Boundary Value Problems for Engineers

Ali Ümit Keskin

Boundary Value Problems for Engineers

with MATLAB Solutions

 Springer

Ali Ümit Keskin
Department of Biomedical Engineering
Yeditepe University
Istanbul, Turkey

ISBN 978-3-030-21082-3   ISBN 978-3-030-21080-9   (eBook)
https://doi.org/10.1007/978-3-030-21080-9

MATLAB® is registered trademark of The MathWorks, Inc. Used with permission. For MATLAB® product information, please contact: The Mathworks, Inc., 3 Apple Hill Drive, Natwick, MA, 01760-2098 USA, Tel: 508-647-7000, Fax: 508-647-7001, E-mail: info@mathworks.com, Web: https://www.mathworks.com

This Springer imprint is published by the registered company Springer Nature Switzerland AG
The registered company address is: Gewerbestrasse 11, 6330 Cham, Switzerland

Preface

This book is a brief yet comprehensive introduction to numerical solution of two-point boundary value problems (BVPs) for ordinary differential equations (ODEs) using the MATLAB® programming language for students and professionals in engineering. It can be used as a supplementary textbook or for self-study, including the solution of common problem types in engineering and applied mathematics, which aims to teach through numerous solved problems and exercises. The book was evolved from the courses taught at Yeditepe University, and endeavors to prepare the reader to solve realistic problems, answer the needs in the field and it is expected to be helpful for undergraduate students as well as to graduates and experts. It is assumed that the reader is comfortable with fundamental mathematical principles and basic MATLAB® use.

It is the continuation of my earlier published book titled "Ordinary Differential Equations for Engineers", and has almost the same objectives. The book is written from the viewpoint of the applied mathematics in engineering to teach fundamentals, new solution methods and applications of boundary value problems (BVPs) in different engineering fields. However, the style of this book is centered on the learning procedure that is based on improving problem-solving techniques using a modern software tool. In most of these, each relevant concept is introduced within the problem statement and solutions are illustrated computationally with the help of short software scripts in MATLAB®. This kind of approach to learn and apply BVPs in particular cases makes everything practical, fast, and easy to grasp, demystifying existing theoretical complexities of the subject via the numerical experiments, rather than delving into mathematical proofs, leaving those in the referenced sources.

As the advent of computers has changed many things in the world, they also influenced the perceptions of many engineers. Because modern computers crunch numbers of big data, carry out symbolic manipulations and put the results of these computations into graphical form easily, many of the earlier difficult and hard-to-solve problems of differential equations and BVPs are best approached with computational techniques.

First two chapters of the book are dedicated to the study of nonlinear root finding methods including Secant, Newton, and Brent methods, their derivative methods and modifications, as well as homotopy and gradient search methods for univariate functions and systems of equations, their stability and convergence, because they play an essential role in the study of BVPs. Fundamental theory and principles of two- point boundary value problems (BVPs), Green's functions, Sturm–Liouville Problems, Eigenvalue problems are the subjects of the following chapters. Shooting methods (including multiple shooting method) for both linear and nonlinear and also for singular problems are studied in some detail. Finite difference methods (again for both linear and nonlinear problems) are exemplified in the next chapter. Solution of linear and nonlinear BVPs using Adomian Decomposition Method and Adomian polynomials, Rayleigh–Ritz, Collocation and Galerkin methods, cubic B-spline techniques, and their convergence issues are described by MATLAB® solutions. The final chapter is devoted to description and applications of built-in MATLAB® BVP solvers, bvp4c, and bvp5c in the solution of BVPs.

Meanwhile, many interesting topics had to be omitted here for space-saving reasons, including invariant embedding, stabilized march, variational iteration methods, homotopy analysis and perturbation methods (HAM and HPM), spectral methods, finite element methods, multipoint and parallel computational techniques, problems involving irregular singularities are left as the subjects of a future book.

Most of the problems concentrate on various engineering projects that are aimed to engage students in the understanding and application of differential equations and BVPs using MATLAB®. These may illustrate direct numerical applications of the explicit formulas or more complicated methods using iterative algorithms or symbolic MATLAB® solutions.

The originator of each of the main ideas were cited as a historical footnote. Almost all of the chapters in the book include a sufficient amount of useful references that have been cited in the related problems of the book.

This book presents complete MATLAB codes in the form of script files as they provide better visibility of variable dimensions, types, and values than the external function files, although the latter tend to be more flexible and modularize computation for most other purposes. When necessary, the use of anonymous functions is preferred.

An outstanding feature of the book is the large number and variety of the all-solved problems that are included in it. Some of these problems can be found relatively simple, while others are more challenging and used for research projects. All solutions to the problems and scripts introduced in the book have been tested using MATLAB®.

Acknowledgements: I would like to thank Dr. Jacek Kierzenka of Mathworks Inc. Natwick, MA, Prof. Dr. A. Okay Çelebi of Yeditepe University Head of Mathematics Department, and Assist. Prof. Namik Ciblak of Mechanical Engineering Department of Yeditepe University, for their fruitful discussions; Prof. Emeritus Lawrence F. Shampine of Southern Methodist University, Dallas, Texas, for his sincere communication related to the solution of some nonlinear systems problems. I also thank our research assistants and graduate students in Biomedical

Engineering Department; Sibel Ozbal, Ibrahim Kapici, Ilayda Hasdemir, Hayrettin Can Sudor, and Ahmet Yetkin who all have offered valuable opinions and suggestions, worked with the problems and proofreading.

Finally, I would like to express my special thanks to my wife Naciye, for her endless patience, encouragement and support.

Istanbul, Turkey   Ali Ümit Keskin

Disclaimer

The software presented in this book is provided "as is" and for academic purpose. Any express or implied warranties, including, but not limited to, the implied warranties of merchantability and fitness for a particular purpose are disclaimed. In no event shall the copyright owner or publisher be liable for any direct, indirect, incidental, special, exemplary, or consequential damages (including, but not limited to, procurement of substitute goods or services; loss of use, data, or profits; or business interruption) however caused and on any theory of liability, whether in contract, strict liability, or tort (including negligence or otherwise) arising in any way out of the use of this software, even if advised of the possibility of such damage.

Contents

About the Author

Ali Ümit Keskin received his BSEE from Bogazici University, MSEE from Yildiz Technical University, and PhD from the Institute of Science and Technology, Istanbul Technical University, and then he joined Siemens AG. He was actively engaged in the fields of Medical Instrumentation, Diagnostic Imaging, and Radiotherapy throughout his professional career.

Since 2002, he is a staff member of Yeditepe University, Faculty of Engineering. Prof. Keskin is one of the founders and recently the Head of Biomedical Engineering Department at Yeditepe University.

He is the author of various patents, numerous research papers, and two books, "Electrical Circuits in Biomedical Engineering, Problems with Solutions" (Springer, 2017, ISBN: 978-3-319-55100-5) and "Ordinary Differential Equations for Engineers, Problems with MATLAB Solutions" (Springer, 2019, ISBN: 978-3-319-95242-0).

His main research interests are applied numerical methods and differential equations, analog circuits, and signal processing, sensors, transducers, and their computer simulations and medical instrumentation.

Chapter 1
Computing Zeros of Nonlinear Univariate Functions

Solving nonlinear equations is one of the most important problems that has many applications in all fields of engineering, and it is one of the oldest and most basic problems in mathematics.

A simple way for obtaining an estimate of the root of a function is to make a plot of the function and observe where it crosses the x axis. This point provides a rough approximation of the root, and can be employed as starting guesses for numerical methods. However, functions that are tangential to the x-axis and discontinuous functions may introduce difficulties in finding all roots of a function.

In the solution of nonlinear univariate equations, one can use iterative methods such as Newton's method and its variants. Newton's method still remains one of the best and most used root-finding methods for solving scalar nonlinear equations. Recently, the order of many variants of Newton's method have been improved by using weight functions. There has been some progress on iterative methods with higher order of convergence that do require the computation of as lower-order derivatives as possible.

Numerous methods of solving nonlinear equations are exemplified in this chapter. Among these are the bisection method, fixed point iteration (FPI), Aitken's Δ^2 Method, Secant Method, Newton-Raphson (or Newton method), modified versions of Newton method, Brent's method, the Steffenson algorithm, Halley's method, Olver's method and improved Ostrowski's method.

Homotopy and Gradient Search Methods for the solution of univariate nonlinear functions are included in Chap. 2, while Adomian Decomposition Method is given in Chap. 5.

MATLAB has its own built-in algorithms to compute the roots of nonlinear functions. This chapter contains various applications of them.

© Springer Nature Switzerland AG 2019
A. Ü. Keskin, *Boundary Value Problems for Engineers*,
https://doi.org/10.1007/978-3-030-21080-9_1

1.1 Bisection and Fixed Point Iteration Methods

Problem 1.1.1 Use bisection method to compute a root of function,

$$f(x) = x^2 - 3x + 1, \quad 1.5 \leq x \leq 3,$$

with an error tolerance of 10^{-6}. Plot the function and the root.

Solution
The bisection method is a bracketing type root finding method in which the interval is always divided in half. If a function changes sign over an interval, the function value at the midpoint is evaluated. The location of the root is then determined as lying within the subinterval where the sign change occurs. The subinterval then becomes the interval for the next iteration. The process is repeated until the root is known to the required precision [1]. Although bisection is generally slower than other methods, the neatness of the procedure (its error analysis, in particular) is a positive feature that makes bisection attractive. This method is often used in the built-in root-finding routine of programmable calculators.
One disadvantage of bisection is that it cannot be used to find roots when the function is tangent to the axis and does not pass through the axis.

The root and the function value at the root are computed by the m-file (bisect1. m), and Fig. 1.1 displays the graph of the polynomial and the location of the root. Bisection method used in this case (with error tolerance of 10^{-6}) stops at the end of 19th iteration.

Note that the code uses $a + 0.5(b - a)$ as the basic equation of bisection. The reason is that for very large values of a, b, $(a + b)/2$ can lead to a computational overflow [2].

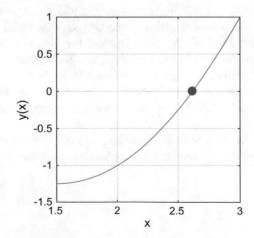

Fig. 1.1 One root of the function, $f(x) = x^2 - 3x + 1, \quad 1.5 \leq x \leq 3$

```
%bisect1.m bisection method to compute a root of f(x)
clc;clear;close;
f =@(x) x^2-3*x+1;
xL=1.50     % lower   guess
xU=2.75     % upper   guess
tol = 1e-6 % prescribed error tolerance
imax = 100 % maximum number of iterations
chk = f(xL)*f(xU);
if chk>0, error('no sign change'),end
i = 0; xr = xL;
while (1)
xrold = xr; xr = xL + 0.5*(xU-xL); i = i + 1
if xr ~= 0, err = abs((xr - xrold)/xr);end % err = approximate error
chk = f(xL)*f(xr);
if chk < 0, xU = xr; elseif chk > 0 ; xL = xr; else, err = 0; end
if err <= tol || i >= imax, break, end
end
root = xr
fx = f(xr) %  function value at root
x=linspace(1.5,3);plot(x, x.^2-3*x+1);hold on;
plot(root,fx,'.','markersize',24);xlabel('x');ylabel('y(x)');grid;
```

Problem 1.1.2 (Fixed point iteration, FPI). An open method to compute a root of a one dimensional function is fixed point iteration (or successive substitution, one-point iteration) method. In this method, unknown is removed from $f(x) = 0$ to the other side of the equation to get

$$x = g(x) \tag{1.1}$$

which can be done either by algebraic manipulation or by adding to both sides of original equation,

$$g(x) = f(x) + x = x \tag{1.2}$$

For a given starting guess value of the root, x_i, Eq. (1.1) is used to determine the estimate x_{i+1},

$$x_{i+1} = g(x_i) \tag{1.3}$$

Fixed point itcration method has linear convergence property. Convergence requires that

$$|g'(x)| < 1. \tag{1.4}$$

It is called FPI because the root r is a fixed point of the function $g(x)$, that is, r is a number for which $r = g(r)$. Note that it is not always clear how to construct $g(x)$ for a fast convergence. If the range of $g'(x)$ is not known, then the convergence is not guaranteed.

Use above given information, and find the positive root of the function $f(x) = x^2 - x - 1$. Let $x_1 = 2$, maximum number of iterations $= 40$, and tolerance $= 10^{-4}$. Note that exact solution is $x_r = 1.618000$ (fixedpointiter.m).

Solution

The solution depends in which manner $g(x)$ is obtained. It is specified that $x > 0$.

Case 1: $x^2 = x + 1$

Divide both sides by x,

$$x = 1 + \frac{1}{x} = g(x) \rightarrow |g'(x)| = \frac{1}{x^2} < 1, \quad x > 1$$

Converges to the root.

Case 2: $x^2 = x + 1 \rightarrow x = \sqrt{x+1} = g(x)$

$$|g'(x)| = \frac{1}{2(x+1)^2} < 1$$

Converges, $x \geq 0$.

Case 3: $x^2 - x - 1 = 0 \rightarrow x(x-1) = 1 \rightarrow x = 1/(x-1) = g(x)$

$$|g'(x)| = \frac{1}{(x-1)^2} < 1,$$

Diverges.

Case 4: $x = x^2 - 1 = (x-1)(x+1) = g(x)$

$$|g'(x)| = |2x| > 1$$

Diverges for $x > 1/2$.

We work on the first case and demonstrate the first seven calculations;

Since $x_0 = 2$,

$$x_1 = 1 + \frac{1}{2} = 1.5; \qquad\qquad x_2 = 1 + \frac{1}{1.5} = 1.666666666666667$$

$$x_3 = 1 + \frac{1}{1.666666666666667} = 1.6; \quad x_4 = 1 + \frac{1}{1.6} = 1.625$$

$$x_5 = 1 + \frac{1}{1.625} = 1.615384615384615;$$

$$x_6 = 1 + \frac{1}{1.615384615384615} = 1.619047619047619$$

Using MATLAB script (fixedpointiter.m), we obtain following results;

```
                            iterations=10
                                r =
                        1.500000000000000
                        1.666666666666667
                        1.600000000000000
                        1.625000000000000
                        1.615384615384615
                        1.619047619047619
                        1.617647058823529
                        1.618181818181818
                        1.617977528089888
                        1.618055555555556
```

```
%fixpointiter.m    fixed point iteration to compute a zero of f(x)=0
clc;clear;format long
g=@(x) 1+1/x ; %function
x0=2; maxIt=40; tol=1e-4;
x=x0; xold=x;
for i =1:maxIt
    x=g(x); % g(x)
    err =abs(x-xold); %absolute error
    xold=x;
    X(i)=xold; r=X'; I=i;
    if(err<tol)||(i==maxIt)
        break
    end
end
fprintf('iterations=%i',I),r
%Root=xold
```

Problem 1.1.3 Use Fixed Point Iteration method to find a positive root of the function

$$f(x) = \sinh(x)\cos(x) - 0.9x.$$

Let the starting (initial guess) value of solution be $x_0 = 1$, maximum number of iterations = 100, and tolerance = 10^{-6}.

Note that exact solution is $x_r = 0.540013984959211$ (fixedpointiter2.m).

Solution
Results are given below.

```
  iterations=46
        r =
0.734963914784736
0.669162880971041
0.631824316753015
      ...
0.540019403629145
0.540018290021264
0.540017403277271
```

Fig. 1.2 Graph of the
function
$f(x) = \sinh(x)\cos(x) - 0.9x$

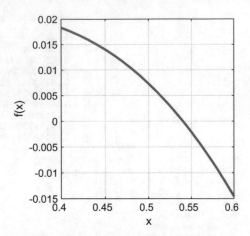

Following MATLAB script (fixedpointiter2.m) is a slightly modified version of the code (fixedpointiter.m) given before. Remark the equation (anonymous function) = $g(x)$, while $f(x) = 0$. Graph of the function $f(x) = \sinh(x)\cos(x) - 0.9x$ is shown in Fig. 1.2.

```
%fixpointiter2.m    fixed point iteration to compute a  zero of f(x)=0
clc;clear;format long
f=@(x) cos(x).*sinh(x)-0.9*x; %function
%Graph
x=0.4:0.01:0.6; y=feval(f,x);
plot(x,y,'linewidth',2);grid;xlabel('x');ylabel('f(x)');
g=@(x) cos(x)*sinh(x)+0.1*x; g(x) = x = f(x)+ x
x0=1; maxIt=100; tol=1e-6;
x=x0; xold=x;
for i =1:maxIt
    x=g(x); % g(x)
    err =abs(x-xold); %absolute error
    xold=x;
    X(i)=xold;r=X';I=i;
    if(err<tol)||(i==maxIt)
        break
    end
end
fprintf('iterations=%i',I),r
```

1.2 Aitken's Δ^2 Method

Problem 1.2.1 When a sequence is slowly converging, it can be transformed into a faster converging sequence by a sequence transformation.

Aitken's[1] Delta-Squared Method can be used to accelerate (the convergence of) a linearly convergent sequence. In fact, Aitken's delta squared method is similar in spirit to Richardson[2] extrapolation, as both methods use assumptions about the convergence of a sequence of approximations to "solve" for the exact solution, resulting in a more accurate method of computing approximations.

Let $\{p_n\}$ be a linearly convergent sequence having the limit of p. For a sufficiently large value of n, to construct a sequence $\{q_n\}$ that converges more rapidly to p than does $\{p_n\}$, we use three consecutive elements of the sequence to determine p,

$$p \approx \frac{p_{n+2}p_n - p_{n+1}^2}{p_{n+2} - 2p_{n+1} + p_n}$$

After some operations and grouping terms appropriately gives an element of the new fast converging sequence, q_n, using the three consecutive elements of the sequence $\{p_n\}$,

$$q_n = p_n - \frac{(p_{n+1} - p_n)^2}{p_{n+2} - 2p_{n+1} + p_n} = p_n - \frac{(\Delta p_n)^2}{\Delta^2 p_n}, \quad \lim_{n \to \infty} \frac{q_n - p}{p_n - p} = 0$$

As an application of this method consider the sequence $\{p_n\}$, $1 \le n \le \infty$, where $p_n = \cos(1/n)$, which converges linearly to p = 1. Show that a sequence $\{q_n\}$ determined by Aitken's Δ^2 Method[3] converges more rapidly to p = 1 than the first one (Aitkendelta2.m).

Solution
First nine values of both sequences are given below.

---

[1]Alexander Craig Aitken (1895–1967) Scottish-New Zealand origin. He was appointed to Edinburgh University in 1925 where he spent the rest of his life. Aitken wrote several books, "The theory of canonical matrices" (1932), "Determinants and matrices (1939)", and "Statistical Mathematics (1939)".

[2]Lewis Fry Richardson, (1881–1953) was an English mathematician who pioneered mathematical techniques of weather forecasting. He is also known for his work for solving a system of linear equations (Modified Richardson Iteration).

[3]It has been pointed out that the delta squared process was first proposed by a Japanese mathematician Takakazu Seki (1642–1708) who used the method for computing π in 1712 [3, 4].

Iteration	pn	qn
0	0	0
1	0.540302305868140	0.961775060161943
2	0.877582561890373	0.982129354477605
3	0.944956946314738	0.989785513554601
4	0.968912421710645	0.993415649669202
5	0.980066577841242	0.995409941655287
6	0.986143231562925	0.996619958741847
7	0.989813260446615	0.997408315133709
8	0.992197667229329	0.997950172708184
9	0.993833508538892	0.998338438634858

```
%aitkendelta2.m  Aitken's delta^2 Method (Accelerating Convergence )
clc;clear; format longg
p=@(n)(cos(1/n)); % n>=1
a=[0 0 0]; % initialization
for n=1:9
pn=p(n);    pn1=p(n+1);    pn2=p(n+2);
qn=pn-((pn1-pn)^2)/(pn2-2*pn1+pn);
a=[a; n, pn, qn];
end
disp('        Iteration       pn        qn');  disp(a)
```

Note References [5–14] can be studied for more detailed information and recent developments on the concepts of sequence transformations, convergence acceleration and extrapolation.

Problem 1.2.2 Use Aitken's delta square method to obtain an accelerated sequence approximating the value of π after 20 iterations. Determine and plot the relative errors (in percent) for the linearly converging sequence and the accelerated sequence on the same figure (aitkendelta4.m).

Solution
The sequence that slowly converges to π is given by

$$\pi = 4 \sum_{n=0}^{\infty} \frac{(-1)^n}{2n+1}$$

For 20 iterations, original and accelerated approximations for the value of π are,

i	original	accelerated
1	4	0
2	2.66666666666667	0
3	3.46666666666667	3.16666666666667
4	2.89523809523810	3.13333333333333
5	3.33968253968254	3.14523809523810
16	3.07915339419743	3.14151898559528
17	3.20036551540955	3.14165339419743
18	3.08607980112383	3.14154198599778
19	3.19418790923194	3.14163535667939
20	3.09162380666784	3.14155633028457

respectively.

Fig. 1.3 Graph of the relative errors, in percent, for both (linearly converging and the accelerated) sequences

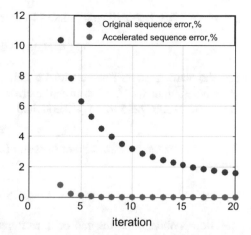

The first and the last five approximation errors, in percent, for both (linearly converging and the accelerated) sequences (each having 20 elements) are listed below in edited form. Approximation errors (in percent) are displayed in Fig. 1.3.

i	Err.orig.%	Err.accel.%
1	27.324	0
2	15.117	0
3	10.347	0.7981300
4	7.8417	0.2629000
5	6.3054	0.1160400
16	1.9875	0.0023449
17	1.8708	0.0019334
18	1.767	0.0016128
19	1.6742	0.0013593
20	1.5906	0.0011562

```
% %aitkendelta4.m  Aitken delta^2 Method
clc;clear;format shortg
sum=0;
for k=0:20;sum=sum + 4*(-1).^k/(2*k+1);X(k+1)=sum;end
x=X';
for i=1:length(x)-2
x1=x(i);x2=x(i+1); x3=x(i+2);  delx1=x2-x1;delx2=x3-x2;
del2x1=delx2-delx1; r(i+2)=x1-delx1.^2/del2x1;
end
%r'
% format longg
% for i=1:20
% disp([i   x(i)    r(i)])
% end
for i=1:20
y(i)= abs(100*(x(i)-pi)/pi); a(i)= abs(100*(r(i)-pi)/pi);
 a(1)=0;a(2)=0; disp([i    y(i)     a(i)])
%fprintf('i= %i y= %4.4f  a= %4.4f \n',i,y(i),a(i))
end
i=3:20;
plot(i,y(i),'.r',i,a(i),'.','markersize',20);xlabel('iteration');grid;
legend('Original sequence error,%','Accelerated sequence error,%')
```

Problem 1.2.3 Use Aitken's process to compute the root for the function

$$f(x) = \sin(x) + 1 - 4x = 0$$

Let the starting guess for the root be unity. Perform iterations until the absolute error is less than 10^{-16}. Determine the error at the end of iterations (actual solution is $r = 0.331323772154633$) (aitkendelta3.m).

Solution
This is an accelerated Fixed Point Iteration (accelerated with Aitken's delta squared process).

$$g(x) = x = 0.25(\sin(x) + 1)$$

Solution is obtained at the end of third iteration, as shown below;

iterations	root_estimate
0	1
1	0.331557005991904
2	0.331323772192903
3	0.331323772154633

Abserror = 3.33066907387547e-16

```
%aitkendelta3.m  Aitken delta^2 Method
clc;clear;
R=0.331323772154633;% best approximation to root value
format longg
g=@(x) 0.25*(sin(x)+1); % g(x) = x = f(x)+ x
x0=1;          % starting (guess) solution
%X(1)=x0;      % initialization
a=[0   x0 ]; % initialization [iteration    root_estimate]
for i=1:3
x1=g(x0);x2=g(x1); x3=g(x2); delx1=x2-x1; delx2=x3-x2;
del2x1=delx2-delx1;  r=x1-delx1.^2/del2x1;x0=r;
if abs(x0-R)<=1e-16; return, end
a=[a; i,  x0];
end
disp('         iteration      root_estimate(i)')
disp(a)
Abserror=abs(a(end,2)-R)
```

1.3   Secant Method

Problem 1.3.1

(a) Describe the Secant Method to compute an approximate root of one dimensional functions.

(b) What are the advantages and limitations of this method?
(c) Use the Secant Method to compute the root of $f(x) = \cos(x) - 2\sin(x)$ around $x = 1$ (secant1.m).

Solution

(a) Secant method is an iterative open method to compute an approximate root of f $(x) = 0$. Note that the Bisection method always converges but its speed of convergence is slow. Secant method chooses the x-intercept of the secant line to the function (which assumes the function to be almost linear in the region of interest). This places the approximation closer to the endpoint of the interval for which function has smaller absolute value. The sequence of approximations generated by the Secant method is started by setting values for two initial guessed zeros of the function which do not need to be bracketing the actual root (consequently, the method may not always converge). Equation for the iterations to compute an approximate root of one dimensional functions is given as

$$x_n = x_{n-1} - f(x_{n-1}) \frac{x_{n-1} - x_{n-2}}{f(x_{n-1}) - f(x_{n-2})}$$

(b) Advantages of Secant Method: Its rate of convergence is more rapid than that of bisection method. It is a fast root finding method. The convergence rate (=1.62) of Secant Method is slightly lower than the quadratic convergence rate of Newton method, but much higher than that of the Bisection Method. It does not need to find the derivative of the function. It is often considered to be a finite difference approximation of Newton method and used as an alternative to Newton's method without requiring any derivatives.
Limitations: The method fails to converge when $f(x_n) = f(x_{n-1})$. Additionally, if x-axis is tangential to the curve, it may not converge to the solution.
(c) The program inputs, $x0 = 1$, $x1 = 2$, $e = 0.000001$ and maximum number of iterations $N = 50$ produces the following results for the computation of the zeros for $f(x) = \cos(x) - 2\sin(x)$,

Iterations	Root	Func
1	0	1
2	1	-1.14263966374765
3	0.466714033591126	-0.00685672308623742
4	0.463494585520077	0.000342170903728456
5	0.463647609228651	-5.09477238175293e-10
6	0.463647609000806	-1.11022302462516e-16

Fig. 1.4 Plot of the function $f(x) = \cos(x) - 2\sin(x)$ and its root, $x = 0.463647609$.

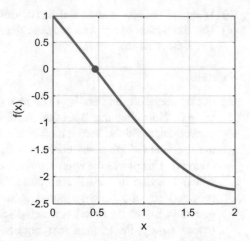

There are two stopping conditions in MATLAB code for Secant method (secant1.m). It is assumed that x_n is sufficiently accurate when the absolute value of $(x_n - x_{n-1})$ is within the given tolerance limits. Also, a safeguard exit based upon a maximum number of iterations is given if the method fails to converge. Figure 1.4 shows the graph of the function $f(x) = \cos(x) - 2\sin(x)$ and its root, $x = 0.463647609$.

```
%secant1.m       secant method
clc;clear;close;tic;format long
f=@(x) cos(x)-2*sin(x)  %1D function
x=[0 1];                %two guesses
e=1e-6;                 %tolerance
N=50;                   %iterations limit
it=0;   fx(1)=f(x(1)); fx(2)=f(x(2));
for j=3:N
x(j) = x(j-1) - (f(x(j-1)))*((x(j-1) - x(j-2))/(f(x(j-1)) - f(x(j-2))));
fx(j)=f(x(j)); it=it+1;
if (abs((x(j)-x(j-1)))<e)
zero=x(j);
%Iterations=j
%fprintf('Iterations =%3i\n',j);
break,end
if(j>(N-1)),disp('iteration limit is reached'),break,end
end
%   Table of results
Iter = 1:j;Iterations = Iter'; Root = x';Func = fx';
T = table(Iterations,Root,Func)
x=linspace(0,2); plot(x,f(x),'linewidth',2); hold on;
plot(zero,0,'.','markersize',22);xlabel('x'), ylabel('f(x)'); grid;
```

1.4   Newton Method

Problem 1.4.1 (Newton-Raphson[4] or Newton Method of finding a root of functions). In convergent (bracketing) methods of root finding, the root is bracketed in an interval prescribed by a lower and an upper bound. Repeated application of these methods always results in closer estimates of the true value of the root.

In contrast, the Newton-Raphson method[5] (which is an open method) requires only a single starting value that does not bracket the root. When it converges, it usually requires less steps (and therefore less computational time) than the bracketing methods.

Use Newton's Method to compute the root of the equation $e^{0.7x} = 0.7x + 7$ with at least 9 digit accuracy. Employ MATLAB Symbolic Math[6] commands (newton1. m).

Solution
MATLAB m-file (newton1.m) uses Symbolic commands to compute the root of f (x),

$$f(x) = e^{0.7x} - 0.7x - 7 = 0$$

Following is the iteration we perform in Newton's Method,

$$x_{i+1} = x_i - \frac{f(x_i)}{f'(x_i)}$$

where denominator in right-hand side of the equation is the first derivative of $f(x)$.

The solution (root value) is obtained at the end of tenth iteration, $x = 3.173631859$.

Given function and its root are displayed in Fig. 1.5.

---

[4]The method was first proposed by Newton in 1669 and later by Raphson in 1690.

[5]For more detailed study on this method, readers may refer to basic textbooks on numerical analysis, such as [15–19].

[6]Algorithms involving Symbolic Math commands are usually slower than those using double precision operations.

Fig. 1.5 The function, $f(x) = e^{0.7x} - 0.7x - 7 = 0$, and its root, $x = 3.173631859$

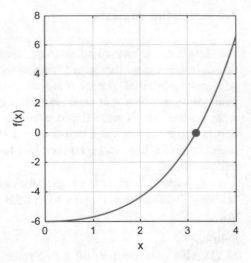

```
%newton1.m    Newton-Raphson (or Newton's) method
clc;clear;close;tic;
syms f(x)
f=exp(0.7*x)-0.7*x-7; df=diff(f);
tol=1e-6; % error tolerance
imax=50;  % maximum allowable iterations
x0=1;     % initial guess
i = 0;    % initialization
while (1)
xold = x0; x0 = x0 - subs(f,x,x0)/subs(df,x,x0); x0=vpa(x0,10);
i = i + 1;
if x0 ~= 0, err = abs((x0 - xold)/x0); end
if err <= tol || i >= imax, break, end
end
i, root = x0
x = linspace(0,4); f=eval(vectorize(f)); % Symbolic to double conversion
plot(x,f,'linewidth',1); hold on; plot(root,0,'.','markersize',22);
xlabel('x'), ylabel('f(x)'); grid; toc
```

Problem 1.4.2 In Newton Method, two difficulties may arise if initial guess is not sufficiently close to the isolated solution. One is that step size may be too long. The second difficulty is that (in selected descent direction) no step size leads to an isolated solution. This can be linked to the Jacobian's being effectively singular (a matrix is said to be "effectively singular" if its condition number is too large). We usually assume that Jacobian matrix is nonsingular. However, in some applications Jacobian matrices are effectively singular. In such cases, altering the Newton direction can be a way out. If the Jacobian matrix is sparse, it is more challenging task than damping the Newton step size alone [19].

For the function, $f(x) = x^{10} - 1$, choose trial values of 0, 0.5, 0.8, and 2 and record the iteration number below which the method converges to the root of the function, for an error tolerance of 10^8 (newton1.m).

Solution

When $x_0 = 0.5$, iteration number below which the method converges to unity (for an error tolerance of 10^8) is 43. When $x_0 = 0.8$, this figure of merit is 10, while it is 11 for $x_0 = 2$. When $x_0 = 0$, method collapses. Figure 1.6 shows the graph and zero of the function, $x \geq 0$.

Problem 1.4.3 Newton-Raphson like sequence without involving a derivative term, such as the sequence

$$x_{n+1} = x_n - \frac{f(x_n)}{f(x_0)}$$

may also converge (but not quadratically) if initial estimate x_0 is quite close to the actual value of a zero of the function [20].

Use $f(x) = x^{10} - 1$, $x_0 = 0.95$, $\epsilon_{rel} = 10^{-8}$ and determine the number of iterations required to compute the zero of $f(x)$, if the equation given above is employed instead of familiar Newton-Raphson equation (newton1.m).

Solution

$$f(x) = x^{10} - 1 = 0, \quad x_0 = 0.95, \quad \epsilon_{rel} = 10^{-8}$$

It takes 30 iterations to compute the zero of $f(x)$, if the equation given above is employed instead of Newton-Raphson formula which requires only five iterations.

Problem 1.4.4 Prove that the Newton-Raphson method does not converge to a solution for the function $f(x) = \arctan(x)$ if initial guess point is $|x_0| > 1.39175$ (newton1.m).

Fig. 1.6 Graph and zero of the function $f(x) = x^{10} - 1$

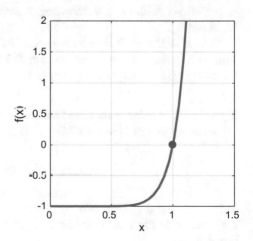

Fig. 1.7 The function $f(x) = \arctan(x)$ and the zero of the function at $x = 0$

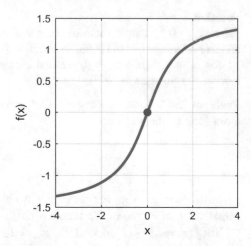

Solution

We use MATLAB script (newton1.m) for different values of x_0 around given point, and note that $f(x) = \arctan(x) = 0$ function can not be computed using Newton-Raphson method if $|x_0| > 1.39175$, but the value of $|x_0| = 1.39174 < 1.39175$ result in convergence to solution at $x = 0$.

The limiting accuracy can be expanded to $|x_0| < 1.391745201$. Beyond this value, an error message "division by zero" is received.

Figure 1.7 displays the function $f(x) = \arctan(x)$ and its zero at $x = 0$.

Problem 1.4.5 Deciding whether a univariate function has a multiple root is an ill-posed problem. For example, a small perturbation of a polynomial coefficient may change the answer from yes to no. In particular a real double root may change into two simple (real or complex) roots. Therefore it is hardly possible to verify that a polynomial or a nonlinear function has a double root if not the entire computation is performed without any rounding error, i.e. using methods from Computer Algebra [21].

Determine the number of steps needed for the Newton's Method to get the error within six decimal digits, or smaller than 0.5×10^{-6} (the rate of convergence) if the function is $f(x) = \sin(x) + 2\cos(x) - x - 2$. How can it be improved?

Solution

If $f'(r) \neq 0$, r being a root, and assuming that there is no multiplicity of a root, convergence rate of Newton's Method is quadratic,

$$e_{i+1} \approx M e_i^2, \quad M = 0.5 \frac{f''(r)}{f'(r)}$$

In this problem (by differentiating given function three times), we see that there are two (multiple) roots of given function at $x = 0$, although this is not apparent in Fig. 1.8.

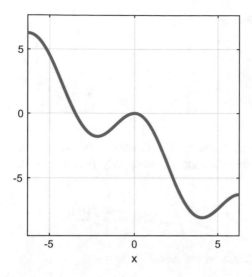

Fig. 1.8 Graph of the function, $f(x) = \sin(x) + 2\cos(x) - x - 2$

```
clc;clear; syms f(x)
f=sin(x)+2*cos(x)-x-2; df1=diff(f);df2=diff(f,2);df3=diff(f,3);
f0=subs(f,x,0), f01=subs(df1,x,0), f02=subs(df2,x,0)
```

If there is a multiplicity of a root, the Newton's Method is locally convergent to the root value and error at step $i(= e_i)$ satisfies

$$\lim_{i \to \infty} \frac{e_{i+1}}{e_i} = \frac{m-1}{m}$$

For two multiple roots, $m = 2$, then classical method converges linearly with $e_{i+1} \approx e_i/2$, in other words the error will decrease by a factor of $1/2$ on each step. The number of steps needed to get the error within six decimal digits, or smaller than 0.5×10^{-6} is

$$(1/2)^n < 0.5 \times 10^{-6} \ \rightarrow \ n > \frac{\log_{10}(0.5) - 6}{\log_{10}(0.5)} \approx 21$$

In this case, convergence of Newton Method can be improved with a small modification.

Fig. 1.9 Diode rectification
circuit

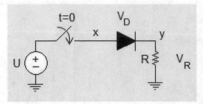

Modified Newton method: If a function f is $(m + 1)$-times differentiable on $[a, b]$, which contains a root of multiplicity $m > 1$, then the Modified Newton Method

$$x_{i+1} = x_i - \frac{m \cdot f(x_i)}{f'(x_i)}$$

converges quadratically to the root.

Problem 1.4.6 Determine current and voltage drops across each element for $t > 0$ in rectification circuit[7] of Fig. 1.9. $U = 5$ V, $R = 1000 \ \Omega$, diode characteristic equation,

$$i_D = Is\left(\exp\left(\frac{v_D}{nV_T}\right) - 1\right), \quad Is = 10^{-15} \text{ A}, \quad n = 1, \quad V_T = 26 \text{ mV}$$

What happens if polarity of voltage source is reversed? If $U = 0$ V? (newtonDiode. m).

Solution

(a) Because explicit analytic solution is not possible in this circuit, iterative methods can be employed.

We source transform the circuit and then apply Kirchhoff's Current Law [22],

$$\frac{U}{R} - i_R - i_D = 0$$

where v is the node voltage.

---

[7]This is a simpler variant of problem described in [23].

Substituting current flows through each element in KCL equation, we obtain

$$\frac{U}{R} - \frac{v}{R} - Is\left(\exp\left(\frac{v_D}{nV_T}\right) - 1\right) = 0$$

$$U - v - Is\,R\left(\exp\left(\frac{v_D}{nV_T}\right) - 1\right) = 0 = f(v)$$

(1.5)

Newton-Raphson iterative method can be applied in (1.5). Following edited results are obtained using MATLAB script (newtonDiode.m) with relative error tolerance of 10^{-8}, and initial (guessed) value of diode (node) voltage, $v_{D0} = 0$ V.

```
U = 5
iterations = 14
Diode  voltage  = 0.755990 V
Resistor voltage  = 4.244010 V
Diode  current  = 0.004244 A
Elapsed time is 0.549887 s.

U = -5
iterations = 13
Diode  voltage  = -5.000000 V
Resistor voltage  = -0.000000 V
Diode  current  = -0.000000 A
Elapsed time is 0.515261 s.

U = 0
iterations = 18
Diode  voltage  = 0.000000 V
Resistor voltage  = 0.000000 V
Diode  current  = 0.000000 A
Elapsed time is 0.957809 s.
```

MATLAB script (newtonDiode.m) is listed below.

```
%newtonDiode.m   Newton-Raphson method applied to diode circuit
clc;clear;close;tic;
syms  f(x)
 VT=0.026; n=1; Is=1e-15; R=1000;
%U=5 % Forward bias
%U=-5 % Reverse bias
U=0 % zero bias
f=U+Is*R-x-Is*R*exp(x./(n*VT));
x0=1;       % initial guess
tol=1e-8;    % error tolerance
imax=50;     % maximum allowable iterations
df=diff(f);
i = 0;       % iteration counter initialization
while (1)
xold = x0;
x0 = x0 - subs(f,x,x0)/subs(df,x,x0); x0=vpa(x0,6);
i = i + 1;
if x0 ~= 0, err = abs((x0 - xold)/x0); end
if err <= tol || i >= imax, break, end
end
iterations=i
Vd = x0; if Vd<1e-8, Vd=0; end
VR=U-Vd; id=VR/R;
fprintf('Diode voltage = %6.6f V \n',Vd);
fprintf('Resistor voltage = %6.6f V \n',VR);
fprintf('Diode current = %6.6f A \n',id);toc
```

1.5 Brent's Method

Problem 1.5.1 (Brent's Method). Brent's Method[8] does not require the evaluation of the derivatives [24]. It is a hybrid method (for finding a root of functions) which consists of a combination of bisection method and two different open methods. The first is the secant method, the second is inverse quadratic interpolation. The method is usually faster than bisection alone for smooth functions. Brent's Method uses quadratic interpolation which requires three points: u1, u2 and u3, the last one is provided by a bisection step [27–29].

MATLAB's command `fzero` implements a version of Brent's Method, along with a preprocessing step, to discover a good initial bracketing interval if one is not provided by the user.

Given the function $f(x) = x^{10} - 1$, $[a, b] = [0, 5]$, $\epsilon = 10^{-10}$ determine the number of iterations required to compute the zero of $f(x)$.

---

[8]Brent's method builds on an earlier algorithm by Dekker [25]. The method is also known as the Van Wijngaarden–Dekker–Brent Method [26].

Solution

After 12 iterations the solution (value for the zero of the function) is reached;

```
0.000262399722939
2.500000000000000
1.250131199861469
0.134442147776487
0.692286673818978
0.971208936840224
1.059755426947500
0.994168275015082
1.000264601642619
0.999997740782234
1.000000000100887
1.000000000000000
```

```
%brent2.m      Finds a root of f(x)=0 by Brent's method
clc;clear;format long;
fun=@(x) x^10-1; %   function
a=0;b=5;  %   interval bracketing the root
X=0:0.01:5; Y = feval(fun,X); plot(X,Y,'linewidth',2);
xlabel('x');ylabel('f(x)');grid;hold on
x1 = a; f1 = feval(fun,x1); if f1 == 0; root = x1; return; end
x2 = b; f2 = feval(fun,x2); if f2 == 0; root = x2; return; end
if f1*f2 > 0.0, error('Root is not bracketed in (a,b)'), end
x3 = 0.5*(a + b);
for i = 1:100
f3 = feval(fun,x3);if abs(f3) < 1e-10;
Root = x3, plot(Root,0,'.','markersize',22); return,end
if f1*f3 < 0.0; b = x3; else a = x3; end
if (b - a) < 1e-10*max(abs(b),1.0); root = 0.5*(a + b); return,end
den = (f2-f1)*(f3-f1)*(f2-f3);
num = x3*(f1-f2)*(f2-f3+f1)+ f2*x1*(f2-f3) + f1*x2*(f3-f1);
if den == 0; dx = b - a; else dx=f3*num/den; end
x = x3 + dx; if (b - x)*(x - a) < 0, dx =(b - a)/2;  x = a + dx; end
if x < x3; x2 = x3; f2 = f3; else x1 = x3; f1 = f3; end
x3 = x;R(i)=x3; r=R'
end
root = []; % root does not converge.
```

Problem 1.5.2 Compute the zero of $f(x) = x.\sin^2(x) - 1$ that lies in the interval (3, 5) using Brent's Method, with $\epsilon = 10^{-10}$ (brent2.m).

Solution

The graph of the given function is shown in (Fig. 1.10). Brent's Method converges to the root in four iterations as given below. It is seen that the root is somewhere between 3 and 4. We select a tight bracketing interval of (3.5, 3.7) and obtain the following result;

Fig. 1.10 Graph of the
function and its zero,
$f(x) = x.\sin^2(x) - 1$

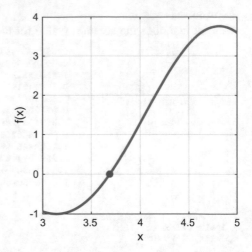

```
        Root:
3.689625855891553
3.689188133907209
3.689188865530172
3.689188865479850
```

Following is the MATLAB script to be patched onto the beginning of file
(brent2.m).

```
%brent2.m      Finds a root of f(x)=0 by Brent's method
clc;clear;format long;
fun=@(x) x.*sin(x).^2-1; %   function
X=3:0.01:5; Y = feval(fun,X); plot(X,Y,'linewidth',2);
xlabel('x');ylabel('f(x)');grid;hold on
a=3.5;b=3.7;   %  interval bracketing the root
```

Problem 1.5.3 If a function has multiple roots, some care must be taken.

  i. It is not possible to use root-bracketing algorithms on even-multiplicity roots.
     Functions having roots with even-multiplicity do not cross horizontal x-axis,
     but only become tangential to it instantaneously.
 ii. A root bracketing algorithm still works for odd-multiplicity roots. Open type
     algorithms work with higher multiplicity roots, but at a reduced rate of
     convergence.[9]

---

[9]The Steffenson algorithm can be used to accelerate the convergence of multiple roots.

Check where does the Brent's Method stand for the case of multiple roots. (brent2.m).

Solution

Since Brent's Method includes bracketing operation, it can be expected that above given conditions should apply. We test Brent's Method for the first case and try to see its performance using functions having even-multiplicity roots. For example, applying the function $f(x) = (x-3)^2$ in Brent's algorithm with $x \in [2,4]$, $\epsilon = 10^{-10}$ gives a message,

```
Error: Root is not bracketed in (a,b)
```

although it is bracketed, as shown in Fig. 1.11.

However, Brent's algorithm works for odd-multiplicity roots. For example, testing the function $f(x) = (x-1)^3(x^2 - 5x + 6)$, $x \in [0.6, 1.3]$ with $\epsilon = 10^{-10}$, we obtain the root of the function at the end of 13th iteration as 0.998430628208406.

Following is the MATLAB script to be patched onto the beginning of file (brent2.m).

```
%brent2.m     Finds a root of f(x)=0 by Brent's method
clc;clear;format long;
%fun=@(x) (x-3).^2;
fun=@(x) (x-1).^3. * (x.^2-5*x+6);
X=0.5:0.001:1.5; Y = feval(fun,X); plot(X,Y,'linewidth',2);
xlabel('x');ylabel('f(x)');grid;hold on
a-0.6;b=1.3;  % interval bracketing the root
```

1.6 Halley's, Olver's, and Improved Ostrowski's Methods

Problem 1.6.1 Halley's[10] method, is similar to Newton method, but converges more rapidly in the neighborhood of a root. Halley's method requires an initial guess for the root and evaluates the function, as well as its first two derivatives of the root and uses that information to iteratively improve the approximation.

$$x = x_0 - \frac{f(x_0)}{f'(x_0)} \left(1 - \frac{f(x_0)f''(x_0)}{(f'(x_0))^2}\right)^{-1}$$

---

[10]Edmond Halley (1656–1742) is best known for predicting the orbital period of the comet that bears his name.

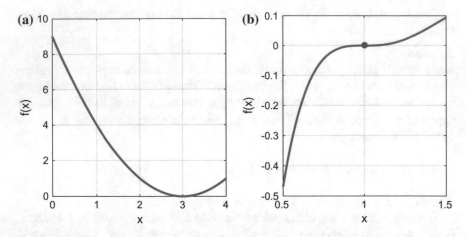

Fig. 1.11 a Brent's method fails to compute even-multiplicity roots, but works for functions having odd-multiplicity of roots, b although convergence rate is relatively slow

Halley's method can be preferred to Newton's method when the ratio $f''(x)/f'(x)$ can be evaluated cheaply. For example, this ratio simplifies if the function is the product of a simple polynomial and an exponential.

Use Halley's Method to compute one root of $f(x) = e^x(0.5x - x^3)$, $x_0 = 1$, $\epsilon = 10^{-6}$ (halley.m).

Solution

Following is a list of the output of MATLAB script (halley.m). Graph of the function is shown in Fig. 1.12.

Fig. 1.12 Graph of the function,
$f(x) = e^x(0.5x - x^3)$

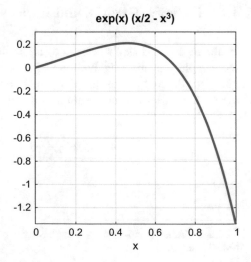

```
                        x0 = 1
                  0.5384615384615
                  0.5962806247542
                  0.6620547510109
                  0.7003103614297
                  0.7069610301063
                  0.7071067148637
                  0.7071067811865
                Zero of f(x) = 0.707107
```

```
%Halley.m  %Halleys Method of finding zeros of f(x)=0
%  f(x) = x^2 - 9
clc;clear;syms f(x);tic
x0 = 1              %initial (guess) value of the root
N = 10;             %maximum number of iterations
tol =1e-6;          %tolerance
f(x)=exp(x)*(x./2- x.^3) %function
p=ezplot(f(x), [0, 1]);grid
%-------------------------------------------------------------
dx = diff(f);   d2x = diff(dx); % first & second derivative of f(x)
for i=1:N
x = x0 - (f(x0)/dx(x0))*(1 - (f(x0)*d2x(x0)/dx(x0)^2))^(-1);
fprintf('%13.13f \n', double(x))
if (abs(x - x0)) < tol   %stopping criterion
fprintf('Zero of f(x) = %f \n', double(x));toc
return
end
x0 = x;
end
fprintf('No covergence, %d iterations at precision  %d \n', N, tol);
```

New modifications and recent studies on this method are given in [30–37].

Problem 1.6.2 Olver's Method employs a cubically convergent variant of the Newton's Method, with the following iterative form;

$$x_{n+1} = x_n - \frac{f(x_n)}{f'(x_n)} - \frac{1}{2}\left(\frac{|f(x_n)|^2 f''(x_n)}{|f'(x_n)|^3}\right), \quad n = 0, 1, \ldots$$

Use this method to solve the nonlinear equation,

$$f(x) = e^{(x^2-3x)} \cdot \sin(x) + \ln(x^2 + 1) = 0, \quad x_0 = 3.6, \quad \epsilon_{rel} = 10^{-12}$$

(olver.m).

Solution
First iteration:

$$f(3.6) = -1.200959,$$

$$f'(x)|_{x_0=3.6} = e^{x^2-3x} \cdot \cos(x) + (2x-3)e^{x^2-3x} \cdot \sin(x) + \frac{2x}{x^2+1}\bigg|_{x_0=3.6} = 11.45644046$$

$$f''(x)|_{x_0=3.6} = (2x-3)^2 e^{x^2-3x} \cdot \sin(x) + e^{x^2-3x} \cdot \sin(x)$$

$$+ 2(2x-3)e^{x^2-3x} \cdot \cos(x) + \frac{2}{x^2+1} - \frac{4x^2}{(x^2+1)^2} = -1.369650$$

$$x_2 = x_1 - \frac{f(x_1)}{f'(x_1)} - \frac{1}{2}\left(\frac{|f(x_1)|^2 f''(x_1)}{|f'(x_1)|^3}\right)$$

$$x_2 = 3.6 - \frac{-1.200959}{11.45644046} - \frac{1}{2}\left(\frac{(1.200959)^2 \times (-1.369650)}{11.45644046^3}\right) = 3.556357$$

Second iteration:

$$f(3.556357) = -1.200959, \quad f'(3.556357) = -18.085010,$$

$$f''(3.556357) = -1.067851,$$

$$x_3 = x_2 - \frac{f(x_2)}{f'(x_2)} - \frac{1}{2}\left(\frac{|f(x_2)|^2 f''(x_2)}{|f'(x_2)|^3}\right)$$

$$x_3 = 3.556357 - \frac{-1.200959}{-18.085010} - \frac{1}{2}\left(\frac{(1.200959)^2 \times (-1.067851)}{(-18.085010)^3}\right)$$

$$x_3 = 3.540531$$

Proceeding in this way, the results of iterations are listed below;

Iteration	Zero
0	3.6
1	3.5563571064855
2	3.54053129149807
3	3.5388534778617
4	3.53883657157615
5	3.53883656988142

```
%olver.m
clc; clear; format longg
syms f(x)
x0=3.6; f(x)=exp(x.^2-3*x).*sin(x)+log(x.^2+1); df=diff(f);d2f=diff(df);
i=1;x=0; err=abs(x-x0)/abs(x);
while err>=1e-12
X(i)=x0;
x=x0-(f(x0)/df(x0))-0.5*((abs(f(x0)))^2*d2f(x0)/(abs(df(x0)))^3);
i=i+1; I=i-1;
x=double(x); err=abs(x-x0)/abs(x); x0=x;
end
disp(' Iteration   Zero');[(0:I-1)'  X']
```

Problem 1.6.3 An improved Ostrowski's method presented by Grau and Díaz-Barrero [38] to solve the nonlinear equation $f(x) = 0$ consists of adding the evaluation of the function at another point in the procedure which is iterated by Ostrowski's[11] method [39]. As a consequence, the order of convergence of the method increases.

Computational procedure is the following:

$$y_n = x_n - \frac{f(x_n)}{f'(x_n)} \tag{1.6}$$

$$\mu = \frac{x_n - y_n}{f(x_n) - 2f(y_n)} \tag{1.7}$$

$$z_n = y_n - \mu f(y_n) \tag{1.8}$$

$$x_{n+1} = z_n - \mu f(z_n) \tag{1.9}$$

Use improved Ostrowski's method to compute the solution of $f(x) = e^x - 2x^2 = 0$, let initial estimate be $x_0 = 1.0$ (ostrowski_improved1.m).

---

[11]Alexander M. Ostrowski (1893–1986) was an Ukrainian born mathematician.

Solution

All calculations for the solution of given equation using Eq. (1.6–1.9) are given below.

First iteration,

$$x_0 = 1, \ f(x_0) = 0.718282, \ f'(x_0) = -1.281720,$$

$$y_0 = 1 - \frac{0.718282}{(-1.281720)} = 1.56041$$

$$\mu = \frac{x_0 - y_0}{f(x_0) - 2f(y_0)} = \frac{1 - 1.56041}{0.718282 - 2(-0.108979)} = -0.59857$$

$$z_0 = y_0 - \mu f(y_0) = 1.56041 - (-0.59857)(-0.108979) = 1.49517$$

$$x_1 = z_0 - \mu f(z_0) = 1.49517 - (-0.59857)(-0.010977) = 1.4886$$

Second iteration,

$$f(x_1) = -0.000976, \ f'(x_1) = -1.52351, \ y_1 = 1.4886 - \frac{(-0.000976)}{(-1.52351)} = 1.48796$$

$$\mu = \frac{x_1 - y_1}{f(x_1) - 2f(y_1)} = \frac{1.4886 - 1.48796}{-0.000976 - 2(0.000000088)} = -0.65626$$

$$z_1 = y_1 - \mu f(y_1) = 1.48796 - (-0.65626)(0.000000088) = 1.48796$$

$$x_2 = z_1 - \mu f(z_1) = 1.48796 - (-0.59857)\left(1.82 \times 10^{-14}\right) = 1.48796$$

Third iteration,

$$f(x_2) = 3.9357 \times 10^{-21}, \ f'(x_2) = -1.52379, \ y_2 = 1.48796 - 0 = 1.48796$$

$$\mu = \frac{x_2 - y_2}{f(x_2) - 2f(y_2)} = -0.65626$$

$$z_2 = y_2 - \mu f(y_2) = 1.48796 - (-0.65626)(0) = 1.48796$$

$$x_3 = z_2 - \mu f(z_2) = 1.48796 - (-0.65626)(0) = 1.48796$$

```
%ostrowski_improved1.m      Grau-Diaz Barrero method, 2006
clc;clear;tic;
syms f(x)
f=exp(x)-2*x^2; df=diff(f);
tol=1e-6; % error tolerance
imax=5;    % maximum allowable iterations
x0=1;      % initial guess
i = 0;     % initialization
while (1)
xold = x0
y0 = x0 - subs(f,x,x0)/subs(df,x,x0);y0=vpa(y0,6);
mu=(x0-y0)/(subs(f,x,x0)-2*subs(f,x,y0)); mu=vpa(mu,6);
z0=y0-mu*subs(f,x,y0);z0=vpa(z0,6);x0=z0-mu*subs(f,x,z0);x0=vpa(x0,6);
i = i + 1
if x0 ~= 0, err = abs((x0 - xold)/x0); end
if err <= tol || i >= imax, break, end
end
i, root = x0
x = linspace(0,2); f=eval(vectorize(f)); % Symbolic to double conversion
plot(x,f,'linewidth',1); hold on; plot(root,0,'.','markersize',22);
xlabel('x'), ylabel('f(x)'); grid; toc
```

1.7   Steffensen's Method

Problem 1.7.1 Steffensen's[12] root finding method does not require evaluation of first derivative of function $f'(x)$ and it converges quadratically.[13]

It can be a combination of a linearly converging root finding method with Aitken's delta-squared acceleration. The method generates a new sequence q_i, if the original iterates are x_i,

$$q_i = x_i - \frac{(x_{i+1} - x_i)^2}{x_{i+2} - 2x_{i+1} + x_i}$$

Algorithm Devise a fixed-point iteration function $g(x)$ from given $f(x)$. Use an error tolerance value ϵ, a = starting value, $b = g(a)$ and $c = g(b)$,

$$q = a - \frac{(b-a)^2}{a - 2b + c}$$

---

[12]Johan Frederik Steffensen was a Danish mathematician, who worked in the fields of finite differences and interpolation. He was professor at the University of Copenhagen from 1923 to 1943, and described this method in [40].

[13]This method can be unstable if the function is not well-behaved. It can diverge or divide by zero, depending upon f(x). It is much more demanding with respect to the starting points than Newton's method which justifies that Steffensen's method is less used than Newton's method to approximate solutions of equations [41].

Then, restart with $a = q$ until $|q - a| < \epsilon$.

Note that every iteration in this algorithm requires two function cells, $g(a)$, $g(b)$.

(a) Use Steffensen Method to compute the square root of π with tolerance $\epsilon = 10^{-7}$
   (hint: $\sqrt{\pi} = 1.7724538509055$).

(b) Compute one root with a multiplicity 2 of the nonlinear function [42]

$$f(x) = e^x - 1 - x + \frac{x^2}{2} = 0, \quad -5 < x < 2, \quad \epsilon = 10^{-7}$$

(hint: multiple roots at $x = 0$) (steffensen.m).

Solution

(a) We set up fixed point iteration function from $f(x) = x^2 - \pi$,

$$g(x) = f(x) + x = x^2 - \pi + x$$

We define this as an anonymous function in the MATLAB script (steffensen.m). Resulting output is listed as shown below.

iteration	root_estimate(i)
0	2
1	1.82331507319071
2	1.77555832439313
3	1.77246615607497
4	1.77245385109964

Abserror = 1.94128046970832e-10

(b) We first find the function $g(x)$,

$$g(x) = f(x) + x = e^x - 1 - x + \frac{x^2}{2} + x = e^x - 1 + \frac{x^2}{2}, \quad -5 < x < 2$$

We define this as an anonymous function in the MATLAB script (steffensen.m). Truncated resulting output is listed as shown below.

```
      iteration          root_estimate(i)
          0                        1
          1               0.794298129201741
          2               0.580145245153757
          3               0.386542703535052
          4               0.235948764112433
         17              3.83419616269908e-05
         18              1.91577986843151e-05
         19              9.55933537867502e-06
         20              5.26967293986357e-06
         21              6.02600619797562e-07

      Abserror = 6.02600619797562e-07
```

```
%steffensen.m    Steffensen method for root finding
clc;clear;format longg
%g=@(x) x.^2-pi+x;          % f=function
g=@(x) exp(x)-1+x.^2/2      % f=function
%a=2 ;          % first guess for the root of g= x.^2-pi+x
a=1 ;           % first guess for the root of g=exp(x)-1+x.^2/2
tol=1e-7 ;      %tolerance
itmax=40;       % max. number of iter.
A=[0   a ]; % initialization [iteration    root_estimate]
for i=1:itmax
    b=g(a);  % calculate the next two guesses for the fixed point.
    c=g(b);
q=a-(b-a)^2/(a-2*b+c); %  Aitken's delta squared method
    if abs(q-a)<tol    %  Note that a=q(i-1).
    break
    end
    a=q ;              % update q
A=[A; i,   a];
end
%A function can be written as function   q= steffensen(g,a,tol,itmax)
disp('                    iteration        root_estimate(i)')
disp(A)
%Abserror=abs(A(end,2)-sqrt(pi))
Abserror=abs(A(end,2)-0)
```

1.8   MATLAB Built-in Functions, Fzero, Roots, and Fsolve

Problem 1.8.1

(a) Describe basic properties of MATLAB's built-in function `fzero`. Give simple examples. What happens if we place an initial guess in the middle of the two zeros of a function? What are the limitations of this function?

(b) Use MATLAB's `fzero` function to find a root of $f(x) = 2\sin(3x)\sin(5x)$, within the following intervals;

$$(i) \; x \subset [0.5 \; 1], \quad (ii) \; x \in [0.8 \; 1.15], \quad (iii) \; x \in [1.2 \; 1.8],$$
$$(iv) \; x \in [1.5 \; 2], \quad (v) \; x \in [1 \; 2], \quad (vi) \; x \in [2 \; 3]$$

(fzero1.m)

Solution

(a) The `fzero` built-in function finds the real root of a single equation using a simple syntax as `fzero(function,a)`, where scalar a is the initial guess of the zero. A more complete representation of the syntax is `[x, fx] = fzero(function,x0,options,p1,p2,...)`, where `[x, fx]` is a vector containing the root and the function evaluated at the root, "options" is a data structure created by the "optimset" function, and p1, p2 ... are any parameters that the function requires.

Two guesses that bracket the root can also be used: `fzero(function,[a b])` where a and b are guesses that bracket a zero.[14]

Example: Let the function be $x^2 - 16$. The negative root is:

x = fzero(@(x) x^2-16, -5)

x = -4

The positive root, using a guess that is near the positive zero:

x = fzero(@(x) x^2-16,6)

x = 4

If an initial guess is placed in the middle of the two zeros of a function, `fzero` finds the negative root:

x = fzero(@(x) x^2-16,0)

x = -4

Following is an alternative (but an earlier MATLAB version) single line expression `fzero`;

```
x=fzero(inline(' x^2-16','x'),0)
x = -4
```

The command finds a point where the function changes sign. If the function is discontinuous, fzero may return values that are discontinuous points instead of zeros.

Since a zero is defined as a point where the function crosses the x-axis, points where the function touches, but does not cross the x-axis are not valid zeros. For functions with no valid zeros, fzero executes until Inf, NaN, or a complex value is detected. The initial interval must be finite; it cannot contain ± Inf.

(b) The graph of the function is shown in Fig. 1.13.

(i) $root1 = 0.6283$, (ii) $root2 = 1.0472$, (iii) $root3 = 1.2566$,

(iv) $root4 = 1.8850$, (v) $root5 = 1.8850$, (vi) $root6 = 2.5133$

---

[14]MATLAB's `fzero` is based on the work described in [43].

Fig. 1.13 The graph of the function,
$f(x) = 2\sin(3x)\sin(5x)$

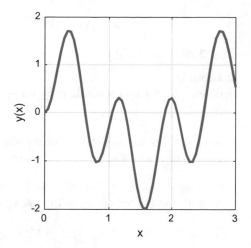

Edited results are shown below.

```
z = 0.6283 z = 1.0472 z = 1.2566 z = 1.8850 z = 1.8850
```

```
Error using fzero (line 290).The function values at the interval end-
points must differ in sign. Error in fzero1 (line 12) case 6, z = fzero
(fun,[2 3])
```

Here, the fifth case gives the largest one of the roots within the given interval. The sixth case gives an error message for software version 2017a (while prints the largest one of the roots within the given interval, in version 2015a). Tight bracketing is required for accurate results.

```
%fzero1.m  Find the zero of f(x) within given interval
clc;clear;close;
fun=@(x,y) 2*sin(3*x).*sin(5*x)   ;
for j=1:6
    switch j
    case 1, z = fzero(fun,[.5 1]),     case 2, z = fzero(fun,[.8 1.15])
    case 3, z = fzero(fun,[1.2 1.8]), case 4, z = fzero(fun,[1.5 2])
    case 5, z = fzero(fun,[1 2]),      case 6, z = fzero(fun,[2 3])
    end, end, x=0:0.05:3;plot(x,   2*sin(3*x).*sin(5*x));grid;
```

Problem 1.8.2 Use fzero and classical Newton's methods to solve the following function,

$$f(x) = (x-1)^{2n+2} = 0, \quad n = 0, 1, 2, \ldots$$

(fzero3.m), (newton1.m).

Solution

Following are the results obtained by running fzero3.m and newton1.m MATLAB scripts.

fzero: Exiting fzero: aborting search for an interval containing a sign change because NaN or Inf function value encountered during search.

Function value at -1.7162e + 154 is Inf.).Check function or try again with a different starting value. Root = NaN.

Newton's Method: i = 34,   Root = 1.0

```
%fzero3.m
clc;clear;
f=@(x) (x-1).^2 ;
Root= fzero(f,2)
x=0:0.01:2 ; y=feval(f,x); plot(x,y);grid; %Graph

%newton1.m    Newton-Raphson method
clc;clear; syms f(x)
f=(x-1).^2;%multiple root
df=diff(f);
tol=1e-10; % error tolerance
imax=50;   % maximum allowable iterations
x0=2;      % initial guess
i = 0;     % initialization
while (1)
xold = x0;
x0 = x0 - subs(f,x,x0)/subs(df,x,x0); x0=vpa(x0,10);
i = i + 1;
if x0 ~= 0, err = abs((x0 - xold)/x0); end
if err <= tol || i >= imax, break, end
end
i, root = x0
```

Problem 1.8.3 (Roots of a polynomial). Although the techniques such as bisection and the Newton-Raphson method can have utility to determine a single real root of a polynomial, one may wish to determine all the roots, both real and complex, which Bisection, Secant, Newton-Raphson and Brent methods are not available for determining all the roots of higher-order polynomials. MATLAB has a built-in roots function for this task which has the syntax, x = roots(f). Polynomials are entered into MATLAB by storing the coefficients as a row vector.

Use MATLAB to find the roots of polynomial, $f(x) = 3x^4 - x^2 + 1.25x - 7.5$.

Solution
Command Window Output list is given below.

```
f = 1.0000    0   -1.0000    1.2500    -7.5000
```

```
          x =
        -1.9199 + 0.0000i
         1.6950 + 0.0000i
         0.1124 + 1.5140i
         0.1124 - 1.5140i
```

MATLAB script is the following;

```
clc;clear;
format short
% Roots of a polynomial
% f is the vector of polynomial coefficients,
% coefficient of highest power comes first
c=[1   0   -1   1.25   -7.5]
x=roots(c)
```

Problem 1.8.4 MATLAB commands `fzero`, `fsolve` and `roots` are three different means of finding zeros of functions. Use these commands (in their default form) to compute the cube-root of 10. Comment on the results (rootFind.m).

Solution

$$f(x) = x^3 - 10 = 0$$

Let the initial guess for the zero of this function be 2. Following are the edited results of using MATLAB script (rootFind.m);

```
X =   2.154434692237897   (fsolve)
x =   2.154434690031884   (fzero)
y =   2.154434690031883   (roots)
```

The results for `fzero` and `roots` are alike, but zero obtained by the application of `fsolve` differs from the others.

If the initial guess for the zero of $f(x)$ is 0, following are the edited results of using MATLAB script (rootFind.m);

No solution found. fsolve stopped because the problem appears regular as measured by the gradient, but the vector of function values is not near zero as measured by the default value of the function tolerance.

```
X =   0                        (fsolve)
x =   2.154434690031884        (fzero)
y =   2.154434690031883        (roots)
```

```
%rootFind.m
clc; format long
f=@(x) x^3-10, X = fsolve(f,2), x=fzero(f,2), y=roots([1 0 0 -10])
```

Problem 1.8.5 Use MATLAB `solve` and `fzero` functions to find the real roots of following nonlinear equations [44–46] each having the form $f = 0$, and plot the solution curves in the same figure (solveNL1.m).

$$(a)\, f_1 = \sin^2 x - x^2 + 1,$$

$$(b)\, f_2 = x^2 - e^x - 3x + 2,$$

$$(c)\, f_3 = \cos(x) - x,$$

$$(d)\, f_4 = (x - 1)^3 - 1,$$

$$(e)\, f_5 = e^{x^2 + 7x - 30} - 1,$$

$$(f)\, f_6 = xe^{x^2} - \sin^2 x + 3\cos x + 5,$$

$$(g)\, f_7 = \sin(2\cos(x)) - 1 - x^2 + exp(\sin(x^3)),$$

$$(h)\, f_8 = \sqrt{x} - \cos(x).$$

Solution
The roots of nonlinear equations are given below.

$(a) \pm 1.4044916,$ $(b)\, 0.25753029,$ $(c)\, 0.73908513,$ $(d)\, 2.0000000,$
$(e) - 1.000000,$ $0.500000,$ $(f) - 1.2076478,$
$(g) - 0.78489599, 1.3061752018,$ $(h)\, 0.6417143708.$

Solution curves are all plotted in Fig. 1.14. Note that using `solve` function gives only a single root for the solutions of equations f_1, f_7, therefore we apply `fzero` function to search for the other root of f_7 about $x = 1$, and use symmetry to get the value of second root for f_1.

On the other hand, using `solve` function cannot find explicit solution for f_8. Application of `fzero` function to search for a root of f_8 about $x = 1$ yields the value of the root as $x = 0.6417143708$.

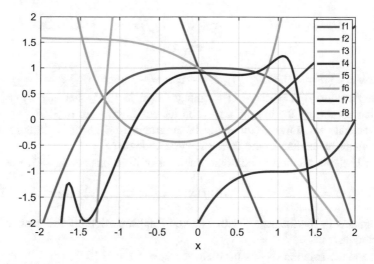

Fig. 1.14 Solution curves of nonlinear equations

```
%solveNL1.m  solve for real roots of scalar nonlinear equations
clc;clear;close;syms x
format long
f1=(sin(x)).^2-x.^2+1;F1=solve(f1 ,'Real',true);   F1=vpa(F1,8)
f2-x.^2-exp(x)-3^x+2; F2=solve(f2 ,'Real',true);  F2=vpa(F2,8)
f3=cos(x)-x ;  F3=solve(f3,'Real',true);          F3=vpa(F3,8)
f4=(x-1).^3-1; F4=solve(f4 ,'Real',true);         F4=vpa(F4,8)
f5=exp(x.^2+0.5*x-0.5)-1;F5=solve(f5,'Real',true);   F5=vpa(F5,8)
f6=x*exp(x.^2)- (sin(x)).^2+3*cos(x)+5;
F6=solve(f6,'Real',true);F6=vpa(F6,8)
f7=sin(2*cos(x))-1-x.^2+ exp(sin(x.^3));
F7=solve(f7, 'Real',true); F7=vpa(F7,8)
f8=x.^(1/2)-cos(x);F8=solve(f8,'Real',true);  F8=vpa(F8,8)
a=ezplot(f1);hold on; b=ezplot(f2);hold on; c=ezplot(f3);hold on;
d=ezplot(f4);hold on; e=ezplot(f5);hold on; f=ezplot(f6);hold on;
g=ezplot(f7);hold on;
h=ezplot(f8);hold on;
set(a,'linewidth',2);set(b,'linewidth',2);set(c,'linewidth',2);
set(d,'linewidth',2);set(e,'linewidth',2);set(f,'linewidth',2);
set(g,'linewidth',2);set(h,'linewidth',2);
axis([-2 2 -2 2]); grid;
legend('f1','f2','f3','f4','f5','f6','f7','f8');
F77=fzero(@(x)  sin(2*cos(x))-1-x.^2+ exp(sin(x.^3)),1)
F88=fzero(@(x) x.^(1/2)-cos(x) ,1)
```

1.9 Exercises

1. Use bisection, secant, Newton, Halley and Brent methods to find one real zero
   of the following polynomials with constant coefficients:

$$(i)\ f(x) = 7x^5 - x^3 + 5x - 1; \quad (ii)\ f(x) = x^4 - 2x^2 + 3x + 4;$$
$$(iii)\ x^3 - x^2 + x - 1$$

Answers: (i) $x = 0.2012$; (ii) $x = -1$, $x = -1.4856$; (iii) $x = 1$.

2. The function $f(x) = (5x - 1)^{-1}$ changes sign in [0,1], that is $f(0) \cdot f(1) < 0$. What point does the bisection method locate? Is it a zero of $f(x)$?

3. Investigate on how circuit analysis programs (e.g., SPICE) handle similar problems as studied in the code (newtonDiode.m)?

4. Use Halley's Method to compute one root of functions for $x_0 = 1$, $\epsilon = 10^{-6}$,

$$(a)\ f(x) = e^x - 2x - 0.9, \quad (b)\ f(x) = \arctan(x), \quad (c)\ f(x) = xe^x - 1 = 0$$

Compare its performance with Newton-Raphson Method for each of these functions.
Repeat the solution of this problem for $x_0 = 10$, $\epsilon = 10^{-6}$.

5. Find the critical "initial guess value" x_0, below which the zero of function $f(x) = \arctan(x)$
can be computed using the Halley's Method (Hint: For the convergence of Newton-Raphson method, critical initial guess value threshold is $|x_0| < 1.391745201$).

6. Test the performance of Halley's method for finding multiple roots of nonlinear equations.

7. Detemine the rate (order) of convergence for the Olver's method, using the results of iterations listed in the solution of sample problem presented in this chapter.

8. Find the critical "initial guess value" x_0, below which the zero of function $f(x) = \arctan(x)$ can be computed using the Olver's Method (Hint: For the convergence of Newton-Raphson method, critical initial guess value threshold is $|x_0| < 1.391745201$).

9. Test the performance of Olver's method for finding multiple roots of nonlinear equations.

10. Use Halley's and Olver's Methods to solve the same problem presented in this chapter for the improved Ostrowski's method.

11. Detemine the order of convergence for the improved Ostrowki's method presented here, using the results of iterations listed in the solution of this problem.

12. Find the critical "initial guess value" x_0, below which the zero of function $f(x) = \arctan(x)$ can be computed using the improved Ostrowski's Method (Hint: For the convergence of Newton-Raphson method, critical initial guess value threshold is $|x_0| < 1.391745201$).

13. Discuss common properties of Halley's, Olver's and improved Ostrowski's Methods.

14. Test the performance of improved Ostrowski's method for finding multiple roots of nonlinear equations.

15. Use Halley's, Olver's and improved Ostrowski's methods to solve the same problem presented in this chapter for the Steffensen's method.
16. Detemine the order of convergence for the Steffensen's method using the results of iterations listed in the solution of this problem presented in this chapter for the Steffensen's method.
17. Find the critical "initial guess value" x_0, below which the zero of function $f(x) = \arctan(x)$ can be computed using the Steffensen's Method.
18. Use MATLAB `roots` function to find all the roots of following polynomials:

$$(i)\, f(x) = 7x^5 - x^3 + 5x - 1; \quad (ii)\, f(x) = x^4 - 2x^2 + 3x + 4;$$
$$(iii)\, x^3 - x^2 + x - 1$$

Find real zero of the following nonlinear equations [47] by using one of the methods described in this chapter, correct to three significant digits.

19. $10^{150-5x^2} - 1, \quad (A : x = 5.477)$
20. $e^{x^2 + 7x - 30} - 1, \quad (A : x = 3.000)$
21. $\sqrt{x} - 3 - 1/x, \quad (A : x = 9.633)$
22. $e^x + x - 20, \quad (A : x = 2.842)$
23. $ln(x) + \sqrt{x} - 5, \quad (A : x = 8.309)$

Find the smallest positive zero of the following functions [48] on the interval [0,1] correct to four significant digits using Brent's Method.

24. $f(x) = e^{2x} - e^x - 2$
25. $f(x) = 4\sin(x) - e^x$
26. $f(x) = x - \sin(x)/5 - 1/2$
27. Use Newton and Secant methods to find the roots of

$$f(x) = -8\sin(x) + (36\sin(x) + (-54\sin(x) + 27\sin(x)x)x)x$$

(Answer: Triple root, $x = 2/3$) [21].
28. Use any of the methods described in this chapter to find the smallest nonzero root of the nonlinear function,

$$f(x) = \tanh\left(x^{1/4}\right) - \tan\left(x^{1/4}\right) = 0$$

with at least 8 decimal digits of accuracy. Compute the relative error of approximation.
(Exact solution: $x = 237.72106753$).

References

1. Chapra SC (2012) Applied numerical methods with MATLAB, for Engineers and Scientists (Chap. 5.4), 3rd edn. McGraw-Hill
2. Epperson JF (2007) An introduction to numerical methods and analysis (Chap. 3.1). John Wiley & Sons

3. Brezinski C (1991) History of continued fractions and Padé approximants. Springer, Berlin, p 90
4. Osada N (1993) Acceleration methods for slowly convergent sequences and their applications. Ph.D. thesis
5. Fikioris G (1999) An application of convergence acceleration methods. IEEE Trans Antennas Propag 47:1758–1760
6. Brezinski C, Zaglia MR (1991) Extrapolation methods. Theory and practice. North-Holland, Amsterdam
7. Delahaye JP (1988) Sequence transformations. Springer, Berlin
8. Wimp J (1981) Sequence transformations and their applications. Academic Press, New York
9. Keshavarzzadeh V, Ghanem RG, Masri SF et al (2014) Convergence acceleration of polynomial chaos solutions via sequence transformation. Comput Methods Appl Mech Eng 271:167–184
10. Buoso D, Karapiperi A, Pozza S (2015) Generalizations of Aitken's process for a certain class of sequences. Appl Numer Math 90:38–54
11. Picca P, Furfaro R, Ganapol BD (2016) Application of non-linear extrapolations for the convergence acceleration of source iteration. J Comput Theor Transp 45(5, Special issue, Part: 3):351–367
12. Chang X-K, He Y, Hu X-B et al (2018) A new integrable convergence acceleration algorithm for computing Brezinski-Durbin-Redivo-Zaglia's sequence transformation via pfaffians. Numer Algorithms 78(1):87–106
13. Zhang X, Zou L, Liang S et al (2018) A novel analytic approximation method with a convergence acceleration parameter for solving nonlinear problems. Commun Nonlinear Sci Numer Simul 56:354–364
14. Jamali H, Kolahdouz M (2018) Modified frame-based Richardson iterative method and its convergence acceleration by Chebyshev Polynomials. Univ Politehnica Bucharest Sci Bull-Ser A-Appl Math Phys 80(3):83–92
15. Chapra SC (2012) Applied numerical methods with MATLAB for Engineers and Scientists (Chap. 6.2), 3rd edn. McGraw-Hill
16. Stanton RG (1961) Numerical methods for science and engineering (Chap. 4.7). Prentice Hall
17. Epperson JF (2007) An introduction to numerical methods and analysis (Chap. 3.2). Wiley
18. Sauer T (2012) Numerical analysis (Chap. 1.4), 2nd edn. Pearson
19. Ascher UM, Mattheij RMM, Russell RD (1995) Numerical solution of boundary value problems for ordinary differential equations (Chap. 2.3). SIAM
20. Henrici P (1964) Elements of numerical analysis. Wiley Inc., p 87
21. Rump SM, Oishi S (2009) Verified computation of a disc containing exactly k roots of a univariate nonlinear function. Nonlinear Theory App IEICE 1(1):1–8
22. Keskin AU (2019) Ordinary differential equations for engineers, problems with MATLAB solutions. Springer, Berlin, p 724
23. Keskin AU (2017) Electrical circuits in biomedical engineering, problems with MATLAB solutions. Springer, Berlin, pp 155–165
24. Brent RP (1973) Chapter 4: an algorithm with guaranteed convergence for finding a zero of a function, algorithms for minimization without derivatives. Prentice-Hall, Englewood Cliffs, NJ
25. Dekker TJ (1969) Finding a zero by means of successive linear interpolation. In: Dejon B, Henrici P (eds) Constructive aspects of the fundamental theorem of algebra. Wiley-Interscience, London
26. Press WH, Teukolsky SA, Vetterling WT, Flannery BP (2007) "Section 9.3. Van Wijngaarden–Dekker–Brent Method". Numerical recipes: the art of scientific computing, 3rd edn. Cambridge University Press, New York
27. Kiusalaas J (2005) Numerical methods in engineering with MATLAB. Cambridge University Press, Cambridge, pp 150–155
28. Sauer T (2012) Numerical analysis, 2nd edn. Pearson, pp 64–65

29. Chapra SC (2012) Applied numerical methods with MATLAB for engineers and scientists (Chap. 6.4), 3rd edn. McGraw-Hill
30. Gander W (1985) On Halley iteration method. Am Math Mon 92(2):131–134
31. Ezquerro JA, Hernandez MA (2004) On Halley-type iterations with free second derivative. J Comp Appl Math 170:455–459
32. Ramos H (2014) Some efficient one-point variants of Halley's method, with memory, for solving nonlinear equations. In: International conference on numerical analysis and applied mathematics (ICNAAM), Article number: UNSP 810004, Rhodes, Greece
33. Liu S, Song Y, Zhou X (2015) Convergence radius of Halley's method for multiple roots under center-Holder continuous condition. Appl Math Comput 265:1011–1018
34. Proinov PD, Ivanov SI (2015) On the convergence of Halley's method for multiple polynomial zeros. Mediterr J Math 12(2):555–572
35. Prashanth M, Gupta DK (2015) Semilocal convergence for Super-Halley's method under omega-differentiability condition. Jpn J Ind Appl Math 32(1):77–94
36. Kaltenbacher B (2015) An iteratively regularized Gauss-Newton-Halley method for solving nonlinear ill-posed problems. Numer Math 131(1):33–57
37. Gnang C, Dubeau F (2018) On the rediscovery of Halley's iterative method for computing the zero of an analytic function. J Comput Appl Math 335:129–141
38. Grau M, Diaz-Barrero JL (2006) An improvement to Ostrowski root-finding method. Appl Math Comput 173:450–456
39. Ostrowski AM (1960) Solutions of equations and system of equations. Academic Press, NewYork
40. Steffensen JF (1933). Remarks on iteration. Scand Actuarial J 1:64–72
41. Ezquerro JA, Hernández MA, Romero N, Velasco AI (2013) On Steffensen's method on Banach spaces. J Comput Appl Math 249:9–23
42. Yun BI (2010) Transformation methods for finding multiple roots of nonlinear equations. Appl Math Comput 217(2010):599–606
43. Forsythe, GE, Malcolm MA, Moler CB (1976) Computer methods for mathematical computations. Prentice-Hall
44. Chun C (2005) Iterative methods improving Newton's method by the decomposition method. Comput Math Appl 50:1559–1568
45. Weerakoon S, Fernando GI (2000) A variant of Newton's method with accelerated third-order convergence. Appl Math Lett 17(8):87–93
46. Babajee DKR (2015) Some improvements to a third order variant of Newton's method from Simpson's rule. Algorithms 8:552–561. https://doi.org/10.3390/a8030552
47. Li X, Mu C, Ma J, Wang C (2010) Sixteenth-order method for nonlinear equations. Appl Math Comput 215(10):3754–3758
48. Conte SD, de Boor C (1980) Elementary numerical analysis (Chap. 3). McGraw-Hill

Chapter 2
Solution of Nonlinear Systems of Equations

In previous chapter, we worked on problems related to determine the value of unknown x that satisfied a single scalar equation, $f(x) = 0$. In this chapter, we deal with the problems of finding the unknown values of multiple independent functions that simultaneously satisfy a set of nonlinear equations, $f(x_i) = 0$, $i = 1, 2, \ldots, N$, where N is the number of independent functions and also the total number of nonlinear equations in the system.

Single-parameter nonlinear systems result in a single equation while multi-parameter systems result in a coupled set of nonlinear equations. These equations are coupled because the individual parts of the system are influenced by other parts. These dependencies result in nonlinear simultaneous equations.

In the solution of system of nonlinear equations, mostly one deals with iterative solution techniques, which are similar in nature to the methods discussed in the previous Chapter. In other words, they require guessing solution values and then perform iterations to have an estimate. While linear systems plot as straight lines, equations for nonlinear functions in a system plot as curves on each other, and their solution being the intersection of these curves, the solutions are the values of the x_i that make the equations equal to zero.

We demonstrate the solution of these systems of equations using Newton-Raphson (Newton method). Considering difficulties encountered in employing Newton method, we also present solution techniques of nonlinear systems using other methods, such as Damped Newton Method, Homotopy methods and Gradient Search. Homotopy methods belong to the family of continuation methods and represent a way to find a solution to a problem by constructing a new simpler problem, then gradually deforming the latter into the original one. Major advantage of the Homotopy Methods is that, they provide a globally convergent (always converging) method to solve nonlinear systems of algebraic or differential equations. In other words, one can get a solution for any arbitrary initial solution (guess) x_0.

We also describe how to apply MATLAB's built-in function `fsolve` in the solution of a nonlinear system and demonstrate its advantages.

© Springer Nature Switzerland AG 2019
A. Ü. Keskin, *Boundary Value Problems for Engineers*,
https://doi.org/10.1007/978-3-030-21080-9_2

2.1 Newton's Method for Systems

Problem 2.1.1 (*Newton's Method for systems*) Consider the solution of m-dimensional nonlinear equation[1] of the form $f(x) = 0$. This means that we solve m simultaneous nonlinear equations (in scalar notation),

$$
\begin{aligned}
f_1(x_1, x_2, \ldots, x_m) &= 0 \\
f_2(x_1, x_2, \ldots, x_m) &= 0 \\
&\cdots \\
f_m(x_1, x_2, \ldots, x_m) &= 0
\end{aligned}
\tag{2.1}
$$

Note that in previous problems we had only a single nonlinear equation, $f(x) = 0$. The difficulty of the solution of simultaneous nonlinear equations is that one cannot find an algorithm with a "guaranteed" initial estimate of x (if such a value is not suggested by the nature of the problem).

Newton's (or Newton-Raphson[2]) method is the method of choice for solving problems of this kind, as long as a good starting vector of unknowns is supplied.

Derivations of Newton's method for solving a system of equations start with the Taylor series expansion of $f_i(x)$ about the point x, $(i = 1, 2, \ldots, m)$,

$$
f_i(x + \Delta x) = f_i(x) + \sum_{k=1}^{m} \frac{\partial f_i}{\partial x_k} \Delta x_k + \mathcal{O}(\Delta x^2)
\tag{2.2}
$$

$$
f(x + \Delta x) \cong f(x) + J(x) \Delta x
\tag{2.3}
$$

where $J(x)$ is the $(m \times m)$ Jacobian matrix.

For the iterative solution, we assume that x is the current approximate solution and $(x + \Delta x)$ is the improved one. Setting $f(x + \Delta x) = 0$ to find the new Δx yields a set of linear equations,

$$
J(x) \Delta x = -f(x)
\tag{2.4}
$$

Newton's method for solving simultaneous nonlinear equations consist of the following steps: (i) Guess the solution vector x, (ii) Find $f(x)$, (iii) Compute the $j(x)$,

---

[1] A simple approach for solving nonlinear systems of algebraic equations (which is called successive substitution) is to use the same strategy that is employed for fixed-point iteration, where, each one of the nonlinear equations can be solved for one of the unknowns, and they can then be implemented iteratively to compute new values until they converge on the solutions. Shortcomings of successive substitution method are that the convergence often depends on the manner in which the equations are formulated and divergence can occur if the initial guesses are selected far from the actual solution. Therefore, successive substitution iteration has limited utility for solving nonlinear systems and we do not include sample solutions of problems related to this method.

[2] We use both names interchangeably for the same method that has been employed for solving single and multi-dimensional nonlinear equations.

(iv) Set up Eq. (2.4) and solve for Δx, (v) Let $(x + \Delta x)$ replace x, (vi) Repeat previous steps until $|\Delta x| < \epsilon$, where ϵ is the described error tolerance value.

Once a good starting vector is used, convergence rate of the Newton's method is high.

Note that partial derivatives in the Jacobian matrix are computed using the finite difference approximation of

$$\frac{\partial f_i}{\partial x_k} \cong \frac{f_i(x + e_j h) - f_i(x)}{h} \tag{2.5}$$

where h is a small increment and e_j is the unit vector in the direction of x_k.

(a) What are the main differences between Newton's methods for single variable and for systems solution?

(b) Investigate the major weaknesses of this method for systems solution.

(c) Compute analytically the coordinates of intersections for a circle of unit radius centered at the origin and a cubic polynomial, $y - x^3 = 0$. Plot these curves (intersection1.m).

(d) Use Newton's method to solve numerically these simultaneous equations (system_nonlin2.m), (newton_sys.m), (jac.m).

Solution

(a) Newton's method for systems, (i)—replaces the derivative in the single-variable case with the m × m Jacobian matrix, (ii)—substitutes multiplying by the reciprocal of the derivative with multiplying by the inverse of the Jacobian matrix.

(b) If the system has a singular Jacobian matrix, convergence may not occur within a reasonable number of iterations. Another weakness of this method is the requirement that a Jacobian matrix be computed at each iteration and a linear system solved that involves the Jacobian matrix which considerably increases the amount of computational cost. Since the exact evaluation of the partial derivatives is inconvenient, or impossible, this difficulty can generally be overcome by using finite-difference approximations to the partial derivatives. Even if this approximation is used, it still requires that at least m^2 scalar functional evaluations be performed to approximate the Jacobian matrix, and does not decrease the amount of calculations. This amount of computational effort is prohibitive except for relatively small values of m and easily evaluated simple scalar functions [1].

A class of methods known as Quasi-Newton[3] methods can be employed in more challenging problems. On the other hand, methods such as "Homotopy

---

[3]For example, Broyden's method replaces the Jacobian matrix in Newton's method with an approximation matrix that is updated at each iteration, requiring only m scalar functional evaluations per iteration at the expense of the lost quadratic convergence of Newton's method.

continuation" or "Steepest Descent" can be used whenever there is a difficulty in estimating an initial guess vector.

(c) Analytical solution:

$$f_1 = x^2 + y^2 - 1 = 0, \quad f_2 = x^3 - y = 0, \quad y = x^3 = \pm\sqrt{1 - x^2}$$
$$p = x^6 + x^2 - 1 = 0$$

The roots of this polynomial equation are $x = \pm0.8260$, which correspond to ordinate values at $y = \pm0.5636$. MATLAB m-file (intersection1.m) computes the roots and plots the two curves as shown in Fig. 2.1.

(d) We apply Newton's method to solve these simultaneous equations using MATLAB m-file (system_nonlin2.m) and two external function files (newton_sys.m) and (jac.m). The latter function file computes the jacobian matrix of the system equations in double format and does not employ Symbolic Math built-in function called "jacobian". The system of equations is defined using an anonymous function in the main m-file (system_nonlin2.m).

Initial estimates for the solution vector, [1; 0] and [1; −1] yield solutions of $x = 0.8260$, $y = 0.5636$. Initial estimates for the solution vector, [−1; 1] yields solutions of $x = -0.8260$, $y = -0.5636$, while the choices of initial solution vector estimates [0; 0] and [0; 1] cause the code to exceed preset number of iterations. The code (intersection1.m) uses the command fimplicit which has been introduced in MATLAB version 2016b.

Fig. 2.1 A cubic polynomial intersecting a circle of unit radius centered at the origin

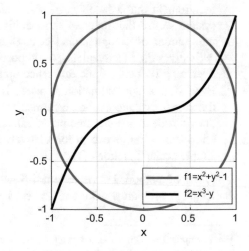

```
%intersection1.m
clc;clear;format short
fimplicit(@(x,y) x.^2 + y.^2 - 1, [-1  1]);hold on;
fimplicit(@(x,y) x.^3 - y , [-1  1]);
legend('f1=x^2+y^2-1','f2=x^3-y');grid;
x = fzero(@(x) x.^6 + x.^2 - 1, 1), y=x^3, p=x^2+y^2-1  % p --> 0

% system_nonlin2.m  solves the nonlinear system,  x^2+y^2-1=0; x^3-y=0
clc;clear;
u = [1 ; 0];  % starting (guessed) solution vector [x, y].
eq=@(x) [x(1)^2+x(2)^2-1; x(1)^3-x(2)];
x=newton_sys(eq,u)

function r = newton_sys(f,u)
% Newton's method of finding  solution set of simultaneous
% f = function handle  returning [f1,f2,...,fn]
% u = starting solution vector [x,y,z,..]
tol=1e-6;
for i = 1:25
[J,fx] = jac(f,u); DP=dot(fx,fx);
if (DP/length(u))^(1/2) < tol, r = u; return, end
dx = J \(-fx); u = u + dx;
if sqrt(dot(dx,dx)/length(u)) < tol*norm(u,Inf), r = u; return, end
end, error(' iterations exceeded allowable count')

function [J,fx] = jac(f,x)
% J= Jacobian matrix
J = zeros(length(x)); fx = feval(f,x);
for k =1:length(x)
g = x(k); x(k) = g + 1.0e-5;  fa = feval(f,x); x(k) = g;
J(:,k) = (fa - fx)/1.0e-5;
end
```

Problem 2.1.2 Use Symbolic Math and Newton-Raphson Method to solve the following system of nonlinear equations [2],

$$x - \sinh(y) = 0$$
$$2y - \cosh(x) = 0$$

with the following conditions:
Maximum number of iterations = 10,
Error tolerance (norm2) = 10^{-4},
$x_0 = [0.6 \quad 0.6]$.
    Display solutions at each step (NewtonNL2.m), (plotNL2.m).

Solution
Graph of given functions are shown in Fig. 2.2.
    Following is an edited list from MATLAB Command Window. Solution vector is

$$x = [0.646280215284669 \quad 0.608103070221628]$$

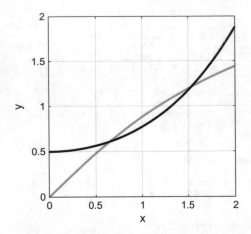

Fig. 2.2 Graph of functions, $x - \sinh(y) = 0, 2y - \cosh(x) = 0$. It is found that the intersection point on the left has coordinate values of $[0.6463 \quad 0.6081]$

```
MaxIter = 10,   tol = 1e-04,    F =[ x - sinh(y),  2*y - cosh(x)]
x0 = 0.600000000000000   0.600000000000000
J =[         1, -cosh(y)]
   [ -sinh(x),        2]

iteration=1
f' = -0.036653582148241    0.014534781757732
J =    [1.000000000000000  -1.185465218242268
       -0.636653582148241   2.000000000000000]
df' = [-0.045031773499928  -0.007067429075743]
X' =    0.645031773499928   0.607067429075743

iteration=2
f' =    -0.000015969775611  -0.001211872907992
J =       [1.000000000000000  -1.185465218242268
          -0.636653582148241   2.000000000000000]
df' =  [-0.001179321377540  -0.000981346043753]
X' =    0.646211094877468   0.608048775119496

iteration=3
f' =    1.0e-04 *   [-0.047554521027702  -0.645844494196415]
J =    [1.000000000000000  -1.185465218242268
       -0.636653582148241   2.000000000000000]
df' =    1.0e-04 *   [-0.691204072009869  -0.542951021318475]
X' =    [0.646280215284669   0.608103070221628]
```

```
%NewtonNL2.m    Newton-Raphson method using Symbolic Math
clc;clear;
syms x y ;format long
MaxIter=10
tol=1e-4
F=[x-sinh(y);2*y-cosh(x)],x0=[0.6 0.6]
%------------------------------------------
 if size(x0,1) == 2; x0 = x0';end
J=jacobian(F,[x y])
for i=1:MaxIter
fprintf('iteration=%i',i);f=double((subs(F,[x y],x0)))
J=double(subs(J,[x,y],x0)), df=J\f,  X=x0'-df
if norm(df)<tol,return,end
x0=X'
end
```

```
%plotNL2.m
clc;clear;close; syms x y
f1=x-sinh(x); f2=2*y-cosh(x); xL=[0, 2];
g=ezplot(f1,xL); set(g,'linewidth',2); hold on;
h=ezplot(f2,xL); grid; set(h,'color','k','linewidth',2); hold off;
```

Problem 2.1.3 Use Newton's method (without using Symbolic Math) to compute the coordinates of intersection points (for positive x-axis values) of an ellipse centered at the origin having horizontal semi-major axis length of $\sqrt{2}$, semi-minor axis length of 1, with the function $10^x \cos(\pi x/2) + 10^y \cos(\pi y/2) = 0$. Let tolerance value be $\epsilon = 10^{-4}$. Plot these curves in the same figure. (plot_implicit2.m), (system_nonlin3.m), (newton_sys2.m), (jac.m).

Solution
Curves of both functions and their intersections are displayed in Fig. 2.3.

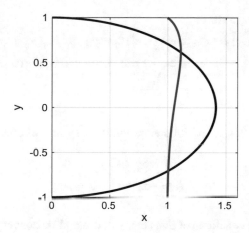

Fig. 2.3 Curves of ellipse and the function $10^x \cos(\pi x/2) + 10^y \cos(\pi y/2) = 0$ in the first and fourth quadrants

We run main MATLAB file (system_nonlin3.m) separately to compute intersection points in each quadrant by assigning different starting points. It takes 5
iterations to obtain the intersection point coordinate in first quadrant as $x =$
1.1151, $y = 0.6151$ while 3 iterations are needed to obtain the intersection point
coordinate in fourth quadrant as $x = 1.0056$, $y = -0.7031$.

```
%plot_implicit2.m
clc;syms x y
r=[0 1.6  -1 1]; a=ezplot(x^2/2+y^2-1, r);hold on;
b=ezplot('10^x*cos(x*pi/2)+10^y*cos(y*pi/2)=0',r);
set(a,'color','k','linewidth',2); set(b,'color','b','linewidth',2);grid
hold off;
```

```
% system_nonlin3.m  solves the nonlinear system,  x^2/2+y^2-1=0; x^3-y=0
clc;clear;
global tol
tol=1e-4;
u = [1 ; 1];  % starting solution vector [x, y] in 1st quadrant
%u = [1 ; -1];  % starting solution vector [x, y] in 4th quadrant
eq=@(x) [x(1).^2/2 + x(2).^2-1;
         (10.^x(1))*cos(pi*x(1)./2)+(10.^x(2))*cos(pi*x(2)./2)];
[Iter,x]=newton_sys2(eq,u)
```

```
function [I,r] = newton_sys2(f,u)
% Newton's method of finding  solution set of simultaneous
% equations fi(x1,x2) = 0, i = 1,2.
% f = function handle  returning [f1,f2].
% u = starting solution vector [x1,x2].
global tol
if size(u,1) == 1; u = u';end
for i = 1:50
[J,fx] = jac(f,u); DP=dot(fx,fx); fx=fx';
if (DP/length(u))^(1/2) < tol, r = u; return, end
dx = J \ (-fx); u = u + dx;
if sqrt(dot(dx,dx)/length(u)) < tol*norm(u,Inf), r = u; return, end
I=i;
end, error(' iterations exceeds allowable count')
```

```
function [J,fx] = jac(f,x)
% J= Jacobian matrix
J = zeros(length(x)); fx = feval(f,x);fx=fx';
for k =1:length(x)
g = x(k); x(k) = g + 1.0e-5; fa = feval(f,x); fa=fa'; x(k) = g;
J(:,k) = (fa - fx)/1.0e-5;
end
```

Problem 2.1.4 Given is the following nonlinear system of equations;

$$f_1 = x_1^2 + x_2^2 + 9 = 0$$
$$f_2 = -x_1 + x_2^2 - 3 = 0$$

(a) Determine the equations of lines that will cause slow convergence of Newton's
method if initial points are located on these lines.

(b) If the solutions of this system of nonlinear equations are $A = (2, \sqrt{5})$, $B = (2, -\sqrt{5})$ and $C = (-3, 0)$, which of these will cause linear (slower) convergence of Newton's method then the other points?

Solution

(a) We calculate determinant of the Jacobian matrix, $\det(J) = 0$

$$\begin{vmatrix} 2x_1 & 2x_2 \\ -1 & 2x_2 \end{vmatrix} = 0$$

$$4x_1 x_2 + 2x_2 = 0$$

which gives $x_2 = 0$, $x_1 = -1/2$.

Hence, one should avoid selecting initial points that are located on these lines.

(b) We substitute each point into Jacobian matrix and observe that "C" yields a singular Jacobian matrix,

$$\begin{bmatrix} 2 \times 3 & 2 \times 0 \\ -1 & 2 \times 0 \end{bmatrix} = \begin{bmatrix} -6 & 0 \\ -1 & 0 \end{bmatrix}, \quad \det \begin{bmatrix} -6 & 0 \\ -1 & 0 \end{bmatrix} = 0$$

This means that f_1 is tangential to f_2 at C and this will cause linear convergence of Newton's method instead of quadratic convergence.

2.2 Broyden's Method

Problem 2.2.1 Newton's Method for solving a univariate equation requires knowledge of the derivative. The Secant Method can be used when the derivative is not available. For (nonlinear) systems of equations, Newton's Method is a good choice if the Jacobian is available or it can be evaluated. Otherwise, one may look for an alternative such as Broyden's Method [3]. Broyden's Method uses the Newton's Method matrix solver step,

$$x_{i+1} = x_i - A_i^{-1} F(x_i)$$

where A_i is the best approximation available at step i to the Jacobian matrix, while updating the approximate Jacobian by

$$A_{i+1} = A_i + \frac{[F(x_{i+1}) - F(x_i) - A_i\delta_i]\delta_{i+1}^T}{\delta_{i+1}^T\delta_{i+1}}$$

with $\delta_{i+1} = x_{i+1} - x_i$, and T stands for transpose operation.

The Broyden's algorithm starts with an initial guess x_0 and an initial approximate Jacobian A_0, which can be chosen to be the identity matrix as a first choice. Note that the Newton-type step is carried out by solving $A_i\delta_{i+1} = F(x_i)$, just as for Newton's Method.

Relatively expensive matrix solver step can be relaxed by using an estimate for the matrix inverse of the Jacobian, B_i, which can be taken as the identity matrix in most cases,

$$x_{i+1} = x_i - B_iF(x_i)$$

$$B_{i+1} = B_i + \frac{\{x_{i+1} - x_i - B_i[F(x_{i+1}) - F(x_i)]\}\delta_{i+1}^TB_i}{\delta_{i+1}^TB_i[F(x_{i+1}) - F(x_i)]}$$

Broyden's Method converge slightly slower than the quadratic convergence of Newton's Method. If Jacobian is available, it usually speeds up convergence. Also like Newton's Method, Broyden's Method is not guaranteed to converge to a solution [4].

Use this information, solve the following system of nonlinear equations employing Broyden's method with initial guess vector of $x_0 = [1, 1]$, taking identity matrix as the matrix inverse of the Jacobian, for a maximum number of 10 iterations.

$$y - 2x^2 = 0$$
$$x^2 + y^2 - 1 = 0$$

Solution

We use MATLAB script file (broyden.m) to compute the solution of this system. Following is the edited output list obtained after running this code.

```
N = 10
x0 = 1 1
x = 0.6248 0.7808
Elapsed time is 0.018215 seconds.
```

MATLAB script file (broyden.m) is given below.

```
%broyden.m
clc;clear;tic;
%input: f=system matrix, k= maximum number of steps,x0=initial vector
f=@(x) [x(2)-2*x(1)^2; x(1)^2+x(2)^2-1];%system matrix
k=10        % max number of steps
x0=[1;1]     %initial (guessed) solution vector
n=length(x0);
b=eye(n,n);  % initial guess for inverse Jacobian matrix
 for i=1:k
x=x0-b*f(x0); del=x-x0; delta=f(x)-f(x0);
b=b+(del-b*delta)*del'*b/(del'*b*delta);
x0=x;
end
x=x
toc;
```

2.3   Damped Newton Method

Problem 2.3.1 (*Damped Newton Method*) As an improvement for the local convergence of Newton's Method, following iteration can be used,

$$x^{m+1} = x^m + \lambda z^m \tag{2.6}$$

Here, damping factor is $0 < \lambda \le 1$, and z solves the linear system

$$J(x^m)z = -f(x^m) \tag{2.7}$$

where J is $n \times n$ Jacobian matrix. Here, (2.7) can be written as

$$z = -J(x)^{-1}f(x) \tag{2.8}$$

(we dropped iteration counter m for notational simplicity).

The value of λ changes at each iteration. In order to find its value, an objective function is defined which relates to 2-norm of the residual $f(x)$,

$$g(x) = |f(x)|^2 = \sum_{j=1}^{n} f_j(x)^2$$

and define a function $\phi(\lambda)$,

$$\phi(\lambda) = \frac{g(x^m + \lambda z) - g(x^m)}{\lambda z^T \nabla g(x^m)}$$

Initially, a test value of λ is accepted if $\phi(\lambda) \geq \sigma$, where $0 < \sigma < 1/2$.

Denoting $g_\lambda = g(\widehat{x})$, $\widehat{x} = x + \lambda z$, if $g_\lambda \leq (1 - 2\lambda\sigma)g_0$ then λ is accepted, otherwise λ value is taken the maximum of either $\tau\lambda$ or $\lambda^2 g_0/[(2\lambda - 1)g_0 + g_\lambda]$. Iterations are repeated until λ is accepted or no convergence detected. Stopping criterion is $|z| \leq \epsilon$, until iteration limit is reached. Initial parameters can be $\lambda_{min} = \sigma = 0.01$, $\tau = 0.1$ [5].

(a) Use standard Newton-Raphson method to compute the real zero of $f(x) = x^{10} - 1$ when $x_0 = 0.15$.
(b) Use Damped Newton Method to determine the real zero of $f(x) = x^{10} - 1$ when $x_0 = 0.15$. and find the value of λ for minimum number of iterations (Damped Newtonm.m).

Solution

(a) Standard Newton-Raphson method converges to $x = 1$ at the end of 145th iteration. (error $= 4.78 \times 10^{-10}$). For undamped (standard) Newton-Raphson method, assign $\lambda = 1$, in the code.
(b) Damped Newton method converges to $x = 1$ at the end of 26th iteration (error $= 1.48 \times 10^{-9}$) when $\lambda = 10^{-7}$. For $\lambda = 10^{-6}$ convergence takes place after 33 iterations; while for $\lambda = 10^{-8}$, this happens at the end of 30th iteration. See, Table 2.1.

Table 2.1 Damping coefficient versus number of iterations

λ	n (iterations)
1	145
10^{-1}	126
10^{-2}	107
10^{-3}	88
10^{-4}	70
10^{-5}	51
10^{-6}	33
10^{-7}	26
10^{-8}	30

```
%Damped Newtonm.m
clc;clear;
global eps maxIter display
eps = 1e-8;% tolerance
maxIter=50;% max.iteration count
display='on';
%display='off';
%---------------------------------
% F=@(x)  x^10-1;
% x0=0.15;  % initial guess
%---------------------------------
%[x,residual] = DampedNewton2(Function,initial guess,Lambda)
[x,F] = DampedNewton2(F, x0, 1e-1)

function [x, F] = DampedNewton2(fun, x, L)
global eps maxIter display
%  x = Solution,  F = Value at x,  J= Jacobian at x,  L=lambda
N = length(x); Lmin = 1e-10; %  System order and Minimum damping factor
J = zeros(N);% Allocate Memory for Jacobian (NxN) matrix
F = fun(x); % get first Values for x
for i = 1:N; xtmp = x; h = sqrt(eps)*x(i); xtmp(i) = xtmp(i)+h;
    J(i,:) = (fun(xtmp)-F)./h; % Jacobian
end
s = (J\-F); x = x + L*s; %  step size computation and first newton step
F = fun(x); err = sqrt(sum(F.^2)/N); n = 0; % Iterations counter set
if(strcmp(display,'on'));
    fprintf('n= %i damping= %e error: %e x= ',n,L,err);
    fprintf(repmat('%e ',1,N),x);  fprintf('\n');
end
while( err > eps && n < maxIter)
    n = n+1;
    for i = 1:N; xtmp = x;h = sqrt(eps)*x(i); xtmp(i) = xtmp(i)+h;
 J(i,:) = (fun(xtmp)-F)./h; end % Jacobian
 stmp = (J\-F);
 if(~(max(abs(L * stmp)) < max(abs(s)) )) % optimal step size



% Decrease damping if step size does not get shorter
 while( ( max(abs(s)) < max(abs(L * stmp))))
 L = L / 2;
 if( L < Lmin ); fprintf(' Extreme damping.\n'); return;  end
 end % step size is fine
 end
s = stmp; x = x + L*s;  F = fun(x); % next Newton step
L = min( 1., L * 2);% increase damping factor
err = sqrt(sum(F.^2)/N);  if(strcmp(display,'on'));
fprintf('n=%i  damping=%e  error= %e  x= ',n,L,err);
fprintf(repmat('%e ',1,N),x);  fprintf('\n'); end
 end % the while loop ends here
if(n<maxIter)&&(err<=eps), disp('Solution converged');
else  disp('Solution did not converge');  end; end
```

2.4 MATLAB Built-in Functions, Solve and Fsolve

Problem 2.4.1 Use MATLAB Symbolic Math in finding roots of following simultaneous equations (roots_system_sym1.m).

$$f_1 = 3x_1 - \cos(x_1 x_3) - \frac{1}{2} = 0$$
$$f_2 = x_1^2 - 81(x_2 + 0.1)^2 + \sin(x_3) + 1.06 = 0$$
$$f_3 = e^{-x_1 x_3} + 20x_3 + (10\pi - 3)/3 = 0$$

Solution
We use MATLAB script (roots_system_sym1.m) to compute the roots of given system of equations. However, the solutions are approximate, as indicated by the following message;

```
Warning: Cannot solve symbolically. Returning a numeric approximation
instead.
```

```
x1 = 0.48852564005812573687043534 76182
x2 = -0.0015126675140186151776475601935424
x3 = -0.53864948362082392691720958715253
```

Actual solutions are $x_1 = 0.5$, $x_2 = 0$, $x_3 = -0.5235988$ [6].

```
%roots_system_sym1.m
% Symbolic numeric approximation
clc;clear;
syms x1 x2 x3;
f1 = 3*x1-cos(x1*x3)-1/2;
f2=x1.^2-81*(x2+0.1).^2+sin(x3)+1.06;
f3=exp(-x1*x3)+20*x3+(10*pi-3)/3;
eqns = [f1 == 0, f2 == 0, f3==0]; [x1, x2, x3 ] = solve(eqns)
```

Problem 2.4.2 Two simultaneously ejected electrons at $t = 0$ have the following path equations;

$$y_1 = \sin^2 t - t^2 + 1 = 0,$$
$$y_2 = e^{t^2 + 7t - 30} - 1 = 0.$$

Use MATLAB `solve` built-in function to find the time at which there is a possibility of collision for these particles with each other for $t > 0$ (solveNL2.m).

Solution

Equating above given equations and solving for t gives, $t = 0.83567076$ (time unit).
Collision takes place at $y = 0.851842$ (distance unit).

```
%solveNL2.m   solve for intersection point of two nonlinear equations
clc;clear;syms t; format long
y1=(sin(t)).^2-t.^2+1;   y2=exp(t.^2+0.5*t-0.5)-1;
T=solve(y2-y1,'Real',true);  T=vpa(T,6)
y=subs(f1,t,T);y=vpa(y,6)
```

Problem 2.4.3

(a) Briefly describe MATLAB's built-in function `fsolve`, and its difference from the function `fzero`. What is the goal of solving n nonlinear equations with n unknowns? What is the principle of operation of this function? Name the algorithms it can use.

(b) Solve the following system of nonlinear equations using `fsolve` function.

$$\cos(x) + y\sin(x) = -1$$
$$e^{(x-4y)} + x^2 = 1 + 16y^2$$

Solution

(a) While the `fzero` function solves a single one-dimensional scalar nonlinear equation for a scalar variable, MATLAB built-in function `fsolve` solves system of nonlinear equations. More specifically, it solves a problem specified by $F(x) = 0$, where x is a vector or a matrix; $F(x)$ is a function that returns a vector value. The goal of solving n nonlinear equations with n unknowns is to find a vector x that makes all $F_n(x) = 0$.

The function `fsolve`[4] attempts to solve systems of equations by minimizing the sum of squares of the components. If the sum of squares is zero, the system of equation is solved. It employs three algorithms: Trust-region-reflective, Trust-region dogleg, Levenberg-Marquardt.

Simple form of syntax is x = `fsolve(fun, x0)`, starts at ×0 and tries to solve the equations,
  `fun(x) = 0`.

(b) First, convert the equations to the form $F_n(x) = 0$, $n = 1, 2$

$$\cos(x) + y.\sin(x) + 1 = 0$$
$$e^{(x-4y)} + x^2 - 1 - 16y^2 = 0$$

---

[4]Readers may refer to Optimization Toolbox documentation of Mathworks, Inc., for more examples and how to solve Nonlinear Equations with Analytic Jacobian, Nonlinear Equations with Finite-Difference Jacobian, Nonlinear Equations with Jacobian, and Nonlinear Equations with Jacobian Sparsity Pattern. Details of algorithms for Trust-Region Dogleg Method, Trust-Region Reflective fsolve Algorithm, Levenberg-Marquardt Method are also described in the same documentation.

We can write an anonymous function that computes the left-hand side of these two equations, then solve the system of equations starting at the point $[1, 1]$.

Following is the result of running the code (system_nonlin1.m) $(x \cong \pi, y \cong \pi/4)$,

```
Equation solved.
fsolve completed because the vector of function values is near zero
as measured by the default value of the function tolerance, and
the problem appears regular as measured by the gradient.
                          w =    3.1416    0.7854
```

MATLAB m-file (system_nonlin1.m) is listed below.

```
% system_nonlin1.m  Solves nonlinear system of eqns
clc;clear;
u = [1 1];   % Trial values
eq=@(x,y) [cos(x(1))+x(2)*sin(x(1))+1;
          exp(x(1)-4*x(2))+x(1)*x(1)-1-16*x(2)*x(2)];
w = fsolve(eq, u)
```

Problem 2.4.4 The function `fsolve` can fail to solve an equation for various reasons.[5]

(a) As a first remedy, try changing the starting point. By changing initial points, the chances of success is increased.
(b) Check the definition of the equation to make sure that it is smooth. This function can fail to converge for functions with discontinuities.
(c) Check that dimensions for input and output has the same number of unknowns as values of the equation.
(d) Change tolerances, especially "OptimalityTolerance" and "StepTolerance". If you attempt to get high accuracy by setting tolerances to very small values, `fsolve` can fail to converge. If you set tolerances too high, `fsolve` can fail to solve an equation accurately.
(e) Check the problem definition. Some problems have no real solution.
(f) `fsolve` expects a function handle to your function that you want to solve, not the actual name of the function itself.

Solve following system of nonlinear equations with given starting vector values. (fsolve2.m).

(a) $$x^2 + 2|y| - 1 = 0, \quad 2xy + 1 = 0; \quad [1, 1]$$

(b) $$x/(y - 1) + 1 = 0, \quad 2xy + 1 = 0; \quad [1, 1], [2, 2]$$

---

[5]See, Mathworks web site for more information. (https://www.mathworks.com/help/optim/ug/when-the-solver-fails.html) last time accessed 15th, March 2019.

Solution

Both problems have discontinuities.

(a) $$x^2 + 2|y| - 1 = 0, \quad 2xy + 1 = 0$$

```
f = @(y)[y(1).^2 + 2*abs(y(2))-1;2*y(1)*y(2) + 1]
```

No solution found. fsolve stopped because the problem appears regular as measured by the gradient, but the vector of function values is not near zero as measured by the default value of the function tolerance.

(b) $$x/(y - 1) + 1 = 0, \quad 2xy + 1 = 0$$

```
When initial vector=[1,1],
f2 = @(y)[y(1)./(y(2)-1)+1;y(1)*y(2)-1]
Error using trustnleqn (line 28)
Objective function is returning undefined values at initial point. FSOLVE
cannot continue.
```

```
When initial vector=[2,2],
f2 = @(y)[y(1)./(y(2)-1)+1;y(1)*y(2)-1]
Solver stopped prematurely.fsolve stopped because it exceeded the function
evaluation limit,options. MaxFunctionEvaluations = 200 (the default value).
```

```
%fsolve2.m  solves nonlinear system of equations
%yeq = fsolve(f,[x0,y0]) y(1)=x, y(2)=y
clc;clear;format long
%f1 =@(y) [ y(1).^2+2*abs(y(2))-1;  2*y(1)*y(2)+1]
%yeq = fsolve(f1,[1,1])
f2 =@(y) [ y(1)./(y(2)-1)+1;  y(1)*y(2)-1]
yeq = fsolve(f2,[2,2])
```

2.5   Homotopy Methods

Problem 2.5.1 Two ideas have been used with some success to force convergence of a root finding method, viz., homotopy continuation (or imbedding), and damping. The basic theory for homotopy algorithms was developed in 1976. They are applicable to zero-finding problems, optimization and the discrezations of nonlinear two-point boundary value problems based on shooting, finite differences, collocation, and finite elements. They are closely related to ODE algorithms, and make heavy use of ODE techniques.

Describe the principles of Homotopy continuation[6] methods used to solve nonlinear systems of algebraic or differential equations. Additionally, describe steps of a numerical algorithm for the solution of nonlinear systems of algebraic or differential equations using "Newton homotopy method".

Solution

Homotopy methods belong to the family of continuation methods and represent a way to find a solution to a problem by constructing a new simpler problem, then gradually deforming the latter into the original one. Major advantage of the Homotopy Methods is that, they provide a globally convergent (always converging) method to solve nonlinear systems of algebraic or differential equations. In other words, one can get a solution for any arbitrary initial solution (guess) x_0. Roughly speaking, this method starts from a system of equations of which the solutions are known and goes to the system of equations that is to be solved. The idea is to deform the given system $f(x) = 0$ to another system

$$H(x, \lambda) = \lambda f(x) + (1 - \lambda)g(x)$$

where $g(x)$ has a known solution and λ is the continuation parameter[7] in the closed interval $0 \leq \lambda \leq 1$. $H(x, \lambda)$ is called homotopy or deformation of $f(x)$. Here $g(x)$ and $f(x)$ are homotopic because $H(x, 0) = g(x)$, $H(x, 1) = f(x)$. Starting from the known solution of $g(x) = 0$ at $\lambda = 0$, we can trace the zero curve (homotopy path) until $\lambda = 1$ at which the desired solution is reached.

Homotopy methods give different roots depending upon the selection of the initial guess values for the solution of $g(x) = 0$ (at $\lambda = 0$). One root is obtained for each initial solution. Depending on the smoothness of the path, the best approximation to a root value is obtained.

A good way of choosing $g(x)$ is to implement a function that can easily be solved to get the initial solution. For example, one can choose $g(x) = f(x) - f(x_0)$ for any value of x_0. In this case the "Newton homotopy" takes the form

$$H(x, \lambda) = f(x) + (\lambda - 1)f(x_0)$$

Alternatively, if $g(x) = x - x_0$ the homotopy becomes a "fixed point homotopy",

$$H(x, \lambda) = (1 - \lambda)(x - x_0) + \lambda f(x)$$

A more general homotopy is the linear homotopy,

$$H(x, \lambda) = \lambda f(x) + (1 - \lambda)g(x)$$

---

[6]Detailed presentation of this subject can be found in [7–12].

[7]This parameter is represented in literature using different symbols, t, λ being the most common ones.

This will transform $g(x)$ into $f(x)$.

Other types of homotopies are also possible (constant, identity, etc.) to find good approximation for the solution of a given system of nonlinear equations.

Although difficulties may be encountered if a point of singularity exists during the course of integration, this method can be used as a stand-alone method not requiring a particularly good choice of starting values. The method can also be used to provide a starting approximation for Newton's or Quasi-Newton Methods, such as Broyden's method.

Algorithm for Newton homotopy method, $H(x, \lambda) = f(x) + (\lambda - 1)f(x_0)$ to solve a nonlinear system of algebraic equations consists of following steps:

1. Consider the given system in the form $f(x) = 0$,
2. Compute $f(x(0)) = 0$,
3. Compute the Jacobian matrix $J(x)$ of $f(x)$,
4. Write Newton homotopy differential equation (NHDE),

$$x'(\lambda) = -J(x(\lambda))^{-1}f(x_0) = \varphi(x, \lambda)$$

5. Choose h (for uniform step size numerical ODE solvers) for the solution of NHDE,
6. Solve NHDE by using a numerical method such as Euler's Method or fourth order Runge Kutta Method [13].

Problem 2.5.2 (*Implicit Function Theorem*) Let $G(u) = 0$. Newton's method for solving this nonlinear equation may not converge if the initial guess is not near to a solution. The solution can still be possible if one can put a homotopy parameter λ in the equation [14].

Let $G : \mathbb{R}^n \times \mathbb{R} \to \mathbb{R}^n$ and

$$G_u(u_0, \lambda_0) = \frac{\partial G}{\partial u}(u_0, \lambda_0)$$

If,

(i) $G(u_0, \lambda_0) = 0, \lambda_0 \in \mathbb{R}, u_0 \in \mathbb{R}^n$,
(ii) $G_u(u_0, \lambda_0)$ is nonsingular (u_0 is an isolated solution),
(iii) G and G_u are smooth functions near u_0,

then, there is a unique and smooth solution family $u(\lambda)$, $G(u(\lambda), \lambda) = 0 \forall \lambda$ near λ_0, and $u(\lambda_0) - u_0$.

Let $g(x, \lambda) = (x^2 - 4)(x^2 - 9) + \lambda x^2 e^{x/5} = 0$

(a) How many solutions are there for $g(x, 0)$?
(b) Are they isolated solutions?
(c) Does each one of these solutions persist as $\lambda \to \neq 0$? (continuation1.m).

Solution

(a) When $\lambda = 0$, $g(x, 0) = 0$ has four solutions, namely, $x = \mp 2$, $x = \mp 3$.
(b) We have

$$g_x(x, \lambda)|_{\lambda=0} = \frac{d}{dx}(x, 0) = 4x^3 - 26x \qquad (2.9)$$

Substituting four solution values into (2.9) we get

$$g_x(-2, 0) = -20, \quad g_x(2, 0) = -20, \quad g_x(-3, 0) = -30, \quad g_x(3, 0) = 30$$

which are all non-zero. Hence, each of these four solutions is isolated when $\lambda = 0$.

(c) From part b, since each solution is isolated when $\lambda = 0$, then each of these solutions persists as λ becomes non-zero. Following is the output list of MATLAB script (continuation1.m).

```
gx0 =(x^2 - 4)*(x^2 - 9)
R =   -3     -2      2      3
gxL0(x)  = 4*x^3 - 26*x

ans =[ -30, 20, -20, 30]

BZ =  1      2      3      4
Solutions are  isolated,they persist for nonzero lambda values

%continuation1.m     definition: isolated persistant solutions
clc;clear;syms x L
%H= (x^2-4)*(x^2-9)   + L*x.^2*exp(x./5)
gx0=(x^2-4)*(x^2-9)
R=double(root(gx0,x))' %roots
gxL0(x)= diff(gx0);gxL0(x)=simplify(gxL0(x))% first derivative of g(x,L=0)
gxL0(R)
BZ=find(gxL0(R))
BZR=length(BZ)-length(R);% Any zeros of first derivative of g(x,lambda=0)?
if BZR==0 % If no zero valued derivative, solutions are isolated.
disp(' solutions are isolated,they persist for nonzero lambda values');
else
disp('solutions are not isolated,they do not persist for nonzero lambda ');
return
end
```

Problem 2.5.3 (*Univariate Linear Homotopy*) This problem aims to demontrate how to construct a simple Linear univariate Homotopy solution to compute the zeros of a function.

Find the zeros of the function $f(x) = x^2 + 3x - 4$ using the Linear Homotopy Method. Plot homotopy parameter (λ) versus independent parameter (x) (zero tracing) curves. For the sake of simplifying the solution of this problem, use MATLAB roots function to compute the zeros of $f(x)$, in place of Newton's Method (homotopyLH1.m).

Solution

The Linear Homotopy function $H(x, \lambda)$ is defined in one dimensional case, as

$$H(x, \lambda) = \lambda f(x) + (1 - \lambda) g(x) = 0$$

where $g(x)$ is the start (base) function, $f(x)$ is the target function, with

$$H(x, 0) = g(x); \quad H(x, 1) = f(x)$$

A polynomial of the form $g(x) = x^n - c$ can be used as a starting function, where n is the highest power available in $f(x)$, while c is a constant term, so that the zeros of $g(x) = 0$ can be found easily [15].

In this case, we may construct the homotopy function using given $f(x)$ as

$$H(x, \lambda) = \lambda f(x) + (1 - \lambda)(x^2 - 4) = 0$$

which simplifies to

$$H(x, \lambda) = x^2 + 3\lambda x - 4 = 0$$

We start tracing the zeros of $g(x) = x^2 - 4$ at $\lambda = 0$ (which are computed to be $x = -2, x = 2$) and incrementing λ by 1/5 units each time, we reach to the solution of the problem at $\lambda = 1$ as $x = -4$, $x = 1$.

Following is the output list of MATLAB script (homotopyLH1.m). Each line below, as well as each curve of Fig. 2.4 shows the zeros of homotopy function at a given value of $\lambda = [0, 0.2, 0.4, 0.6, 0.8, 1]$, starting from different zeros of homotopy function (=starting function) at $\lambda = 0$.

```
p =    -2.0000    -2.3224    -2.6881    -3.0932    -3.5324    -4.0000

p =     2.0000     1.7224     1.4881     1.2932     1.1324     1.0000
```

```
%homotopyLH1.m
clc;clear;close;
for k=1:2
L=0:0.2:1;
for i=1:length(L)
h=roots([1 3*L(i) -4]);
if k==1,p(i)=min(h);
else p(i)=max(h);
end
end
p=p
plot(L,p,'linewidth',2);hold on;grid on;xlabel('lambda'),ylabel('x')
end
```

Fig. 2.4 Solution families of $H(x, \lambda) = 0$

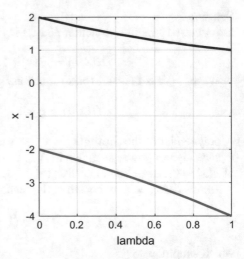

Problem 2.5.4 Find the zeros of the univariate function, $f(x) = \cos(x) - x$ using the Linear Homotopy Method. Plot homotopy parameter (λ) versus independent parameter (x) (zero tracing) curves. For the sake of writing a short and simple code, use MATLAB `fzero` function to compute the zeros of $f(x)$, instead of using the Newton's Method (homotopyLH2.m).

Solution

The Linear Homotopy[8] function $H(x, \lambda)$ is defined in one dimensional case, as

$$H(x, \lambda) = \lambda f(x) + (1 - \lambda) g(x) = 0$$

where $g(x)$ is the start (base) function, $f(x)$ is the target function, with

$$H(x, 0) = g(x); \quad H(x, 1) = f(x)$$

In this case, we may construct the homotopy function using given $f(x)$ as

$$H(x, \lambda) = \lambda(x - \cos(x)) + (1 - \lambda)(x) = 0$$

which simplifies to

$$H(x, \lambda) = x - \lambda \cos(x) = 0$$

We start tracing the zero of $g(x) = x$ at $\lambda = 0$ (which is $x = 0$) and incrementing λ by 1/5 units each time, we reach to the solution of the problem at $\lambda = 1$ as $x = 0.739085$.

---

[8]Some books use the term "Convex homotopy", such as in [16].

Following is the output list of MATLAB script (homotopyLH2.m). Each line below shows the zeros of homotopy function at a given value of $\lambda = [0, 0.2, 0.4, 0.6, 0.8, 1]$.

> 0.000000000000000
> 0.196164281187842
> 0.372559495832110
> 0.520532639238019
> 0.641134282813549
> 0.739085133215161

Alternatively, if the homotopy function is written as $H(x, \lambda) = \cos(x) - \lambda x = 0$, we obtain the following results. Note, however that it starts from $H(x, 0) = 1.570796$ in this case.

> 1.570796326794897
> 1.306440008369511
> 1.110510503581112
> 0.958251898467990
> 0.837060770621973
> 0.739085133215161

Both of these results are plotted in Fig. 2.5.

```
%homotopyLH2.m
clc;clear;close;format long
syms x
dL=.2;   L=0:dL:1;   N=length(L); L=0;
for i=1:N
p(i)=fzero(@(x) x-L*cos(x) ,1);
%p(i)=fzero(@(x) cos(x)-L*x ,1);
L=L+dL;
end
L=0:dL:1;  p'
plot(L,p,'linewidth',2);hold on;grid on;xlabel('lambda'),ylabel('x');
```

Problem 2.5.5 When a function $f(x) = 0$ is a sum of a polynomial $p(x)$ and a transcendental function, $q(x)$, i.e., $f(x) = p(x) + q(x)$, Linear Homotopy (LH) function can be written as

$$H(x, \lambda) = \lambda f(x) + (1 - \lambda)g(x) = p(x) + \lambda q(x) = 0 \qquad (2.10)$$

Use LH Method to determine zeros of the nonlinear univariate function,

$$f(x) = x^2 + 3x + 2 - 1.12e^x\sin(x) = 0,$$

and plot independent parameter x as a function of homotopy parameter (λ).

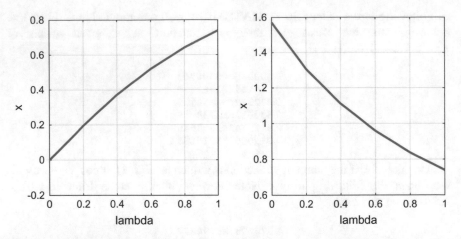

Fig. 2.5 Zero tracing curves for the Linear Homotopy functions $H(x, \lambda) = x - \lambda\cos(x)$ (left), and $H(x, \lambda) = \cos(x) - \lambda x$, (right)

For the sake of simplicity, use MATLAB fzero function to compute the zeros of $f(x)$, instead of using the Newton's Method (homotopyLH3.m).

Solution

Here, $p(x) = x^2 + 3x + 2$, $q(x) = -1.12e^x\sin(x)$;

Using (2.11), one can write the LH function as

$$H(x, \lambda) = x^2 + 3x + 2 - 1.12\lambda e^x\sin(x) = 0$$

Following is part of the output list of MATLAB m-file (homotopyLH3.m), showing only the first and last two values for each zero tracing curves.

```
-1.000000000000000
-1.035460999813394
-1.412682315335193
-1.496930152679000

-2.000000000000000
-1.985725453990008
-1.794253255840044
-1.734439742902888
```

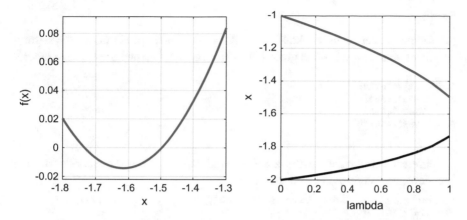

Fig. 2.6 The graph of $f(x)$ (left) and zero tracing curves for the Linear Homotopy function $H(x, \lambda) = x^2 + 3x + 2 - 1.12\lambda e^x \sin(x) = 0$ (right)

The graph of $f(x)$ and zero tracing curves for the LH function are shown in Fig. 2.6.

```
%homotopyLH3.m
clc;clear;format long
syms x
%  h=x.^2+3*x+2-1.12*exp(x)*sin(x);
%  c=ezplot(h, [-1.8 -1.3]);set(c,'linewidth',2);ylabel('f(x)'); grid;
dL=.1;   L=0:dL:1;   N=length(L);   L=0;
for i=1:N
%p(i)=fzero(@(x) x^2+3*x+2-1.12*L*exp(x)*sin(x), -1);
p(i)=fzero(@(x) x^2+3*x+2-1.12*L*exp(x)*sin(x), -3);
L=L+dL;
end
L=0:dL:1;   p'
plot(L,p,'linewidth',2);hold on;grid on;xlabel('lambda'),ylabel('x');
hold on
```

Problem 2.5.6 Find the zeros of the univariate function, $f(x) = (x-1)^3$, using the Linear Homotopy Method. Plot homotopy parameter (λ) versus independent parameter (x) (zero tracing) curves;

(a) Use MATLAB fzero function to compute the zeros of $f(x)$. Chose initial guess, $x_0 = 2$ (homotopyLH4.m).
(b) Repeat part (a) using the Newton's Method. (homotopyLH5.m), (newtonH.m).

Solution

(a) Output numerical list of the MATLAB script (homotopyLH4.m) for the starting
   function $g(x) = x^3 - 3$ is given below (for the first two and the last two terms);

```
1.442249570307408
1.442109677655019
1.226732583482821
0.999988996448478
```

Different zero tracing paths are displayed in Fig. 2.7 using various starting
functions, and steps of $\Delta\lambda = 0.01$.

```
%homotopyLH4.m    Linear Homotopy for multiple roots,    f(x)=(x-1)^3
clc;clear;%close;
format long
syms x
dL=.01;   L=0:dL:1; N=length(L); L=0;
for i=1:N
p(i)=fzero(@(x) (x-1).^3*L +(1-L)*(x.^3-3), 2);
L=L+dL;
end
L=0:dL:1;   p'
plot(L,p,'linewidth',2);hold on;grid on;xlabel('lambda'),ylabel('x');
hold on
```

(b) We obtain the same pattern for the zero tracing paths for three different starting
   functions (as indicated in Fig. 2.7). However, even using larger steps of $\Delta\lambda =$
   0.25 in the MATLAB script (homotopyLH5.m) in relation with the function
   (newtonH.m) gives more accurate result for the zero value of the function than
   the previously obtained one, as listed below. Note that homotopy function and
   its sub-functions are defined in (newtonH.m).

Fig. 2.7 Zero tracing paths
for three different starting
functions using steps of
$\Delta\lambda = 0.01$

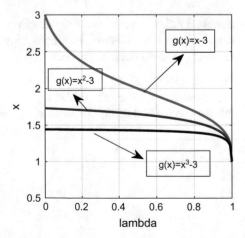

```
1.4422495703074083823216383108015
1.4377546380286991175200764785319
1.4294445359356209065669886447239
1.4086628482322745773354818217425
1.0000000141149569093556276010031
```

```
%homotopyLH5.m    Linear Homotopy for multiple roots,    f(x)=(x-1)^3
clc;clear;%close;
format long
syms x
x0=10; dL=0.25; L=0:dL:1; N=length(L); L=0;
for i=1:N
p(i)=newtonH(L,x0); L=L+dL;
end
L=0:dL:1;  p'
plot(L,p,'linewidth',2);hold on;grid on;xlabel('lambda'),ylabel('x');
hold on

function p = newtonH(L,x0)
%newtonH.m    Newton-Raphson in homotopy method
syms x
f=(x-1).^3; g=x.^3-3; H=f*L+(1-L)*g; %homotopy function
dH=diff(H,x);
imax=50;  % maximum allowable iterations
i = 0;    % initialization
while (1)
xold = x0; x0 = x0 - subs(H,x,x0)/subs(dH,x,x0); x0=(vpa(x0,10));
i = i + 1;
if x0 ~= 0, err =abs((x0 - xold)/x0); end
if err <= 1e-8, break, end % absolute tolerance
if i >= imax, break, end
end
i; p=x0;
```

Problem 2.5.7 Newton's Method fails to compute the zero of $f(x) = \arctan(x)$, if initial guess for zero of the function is $x_0 > 1.39$. Homotopy method can be used to extend this range.

(a) Find the zero of $f(x) = \arctan(x) = 0$ using the "Newton Homotopy Method". Plot homotopy parameter (λ) versus independent parameter (x) (zero tracing) curves. Use the Newton's Method, and initial guess for zero of the function as $x_0 = 8$ (homotopyNH1.m), (newtonN.m).

(b) Find out maximum guess values for the zero of this function if step size is varied between $0.2 \geq \lambda \geq 0.025$.

Solution

(a) The Newton Homotopy function $H(x, \lambda)$ is defined in one dimensional case, as

$$H(x, \lambda) = f(x) - (1 - \lambda)f(x_0) = 0$$

where $g(x) = f(x_0)$ is the start (base) function, $f(x)$ is the target function, with

$$H(x, 0) = g(x); \quad H(x, 1) = f(x)$$

In this case, we may construct the Newton Homotopy function using given $f(x)$ as

$$H(x, \lambda) = \arctan(x) - (1 - \lambda)\arctan(x_0) = 0$$

where $x_0 = 8$.

We start tracing the zero path at $\lambda = 0$ (which corresponds to $x = 8$) and by incrementing λ 1/10 units each time, we reach to the solution of the problem at $\lambda = 1$ as $x = 0.0$.

The output list of MATLAB script (homotopyNH1.m) is plotted in Fig. 2.8.

(b) Maximum guess values for the zero of this function are computed as follows:

When the Homotopy parameter increment is $\Delta\lambda = 0.2$, $\max(x_0) = 4.5$
When the Homotopy parameter increment is $\Delta\lambda = 0.1$, $\max(x_0) = 8.2$
When the Homotopy parameter increment is $\Delta\lambda = 0.05$, $\max(x_0) = 15.3$
When the Homotopy parameter increment is $\Delta\lambda = 0.025$, $\max(x_0) = 28.9$

Fig. 2.8 Plot of the homotopy parameter (λ) versus independent parameter (x) (zero tracing path) for the function, $f(x) = \arctan(x) = 0$, $x_0 = 8$, $\Delta\lambda = 0.1$; Newton Homotopy is used

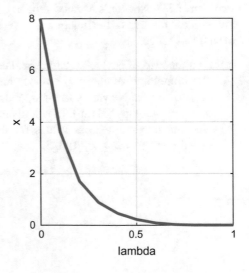

```
%homotopyNH1.m
clc; clear; format long
x0=8; dL=0.1; L=0:dL:1; N=length(L); L=0;
p(1)=x0;%initial guess for the zero
for i=1:N
p(i+1)=newtonN(L,p(i));
L=L+dL;
end
L=0:dL:1; p(1)=[]; p'
plot(L,p,'linewidth',2);hold on;grid on;xlabel('lambda'),ylabel('x');


function p = newtonN(L,x0)
%newtonN.m    Newton-Raphson in Newton homotopy method
syms x
g0=atan(x0);
H=atan(x)-(1-L)*g0;   % Newton homotopy function
dH=diff(H,x);
imax=50;              % maximum allowable iterations
j = 0;                % initialization
while (1)
xold = x0;
x0 = x0 - dx=vpa(subs(H,x,x0)/subs(dH,x,x0)); x0=(vpa(x0,6));
j = j + 1;
if x0 ~= 0, err =abs((x0 - xold)/x0); end
if err <= 1e-6, break, end %  tolerance
if j >= imax, break, end
end
p=x0;
```

Problem   2.5.8 Use   Homotopy   Continuation   method   to   solve
$f(x) = x - \sin(x) - 1/2 = 0$.

Plot homotopy parameter (λ) versus independent parameter (x) (zero tracing path) curves. Let the incremental value of the homotopy parameter be $\Delta\lambda = 1/100$. (homotopy4.m).

Solution

$$H(x, \lambda) = x - \sin(x) - \lambda/2 = 0,$$

Here the base equation is $x - \sin(x) = 0$. Since $\lambda = 0$ is a singular solution, iterations must start at $\lambda/100$. Corresponding value of x_0 is to be calculated;

$$x - \sin(x) = x - \left(x - \tfrac{1}{6}x^3 + \tfrac{1}{120}x^5 - \ldots\right) \cong \tfrac{1}{6}x^3,$$
$$\tfrac{1}{6}x^3 = \tfrac{1}{100} \cdot \tfrac{1}{2} \quad \rightarrow \quad x^3 = \tfrac{6}{200} \quad \rightarrow \quad x = \left(\tfrac{3}{100}\right)^{1/3}$$

Following is part of the output list of MATLAB m-file (homotopy4.m), showing only the first and last three values for the zero tracing path. The graph of zero tracing curve for the given function is shown in Fig. 2.9.

Fig. 2.9 Zero tracing path for $f(x) = x - \sin(x) - 1/2 = 0$, using steps of $\Delta\lambda = 0.01$

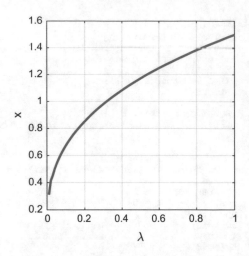

Lambda	x	x-sin(x)
0.01	0.311226135830411	0.00500003868094584
0.02	0.415302687647597	0.01183578132996460
0.03	0.452526266311338	0.01528735341884760
0.98	1.486460802486950	0.49001493550920700
0.99	1.491904412679860	0.49501476601455800
1.00	1.497316145920680	0.50001459992827200

```
%homotopy4.m        x-sin(x)=1/2
clc;clear;
m=100;              % increment
format longg
x(1) = (3/m)^(1/3); % initial guess
for k=1:m
x(k+1) = x(k) - ( x(k)-sin(x(k))- k/(2*m) ) / (1-cos(x(k)));
end
k=0:m;k=k'; x=x';  A= [k/m  x  (x-sin(x))];  A(1,:)=[]
plot(A(:,1),A(:,2),'linewidth',2);grid;xlabel('\lambda'),ylabel('x');
```

Problem 2.5.9 Use Newton homotopy method combined with Forward Euler Method to compute the solution of following system of nonlinear equations,

$$f_1 = x_1^2 - \ln(2 + x_2) - 1 = 0$$
$$f_2 = 2x_1 - \exp(x_1 + x_2) - 3 = 0$$

using the starting vector $x = [5,5]$, $N = 2$ (number of equal Euler steps), and differential homotopy parameter of $\Delta\lambda = 0.125$, $0 \le \lambda \le 1$.

Repeat the computation for the solution of nonlinear equations using $N = 4$ equal step sizes (homotopyEuler1.m).

Solution

Using Newton homotopy differential equation (NHDE)

$$x'(\lambda) = -J(x(\lambda))^{-1}f(x_0) = \varphi(x, \lambda)$$

in Forward Euler equation,

$$\begin{aligned}x^{(k+1)} &= x^{(k)} + h\varphi(x, \lambda)\\ x^{(k+1)} &= x^{(k)} - hJ(x(\lambda))^{-1}f(x_0)\end{aligned} \tag{2.11}$$

where h is the step size within the interval of continuation, $0 \le \lambda \le 1$.

Following is the edited output list of MATLAB m-file (homotopyEuler1.m) which uses Eq. (2.11) to solve the given system of nonlinear equations using 2 equal Euler step sizes and $\Delta\lambda = 0.125$.

```
         x1                    x2
5.000000000000000      5.000000000000000
2.503886978924106      6.171591637152361
1.807385561423441      5.543548507432006
1.691420526284307      4.334724380970358
1.624713421925218      3.076017124478303
1.539066359542077      1.836031792844766
1.419368101359241      0.640731820613131
1.244887921616446     -0.399274962723358
1.056260945059911     -0.942538590785312
 Elapsed time is 1.608167 seconds.
```

When N = 4 and $\Delta\lambda = 0.125$, following results are obtained:

```
5.000000000000000      5.000000000000000
2.246363123222093      6.051086485011235
1.731298130268482      4.863460630376130
1.634816211992216      3.256473136279737
1.523694417193779      1.664046564481559
1.350844218307343      0.170278769725853
1.106070623044144     -0.833713768710677
1.003199702915785     -0.999143203593652
1.000000942220712     -1.000000765417626
 Elapsed time is 2.757970 seconds.
```

Actual solution values are $x_1 = 1, x_2 = -1$.

The meaning of symbols are defined in the MATLAB script given below.

Although this method can be used as a stand-alone nonlinear systems solver, its application for the lower number of (equal) steps can be useful to provide a starting approximation for Newton or Quasi-Newton Methods, since homotopy methods do not require a particularly good choice of starting values. Figure 2.10 displays the

Fig. 2.10 Zero tracing paths
for the Newton Homotopy
Method with two unknowns
and $N = 4$, $\Delta\lambda = 0.125$

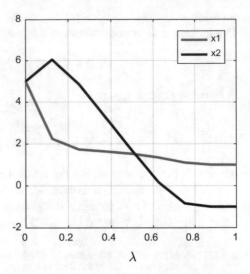

zero tracing paths for the Newton Homotopy Method with two unknowns and
$N = 4$, $\Delta\lambda = 0.125$.

```
%homotopyEuler1.m
% HOMOTOPY & EULER METHOD FOR SOLVING 2 SIMULTANEOUS NONLINEAR EQNS
%  f(x)=0 & initial approximation  u=x=(x1,x2) are given
% Output:  Approximate solution x=(x1,x2)
clc;clear;format long;tic
syms    x1 x2 f
dL=0.125  % delta Lambda
u=5*ones(1,2);% initial guess values
N = 4 ;  % N= number  of steps in Euler equation
k=1/dL;
x=zeros(k+1,2);%initialization
x(1,:)=u; n=1; h=1/N; %Euler step size
for L=dL:dL:1
n=n+1;
f(1) = 'x1*x1-log(2+x2)-1'; % Define components of f
f(2) = '2*x1+exp(x1+x2)-3';
f0=subs(f,[x1,x2],u); f0=vpa(f0',6); % f(x(0))
for i=1:N
J=jacobian(f); J=subs(J,[x1,x2],u); J=vpa(J,8);
dxdL=-J\f0; dxdL=vpa(dxdL',8); u=u+h*dxdL; u=vpa(u,8);
end
x(n,:)=u;
end
disp('x1     x2'); x
L=0:dL:1;plot(L,x,'linewidth',2);grid;xlabel('\lambda');
legend('x1','x2'); toc
```

Problem 2.5.10 Use Newton homotopy method combined with Forward Euler
Method to compute the solution of following system of nonlinear equations,

$$f_1 = x_1 - \exp(2 + x_2 - x_3) - \tfrac{1}{2} + \exp(4) = 0$$
$$f_2 = 2x_1 + \exp(x_2 + x_3) - 2 = 0$$
$$f_3 = \exp(2 + x_1) + x_2 + x_3 - \exp\left(\tfrac{5}{2}\right) = 0$$

using the starting vector $x = [5, 5, 5]$, $N = 2$ (number of equal Euler steps), and differential homotopy parameter of $\Delta\lambda = 0.125$, $0 \le \lambda \le 1$.

Repeat the computation for the solution of nonlinear equations using $N = 4$ equal step sizes (homotopyEuler2.m).

Solution

Using NHDE, $x'(\lambda) = -J(x(\lambda))^{-1} f(x_0) = \varphi(x, \lambda)$ in Forward Euler equation,

$$x^{(k+1)} = x^{(k)} + h\varphi(x, \lambda)$$
$$x^{(k+1)} = x^{(k)} - hJ(x(\lambda))^{-1} f(x_0) \tag{2.12}$$

where h is the step size within the interval of continuation, $0 \le \lambda \le 1$.

Following is the edited output list of MATLAB m-file (homotopyEuler2.m) which employs Eq. (2.1) to solve the given system of nonlinear equations using $N = 2$ equal Euler step sizes and $\Delta\lambda = 0.125$. The meaning of symbols are defined in the MATLAB script given below.

Actual solution values are $x_1 = 0.5$, $x_2 = 1$, $x_3 = -1$.

x1	x2	x3
5.000000000000000	5.000000000000000	5.000000000000000
3.681162182022677	6.108165787441236	2.567048220490083
2.385181674636006	4.963760568886030	2.386261811623403
1.190786023255577	4.059585044213253	1.965620763836495
0.354676579420836	3.352945490501726	1.353257066665834
0.180539787304607	2.700814566839136	0.706675104214859
0.308698121641027	2.079117072305270	0.082625588184164
0.411210087186680	1.531155163541623	-0.467218190599332
0.475565684618430	**1.152903626608252**	**-0.846649099604922**

Elapsed time is 5.039624 seconds.

When $N = 4$ and $\Delta\lambda = 0.125$, following results are obtained;

x1	x2	x3
5.000000000000000	5.000000000000000	5.000000000000000
3.308236709042605	5.359272570076519	2.937745736732251
1.687174090105897	4.318704243552109	2.274887958177755
0.464032411802619	3.447560140761540	1.447963613090893
0.202356209389854	2.611773964346014	0.617237125736976
0.355110673688173	1.836400091091197	-0.160934948205948
0.459540268902229	1.249357206612229	-0.749901967396770
0.496803522837259	1.020462171314659	-0.979479340625693
0.499980681462923	**1.000125170084395**	**-0.999874476506713**

Elapsed time is 9.113810 seconds.

Fig. 2.11 Zero tracing paths
when the number of Euler
steps N = 4, and $\Delta\lambda = 0.125$

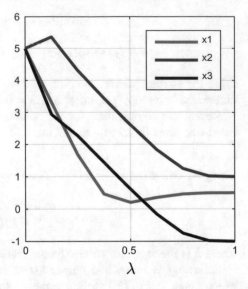

These results are displayed in Fig. 2.11 as functions of the homotopy parameter.
Although this method can be used as a stand-alone nonlinear systems solver, its
application even for smaller number of Euler steps can be useful to provide a
starting approximation for Newton or Quasi-Newton Methods, since homotopy
methods do not require a particularly good choice of starting values.

```
%homotopyEuler2.m
% HOMOTOPY & EULER METHOD FOR SOLVING 3 SIMULTANEOUS NONLINEAR EQNS
%   f(x)=0 & initial approximation  u=x=(x1,x2,x3) are given
% Output:  Approximate solution x=(x1,x2, x3)
clc; clear; format long; tic
syms    x1 x2 x3   f
dL=0.125  % delta Lambda
u=5*ones(1,3);% initial guess values
N = 2 ;  % N= number  of steps in Euler equation
k=1/dL;  x=zeros(k+1,3);%initialization
x(1,:)=u; n=1; h=1/N; %Euler step size
for L=dL:dL:1
n=n+1;
f(1) = 'x1-exp(2+x2-x3)-1/2+exp(4)'; % Define components of f
f(2) = '2*x1+exp(x2+x3)-2';
f(3) = 'exp(2+x1)+x2+x3-exp(5/2)'
f0=subs(f,[x1,x2,x3],u); f0=vpa(f0',6)
for i=1:N
J=jacobian(f);%c=cond(J), DJ=det(J)
J=subs(J,[x1,x2,x3],u); J=vpa(J,8)
dxdL=-J\f0; dxdL=vpa(dxdL',8)
u=u+h*dxdL; u=vpa(u,8)
end
x(n,:)=u;
end
disp('x1    x2     x3'); x
L=0:dL:1;plot(L,x,'linewidth',2);grid;xlabel('\lambda');
legend('x1','x2','x3'); toc
```

2.6   Steepest Descent (Gradient Search) Method

Problem 2.6.1

(a) What is the idea behind the Steepest Descent Method? Describe the principles of this method.
(b) Use Gradient Search (Steepest Descent) method to find (manually) the minimum of the function $f(x, y) = x^2 + y^2 + 2y + 1$. Apply point $u^{(0)} = \left[x^{(0)} y^{(0)} \right]^T = \left[1 \quad 1 \right]^T$ as the initial guess for the minimum point.

Solution

(a) The idea behind this method is to search for a minimum of the function by moving in the direction of steepest decline from the current point. Since the gradient of the function points in the direction of steepest growth of the function, the opposite direction of the gradient of the function is the line of steepest descent. After the new minimum along the line of steepest descent is located, process is repeated, starting at that point. That is, we find the gradient at the new point, and perform minimization in the new direction [17]. Steepest descent method guarantees a decrease in function value from step to step. This fact can be made the basis for a convergence proof of the method, although convergence is slow [18].

(b) $f(x, y) = x^2 + y^2 + 2y + 1$, $f^{(0)}_{min}(1, 1) = 1^2 + 1^2 + 2(1) + 1 = 5$

First iteration: $(i = 1)$,

$$\frac{\partial f}{\partial x} = 2x|_{x=1} = 2, \frac{\partial f}{\partial y} = 2y + 2|_{y=1} = 4, \nabla f = 2i + 4j$$

$$u^{(1)} = u^{(0)} h \nabla f = \begin{bmatrix} 1 \\ 1 \end{bmatrix} + h \begin{bmatrix} 2 \\ 4 \end{bmatrix} = \begin{bmatrix} 1 + 2h \\ 1 + 4h \end{bmatrix} \qquad (2.13)$$

Here, h determines the "step size" in the gradient direction. Substituting of x and y into the objective function gives,

$$f\left(u^{(1)}\right) = (1 + 2h)^2 + (1 + 4h)^2 + 2(1 + 4h) + 1 = 20h^2 + 20h + 5 = g(h)$$
$$g_{min} = \frac{dg}{dh} = 40h + 20 = 0 \quad \rightarrow \quad h = -1/2$$

Replace h in Eq. (2.13),

$$u^{(1)} = \begin{bmatrix} 1 + 2(-1/2) \\ 1 + 4(-1/2) \end{bmatrix} = \begin{bmatrix} 0 \\ -1 \end{bmatrix}$$

Substitute this into objective function, f,

$$f_{min}^{(1)} = 0^2 + (-1)^2 + 2(-1) + 1 = 0$$

$$i = 2:$$

The gradient value at $[0 \quad -1]^T$ is

$$\frac{\partial f}{\partial x} = 2x|_{x=0} = 0, \quad \frac{\partial f}{\partial y} = 2y + 2|_{y=-1} = 0, \quad \nabla f = 0i + 0j$$

We reach the optimum solution since we expect $\nabla f = 0$. No more improvement is possible along this gradient. Hence, the minimum of the function is at point $[0 \quad -1]^T$ with $f_{min}^{(1)} = 0$

Problem 2.6.2 Determine the minimum of the function $f(x)$ with starting point x_0 (steepestdescent3.m).

$$f(x) = x_1^2 + 2x_2^2, \quad x_0 = (5,3) = 5i + 3j, \quad \varepsilon = 10^{-4}$$

Solution

$$\nabla f = (2x_1, 4x_2) - h\nabla f = \begin{bmatrix} x_1 \\ x_2 \end{bmatrix}_{x=x_0} - h \begin{bmatrix} 2x_1 \\ 4x_2 \end{bmatrix} = \begin{bmatrix} x_1(1-2h) \\ x_2(1-4h) \end{bmatrix}$$

$$g = f(x) = x_1^2(1-2h)^2 + 2x_2^2(1-4h)^2$$

$$\frac{\partial g}{\partial h} = f'(x) = 2x_1(-2)(1-2h) + 4x_2^2(1-4h)(-4)$$

$$\frac{\partial g}{\partial h} = 0 = -4x_1^2(1-2h) - 16x_2^2(1-4h)$$

Solve for h,

$$h = \frac{4x_2^2 + x_1^2}{2(8x_2^2 + x_1^2)}$$

starting from x_0, we compute the values as shown below (tolerance $= 10^{-4}$),

n	x_1	x_2	h
1	5.0	3.0	0.3144
2	1.86	−0.733	0.3547
10	0.000251	−0.000105	0.3547
12	0.000027	0.000011	0

The solution converges at the origin, as shown in Fig. 2.12.

Fig. 2.12 Contour diagram
and convergence to minimum
for $f(x) = x_1^2 + 2x_2^2$,
$x_0 = [5,3]$.

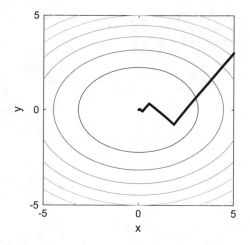

Problem 2.6.3 Use steepest descent method to find the minimum of the following function,

$$f(x_1, x_2) = 3x_1^2 - x_2^2 + x_1 - x_2 - 1/6 = 0$$

Plot a contour diagram to show the convergence. Note that the coordinates of minimum point yield $f(x_1, x_2) = 0$. Let the error tolerance $\varepsilon = 10^{-4}$, and the initial guess be $p_0 = [5,4]$.

How many iterations are needed to obtain minimum of the function if initial guess point is set at $p_0 = [5,1]$? Actual solution is $p = [-1/6, -1/2]$ (steepestdescent3.m).

Solution

Using the initial guess $p_0 = [5,4]$, and the error tolerance of $\varepsilon = 10^{-4}$, we obtain following edited list at the end of 52 iterations;

i	x	y
1.000000000000000	5.000000000000000	4.000000000000000
10.000000000000000	-0.254413323502561	0.406715453970905
20.000000000000000	-0.158688379779909	-0.582442297829826
30.000000000000000	-0.167392085320125	-0.492504007247593
40.000000000000000	-0.166600708619500	-0.500681566487388
50.000000000000000	-0.166672663843559	-0.499938029172109
51.000000000000000	-0.166698774115100	0.500027964551861
52.000000000000000	-0.166662954028090	-0.500038363931961

Contour diagram showing the convergence of the solution for $p_0 = [5,4]$ is displayed in Fig. 2.13. If initial point s set at $p_0 = [5,1]$, it takes 10 iterations to get the minimum point.

Fig. 2.13 Contour diagram
showing the convergence of
the solution for $p_0 = [5, 4]$

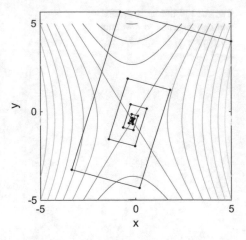

```
%steepestdescent3.m
clc,clear,format long;tic
syms x1 x2
f=3*x1^2-x2^2+x1-x2-1/6;
p=[5,4]; % Initial Guess
x(1) = p(1); y(1) = p(2); tol = 1e-4; % tolerance
n = 1;
fx1 = diff(f, x1); fx2 = diff(f, x2);% Gradients
J = [subs(fx1,[x1,x2], [x(1),y(1)]) subs(fx2, [x1,x2], [x(1),y(1)])];
S = -(J);
while norm(J) > tol
syms h;% Step size
A = [x(n),y(n)]';
g = subs(f, [x1,x2], [x(n) + S(1)*h,  y(n) + S(2)*h]);
h = solve(diff(g,h), h); h=h(1);% Step size
x(n+1) = A(1)+h*S(1); y(n+1) = A(2)+h*S(2); n = n+1;
J = [subs(fx1,[x1,x2], [x(n),y(n)])  subs(fx2, [x1,x2], [x(n),y(n)])];
S = -(J);
end
i = (1:n)'; [i  x'  y']
fcontour(f); % this command is available by MATLAB version 2016a
hold on;  plot(x,y,'.-k'); xlabel('x'); ylabel('y'); toc
```

Problem 2.6.4 Apply the steepest descent method to following functions using the
initial guess point of $x_0 = (5, 3)$. Plot contour diagrams and the path of descent in
the same figure (steepestdescent3.m).

(a) $f(x) = 2x_1^2 + 4x_1 + x_2^2 - 2x_2$

(b) $f(x) = x_1^2 + x_2^2 - x_1 + 4x_2$

(c) $f(x) = x_1^2 + x_2^2/5 - x_1/50$

Solution

(a) $f(x) = 2x_1^2 + 4x_1 + x_2^2 - 2x_2$

```
[ 1.0,      5.0,    3.0, 0.2534]
[ 2.0,   -1.082, 1.986, 0.4868]
[ 3.0,  -0.9221, 1.026, 0.2534]
[ 4.0,   -1.001, 1.013, 0.4868]
[ 5.0,   -0.999,   1.0, 0.2534]
[ 6.0,     -1.0,   1.0,      0]
```


(b) $f(x) = x_1^2 + x_2^2 - x_1 + 4x_2$

```
[ 1.0, 5.0,  3.0, 0.5]
[ 2.0, 0.5, -2.0,   0]
```


(c) $f(x) = x_1^2 + x_2^2/5 - x_1/50$

```
[ 1.0,       5.0,        3.0, 0.5058]
[ 2.0, -0.04755,      2.393,  2.365]
[ 3.0,   0.2247,     0.1291, 0.5058]
[ 4.0, 0.007524,      0.103,  2.365]
[ 5.0,  0.01924,   0.005553, 0.5058]
[ 6.0, 0.009893,   0.004429,  2.365]
[ 7.0,   0.0104, 0.0002389,       0]
```


Contour diagrams and the descent paths are displayed in Fig. 2.14.

Problem 2.6.5 (*Starting approximation to the solution of a nonlinear system*). Let a system of the nonlinear functions be,

$$f_1(x_1, x_2, \ldots, x_n) = 0$$
$$f_2(x_1, x_2, \ldots, x_n) = 0$$
$$\vdots$$
$$f_n(x_1, x_2, \ldots, x_n) = 0$$

which has a solution of $x = [x_1, x_2, \ldots, x_n]^T$. when the function g has the minimum of zero when it is defined by

$$g(x_1, x_2, \ldots, x_n) = \sum_{i=1}^{n} [f_i(x_1, x_2, \ldots, x_n)]^2$$

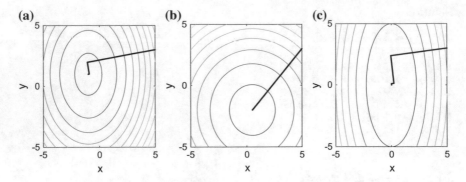

Fig. 2.14 Contour diagrams and the paths of descent for **a** $f(x) = 2x_1^2 + 4x_1 + x_2^2 - 2x_2$,, **b** $f(x) = x_1^2 + x_2^2 - x_1 + 4x_2$, **(c)** $f(x) = x_1^2 + x_2^2/5 - x_1/50$.

The Steepest Descent (or Gradient Search) method of finding a local minimum of a function g uses following steps: [19].

1. Evaluate g at $p^{(0)} = \left[p_1^{(0)}, p_2^{(0)}, \ldots, p_n^{(0)} \right]^T$
2. Determine a direction from $p^{(0)}$ that results in a decrease in the value of g
3. Move in this direction and call the new value $p(1)$.
4. Repeat above steps, replacing $p(0)$ by $p(1)$.

We define the gradient of g at x as

$$\Delta g(x) = \left[\frac{\partial y}{\partial x_1}, \frac{\partial y}{\partial x_2}, \ldots, \frac{\partial y}{\partial x_n} \right]^T = 2J^T F, \quad F = [f_1, f_2, f_3, \ldots, f_n]$$

The direction of maximum decrease of g at x is the direction given by $(-\Delta g)$. The objective is to reduce $g \to 0$, therefore we choose,

$$p^{(1)} = p^{(0)} - \alpha \Delta g \left(p^{(0)} \right), \quad \alpha > 0 \tag{2.14}$$

The value of α that minimizes a single-variable function, h,

$$h(\alpha) = g \left(p^{(0)} - \alpha \Delta g \left(p^{(0)} \right) \right)$$

is the value needed for (2.14).

We choose three numbers $\alpha_1 < \alpha_2 < \alpha_3$ and construct the second order polynomial $p(x)$ interpolating h at these three points. Define $p(\hat{\alpha})$ so that it is minimum in $[\alpha_1, \alpha_3]$ and use $p(\hat{\alpha})$ to approximate $\min(h(\alpha))$. Then, use $\hat{\alpha}$ to determine the new $p^{(1)}$,

$$p^{(1)} = p^{(0)} - \hat{\alpha} \Delta g\left(p^{(0)}\right)$$

The minimum value $\hat{\alpha}$ of $p(x)$ occurs either at the only critical point of p or at α_3 where $p(\alpha_3) = h(\alpha_3)$.

At the beginning, $\alpha_1 = 0, \alpha_3 = 1$. If $h(\alpha_3) \geq h(\alpha_1)$, then successive halving of α_3 are performed and the value of α_3 is reassigned until $h(\alpha_3) < h(\alpha_1)$. The process is repeated starting at p(1) instead of p(0).

Use Steepest Descent (Gradient Search) Method to find a starting approximation to the solution of the following nonlinear system,

$$f_1 = 2x_1 - \sin(x_1 x_3) = 0$$
$$f_2 = x_2^2 - 10(x_1 + 0.2)^3 + \sin(x_3) - 0.4174039592 = 0$$
$$f_3 = e^{-x_2 x_3} + 10(x_1 + x_2) - 6.1324969025 = 0$$

Perform first two iterations by manual calculations. Then, list the result of first 10 iterations (steepestdescent4.m), (steepestdescent5.m).

Solution

$$i = 1 : p(0) = [0, \quad 0, \quad 0]^T$$

$$g = |f_1|^2 + |f_2|^2 + |f_3|^2$$
$$\nabla g = 2J^3 F$$
$$g\left(p^{(0)}\right) = |2 \times 0 - 0|^2 + |0 - 10(0.2)^3 + 0 - 0.4174039592|^2$$
$$\qquad + |1 + 0 - 6.1324969025|^2$$
$$= 0 + |-0.4874039592|^2 + |1 - 6.1324969025|^2 = 26.5899351527$$
$$z_0 = \nabla g(p_2^{(0)} = 2 \cdot J^T \cdot F_2 = 144.76583524$$
$$z = \frac{1}{z_0} \nabla g\left(p^{(0)}\right) = [-0.70082950, -0.70907571, -0.07777941]^T$$

$$\alpha_1 = 0 \quad \rightarrow \quad g_1 = g\left(p^{(0)} - \alpha_1 z\right) = g\left(p^{(0)}\right) = 26.5899351527$$

$$\alpha_3 \quad \rightarrow \quad g_3 = g\left(p^{(0)} - z\right) = 133.728043384$$

$$\alpha_2 = \frac{1}{2}, \quad \rightarrow \quad g_2 = g\left(p^{(0)} - \alpha_2 z\right) = 7.9322074196$$

$$p(\alpha) = g_1 + h_1\alpha + h_3\alpha(\alpha - \alpha_2)$$

$$h_1 = \frac{g_2 - g_1}{\alpha_2 - \alpha_1} = -37.31545546, h_2 = \frac{g_3 - g_2}{\alpha_3 - \alpha_2} = 251.59167192,$$

$$h_3 = \frac{h_2 - h_1}{\alpha_3 - \alpha_1} = 288.90712739$$

$$p(\alpha) = 26.5899 - 37.3154\alpha + 288.9071\alpha(\alpha - 0.5)$$

$$p'(\alpha) = 0 \rightarrow \alpha = \alpha_0 = 0.31458$$

$$g_0 = g\left(p^{(0)} - \alpha_0 z\right) = 1.8301962,$$

$$g_0 < g_1 \quad \text{and} \quad g_0 < g_3$$

$i = 2:$
$$p^{(1)} = p^{(0)} - \alpha_0 z = p^{(0)} - 0.31458z = [0.220467, 0.223061, 0.024467]^T$$

$$g\left(p^1\right) = 1.830196$$

$$z_0 = \left\|\nabla g(p^{(1)})\right\|_2 = \left\|2 \cdot J^T \cdot F\right\|_2 = 14.9014155$$

$$z = \frac{1}{z_0}\nabla g\left(p^{(1)}\right) = [-0.021298, \quad -0.9730746, \quad -0.229503]^T$$

$$\alpha_1 = 0 \quad \rightarrow \quad g_1 = g\left(p^{(1)} - \alpha_1 z\right) = g\left(p^{(1)}\right) = 1.830196$$

$$\alpha_3 \quad \rightarrow \quad g_3 = g\left(p^{(1)} - z\right) = 85.69034471$$

$$\alpha_2 = \frac{1}{2}, \quad \rightarrow \quad g_2 = g\left(p^{(1)} - \alpha_2 z\right) = 19.177914$$

$$h_1 = 34.6954368, \quad h_2 = 133.024860, \quad h_3 = 98.3294235$$

$$\alpha = \alpha_0 = 0.0735755$$

$$g_0 = g\left(p^{(1)} - \alpha_0 z\right) = 1.27672827$$

$$g_0 < g_1 \quad \text{and} \quad g_0 < g_3$$
$$p^{(2)} = p^{(1)} - \alpha_0 z = p^{(1)} - 0.0735755z = [0.222034, \quad 0.294655, \quad 0.041353]^T$$

Actual solution set is [0, 0.5, 0.25]. Following edited list displays the result of first 10 iterations (Convergence rate is too slow). MATLAB code for the algorithm (steepestdescent4.m) and numerical approximation code (steepestdescent5.m) using Symbolic Math are given below.

```
i = 1,    [ 0.2204672, 0.2230613, 0.0244678]
i = 2,    [ 0.2220342, 0.2946557, 0.0413537]
i = 3,    [ 0.1824781, 0.2945411, 0.0474775]
i = 4,    [ 0.1603233, 0.3699583, 0.0702940]
i = 5,    [ 0.1294909, 0.3606133, 0.0736623]
i = 6,    [ 0.1204501, 0.4009605, 0.0866658]
i = 7,    [ 0.0913851, 0.3928208, 0.0905429]
i = 8,    [ 0.0963660, 0.4068275, 0.0936000]
i = 9,    [-0.0321851, 0.5060644, 0.1523278]
i = 10,   [ 0.0178883, 0.5553775, 0.1558418]
```

Elapsed time is 2.996598 seconds.

```
% steepestdescent4.m
clc,clear;tic;
syms x1 x2 x3 a
f1 = 2*x1-sin(x1*x3);    % Function Definition
f2=x2.^2-10*(x1+0.2).^3+sin(x3)-0.4174039592;
f3=exp(-x2*x3)+10*(x1+x2)+x3-6.1324969025;
F=[f1 f2 f3]'; x=[x1 x2 x3];
p=[0,0,0];
for i=1:10
i
p0=p; g=vpa(subs(f1,x,p0)^2 + subs(f2,x,p0)^2 + subs(f3,x,p0)^2);
J=jacobian([f1;f2;f3],[x1 x2 x3]);   Jt=J';Gg=2*Jt*F;
Gg=vpa(subs(Gg,x,p0)); z0=norm(Gg); z=(1/z0)*Gg ; a1=0;a3=1;a2=1/2;
P=p0-z'; g1=g;
g3=vpa(subs(f1,x,P)^2 + subs(f2,x,P)^2 + subs(f3,x,P)^2); R=p0-a2*z';
g2=vpa(subs(f1,x,R)^2 + subs(f2,x,R)^2 + subs(f3,x,R)^2);
h1=(g2-g1)/(a2-a1); h2=(g3-g2)/(a3-a2); h3=(h2-h1)/(a3-a1);
Pa=g1+h1*a+h3*a*(a-a2); dPa=diff(Pa); a0= vpa(root(dPa,a));
S=p0-a0*z';
g0=vpa(subs(f1,x,S)^2 + subs(f2,x,S)^2 + subs(f3,x,S)^2);
p=p0-a0*z'; vpa(p,8)
end
toc

%steepestdescent5.m    Symbolic numeric approximation
clc,clear, syms x1 x2 x3
f1 = 2*x1-sin(x1*x3), f2=x2.^2-10*(x1+0.2).^3+sin(x3)-0.4174039592
f3=exp(-x2*x3)+10*(x1+x2)+x3-6.1324969025;
eqns = [f1 == 0, f2 == 0, f3==0]; [x1, x2, x3 ] = solve(eqns)
```

2.7 Exercises

1.(a) Solve the following system of equations by simple algebra,

$$f_1(x, y) = x^2 - 2x - y + 0.5$$
$$f_2(x, y) = x^2 + 4y^2 - 4$$

What do the equations $f_1 = 0, f_2 = 0$ implicitly define curves in the xy-plane? Plot these curves.

1.(b) Apply (manually) two steps of Newton's method to this system with starting point (1,1).

1.(c) Implement Newton's Method (use a computer) with appropriate starting point to find the solution of this problem with an accuracy of 10^{-6}.

2. Use Newton's Method to compute four solutions of the system of nonlinear equations in two unknowns [20], by choosing appropriate starting points.

$$x^2 + y^2 - 1 = 0$$
$$5x^2 + 21y^2 - 9 = 0$$

3. A global nonlinear transformation may create an algebraically equivalent system on which Newton's method does better because the new system is more linear, although there is no general way to apply this idea [21]. Consider the system,

$$x^{0.2} + y^{0.2} = 2$$
$$x^{0.1} + y^{0.4} = 2$$

Apply Newton's method to this system with initial guess [2, 2] and record the results at the end of each iteration. Then, apply Newton's method to the equivalent system,

$$\left(x^{0.2} + y^{0.2}\right)^5 = 32$$
$$\left(x^{0.1} + y^{0.4}\right) = 16$$

with the initial guess [2, 2], and record the results at the end of each iteration.
Comment on these solutions.

Repeat these solutions using initial guess vector [3, 3]. Which solution method converge more rapidly to true solution [1, 1]?

4. Solve the following nonlinear system of equations [22]

$$\frac{1}{2}x^{-1/2}y^{1/3} - 0.5 = 0$$
$$\frac{1}{3}x^{1/2}y^{-2/3} - \frac{1}{3} = 0$$

Let initial guess vector be $[x(0) \quad y(0)] = [1.2 \quad 1.1]$
Exact solution: [1, 1].

5. Solve the following system of equations using Broyden's Method,

$$f_1(x,y) = x^2 - 2x - y + 0.5$$
$$f_2(x,y) = x^2 + 4y^2 - 4$$

with starting point (1,1) and maximum number of 10 iterations.

6. Apply the steepest descent method to following functions. Use initial guess of $x_0 = (5,3)$. Plot contour diagrams and the path of descent in the same figure.

(a) $f(x) = 2x_1^4 + x_2^4$, (b) $f(x) = 2x_1^2 + x_2^2 - 2x_2^4$ (c) $f(x) = x_1^6 + 2x_2^6 - x_2^4$.

References

1. Faires JD, Burden RL (2003) Numerical methods (Chap. 10.2). Thomson/Brooks/Cole
2. Conte SD, de Boor C (1980) Elementary numerical analysis: an algorithmic approach, 3rd edn. McGraw-Hill Book Company, p 233
3. Broyden CG (1965) A class of methods for solving nonlinear simultaneous equations. Math Comput 19:577–593
4. Sauer T (2012) Numerical analysis (Chap. 2), 2nd edn. Pearson
5. Ascher UM, Mattheij RMM, Russell RD (1995) Numerical solution of boundary value problems for ordinary differential equations (Chap. 8.1.1). SIAM
6. Faires JD, Burden RL (2003) Numerical methods. Thomson/Brooks/Cole, p 509
7. Ortega JM, Rheinbolt WC (1970) Iterative Solution of nonlinear equations in several variables (Chap. 10.4). Academic Press, New York
8. Na TY (1979) Computational methods in engineering boundary value problems (Chap. 10.6). Academic Press, New York
9. Allgower L, Georg K (2003) Introduction to numerical continuation methods. SIAM, Philadelphia, PA, USA
10. Judd KL (1998) Numerical methods in economics (Chap. 5.9). MIT Press
11. Park H, Shim H (2005) What is the homotopy method for a system of nonlinear equations (survey)? Appl Math Comput 17:689–700
12. Watson LT (1989) Globally convergent homotopy methods: a tutorial. Appl Math Comput 31:369–396
13. Keskin AU (2019) Ordinary differential equations for engineers, problems with MATLAB solutions (Chap. 8)
14. Krauskopf B, Osinga HM, Vioque JG (eds) (2007) Numerical continuation methods for dynamical systems, path following and boundary value problem 2.s. Springer, p 3

15. Palancz B, Awange JL, Zaletnyik P, Lewis RH (2010) Linear homotopy solution of nonlinear systems of equations in Geodesy. J Geodasy 84(1):79–95
16. Allgower L, Georg K (2003) Introduction to numerical continuation methods. SIAM, Philadelphia, PA, USA, p 3
17. Sauer T (2012) Numerical analysis (Chap. 13.2.2), 2nd edn. Pearson
18. Conte SD, de Boor C (1980) Elementary numerical analysis: an algorithmic approach, 3rd edn. McGraw-Hill Book Company, p 212
19. Faires JD, Burden RL (2003) Numerical methods (Chap. 10.4). Thomson/Brooks/Cole
20. Süli E, Mayers DF (2003) An introduction to numerical analysis (Chap. 4). Cambridge University Press
21. Judd KL (1998) Numerical methods in economics (Chap. 5.7). The MIT Press, Cambridge MA
22. Kalaba R, Tesfetsion L (1991) Solving nonlinear equations by adaptive homotopy continuation. Appl Math Comput 41:99–115

Chapter 3
Boundary Value Problems (BVPs) for ODEs

In many engineering problems there are more than one endpoints or boundary values, so the corresponding mathematical models involve ordinary differential equations that must be solved subject to given boundary conditions.

While an initial value problem (IVP) consist of an equation together with its initial conditions at a given point, some applications often lead to differential equations in which the dependent variable or its derivative is specified at two different points. The conditions are called boundary conditions. A differential equation and its boundary conditions form a two-point BVP which usually involves a space coordinate as the independent variable. To solve the boundary value problem we first look for the general solution of the ODE and then use the boundary conditions to determine the values of the coefficients for the ODE.

Although the IVP and the BVP may seem similar to each other, they differ significantly. For example, while an IVP may have a unique solution, BVPs under similar conditions may also have a unique solution, but they may also have no solution or infinitely many solutions, as it happens in the solution of systems of linear algebraic equations.

In the BVPs with higher order equations, sometimes a single boundary condition involves values of the solution and/or its derivatives at both ends.

The first section of this chapter deals with basic problems of two-point BVPs. A more general discussion of linear BVPs is given by the introduction of Sturm—Liouville problems later in this chapter. We illustrate fundamentals and some applications of BVPs using MATLAB based solutions. For the proofs of theorems related to the theory of BVPs for ODEs, as well as for more detailed study of a particular subject, one may refer to textbooks [1–20].

© Springer Nature Switzerland AG 2019
A. Ü. Keskin, *Boundary Value Problems for Engineers*,
https://doi.org/10.1007/978-3-030-21080-9_3

3.1   Two Point Boundary Value Problems

Problem 3.1.1

(a) Define the standard (regular) form of a second order boundary value problem (BVP).

(b) Which of the following BVPs are homogenous?

(i) $$y'' + 4y' - 6y = 0; \quad y(0) = y'(1) = 0$$

(ii) $$y'' + y' - 2y = 3x; \quad y(0) = 1, y'(1) = 2$$

(iii) $$y'' = 4; \quad y(-2) = 3, \quad y(2) - 2y'(2) = 0$$

(iv) $$y'' = 1; \quad y(-2) = 0, \quad y(2) - 2y'(2) = 0$$

Solution

(a) A BVP in standard (regular) form is a second order linear ODE and the boundary conditions,

$$y'' + P(x)y' + Q(x)y = f(x) \tag{3.1a}$$

$$A_1 y(a) + B_1 y'(a) = g_1 \tag{3.1b}$$

$$A_2 y(b) + B_2 y'(b) = g_2 \tag{3.1c}$$

P, Q, f are continuous in the interval $[a, b]$ and $A_i, B_i, g_i, i = 1, 2, \in \mathbb{R}$ (A_1, B_1 are not both zero, a_2, b_2 are not both zero).

(b) A BVP is homogenous if both the differential equation and the boundary conditions are homogenous ($f(x) = g_1 = g_2 = 0$).

A more general homogenous BVP is defined using (3.1a–3.1c) with $f(x) = g_1 = g_2 = 0$ and $P(x), Q(x)$ includes an arbitrary constant λ,

$$y'' + P(x, \lambda)y' + Q(x, \lambda)y = 0$$
$$A_1 y(a) + B_2 y(b) = 0, \quad A_2 y(b) + B_2 y(0)$$

where $y(x) = 0$ is a trivial solution.

(i) $y'' + 4y' - 6y = 0, \quad P(x) = 4, Q(x) = -6$
     $y(0) = y'(1) = 0, \quad a_1 = 1, \quad b_1 = 0, \quad a_2 = 0, \quad b_2 = 1,$
     This is a homogenous BVP.

(ii) $y'' + y' - 2y = 3x$; $y(0) = 1$, $y'(1) = 2$, non-homogenous BVP.
($f(x) \neq 0$)

(iii) $y'' = 4 \neq 0$, $f(x) = 4$, $g_1 = 3$, $g_2 = 0$, non-homogenous BVP

(iv) $y'' = 1 \neq 0$ non-homogenous BVP.

Problem 3.1.2

(a) What are the two main types of BVPs (in matrix form)?

(b) What do unseperated and seperated boundary conditions mean? Which one is a more general statement of boundary conditions? Express their forms in linear case.

Solution

(a) BVPs have different forms:

(i) A first order system of equations

$$y' = f(x, y) \tag{3.2}$$

where $y' = dy/dx$ and y, f are n-dimensional vectors,

(ii) An n-th order ODE,

$$y^{(n)} = g\left(x, y, y', \dots, y^{(n-1)}\right) \tag{3.3}$$

which can be reduced to (3.2).

(b) If (3.4) is to be solved on the interval $x \in (a, b)$, then n boundary condition equations uniquely determine the solution of the BVP, which may be of the unseparated general form

$$g(y(a), y(b)) = 0 \tag{3.4}$$

while separated boundary conditions have simpler forms,

$$g_L(y(a)) = 0, \quad g_R(y(b)) = 0 \tag{3.5}$$

where g_L and g_R have dimensions of p and $n - p$, respectively.

For linear boundary conditions, unseparated one take the form of

$$Ly(a) + Ry(b) = c \tag{3.6}$$

L and R have $n \times n$ matrices and c is n-dimensional vector; while separated linear boundary conditions have the form of

$$L_L y(a) = c_L, \quad R_R y(b) = c_R \tag{3.7}$$

L_L, R_R, c_L, c_R are $(p \times n), (n - p) \times n, p$ and $(n - p)$ dimensional.

Problem 3.1.3 What are the Dirichlet, Neumann and Robin boundary conditions for BVPs?

Solution

  (i) Dirichlet[1] Condition; The solution has some particular value at the end point or along a boundary. For example, a Dirichlet BC means fixing the value for the temperature for the solution in one edge of the one-dimensional domain (a metal bar).
 (ii) Neumann[2] Condition; The derivative of the solution equals a particular value at the end point or in the normal direction along a boundary, such as imposing the flux of heat through one edge of the one-dimensional domain (a metal bar). If Newman BC is set to zero, it means that the edge of the bar is isolated and no flux of heat enters or outputs the bar.
(iii) Robin[3] Condition; A linear combination of Dirichlet and Neumann conditions.

Problem 3.1.4 True (T) or False (F) ?

(a) It is impossible for the homogenous system of ODEs to have no solution.
(b) A homogenous system, $Ax = 0$, never has the trivial solution.
(c) In a homogenous system, $Ax = 0$, if A is singular, then there are infinitely many nontrivial solutions.
(d) Trivial solution means $x = 0$.
(e) The nonhomogeneous system $Ax = b$ has either no solutions or infinitely many solutions if and only if the homogenous system $Ax = 0$ has nonzero solutions.
(f) The two point boundary value problem $y'' + 3y' + 5y = 0$; $y(0) = y(1) = 1$ is a homogenous linear BVP.
(g) The two point BVP $y'' + 3y' + 5y = 0$; $y(0) = 0$, $y(1) = 1$ is a homogenous linear BVP.
(h) In a two point BVP, we want to know if the problem has nonzero solutions, because trivial solution is rarely of interest.

Answers
(a) T, (b) F, (c) T, (d)T, (e) T, (f) F, boundary conditions not zero, (g) F, $y(1) = 1$, (h) T.

---

[1]Peter Gustav Lejeune Dirichlet (1805–1859) German mathematician who made contributions to number theory, analysis, and mechanics.
[2]Carl Gottfried Neumann (1832–1925), German mathematician who worked in mathematical physics, potential theory and electrodynamics, as well as in pure mathematics.
[3]Gustave Robin (1855–1897), French mathematician.

Problem 3.1.5 Show that nonlinear BVP, $y'' = e^{-2y}$ with Neumann boundary conditions $y'(0) = 1$, $y'(1) = 1/2$ has the solution of $y = \ln(1+x)$.

Solution

$$y = \ln(1+x), \quad y' = \frac{1}{1+x}, \quad y'' = -\frac{1}{(1+x)^2}$$

Both boundary conditions check for $y'(0)$, $y'(1)$.

$$-\frac{1}{(1+x)^2} = -e^{-2y} \quad \rightarrow \quad \frac{1}{(1+x)^2} = e^{-2y}$$

Taking the logarithm of both sides and simplifying,

$$0 - \ln(1+x)^2 = -2y \quad \rightarrow \quad y = \ln(1+x) \quad QED.$$

Problem 3.1.6 Solve the two point BVP, $y'' + 9y = 0$, $y(0) = 1$, $y(\pi) = 0$.

Solution

Characteristic equation: $s^2 + 9 = 0 \rightarrow s = \pm 3i$,

The general solution is $y = c_1 \cos(3x) + c_2 \sin(3x)$. Using boundary conditions, we get

$$y(0) = c_1 \cos(0) + c_2 \sin(0) = c_1 = 1$$
$$y(\pi) = 1 = c_1 \cos(3\pi) + c_2 \sin(3\pi) = 0$$
$$\cos(3\pi) + c_2 \sin(3\pi) = 0 \quad \rightarrow \quad c_2 = -\cos(3\pi)/\sin(3\pi) = -\cot(3\pi)$$

The solution is $y = \cos(3x) - \cot(3\pi) \sin(3x)$.

Problem 3.1.7 Solve the two point BVP $y'' + y = 0$, $y(0) = y(\pi) = 1$.

Solution

$$s^2 + 1 = 0 \quad \rightarrow \quad s = \pm i \quad \rightarrow \quad y = c_1 \cos x + c_2 \sin x$$
$$y(0) = 1 = c_1 \cos(0) + c_2 \sin(0) = c_1$$
$$y(\pi) = 1 = c_1 \cos(\pi) + c_2 \sin(\pi) = -c_1$$

These two conditions are inconsistent, therefore given BVP has no solution.

Problem 3.1.8 Solve the two point BVP $y'' + y = 0$; $y(0) = y(\pi) = 0$.

Solution

General solution has the form $y = c_1 \cos(x) + c_2 \sin(x)$

$$0 = c_1 \cos(0) + c_2 \sin(0) = c_1$$
$$0 = c_1 \cos(\pi) + c_2 \sin(\pi) = -c_1$$

The secondary boundary condition is satisfied regardless of the value of c_2. Therefore the solution is $y = c_2 \sin(x)$, c_2 is arbitrary. Hence, this homogenous BVP has infinitely many solutions.

Problem 3.1.9 Solve the two point BVP, $y'' + y = 0$, $y(0) = 0$, $y'(2\pi) = 1$.

Solution

Characteristic equation: $s^2 + 1 = 0$, $\quad s = \pm i$, $\quad \rightarrow \quad y = c_1 \cos(x) + c_2 \sin(x)$.
   Using boundary conditions, we obtain

$$y(0) = 1 = c_1 \cos(0) + c_2 \sin(0) = c_1 = 0$$
$$y'(x) = -c_1 \sin(x) + c_2 \cos(x) \quad \rightarrow \quad y'(2\pi) = -c_1 \sin(2\pi) + c_2 \cos(2\pi) = c_2 = 1$$

Therefore solution is $y = \sin x$.

Problem 3.1.10 Solve the BVP, $y'' + (1/2) \cdot y = 0$, $\quad y(0) = 1$, $\quad y(\pi) = 0$

Solution

$$s^2 + \frac{1}{2} = 0 \quad \rightarrow \quad s = \mp \frac{1}{4}j$$
$$y = c_1 \cos(t/4) + c_2 \sin(t/4)$$
$$y(0) = 1 = c_1 \cdot 1 + 0 \quad \rightarrow \quad c_1 = 1$$
$$y(\pi) = 0 = \cos(\pi/4) + c_2 \sin(\pi/4) \quad \rightarrow \quad c_2 = -\cos(\pi/4)/\sin(\pi/4)$$
$$c_2 = -\cot(\pi/4) = -1$$
$$y = \cos(t/4) - \sin(t/4)$$

Problem 3.1.11 Solve the two point BVP, $y'' - 4y = 0$; $y(0) = 0$, $y(1) = 1$.

Solution

Characteristic equation: $\quad s^2 - 4 = 0 \quad \rightarrow \quad s = \pm 2$, the general solution is $y = c_1 e^{2x} + c_2 e^{-2x}$.

$$y(0) = c_1 + c_2 = 0 \quad \rightarrow \quad c_1 = -c_2$$
$$y(1) = c_1 e^2 + c_2 e^{-2} = 1 \quad \rightarrow \quad -c_2 e^2 + c_2 e^{-2} = 1 \quad \rightarrow \quad c_2\left(-e^2 + e^{-2}\right) = 1$$
$$c_2 = 1/\left(-e^2 + e^{-2}\right), \quad c_1 = -1/\left(-e^2 + e^{-2}\right)$$
$$y = \frac{e^{-2x} - e^{2x}}{e^{-2} - e^2} = \left(\frac{1}{e^{-2} - e^2}\right)\left(e^{-2x} - e^{2x}\right)$$

Problem 3.1.12 Find the solution for the two point BVP, $y'' + y = 0$, $y(0) = 0$, $y(4) = -2$.

Solution

$$s^2 + 1 = 0 \quad \rightarrow \quad s = \pm i \quad \rightarrow \quad y = c_1 \cos(x) + c_2 \sin(x)$$
$$0 = c_1 \cos(0) + c_2 \sin(0) = c_1$$
$$-2 = c_1 \cos(4) + c_2 \sin(4) = c_2 \sin(4) \quad \rightarrow \quad c_2 = -2/\sin(4)$$

Therefore, general solution of the given BVP is $y = -[2/\sin(4)]\sin(x)$.

Problem 3.1.13 Show that $y = x \cos(x)$ is a solution of the BVP,

$$y'' + y = -2\sin(x); \quad y(0) = 0, y(\pi) = -\pi$$

Solution
This function solves the ODE because

$$y'' = [\cos(x) - x \sin(x)]' = -2\sin(x) - x \cos(x) = -2\sin(x) - y$$
$$y'' + y = -2\sin(x)$$

Checking the boundary conditions gives

$$y(0) = 0 \cos(0) = 0, \quad y(\pi) = \pi \cos(\pi) = -\pi$$

Problem 3.1.14 A homogenous two point BVP of the form $-y'' = f(x)$; $y(0) = y(1) = 0$ has the solution $y = y_p + Ax + B$ where A, B are constants that are calculated using given boundary conditions, $y(0) = 0 = y_p(0) + B$, $y(1) = 0 = y_p(1) + A + B$. Here, y_p is the particular solution.
Find the solution of the BVP, $-y'' = 2$, $y(0) = y(1) = 0$.

Solution
Integrating both sides of the equation, we obtain

$$y = -x^2 + Ax + B, \quad y(0) = B = 0, \quad y(1) = 0 = -(1)^2 + A = A - 1$$

$A = 1$, therefore $y = -x^2 + x = x(1 - x)$ is the solution.

Problem 3.1.15 Nonhomogeneous BVP, $y'' - x$, $y(0) = y(\pi/2) - 0$, has the solution within the interval $x \in [0, \pi/2]$,

$$y(x) = \frac{x}{24}\left(4x^2 - \pi^2\right)$$

Is the solution for $\pi/2 \le x \le 2$ guaranteed for its correctness? Under which conditions this solution is valid for $x \in [0, 2]$?

Solution

The boundary values are $y(0) = y(\pi/2) = 0$. The solution for $x \geq \pi/2$ cannot be guaranteed for its correctness, since it remains outside the interval of definition for the BVP (this is a fundamental property that relates to definition of a Two-Point BVP, existence, uniqueness and continuity of solutions).

However, this solution will be valid for $x \in [0, 2]$ provided that BVP is

$$y'' = x, \quad y(0) = 0, \quad y(2) = 0.5108663, \quad x \in [0, 2].$$

Problem 3.1.16 Let $y_1(x), y_2(x)$ be two solutions of $y'' + P(x)y' + Q(x)y = 0$ Nontrivial solutions to the homogenous BVP exists if

$$\Delta = \begin{vmatrix} a_1 y_1(a) + b_1 y_1'(b) & a_1 y_1(a) + b_1 y_2'(b) \\ a_2 y_1(a) + b_2 y_1'(b) & a_2 y_1(a) + b_2 y_2'(b) \end{vmatrix} = 0$$

If $\Delta \neq 0$, the only solution is the trivial solution, $y(x) = 0$
Check if the following BVPs have non-trivial solution:

(i) A BVP with the general solution $y = c_1 e^{-x} + c_2 e^x, y(0) = 0, y'(1) = 0$
(ii) $y'' - y' - 2y = 0; \quad y_1(0) = 1, \quad y_2(0) = -1, \quad y_1'(1) = y_2'(1) = 0$

Solution

(i) Applying boundary conditions, $0 = c_1 + c_2, \quad y' = -c_1 e^{-x} + c_2 e^x$

$$0 = -c_1 e^{-1} + c_2 e^1, 0 = -c_1 e^{-1} + (-c_1)e = -c_1(e^{-1} + e) \rightarrow -c_1 = c_2 = 0$$

Hence, the solution is $y = 0$
Alternatively, applying above-given determinant, we obtain

$$y_1 = c_1 e^{-x}, \quad y_2 = c_2 e^x, \quad a_1 = 1, \quad b_1 = 0, \quad a_2 = 0, \quad b_2 = 1$$
$$a = 0, \quad b = 1, \quad g_1 = g_2 = 0, \quad y_1'(1) = -c_1 e^{-1}, \quad y_2'(1) = c_1 e^1$$
$$\Delta = \begin{vmatrix} 1c_1 + 0 & 1c_2 + 0 \\ 0 + 1(-c_1 e^{-1}) & 0 + 1c_1 e^1 \end{vmatrix} = \begin{vmatrix} c_1 & c_2 \\ -c_1 e^{-1} & c_1 e^1 \end{vmatrix} = c_1^2 e + c_1 c_2 e^{-1} \neq 0$$

Solution is trivial, $y = 0$

(ii) $y'' - y' - 2y = 0; \quad y_1(0) = 1, \quad y_2(0) = -1, \quad y_1'(1) = y_2'(1) = 0$
$$a_1 = 1, \quad b_1 = 0, \quad a_2 = 0, \quad b_2 = 1, \quad g_1 = g_2 = 0,$$
$$\Delta = \begin{vmatrix} 1 & -1 \\ 0 & 0 \end{vmatrix} = 0$$

It has non-trivial solution, $y(x) \neq 0$.

Problem 3.1.17 Solve the two-point BVP, $y'' - 2y' = 0$, $y(0) = 0$, $y(L) = 1$, $L > 0$.

Solution

Characteristic equation: $s^2 - 2s = s(s - 2) = 0$, $s_1 = 0$, $s_2 = 2$.

General solution is of the form $y = c_1 e^0 + c_2 e^{2x} = c_1 + c_2 e^{2x}$

Using boundary conditions, we obtain

$$y(0) = 0 = c_1 + c_2 \rightarrow c_1 = -c_2$$

$$y(L) = 1 = -c_2 + c_2 e^{2L} = c_2 (e^{2L} - 1) \rightarrow c_2 = 1/(e^{2L} - 1)$$

Therefore, the solution of given BVP is $y = (e^{2x} - 1)/(e^{2L} - 1)$.

Problem 3.1.18 (Poiseuille[4] flow). Fluid flow in a tube (pipe) can be laminar or turbulent. While turbulent fluid flow relies on semi-empirical theories and on the experimental data, laminar fluid flow problems may have analytic solutions. In a two-dimensional single phase forced convection flow in a pipe, it is assumed that the flow is steady, fluid is incompressible and Newtonian. With these assumptions, axial laminar flow velocity profile of the fluid in a pipe with circular cross-section can be obtained using continuity, momentum and energy equations. This gives

$$\frac{1}{r}(ru')' = -\frac{1}{\eta}p \tag{3.8}$$

where u gives the axial fluid velocity at a point, r is the radial distance from the center of the pipe, p is the pressure gradient and η is the viscosity of the fluid (derivatives are taken with respect to r).

Find the maximum flow velocity, the total flow rate of the fluid and the absolute viscosity of the fluid, using boundary conditions for the velocity profile, $u(R) = 0, 0 < r < R$, where R is the radius of the pipe.

Solution

$$\frac{1}{r} \cdot \frac{d}{dr}\left(r\frac{du}{dr}\right) = -\frac{p}{\eta} \rightarrow \frac{1}{r}\left(\frac{du}{dr} + \frac{d^2 u}{dr^2}r\right) = -\frac{p}{\eta}$$

$$\frac{1}{r}u' + u'' = -\frac{p}{\eta} \tag{3.9}$$

$$u = u_h + u_p \tag{3.10}$$

$u = kr^2$, substitute into (3.9),

---

[4]Jean Louis Marie Poiseuille (1797–1869), French doctor who studied fluid mechanics of blood flow. The unit of viscosity (1 poise = 1 dyne s/cm2) was named in honor of Poiseuille (The SI unit of viscosity is N s/m2).

$$2k + 2k = -\frac{p}{\eta} \quad \rightarrow \quad k = -\frac{p}{4\eta}$$

$$u = u_h - \frac{p}{4\eta}r^2 \tag{3.11}$$

From boundary conditions, $u(R) = 0$, substitute this into (3.11)

$$0 = u_h - \frac{p}{4\eta}R^2 \quad \rightarrow \quad u_h = \frac{p}{4\eta}R^2 \tag{3.12}$$

Using (3.10), we obtain fluid flow velocity profile,

$$u = \frac{p}{4\eta}R^2 - \frac{p}{4\eta}r^2 = \frac{p}{4\eta}(R^2 - r^2) \tag{3.13}$$

This is equation of a parabola and its maximum occurs when $r = 0$, as shown in Fig. 3.1,

$$u_{max} = \frac{pR^2}{4\eta} \tag{3.14}$$

Flow rate is

$$q = \int u \cdot dA \tag{3.15}$$

where A is the differential area perpendicular to the flow direction. With $dA = 2\pi r dr$,

$$q = \int_0^R \frac{p(R^2 - r^2)}{4\eta}dA = \frac{p}{4\eta}\int_0^R 2\pi r(R^2 - r^2)dr = \frac{2\pi p}{4\eta}\int_0^R r(R^2 - r^2)dr$$

$$q = \frac{\pi p}{2\eta}\int_0^R (rR^2 - r^3)dr = \frac{\pi p}{2\eta}\left[R^2\frac{r^2}{2} - \frac{r^4}{4}\right]_0^R q = \frac{\pi p}{2\eta}\left[\frac{R^2}{2} - \frac{R^4}{4}\right] = \frac{\pi pR^4}{8\eta} \tag{3.16}$$

Absolute viscosity is calculated from (3.16),

$$\eta = \frac{\pi pR^4}{8q}$$

Fig. 3.1 Laminar fluid flow velocity profile in a pipe

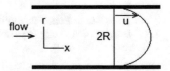

If the pipe length is L, and pressure difference between two ends of the pipe is P, then $p = P/L$,

$$\eta = \pi P R^4 / 8(qL) \tag{3.17}$$

units are $q \to m^3/s, \quad p \to N/m^2, \quad L = m, R = m$
if $L = 1m, R = 1cm, P = 1N/m^2, q = 10^{-8}m^3/s$, From (3.17),

$$\eta = \frac{\pi(10^{-2})^4}{8 \cdot 10^{-8}} = \frac{\pi \cdot 10^{-8}}{8 \cdot 10^{-8}} = \frac{\pi}{8} = 0.3927\left(N/m^2\right).$$

Problem 3.1.19 Write a MATLAB m-file to check if a nontrivial solution for a homogeneous BVP of the form, $y'' + Py' + Qy = 0, \quad A_1 y(a) + B_1 y'(a) = 0, \quad A_2 y(b) + B_2 y'(b) = 0$ exists.

Then, use this code to test the following BVPs for a nontrivial solution.

(i) $y'' + 2y' - 3y = 0; \quad y(0) = 0, \quad y'(1) = 0,$

(ii) $y'' = 0; \quad y(-1) = 0, \quad y(1) - 2y'(1) = 0,$

(iii) $y'' + y = 0; \quad y(0) = 0, \quad y(1) = 0,$

(iv) $y'' + y' - y = 0; \quad y(0) = 0, \quad y(1) = 0,$
            (bvp_nontrivialtest1.m).

Solution
Following MATLAB m-file (bvp_nontrivialtest1.m) can be used to check if a nontrivial solution for a BVP of the form

$$y'' + Py' + Qy = 0, \quad A_1 y(a) + B_1 y'(a) = 0, \quad A_2 y(b) + B_2 y'(b) = 0.$$

A sample output of the edited m-file (for the last BVP) is the following;

```
cp =
   -1.6180
    0.6180
```

```
det = 1.657
only trivial solution (y = 0) exists
```

MATLAB m-file (bvp_nontrivialtest1.m) is given below:

```
%bvp_nontrivialtest2.m        2-point bvp
clc; clear ;
syms x  c1  c2  L
%y"+Py'+Qy=R=0  Homogeneous ODE
% 1st boundary condition: A1*y(a)+B1*y'(a)=e=0
% 2nd boundary condition: A2*y(b)+B2*y'(b)=h=0

% P=2; Q=-3;R=0; a=0; b=1; e=0;h=0; % distinct real  roots: 1,-3
% A1=1; B1=0; A2=0; B2=1;        % Trivial solution

% P=0; Q=0;R=0; a=-1;b=1; e=0;h=0; % repeating roots: 0,0
% A1=1; B1=0; A2=1; B2=-2;       % Nontrivial solution

% P=0; Q=1;R=0;  a=0; b=1; e=0; h=0; % complex roots: +/-i
% A1=1; B1=0; A2=1; B2=0;        % Trivial solution

P=1; Q=-1;R=0; a=0; b=1;  e=0; h=0; % distinct real roots: -1.618, 0.618
A1=1; B1=0; A2=1; B2=0;        % Trivial solution

if (abs(R)+abs(e)+abs(h))~=0
    disp('not a homogeneous BVP')
else
cp=[1 P Q];cp=roots(cp)
r=isreal(cp(1))
 if r==0
 y1=c1*cos(sqrt(cp(1))*x); y2=c2*sin(sqrt(cp(2))*x);
  else if cp(1)==cp(2)
 y1=c1*exp(cp(1)*x);   y2=c2*x*exp(cp(1)*x);
  else
 y1=c1*exp(cp(1)*x);   y2=c2*exp(cp(2)*x);
      end
 end
y1a=subs(y1,x,a); y1b=subs(y1,x,b);y2a=subs(y2,x,a); y2b=subs(y2,x,b);
dy1=diff(y1,1,x); dy1a=subs(dy1,x,a); dy1b=subs(dy1,x,b);
dy2=diff(y2,1,x); dy2a=subs(dy2,x,a); dy2b=subs(dy2,x,b);
d11=A1*y1a+B1*dy1a;    d12=A1*y2a+B1*dy2a;
d21=A2*y1b+B2*dy1b;    d22=A2*y2b+B2*dy2b; D=[d11 d12;d21 d22];
det=det(D);det=simplify(det);det=vpa(subs(det,[c1 c2],[1 1]),4)
if det<=1e-6
disp('Nontrivial solution of given BVP exists')
else
disp('only trivial solution (y = 0) exists')
end
end
```

Problem 3.1.20 Solve the BVP, $y'' - 4y' + 3y = 0, \quad y(0) = 0, \quad y(1) = 1$.
Verify the solution using MATLAB $\texttt{dsolve}$ function.
Plot the solution curves for the function and its first derivative.

Solution
The solution is of the form $y = C_1 \exp(s_1 x) + C_2 \exp(s_2 x)$ where s_1, s_2 are found from

$$s^2 - 4s + 3 = (s-1)(s-3) = 0,$$
$$s_1 = 1, \quad s_2 = 3$$

Applying initial conditions to given ODE we obtain,

$$y(0) = C_1 + C_2 = 0, C_1 = -C_2 \tag{3.18}$$

$$y(1) = C_1 \exp(s_1) + C_2 \exp(s_2) = C_1 \exp(1) + C_2 \exp(3) = 1$$

From (3.18),

$$C_1 \exp(1) - C_1 \exp(3) = C_1(\exp(1) - \exp(3)) = 1$$
$$C_1 = 1/(\exp(1) - \exp(3))$$

Thus, the solution becomes

$$y = \exp(x)/(\exp(1) - \exp(3)) - \exp(3x)/(\exp(1) - \exp(3))$$

Following is the result of applying MATLAB `dsolve` function for this BVP which checks with the result of analytic solution;

$$y = \exp(x)/(\exp(1) - \exp(3)) - \exp(3*x)/(\exp(1) - \exp(3))$$

The solution and its derivative are plotted in Fig. 3.2.

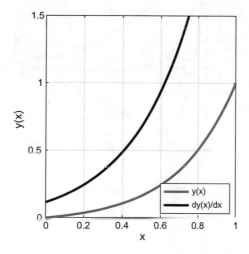

Fig. 3.2 Soluton curves for the BVP, $y'' - 4y' + 3y = 0$, $\quad y(0) = 0$, $\quad y(1) = 1$

```
clc; clear; close; syms y(x)
y=dsolve(diff(y,2,x)-4*diff(y,x)+3*y == 0,y(0)==0, y(1)==1)
p=ezplot(y); set(p,'linewidth',2);hold on;
dy=diff(y,x);
p=ezplot(dy);grid; set(p,'linewidth',2);ylabel('y(x)');
axis([0 1 0 1.5]);legend('y(x)','dy(x)/dx')
```

Problem 3.1.21 Use MATLAB to solve and plot the solution of the following BVP.

$$y'' + 3y' + 2y = 0; \quad y(0) = 1, \quad y'(k) = 0, \quad k = 0,1,2.$$

Solution
MATLAB solution is

$$y(t) = \frac{e^{-t}\left(e^{-t} - 2e^{-k}\right)}{1 - 2e^{-k}}, \quad k = 0,1,2$$

Solution curves for $y'' + 3y' + 2y = 0$, $y(0) = 1$, $y'(k) = 0$, $k = 0,1,2$ are displayed in Fig. 3.3.

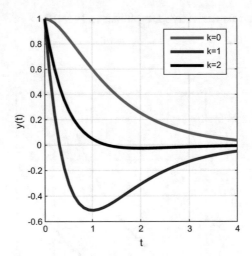

Fig. 3.3 Solution curves for $y'' + 3y' + 2y = 0$; $y(0) = 1$, $y'(k) = 0$, $k = 0,1,2$.

```
% dsolve2_2.m    Solve 2nd order BVP and plot the solutions
clc;clear;close;
syms  y(x)
eq='D2y+3*Dy+2*y=0';    % ODE
ic='y(0)=1,Dy(1)=0';    % boundary conditions
S=dsolve(eq,ic);pretty(S)
t = linspace(0,4,40); y=eval(vectorize(S));
plot(t,y,'linewidth',2.5);xlabel('t'),ylabel('y(t)');grid on;
```

Problem 3.1.22

Solve the BVP, $y'' - y = -x$, $\quad 0 \le x \le 1$, $\quad y(0) = y(1) = 0$

Solution

Let s denote differential operation, then

$$(s^2 - 1)y = -x$$

auxiliary equation: $p^2 - 1 = 0$, $\quad \rightarrow \quad p = \pm 1$,
  The homogenous (complementary) solution is

$$y_h = A_1 e^{p_1 x} + A_2 e^{p_2 x} = A_1 e^x + A_2 e^{-x}$$

Particular solution;

$$y_p = \frac{-x}{p^2 - 1} = \left(1 - p^2\right)^{-1} x = \left(1 + s^2 + s^4 + \cdots\right)x = x$$

$$y = y_h + y_p = A_1 e^x + A_2 e^{-x} + x$$

Boundary Conditions:

$$y(0) = 0 = A_1 + A_2 \quad \rightarrow \quad A_1 = -A_2,$$

$$y(1) = 0 = A_1 e^1 + A_2 e^{-1} + 1 = A_1 e^1 - A_1 e^{-1} + 1 = 0$$

$$A_1 = \frac{-1}{e - e^{-1}}, \quad A_2 = -A_1 = \frac{1}{e - e^{-1}}$$

$$y = \left(-\frac{1}{e - e^{-1}}\right)e^x + \left(\frac{1}{e - e^{-1}}\right)e^{-x} + x = x - \frac{e^x - e^{-x}}{e - e^{-1}}$$

Problem 3.1.23 The existence of a solution of a nonlinear BVP does not guarantee that the general problem can be solved.

Use MATLAB to verify that the nonlinear BVP, $y' = y^2, y(0) = y(1)$ has only the zero solution, while the perturbed problem $y' = y^2 + \epsilon$, $\quad \epsilon > 0$, $\quad y(0) = y(1)$ has no solutions.

(dsolve_perturb1.m).

Solution

MATLAB m-file dsolve_perturb1.m demonstrates that the nonlinear BVP, $y' = y^2, y(0) = y(1)$ has only the zero solution, while the perturbed problem has no solutions.

```
% dsolve_perturb1.m
clc;clear;syms  y(x)
bc='y(0) ==0,y(1)==0';
y=dsolve(diff(y)-y^2 == 0, bc)


y = 0
```

```
% dsolve_perturb1.m
clc;clear;syms  y(x)
eps=1e-30
bc='y(0) ==0,y(1)==0';
y=dsolve(diff(y)-y^2-eps == 0, bc)

Warning: Explicit solution could not be found.
> In dsolve (line 201)
  In dsolve_perturb1 (line 7)
```

Problem 3.1.24 Using string syntax when setting up the equations in the application of Symbolic Math Toolbox 'dsolve' command in solving ODEs has some limitations.

For example assigning numerical values to certain parameters in the equation does not work and one has to modify the ODE manually to replace all appearances of such parameters by numerical values.

Another limitation comes up when checking the solution. At this stage, one needs to plug the solution into the ODE and observe if it gives zero result. However, in that case 'subs' command can not be used directly, since the string input does not permit it to be used, which requires further manual work.

Symbolic equations and functions can be used defining inputs for other Symbolic Math Toolbox functions, such as 'solve'.

Checking the initial values brings about similar problems. In short, use of string syntax for ODE's, such as, Dy, D2y, y(0) = etc., requires some handwork to overcome limitations of string syntax; causing unsmooth workflow. MATLAB introduced symbolic equations and functions in the Symbolic Math Toolbox in R2012a, making the workflow of solving ODEs (and checking solutions and initial conditions) more convenient.

Write MATLAB m-files to illustrate these two cases of solving ODEs using the second order ODE, $y'' + y' + ny = 0$, $y(1) = 1$, $y(2) = n$, and check the solution as well as the initial conditions.

Solution

Following m-files illustrate these two cases using a second order ODE,

$$y'' + y' + ny = 0, \quad y(1) = 1, \quad y(2) = n$$

```
% String syntax example:
clc;clear; syms x
eq = 'D2y + Dy + n*y' ;
f = dsolve(eq, 'y(1)=1', 'y(2)=n', 'x');
n = 1;
f = dsolve(eq, 'y(1)=1', 'y(2)=n', 'x');
eq
f = dsolve('D2y + Dy + n*y', 'y(1)=1', 'y(2)=n', 'x');
check = diff(f,2,x) + diff(f,x) + n*f ;
check = simplify(check)
checkic1 = subs(f,x,1)
checkic2 = subs(f,x,2)
```

```
% Using Symbolic Functions and Equations:
clc;clear;
syms n y(x)
eq = diff(y,2,x) + diff(y,x) + n*y == 0;
f(x) = dsolve(eq, y(1)==1, y(2)==n); f(x)=simplify(f(x))
n=1;
eq = subs(eq);
f(x) = dsolve(eq, y(1)==1, y(2)==n)
check = subs(eq, [y,diff(y),diff(y,2,x)], [f,diff(f),diff(f,2,x)]);
TF = simplify(check(x))
checkic1 = simplify(f(1)--1)
checkic2 = simplify(f(2)==1)
```

Problem 3.1.25 Use MATLAB dsolve function to solve and plot the autonomous system of ODEs,

$$y'' + y = z$$
$$z'' = y + 1$$

Subject to following boundary conditions, $y(0) = y(1) = z(0) = 0, z'(1) = 0$. (dsolvebvp10.m).

Solution
Figure 3.4 displays the two solution curves for the given system of BVP using MATLAB script (dsolvcbvp10.m). Note that, analytical expressions are too complicated and lengthy to include here, additionally simplification process takes too much time. Therefore, these lines are not used, and shown in "commented" form.

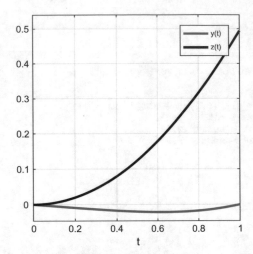

Fig. 3.4 Solution curves for the system of BVP, $y'' + y = z, z'' = y + 1, \quad y(0) = y(1) = z(0) = 0, \quad z'(1) = 0$

```
% dsolvebvp10.m  Solves a system of BVP
clc;clear;close; syms  y(t) z(t)
Dz=diff(z,t);
Y = dsolve( diff(y,2,t) + y== z, diff(z,2,t)== y + 1,...
    y(0)==0, y(1)==0, z(0)==0, Dz(0)==0 );
y=Y.y;  %y=simplify(y)
z=Y.z;  %z=simplify(z)
y=ezplot(y, [0  1]);set(y,'linewidth',2);hold on;
z=ezplot(z, [0  1]);set(z,'linewidth',2);hold off;grid;
legend('y(t)','z(t)');
```

Problem 3.1.26 Use `dsolve` function to solve the following Singular Linear BVP and plot the solution.

$$y'' = \frac{1}{x}y' - \frac{1}{x^2}y + 8x - 2, \quad y(0) = y(1) = 0$$

Solution

The solution is $y(x) = 2x^2(x - 1)$. Its graph is shown in Fig. 3.5. MATLAB script is given below.

```
syms  y(x) % singular BVP
S0=dsolve(diff(y,2,x)-(1/x)*diff(y,1,x)+(1/x^2)*y == 8*x-2,...
y(0)==0,y(1)==0);S=expand(S0)
p=ezplot(S,[0 1]);grid; set(p,'linewidth',2);ylabel('y(x)');
```

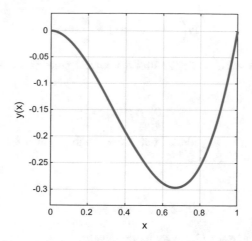

Fig. 3.5 The solution curve for the BVP, $y'' = x^{-1}y' - x^{-2}y + 8x - 2$, $y(0) = y(1) = 0$

Problem 3.1.27 ODE defining the shape of a catenary is

$$y'' = \frac{\rho g A}{T}\sqrt{1 + (y')^2}; \quad y(0) = a, \quad y'(0) = 0 \qquad (3.19)$$

ρ, g, A, T are density, cross sectional area of the cable, gravity and tension in the cable, respectively, and $a = T/(\rho g A)$.

(a) Solve this equation.
(b) Use MATLAB dsolve function to check the result found above.
(c) Plot catenary for $a = T/(\rho g A) = 0.5, \quad 1, \quad 1.5, \quad 2$ (dsolve_catenary1.m).

Solution

(a) Let $y = z$, then (3.19) becomes

$$z' = \frac{\rho g A}{T}\sqrt{1 + z^2}$$

which can be solved using separation of variables,

$$dz = \frac{\rho g A}{T}\sqrt{1 + z^2}\,dx \quad \rightarrow \quad \frac{dz}{\sqrt{1 + z^2}} = \frac{\rho g A}{T}\,dx \quad \rightarrow$$

$$\int \frac{dz}{\sqrt{1 + z^2}} = \frac{\rho g A}{T}\int dx$$

Let $\rho g A/T = 1/a$,

$$\ln\left(z + \sqrt{1 + z^2}\right) = \frac{x}{a} + c_1 \tag{3.20}$$

Arrange the coordinate axes so that the lowest point of catenary occurs when $x = 0$;

$$z(0) = y'(0) = 0$$

This helps us to find out the value of c_1 in (3.20), as $\ln(1) = 0 + c_1 \quad \rightarrow \quad c_1 = 0$
Therefore,

$$z + \sqrt{1 + z^2} = e^{x/a} \tag{3.21}$$

Multiply both sides of (3.21) by its conjugate,

$$\left(z + \sqrt{1 + z^2}\right)\left(z - \sqrt{1 + z^2}\right) = \left(z - \sqrt{1 + z^2}\right)e^{x/a}$$

$$z^2 - \left(1 + z^2\right) = -1 = \left(z - \sqrt{1 + z^2}\right)e^{x/a} \quad \rightarrow \quad z - \sqrt{1 + z^2} = -e^{-x/a}$$

Adding this result to Eq. (3.21), we obtain

$$\left(z + \sqrt{1 + z^2}\right) + \left(z - \sqrt{1 + z^2}\right) = e^{x/a} - e^{-x/a}$$

$$z = \frac{e^{x/a} - e^{-x/a}}{2} = \sinh\left(\frac{x}{a}\right) = y'$$

$$y = \int y'dx = \int \sinh\left(\frac{x}{a}\right)dx = a\cosh\left(\frac{x}{a}\right)$$

Note that in the solution of this BVP, we used $y(0) = a$ and the value of derivative at $x = 0$. The same problem could be defined under the boundary conditions of $0 < x < L$, $y(0) = h_1$, $y(L) = h_2$ where, L, h_1, h_2 are the horizontal distance between the endpoints, height at the first and second end of the cable, respectively.

(b) Edited output of MATLAB m-file is given below. Last line shows the analytical result obtained for the coefficient value of $a = T/(\rho g A) = 1$.

```
eq =   D2y=(1/a)*(1+(Dy)^2)^(1/2)
S = exp(-(x - a*log(a))/a)/2 + (a^2*exp((x - a*log(a))/a))/2
y =   exp(-x)/2 + exp(x)/2
```

Fig. 3.6 Catenary for different values of coefficients

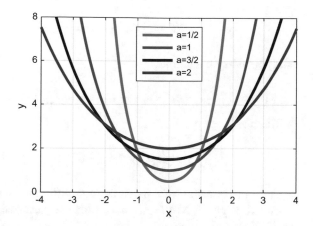

(c) Graphs of catenaries for four different values of $a = T/(\rho g A) =$ 0.5, 1, 1.5, 2 are displayed in Fig. 3.6.

Following is the MATLAB code (dsolve_catenary1.m).

```
%dsolve_catenary1.m   Solve 2nd order BVP
clc;clear;close;
syms  y(x)  a
eq='D2y=(1/a)*(1+(Dy)^2)^(1/2)'
bc='y(0)=a, Dy(0)= 0';    % Boundary conditions
S=dsolve(eq,bc,x); S=simplify(S)
for k=1/2:1/2:2
 y=subs(S,a,k);
h=ezplot(y(1));xlim([-4 4]);ylim([0 8]);set(h,'Linewidth',2.5);hold on
end
xlabel('x'); ylabel('y');;grid on; legend('a=1/2','a=1','a=3/2','a=2');
```

Problem 3.1.28

(a) The shape of the cable supporting a uniform beam is defined by the BVP,

$$y'' = \mu/T, \quad y'(0) = 0, \quad y(L) = h = y(-L) \qquad (3.22)$$

where L is the half the length of the beam, h is the height at both ends of the cable, μ is a constant, T is the tension on the cable (in analogy with the suspension cable of a bridge).

Determine the shape function $y(x)$ of the cable.

(b) Use MATLAB dsolve function to solve this problem analytically. Plot the solution for $\mu = 2, T = 1, L = 1, h = 2$ units (dsolve_suspension1.m).

Solution

Twice integrating Eq. (3.22) we obtain

$$y' = \frac{\mu}{T}x + c_1, \quad y = \frac{\mu}{2T}x^2 + c_1 x + c_2$$

using boundary conditions,

$$y'(0) = 0 \quad \rightarrow \quad c_1 = 0, \quad y(L) = h = \frac{\mu}{2T}L^2 + c_2 \quad \rightarrow \quad c_2 = h - \frac{\mu}{2T}L^2$$

Therefore,

$$y = \frac{\mu}{2T}x^2 + h - \frac{\mu}{2T}L^2 = \frac{\mu}{2T}\left(x^2 - L^2\right) + h$$

This is an equation of a parabola (not cosh function!) opening upward.

Edited output of MATLAB m-file (dsolve_suspension1.m) is given below for $\mu = 2, T = 1$,

$L = 1, h = 2$ units. Figure 3.7 displays the solution (the shape of the cable supporting a uniform beam).

```
eq = D2y = m/T


S = (- m*L^2 + m*x^2 + 2*T*h)/(2*T)


y = x^2 + 1
```

Fig. 3.7 The shape of the cable supporting a uniform beam is a parabola. The BVP is $y'' = \mu/T$, $y'(0) = 0$, $y(L) = h = y(-L)$ where L is the half length of the beam

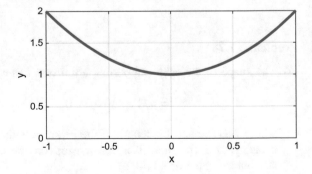

```
% dsolve_suspension1.m    Solve 2nd order BVP
clc;clear;close;
syms  y(x) m T L h
eq='D2y=m/T'
bc='y(L)=h, Dy(0)= 0';    % Boundary conditions
S=dsolve(eq,bc,x); S=simplify(S)
y=subs(S,[m,T,L,h],[2,1,1,2])
h=ezplot(y);xlim([-1 1]);ylim([0 2]);set(h,'Linewidth',2.5);
xlabel('x'); ylabel('y');;grid on;
```

Problem 3.1.29 A cable supports a load uniformly distributed in the x (horizontal) direction, approximating a suspension bridge with different height values at both ends of the cable. The BVP to be solved is

$$y'' = \mu/T; \quad 0 < x < L, \quad y(0) = h_1, \quad y(L) = h_2$$

T = tension, μ = weight, L = length of the beam, h_1, h_2 terms are different height values of the cable at both ends of the beam. Solve this BVP.

Solution

$$y = y_h + y_p,$$

Homogeneous solution: $y_h = c_1 x + c_2$
Particular solution is obtained by double integration of given equation,

$$y_p'' = \frac{\mu x^2}{2T}$$

The general solution is

$$y = \left(\frac{\mu}{2T}\right) x^2 + c_1 x + c_2$$

Using given boundary conditions,

$$y(0) = h_1 \quad \rightarrow \quad c_2 = h_1$$
$$y(L) = h_2 \quad \rightarrow \quad \left(\frac{\mu}{2T}\right) L^2 + c_1 L + c_2 = h_2$$

solving these two equations for c_1, c_2 we obtain

$$y(x) = \frac{\mu}{2T} \left(x^2 - Lx\right) + \frac{h_2 - h_1}{L} x + h_1$$

Edited output of MATLAB m-file (dsolve_suspension2.m) is given below for $\mu = 2, T = 1$,

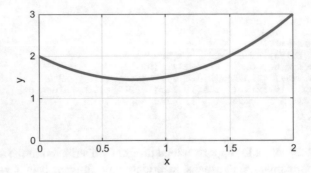

Fig. 3.8 The shape of the cable supporting a uniform beam with different cable heights at both ends of the beam. BVP is $y'' = \mu/T$; $0<x<L$, $y(0) = h_0$, $y(L) = h_1$, where L is the length of the beam

$L = 2, h_1 = 2, h_2 = 3$ units. Figure 3.8 displays the solution (the shape of the cable supporting a uniform beam).

```
eq =D2y=m/T
S = (m*x^2)/(2*T) - ((m*L^2 + 2*T*h1 - 2*T*h2)*x)/(2*L*T) + h1
m=2, T=1, L=2, h1=2, h2=3
y = x^2 - (3*x)/2 + 2
```

```
% dsolve_suspension2.m    Solve 2nd order BVP
clc;clear;close;
syms  y(x)  m T L h1 h2
eq='D2y=m/T'
bc='y(0)=h1, y(L)= h2';    % Boundary conditions
S=dsolve(eq,bc,x); S=simplify(S)
y=subs(S,[m,T,L,h1,h2],[2,1,2,2,3])
h=ezplot(y);xlim([0 2]);ylim([0 3]);set(h,'Linewidth',2.5);
xlabel('x'); ylabel('y');;grid on;
```

Problem 3.1.30 A horizontal metal beam of length L is clamped at both ends. It is subject to a vertical load $f(x)$ per unit length. The resulting vertical displacement in the beam $y(x)$ can be described by the BVP,

$$EI\frac{d^4y}{dx^4} = f(x); \quad y(0) = y'(0) = y(L) = y'(L) = 0$$

$E =$ Young's modulus, $I =$ the moment of inertia of the cross section. Let $k = f(x)/(EI)$, a constant.

(a) Determine the vertical displacement function, $y(x)$.
(b) Compute and plot displacement $y(x)$ for $L = 1, k = -1$ (dsolve_beam1.m) .

Solution

(a) The BVP to be solved is

$$\frac{d^4y}{dx^4} = k; \quad y(0) = y'(0) = y(L) = y'(L) = 0$$

Integrating this equation four times, we get

$$y_4 = kx + c_1, \quad y_3 = k\frac{x^2}{2} + c_1 x + c_2, \quad y_2 = k\frac{x^3}{6} + c_1\frac{x^2}{2} + c_2 x + c_3$$

$$y = k\frac{x^4}{24} + c_1\frac{x^3}{6} + c_2\frac{x^2}{2} + c_3 x + c_4 \tag{3.23}$$

Applying given boundary conditions, we obtain

$$y(0) = c_4 = 0;$$

$$y'(x) = k\frac{x^3}{6} + c_1\frac{x^2}{2} + c_2 x + c_3 = 0, \quad \rightarrow \quad c_3 = 0$$

$$y(L) = k\frac{L^4}{24} + c_1\frac{L^3}{6} + c_2\frac{L^2}{2} = 0 \quad \rightarrow \quad k\frac{L^2}{12} + c_1\frac{L}{3} + c_2 = 0 \tag{3.24}$$

$$y'(L) = k\frac{L^3}{6} + c_1\frac{L^2}{2} + c_2 L = 0, \quad \rightarrow \quad k\frac{L^2}{6} + c_1\frac{L}{2} + c_2 = 0 \tag{3.25}$$

From the last two Eqs. (3.24) and (3.25), we obtain

$$k\frac{L^2}{12} + c_1\frac{L}{3} = k\frac{L^2}{6} + c_1\frac{L}{2} \quad \rightarrow \quad c_1 = -k\frac{L}{2}$$

Substituting this result in (3.24), we get

$$k\frac{L^2}{12} - k\frac{L^2}{6} + c_2 = 0 \quad \rightarrow \quad c_2 = k\frac{L^2}{12}$$

Substituting this result into (3.21), we will have the equation for displacement of the beam,

Fig. 3.9 Deflection profile of
a uniform beam clamped at
both ends, for
$L = 1, \quad k = -1$

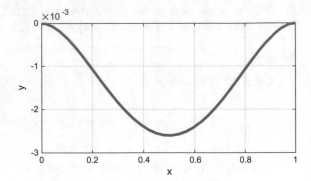

$$y = k\frac{x^4}{24} - \left(k\frac{L}{2}\right)\frac{x^3}{6} + \left(k\frac{L^2}{12}\right)\frac{x^2}{2} = \frac{kx^2}{24}(L-x)^2$$

Output of MATLAB code (dsolve_beam1.m) is given below. Figure 3.9 displays the graph of displacement for $L = 1$ and $k = -1$.

```
eq = D4y=k
S  = (k*x^2*(L - x)^2)/24
y  = -(x^2*(x - 1)^2)/24
```

MATLAB m-file (dsolve_beam1.m) is the following:

```
% dsolve_beam1.m    Solve 4th order BVP
clc;clear;close;
syms  y(x)  k   L
eq='D4y=k'
bc='y(0)=0, y(L)=0, Dy(0)=0, Dy(L)=0';    % Boundary conditions
S=dsolve(eq,bc,x); S=simplify(S)
y=subs(S,[k,L],[-1,1])
h=ezplot(y);set(h,'Linewidth',2.5);xlim([0 1]);ylim([-0.003 0]);
xlabel('x'); ylabel(' y ');;grid on; title('');
```

Problem 3.1.31 Suppose A and q in the linear BVP

$$y' = Ay + q, \quad B_a y(a) + B_b y(b) = \beta$$

are continuous. The BVP has unique solution y if and only if the matrix

$$Q = B_a y(a) + B_b y(b) \tag{3.26}$$

is non-singular [21].

Apply this theorem to the BVP, $y'' + y = 0, y(0) = 0, y(b) = \beta_2$ and determine the conditions that yield unique solutions.

Hint: Fundamental solution matrix which satisfies $Y(0) = I$ for given BVP is

$$Y = \begin{bmatrix} \cos x & \sin x \\ -\sin x & \cos x \end{bmatrix}$$

Solution

The BVP can be put in matrix form as

$$\begin{bmatrix} Y_1' \\ Y_2' \end{bmatrix} = \begin{bmatrix} 0 & 1 \\ -1 & 0 \end{bmatrix} \begin{bmatrix} Y_1 \\ Y_2 \end{bmatrix} \quad 0 < x < b$$

and boundary conditions can be expressed as

$$\begin{bmatrix} 1 & 0 \\ 0 & 0 \end{bmatrix} \begin{bmatrix} Y_1(0) \\ Y_2(0) \end{bmatrix} + \begin{bmatrix} 0 & 0 \\ 1 & 0 \end{bmatrix} \begin{bmatrix} Y_1(b) \\ Y_2(b) \end{bmatrix} = \begin{bmatrix} 0 \\ \beta_2 \end{bmatrix}$$

Substituting boundary value into the fundamental solution matrix and using (3.26) with $Y(0) = I$,

$$Q = B_o I + B_b Y(b) = \begin{bmatrix} 1 & 0 \\ 0 & 0 \end{bmatrix} + \begin{bmatrix} 0 & 0 \\ 1 & 0 \end{bmatrix} \begin{bmatrix} \cos b & \sin b \\ -\sin b & \cos b \end{bmatrix}$$

$$Q = \begin{bmatrix} 1 & 0 \\ 0 & 0 \end{bmatrix} + \begin{bmatrix} 0 & 0 \\ \cos b & \sin b \end{bmatrix} = \begin{bmatrix} 1 & 0 \\ \cos b & \sin b \end{bmatrix}$$

$$\det(Q) = \sin b$$

Hence, Q is singular when $b = k\pi, k = 0, 1, 2, \ldots$(*integer*). Other values of b yield unique solutions of the given BVP.

Problem 3.1.32 Consider the linear BVP (system),

$$y' = Ay + q, \quad a \le x \le b$$
$$B_a y(a) + B_b y(b) = \beta$$

with $a = 0$, $b = 1$, $q = [-\sin x \quad e^{-x}]^T$, $\beta = [1 \quad 0]^T$

$$A = \begin{bmatrix} 1 & -x \\ 0 & 2 \end{bmatrix}, \quad B_a = \begin{bmatrix} 1 & 0 \\ 0 & 0 \end{bmatrix}, \quad B_b = \begin{bmatrix} 0 & 0 \\ 0 & 1 \end{bmatrix}$$

Use MATLAB dsolve function to solve the BVP and plot its solution. (dsolve_systembvp1.m)

Solution

Following are the edited output list print of the MATLAB code.

Fig. 3.10 Solution curves for
the system of first order BVPs

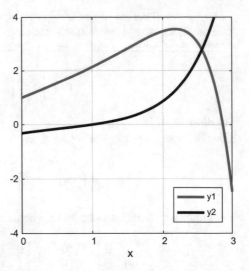

Figure 3.10 displays the two solution curves of this BVP.

```
A = [ 1, -x]    q = -sin(x)
    [ 0,  2]        exp(-x)


Ba = 1   0        Bb = 0   0        beta = 1
     0   0             0   1               0


Y(x) = y1(x)
       y2(x)


S(x) = diff(y1(x), x) == y1(x) - sin(x) - x*y2(x)
       diff(y2(x), x) == exp(-x) + 2*y2(x)


C = y1(0) == 1
    y2(1) == 0


y1 = (7*exp(x))/12 - exp(-x)/12 + (exp(2*x)*exp(-3))/3 - (x*exp(-x))/
6 - (exp(-3)*exp(x))/3 + (2^(1/2)*cos(x - pi/4))/2 - (x*exp(2*x)*exp
(-3))/3


y2 =(exp(-x)*(exp(3*x - 3) - 1))/3
```

```
%dsolve_systembvp1.m    solve system of BVPs in matrix form
clc;clear;close;
syms y1(x) y2(x)
A=[1  -x  ;  0  2 ]; q= [-sin(x); exp(-x)]; Ba=[1 0; 0 0]; Bb=[0 0; 0 1]
beta=[1; 0],a=0; b=1; Y = [y1; y2]; S = diff(Y) == A*Y+q
C = Ba*Y(a)+Bb*Y(b) == beta; [y1, y2] = dsolve(S,C);
y1 = simplify(y1),y2 = simplify(y2), ezplot(y1);hold on;
ezplot(y2);grid on; xlim([0 3]);ylim([-4 4]);legend('y1','y2');
```

Problem 3.1.33 (BVP with Robin boundary condition) Use MATLAB `dsolve` function to solve the following BVP [22],

$$y'' + 4xy' + (4x^2 + 2)y = 0, \quad 0 \leq x \leq 3, \quad y(0) = 1, \quad 1.5y(1) + y'(1) = 0$$

and plot the solutions $y(x), y'(x)$ in the same figure. (dsolveRobin1.m).

Solution

Analytical expression for the solution is computed as $y(x) = (1 + x)\exp(-x^2)$. Figure 3.11 displays these solution curves.

```
%dsolveRobin1.m
syms  y(x)
Dy=diff(y,x);
y=dsolve(diff(y,2,x)+4*x*diff(y,x)+(4*x*x+2)*y == 0,...
     y(0)==1, 1.5*y(1) + Dy(1)==0)
dy=diff(y,x);
p=ezplot(y); set(p,'linewidth',2);hold on;
p=ezplot(dy);grid; set(p,'linewidth',2);ylabel('y(x)');
axis([0 3 -1.2 1.2]);legend('y(x)','dy(x)/dx')
```

Fig. 3.11 Solution curves for
the BVP,
$y'' + 4xy' + (4x^2 + 2)y = 0$,
$y(0) = 1, 1.5y(1) + y'(1) = 0$

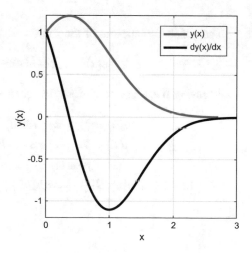

Problem 3.1.34 (Heat conduction in one dimension). A long cylindrical rod of homogeneous material conducts heat along its horizontal (axial) direction in accordance with the differential equation

$$-\mathrm{K} \cdot \frac{d^2 u}{dx^2} = g(x); \quad 0 < x < L \tag{3.27}$$

Where L = the length of the rod, K=heat conductivity of the rod material. The function $g(x)$ describes the rate at which heat enters a differential slice by means other than conduction through the two circular faces of the rod.

If the surrounding temperature of cylindrical rod is at $T = 0$ units for long time while both ends of the rod continuously remaining at temperature T_0,

$$g(x) = -hu(x)C/A \quad 0 < x < L \tag{3.28}$$

where $C =$ the circumference of the rod, $h =$ heat transfer coefficient, $A =$ cross-sectional area of the rod.

(a) What is the physical meaning of minus sign in Eq. (3.28)?
(b) Solve the BVP of Eq. (3.27) with $g(x)$ given by Eq. (3.28).
(c) Use MATLAB to check the solution and plot $u(x)$, if $\mu^2 = hC/KA = 1$, $L = 4$, $T_0 = 40$ units. Compute the temperature in the middle of the rod (dsolve_heat1.m).

Solution

(a) The minus sign indicates that if $u(x) > T$, heat leaves the rod.
(b) Let $\mu^2 = hC/KA$,

$$\frac{d^2 u}{dx^2} - \mu^2 u = 0, \quad 0 < x < L \tag{3.29}$$

The general solution is of the form

$$u(x) = c_1 \cos h(\mu x) + c_2 \sin h(\mu x) \tag{3.30}$$

Applying boundary conditions, we obtain

$$u(0) = T_0 = c_1 \cosh(0) + c_2 \sin h(0) = c_1$$
$$u(L) = T_0 = c_1 \cos h(\mu L) + c_2 \sin h(\mu L)$$
$$T_0 = T_0 \cos h(\mu L) + c_2 \sin h(\mu L)$$
$$c_2 = T_0[1 - \cos h(\mu L)]/\sin h(\mu L)$$

Fig. 3.12 Temperature
distribution in the rod

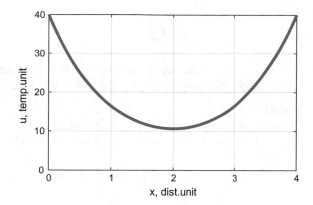

Therefore,

$$u(x) = T_0 \cdot \left[\cos h(\mu x) + \frac{1 - \cos h(\mu L)}{\sin h(\mu L)} \cdot \sin h(\mu x) \right]$$

(c) Edited output of the MATLAB code (dsolve_heat1.m) is given below. The
temperature in the middle of the rod is computed as 10.63 units. Figure 3.12
displays the temperature distribution in the rod.

```
eq = D2u-m^2*u=0
S = (T0*exp(L*m) + T0*exp(2*m*x))/(exp(m*x) + exp(L*m)*exp(m*x))
u = (40*exp(2*x) + 40*exp(4))/(exp(x) + exp(4)*exp(x))
U = 10.63
```

```
% dsolve_heat1.m    Solve 2nd order BVP
clc;clear;close;
syms  u(x) m  T0   L
eq='D2u-m^2*u=0'
bc='u(0)=T0, u(L)= T0';   % Boundary conditions
S=dsolve(eq,bc,x); S=simplify(S)
u=subs(S,[m,T0,L],[1,40,4])
h=ezplot(u);xlim([0 4]);ylim([0 40]);set(h,'Linewidth',2.5);
xlabel('x, dist.unit'); ylabel('u, temp.unit');;grid on; title('');
U=subs(S,[m,T0,L,x],[1,40,4,2]);U=vpa(U,4)
```

Problem 3.1.35 (Singular BVP, radial heat flow). A long cylinder (rod) surrounded
by a fluid at temperature T caries electrical current which heats the rod. The heat
flows in the radial direction and the temperature at an axial distance r in the rod is
described by

$$\frac{1}{r}\frac{d}{dr}\left(r\frac{dy}{dr}\right) = -H, \quad 0 \le r < R, \quad y(R) = T$$

where R is the radius of the rod and H is a constant of proportionality to the electrical power dissipated in the rod. Solve this BVP.

Solution
Here we are only given the physical boundary condition, and the mathematical boundary at $r = 0$ is the singular point. At this boundary, we require $y(0)$ and $y'(0)$ be finite.

Multiplying by r and integrating once we get

$$r\frac{dy}{dr} = -H\frac{r^2}{2} + c_1$$

Then, dividing both sides by r and integrating, we obtain

$$\int \left(\frac{dy}{dr}\right)dr = -\frac{H}{2}\int r\,dr + \int \frac{c_1}{r}dr \quad \rightarrow \quad y = -\frac{Hr^2}{4} + c_1\ln(r) + c_2$$

Since $y(0)$ and $y'(0)$ are finite,

$$y(0) = 0 + c_1\ln(0) + c_2 \rightarrow c_1 = 0$$

$$y(T) = T = -\frac{H}{4}R^2 + c_2 \quad \rightarrow \quad c_2 = T + \frac{HR^2}{4}$$

The solution is

$$y(r) = -\frac{Hr^2}{4} + T + \frac{HR^2}{4} = H\frac{R^2 - r^2}{4} + T$$

3.2 Eigenvalue Problems

Problem 3.2.1 A two-point BVP is described as

$$y'' = f(x, y, y'), \quad a < x < b$$

subject to boundary conditions,

$$A\begin{bmatrix} y(a) \\ y'(a) \end{bmatrix} + B\begin{bmatrix} y(b) \\ y'(b) \end{bmatrix} = \begin{bmatrix} g_1 \\ g_2 \end{bmatrix}$$

where A, B are 2×2 matrices.

The BVP, $y'' + \lambda y = 0, y(0) = y(1) = 0$ has the general solution

$$Y(x) = \begin{cases} c_1 e^{kx} + c_2 e^{-kx} & \text{for } \lambda < 0, \quad k = (-\lambda)^{1/2} \\ c_1 + c_2 x & \text{for } \lambda = 0 \\ c_1 \cos(kx) + c_2 \sin(kx) & \text{for } \lambda > 0, \quad k = \lambda^{1/2} \end{cases}$$

(a) What are the solutions for the first two cases in $Y(x)$?
(b) What is the solution for the third case?

Solution

(a) First two cases for $Y(x)$ imply $c_1, c_2 = 0$ when boundary conditions are applied.
(b) For the third case in $Y(x)$, when we apply boundary conditions, we get $c_1 = 0$,
$c_2 \sin(k) = 0$
where $\sin(k) = 0$ means that

$$k = \pm\pi, \pm 2\pi, \pm 3\pi, \ldots$$
$$\lambda = \pi^2, 4\pi^2, 9\pi^2, \ldots$$

which shows that a nontrivial solution exists for given BVP as
$Y(x) = C \sin(kx), C \neq 0$

A trivial solution $Y(x) = 0$ for other values of k and λ.

Problem 3.2.2 Find the eigenvalues and eigenfunctions of the homogenous two point BVP,

$$y'' + \lambda y = 0; \quad y(0) = 0, \quad y(L) = 0, \quad L > 0.$$

It is known that $\lambda = 0$ and $\lambda < 0$ give trivial solutions.

Solution
We are left with the condition that $\lambda > 0$.
   Characteristic equation: $s^2 + \lambda = 0$, $s = \pm i\sqrt{\lambda}$, general solution,
$y = c_1 \cos\left(\sqrt{\lambda}x\right) + c_2 \sin\left(\sqrt{\lambda}x\right)$.

$$y(0) = c_1 \cos(0) + c_2 \sin(0) = c_1 = 0$$

This leaves as the solution $y = c_2 \sin\left(\sqrt{\lambda}x\right)$.
From second condition, $y(L) = c_2 \sin\left(\sqrt{\lambda}L\right) = 0$
$\sqrt{\lambda}L = n\pi \rightarrow \lambda = n^2\pi^2/L^2$, c_2 arbitrary.
The solution is $y = c_2 \sin[(n\pi/L)x]$.

Problem 3.2.3 Find the eigenvalues and eigenfunctions of the BVP,

$$y'' - 6\lambda y' + 9\lambda^2 y = 0, \quad y(0) = 0, \quad y(1) + y'(1) = 0.$$

Solution

The characteristic equation; $s^2 - 6\lambda s + 9\lambda^2 = (s - 3\lambda)^2 = 0 \quad \rightarrow \quad s = 3\lambda$,

This is a double root, therefore the solution to the ODE is $y = c_1 e^{3\lambda x} + c_2 x e^{3\lambda x}$

Applying boundary conditions, $c_1 = 0$,

$$y' = 3c_1 \lambda e^{3\lambda x} + c_2 e^{3\lambda x} + 3c_2 \lambda x e^{3\lambda x} = 3c_1 \lambda e^{3\lambda x} + c_2 e^{3\lambda x}(1 + 3\lambda x)$$

$$y'(1) = 3c_1 \lambda e^{3\lambda} + c_2 e^{3\lambda}(1 + 3\lambda)$$

$$y(1) = c_1 e^{3\lambda} + c_2 e^{3\lambda}$$

$$y(1) + y'(1) = c_1 e^{3\lambda} + c_2 e^{3\lambda} + e^{3\lambda}[3c_1 \lambda + c_2(1 + 3\lambda)]$$

$$= c_1 + c_2 + 3c_1 \lambda + c_2 + 3c_2 \lambda = c_1(1 + 3\lambda) + c_2(2 + 3\lambda) = 0$$

Thus $c_1 = 0$ and either $c_2 = 0$ or $2 + 3\lambda = 0 \quad \rightarrow \quad \lambda = -2/3$.

The choice $c_2 = 0$ yields a trivial solution, but $\lambda = -2/3$ gives the nontrivial solution, $y = c_2 x e^{-2x}$, thus, the BVP has the eigenvalue $\lambda = -2/3$ and eigenfunction $y = c_2 x e^{-2x}$.

Problem 3.2.4 For the homogenous BVP, $y'' - 6\lambda y' + 9\lambda^2 y = 0$; $y'(1) = y(2) + 2y'(2) = 0$, determine the eigenvalues and corresponding eigenfunctions.

Solution

General solution is of the form,

$$y = c_1 e^{3\lambda x} + c_2 x e^{3\lambda x} \tag{3.31}$$

$$y' = 3\lambda c_1 e^{3\lambda x} + c_2 3\lambda x e^{3\lambda x} + c_2 e^{3\lambda x} \tag{3.32}$$

in

$$y'(1) = 3\lambda c_1 e^{3\lambda} + c_2 3\lambda e^{3\lambda} + c_2 e^{3\lambda} = 0$$
$$= e^{3\lambda}(3\lambda c_1 + 3\lambda c_2 + c_2) = 3\lambda c_1 + c_2(3\lambda + 1) \tag{3.33}$$

$$y(2) = c_1 e^{6\lambda} + 2c_2 e^{6\lambda} \tag{3.34}$$

$$y'(2) = 3\lambda c_1 e^{6\lambda} + 6\lambda c_2 e^{6\lambda} + c_2 e^{6\lambda} \tag{3.35}$$

From (3.34), (3.35),

$$y(2) + 2y'(2) = c_1 e^{6\lambda} + 2c_2 e^{6\lambda} + 6\lambda c_1 e^{6\lambda} + 12\lambda c_2 e^{6\lambda} + 2c_2 e^{6\lambda} = 0$$
$$c_1 + 2c_2 + 6\lambda c_1 + 12\lambda c_2 + 2c_2 = c_1(1 + 6\lambda) + c_2(4 + 12\lambda) = 0 \tag{3.36}$$

From (3.33), (3.36),

$$3\lambda c_1 + (1+3\lambda)c_2 = 0 \tag{3.37a}$$

$$(1+6\lambda)c_1 + (4+12\lambda)c_2 = 0 \tag{3.37b}$$

This system has a nontrivial solution for c_1, c_2 if and only if

$$\Delta = \begin{vmatrix} 3\lambda & 1+3\lambda \\ 1+6\lambda & 4+12\lambda \end{vmatrix} = 0$$
$$= 3\lambda(4+12\lambda) - (1+3\lambda)(1+6\lambda) = (1+3\lambda)[12\lambda - (1+6\lambda)] = 0$$
$$\Delta = (1+3\lambda)(-1+6\lambda) = 0$$

That is, $\lambda = -1/3$ or $\lambda = 1/6$ are the eigenvalues.

If $\lambda = -1/3$, from (3.37a), $-c_1 + 0c_2 = 0 \quad \rightarrow \quad c_1 = 0$, c_2 arbitrary.

If $\lambda = 1/6$, from (3.37b), $2c_1 + (4+2)c_2 = 0 \quad \rightarrow \quad 2c_1 + 6c_2 = 0, c_1 = -3c_2$, c_2 arbitrary.

Then, eigenfunction corresponding to $\lambda_1 = -1/3$, $y_1 = c_2 x e^{3(-1/3)x} = c_2 x e^{-x}$.

Eigenfunction associated with $\lambda_2 = 1/6$, $y_2 = c_1 e^{3(1/6)x} + c_2 x e^{3(1/6)x} = c_2(-3+x)e^{x/2}$.

Problem 3.2.5 Solve the eigenvalue problem $y'' + \lambda y = 0$, $y(0) = 0$, $y'(L) = 0$, $\lambda = 0$, $L > 0$.

Solution

General solution of this BVP is $y = c_1 \cos\left(\sqrt{\lambda}x\right) + c_2 \sin\left(\sqrt{\lambda}x\right)$

$$y(0) = c_1(1) + c_2(0) = c_1 = 0, y = c_2 \sin\left(\sqrt{\lambda}x\right).$$
$$y' = c_2\sqrt{\lambda}\,\cos\left(\sqrt{\lambda}x\right)$$
$$y'(L) = 0 = c_2\sqrt{\lambda}\,\cos\left(\sqrt{\lambda}x\right)$$

This implies that with $c_2 \neq 0$ we choose $\sqrt{\lambda} = (2n-1)\pi/(2L)$.

Then eigenvalues are $\lambda_n = (2n-1)^2\pi^2/(4L^2)$, $n = 1, 2, 3, \ldots$ associated eigenfunction $y_n = \sin[(2n-1)\pi x/(2L)]$.

Problem 3.2.6 Solve the eigenvalue (two point) problem $y' + \lambda y = 0$, $y'(0) = y'(L) = 0$, $\lambda \geq 0$, $L > 0$.

Solution

Characteristic equation: $s^2 + \lambda = 0 \quad \rightarrow \quad s = \pm i\sqrt{\lambda}$

If $\lambda = 0$, $y = c_1 x + c_2$, $y' = c_1$, $y'(0) = 0 \quad \rightarrow \quad c_1 = 0, y = c_2 \quad \rightarrow \quad y_0 = 1$.

If $\lambda < 0$, let $a^2 = \lambda$, $s = \mp ia$, general solution of BVP is $y = c_1 \cos(ax) + c_2 \sin(ax)$

$$y' = -ac_1 \sin(ax) + ac_2 \cos(ax) \quad \rightarrow \quad y'(0) = -c_1 a \sin(0) + c_2 a \cos(0) = 0$$

This implies that $c_2 = 0$ if $a \neq 0$.

$y'(L) = -c_1 a \sin(aL) = 0$, nontrivial solution requires that $c_1 \neq 0$, then $\sin(aL) = 0$, $aL = \pi, 2\pi, 3\pi, \ldots \quad \rightarrow \quad a_n = n\pi/L$, $n = 1, 2, 3, \ldots$

$\lambda_n = a_n^2 = n^2\pi^2/L^2$ are the eigenvalues of the ODE,

$y_n = \cos(n\pi x/L)$, $n = 1, 2, 3, \ldots$ are the eigenfunctions.

Problem 3.2.7 Solve the BVP $y'' + \lambda y = 0$, $y(0) = y(\pi) = 0$, $\lambda > 0$.

Solution

The general solution of the BVP is $y = c_1 \cos(\lambda x) + c_2 \sin(\lambda x)$

$y(0) = c_1 \cos(0) + c_2 \sin(0) \quad \rightarrow \quad c_1 = 0$. Therefore $y = c_2 \sin(\lambda x)$.

$y(\pi) = c_2 \sin(\lambda \pi) = 0$. This is possible either for $c_2 = 0$ or $\lambda = 1, 2, \ldots n$. If $c_2 \neq 0$, the solution is $y = c_2 \sin(\lambda x)$, which has infinitely many nontrivial solutions; in other words, whatever the value of c_2 is, the solution "hits" the target value for $x = \pi$.

Problem 3.2.8 (Whirling string). The BVP, $y'' + \lambda y = 0$, $y(0) = y(L) = 0$, $\lambda = \rho\omega^2/T$ (where L is the length of the string, ρ is density of string, ω is the angular frequency of relation, T is the tension force at a point on the string) is the equation of a whirling string. In an earlier solution of a problem of this kind, it was found that the eigenvalues are $\lambda_n = n^2\pi^2/L^2$, $n = 1, 2, \ldots$ with the eigenfunction $y_n(x) = \sin(n x \pi/L)$.

Describe the physical meaning of these equations.

Solution

We understand from these equations that the only solution is $y(x) = 0$, if λ is not one of the eigenvalues of this BVP (There will be no deflection of the string). However, equating $\lambda = \rho\omega^2/T$ to λ_n and solving for ω_n, we obtain

$$\omega = \sqrt{\lambda_n/\rho} = n\pi\sqrt{T/\rho}/L, \quad n = 1, 2, 3, \ldots$$

this means that we have some set of critical speeds of rotation, at which the string will whirl up, and take the form $y_n = c_n \sin(n x \pi/L)$. At lower and upper angular frequencies, th string will remain in its undeflected position.

Problem 3.2.9 A thin vertical column of length L constrained at both ends is subject to an axial compressive force, P. The ODE modeling the lateral bending of the column is

$$EIy'' = -Py, \quad 0 < x < L, \quad y(0) = y(L) = 0$$

where E, I are Young's modulus and moment of inertia, respectively. Determine $y(x)$ and find the minimum compressive force if $I = b^2/12$, where b^2 is the area of the cross section of vertical column.

Solution

Let $\lambda = P/EI$, then the ODE becomes

$$y'' + \lambda y = 0.$$
$$\lambda_n = P_n/EI = n^2\pi^2/L^2, \quad y_n = c_n \sin(n\pi x/L)$$

The minimum force to bend the column is $P_n = \pi^2(EI/L^2)$

The shape of the vertical column when it bends is $y_1 = c_1 \sin(\pi x/L)$.

The second bending mode is a sinusoidal waveform at a force $P_2 = 4P_1$ (the first mode is not possible when the column is pinned at $x = L/2$).

The minimum compressive forces for the two modes are

$$P_1 = \pi^2(EI/L^2) = \frac{\pi^2}{12} \cdot \frac{b^4 E}{L^2},$$
$$P_2 = 4P_1 = \frac{\pi^2}{3} \cdot \frac{b^4 E}{L^2}$$

Problem 3.2.10 (Semi-Infinite and Infinite Intervals). In some BVPs, the interval $0 < x < \infty$ is called semi-infinite interval and the condition is usually written

$$y(x) \text{ and } y'(x) \text{ bounded as } x \to \infty$$

Solve the BVP $y'' - k^2 y = 0$, $\quad y(x), \quad y'(x)$ bounded as $x \to \infty$ and $y(0) = T_0$.

Solution

The general solution of the ODE is

$$y(x) = c_1 \cosh(kx) + c_2 \sin h(kx),$$

The boundary condition at $x = 0$ requires $y(0) = T_0$,

$$T_0 = c_1 \cos h(0) + c_2 \sin h(0) = c_1$$

The boundedness condition implies that of all the linear combinations of $\cos hx$ and $\sin hx$, the only one that is bounded as x tends to infinity is $\cos h(kx) - \sin h(kx) = e^{-kx}$ (and its multiples), then $c_2 = -c_1$,

$$y(x) = T_0 \cosh(kx) + (-T_0) \sin h(kx)$$
$$y(x) = T_0[\cosh(kx) - \sin h(kx)]$$

Problem 3.2.11 Displacement equation of a beam with homogeneous material properties and constant cross section is

$$\frac{d^4y}{dx^4} = a = \frac{\omega^2 A}{EI} \tag{3.38}$$

where E, I are Young's modulus, area moment of inertia of the beam[5] with density p and cross-sectional area A, ω is the angular frequency. Analytical solution is determined as

$$y(x) = c_1 \sin h(ax) + c_2 \cosh(ax) + c_3 \sin(ax) + c_4 \cos(ax) \tag{3.39}$$

Boundary conditions for "simply supported-simply supported" case are

$$y(0) = y''(0) = y(L) = y''(L) = 0 \tag{3.40}$$

Solve the BVP problem[6] for nontrivial solutions. Determine first three characteristic vibration frequencies of this system (fzero2.m).

Solution
Considering boundary conditions in (3.39), we obtain

$$\begin{bmatrix} \sin h(aL) & \sin(aL) \\ \sin h(aL) & -\sin(aL) \end{bmatrix} \begin{bmatrix} c_1 \\ c_2 \end{bmatrix} = 0 \tag{3.41}$$

The non-trivial solution for the determinant of the matrix is computed by (fzero2.m),

$$\sin(a_k L) \sin h(a_k L) = 0, \quad k = 1, 2, \ldots \tag{3.42}$$

Roots of Eq. (3.42) are also determined using the same MATLAB script as

$$n = 1, aL = \pi; n = 2, aL = 2\pi; n = 3, aL = 3\pi \tag{3.43}$$

$$\text{For } n = 1, \quad \omega_1 = \sqrt{EI\pi/(\rho AL)} \tag{3.44}$$

$$\text{For } n = 2, \quad \omega_2 = \sqrt{2}.\omega_1 \tag{3.45}$$

$$\text{For } n = 3, \quad \omega_3 = \sqrt{3}.\omega_1 \tag{3.46}$$

Graph of $y(x) = \sin x . \sin hx = 0$ is shown in Fig. 3.13.
The roots are, π, 2π, 3π, 4π, $5\pi, \ldots$, $k\pi$

---

[5]As an example, for aluminium, $E = 70 \times 10^9 [N/m^2]$, $\rho = 2700[kg/m^3]$, $I_{square} = d^4/12$, $I_{circle} = \pi r^4/4$, where $d = edgelength, r = radius$.
[6]In the next chapters, we will deal with the higher order BVPs and solve them using numerical methods, directly.

Fig. 3.13 Graph of
$y(x) = \sin x . \sin hx = 0$

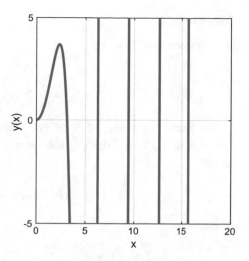

The roots are, π, 2π, 3π, 4π, 5π, ... , $k\pi$

```
%fzero2.m
clc;clear;
% syms x
% A=[sinh(x)  sin(x);  sinh(x)    -sin(x)];
% b=det(A); simplify(b)
% i=0;
% f=@(x) sin(x)*sinh(x); X(i+1) = fzero(f, i);
% for i=2:18; X(i) = fzero(f, i); end; X'
% Graph
x=0:0.01:18; f=sin(x).*sinh(x); plot(x,f);grid;ylim([-5 5]);
```

3.3 Sturm-Liouville Problems

Problem 3.3.1

(a) What is the form of a Sturm[7]-Liouville[8] (SL) ODE?
(b) Obtain the SL form of $y'' + 3x^{-1}y' + (x^2 - \lambda x)y = 0$

---

[7]Jacques Sturm (1803–1855) was a professor of mechanics at the Sorbonne, France.
[8]Joseph Liouville (1809–1882) was a professor of mathematics at the Collége de France.

Solution

(a) Consider the ODE

$$y'' + R(x)y' + [Q(x) + \lambda P(x)]y = 0 \qquad (3.47)$$

We search for values of λ on the interval $[a, b]$ so that there exist nontrivial solutions of this ODE. Multiply each term of above equation by $\exp\left[\int R(x)dx\right]$, we obtain:

$$[ry']' + (q + \lambda p)y = 0 \qquad (3.48)$$

where,

$$r = \exp\left(\int Rdx\right), \quad q = Q \exp\left(\int Rdx\right), \quad p = P \exp\left(\int Rdx\right)$$

The ODE (3.48) is called the Sturm-Liouville ODE, or the SL form of Eq. (3.47), and it has the same solutions as the original ODE on any interval not containing zero.

(b) Assuming an interval not containing zero, this equation is in the form of Eq. (3.47) with

$$R = 3x^{-1}, \quad Q = x^2, \quad P = -x.$$
$$r = \exp\left(\int 3x^{-1}dx\right) = \exp(3lnx) = \exp(lnx^3) = x^3,$$
$$q = Q \cdot x^3 = x^5, \quad p = (-x) \cdot x^3 = -x^4$$
$$(x^3y')' + (x^5 - \lambda x^4)y = 0$$

Problem 3.3.2 Is the BVP $y'' + 9y = 0; y(0) = 0, y(1) = 0$, a Sturm-Liouville (SL) problem?

Solution

A SL problem has the form of $[ry']' + qy + \lambda py = 0$, $r = r(x), p = p(x)$.

In this case, $r = 1, q = 0, p = 1, \lambda = 9$. Here both $r(x), p(x)$ are positive and continuous everywhere, $[0, 1]$ in particular. This is a SL problem.

Problem 3.3.3 Is the BVP $xy'' + y' + (x^2 + 4 + \lambda)y = 0, y(0) = y'(1) = 0$, a SL problem?

Solution

$$[ry']' + qy + \lambda py = 0 \quad \rightarrow \quad [xy']' + (x^2 + 4)y + \lambda y = 0$$
$$r = x, \quad q = x^2 + 4, \quad p = 1;$$

Here $r(x) = 0$ at a point $[0, 1]$, therefore, given ODE is not SL problem.

Problem 3.3.4

(a) Classify the SL problem, $y'' + \lambda y = 0$, $y(0) = y(\pi/2) = 0$, as regular, periodic or singular, specify the relevant interval and find the eigenvalues or an
equation satisfied by the eigenvalues.
(b) Classify the SL problem, $y'' + 4y = 0$, $y(0) = y(\pi/2) = 0$, as regular, periodic or singular, specify the relevant interval and find the eigenvalues.
(c) Use MATLAB to verify the result found in (b) (dsolve2_3.m).

Solution
There are three kinds of problems involving the SL ODE. Assume that p, q are
continuous on [a, b], then we search for the values of λ and nontrivial solutions of
$(ry')' + (q + \lambda p)y = 0$, satisfying the boundary conditions of the form,

(i) $A_1 y(a) + A_2 y'(a) = 0, \quad B_1 y(b) + B_2 y'(b) = 0$
where A_1, A_2 are real numbers and not both zero and B_1, B_2 are given and not
both zero. This is regular SL problem on [a, b].
(ii) $y(a) = y(b), y'(a) = y'(b)$ constitutes the periodic SL problem.
(iii) If the SL ODE satisfies one of the following types of boundary conditions,
the singular SL problem on [a, b] is considered;

  - If $r(a) = 0$, $B_1 y(b) + B_2 y'(b) = 0$, B_1, B_2 are given and not both
    zero,
  - If $r(b) = 0$, $A_1 y(a) + A_2 y(a) = 0$, A_1, A_2 are given and not both
    zero,
  - If $r(a) = r(b) = 0$, no boundary conditions are specified at a or b but
    require that solutions be bounded on [a, b].

For ii and iii, $r(x) > 0, p(x) > 0$ for $a < x < b$.
   Using this information for the given ODE, $y'' + \lambda y = 0$, $y(0) = y(\pi/2) = 0$;
The interval is $[0, \pi/2]$, $r(x) = p(x) = 1$, $q(x) = 0$. This is a Regular SL
problem on $[0, \pi/2]$. Characteristic equation of the ODE is $s^2 + \lambda = 0$, its roots are
$s = \pm - \lambda\sqrt{-\lambda}$

  I. If $\lambda = 0$, the ODE is $y'' = 0$, $y(x) = ax + b$, $y(0) = b =$
    $y(\pi/2) - 0$, $a - 0$
    Zero is not an eigenvalue.

II. If $\lambda > 0$, $\lambda = k^2$, $k > 0$, $y(x) = a\cos(kx) + b\sin(kx)$, $y(0) = a = 0$,

$y(x) = b\sin(kx); y(\pi/2) = b\sin(k\pi/2) = 0$, $b = 0$ gives a trivial solution, but $b \neq 0$ gives $y(\pi/2) = b\sin(k\pi/2) = 0$; $k\pi/2$ should be an integer multiple of π, or $k = 2n$, $n = 1, 2, 3, \ldots$

Therefore, eigenvalues are $\lambda = k^2 = 4n^2$ and eigenfunction corresponding to these eigenvalues is $\varphi(x) = b\sin(2nx), b > 0$.

III. If $\lambda < 0, \lambda = -k^2, k > 0$, $y'' - k^2 y = 0 \rightarrow y(x) = ae^{kx} + be^{-kx}$, $y(0) = a + b = 0$,

$$a = -b, \quad \rightarrow \quad y(x) = ae^{kx} - abe^{-kx} = 2a\sin h(kx)$$
$$y(\pi/2) = 2a\sin h(k\pi/2) = 0 \quad \rightarrow \quad a = 0$$

This case gives only the trivial solution. There are no negative eigenvalues.

(b) In this problem, $\lambda = 4 > 0$, using the results of part (a), This is a Regular SL problem on $[0, \pi/2]$. Therefore, eigenvalues are $\lambda = 4 = 4n^2$, $n = 1$ and eigenfunction corresponding to these eigenvalue is $\varphi(x) = b\sin(2x), b > 0$.

(c) MATLAB gives the following result;

$$S = C17 * \sin(2 * t)$$

MATLAB code is given below;

```
% dsolve2_3.m    Solve 2nd order BVP
clc; clear; syms  y(x)
eq='D2y+4*y=0';   ic='y(0)=0,y(pi/2)=0'; S=dsolve(eq, ic)
```

Problem 3.3.5 Write Legendre's ODE in SL form, and classify the SL problem as regular, periodic, or singular.

Solution
The Legendre's ODE is

$$(1 - x^2)y'' - 2xy' + \lambda y = 0$$

The SL form of Legendre's ODE is

$$[(1 - x^2)y']' + \lambda y = 0$$

Since $r = 1 - x^2$, $q = 0$, $p = 1$, the interval is [-1,1], and $r(-1) = r(1) = 0$. This is the third case of boundary conditions for a singular SL problem.

If n is positive integer, $\lambda = n(n+1)$ is an eigenvalue of this SL problem, the eigenfunction is $P_n(x)$, the nth Legendre polynomial. Legendre polynomials are orthogonal on [-1,1] with weight function 1,

$$\int_{-1}^{1} P_n(x)P_m(x)dx = 0, \quad \text{if } n \neq m$$

Problem 3.3.6 (A periodic SL problem). Consider the BVP,

$$y'' + \lambda y = 0, \quad y(-\pi) = y(\pi), \quad y'(-\pi) = y'(\pi)$$

Show that the SL problem is periodic one, specify the relevant interval, find the eigenvalues and eigenfunctions.

Solution

In a periodic SL problem, for $r(x) > 0$, $p(x) = 0$ in $x \in [a,b]$, we seek λ values and nontrivial solutions of $[ry]' + (q + \lambda p)y = 0$, satisfying boundary conditions $y(a) = y(b)$, $y'(a) = y'(b)$.

This is exactly the case in this problem, thus the SL problem is a periodic one, the relevant interval is $[-\pi \quad \pi]$.

  I. $\lambda = 0$; $y'' = 0$, the solution is $y = ax + b$, $y(-\pi) = -a\pi + b$, $y(\pi) = a\pi + b$, then $a = 0$, $y = b$, zero is an eigenvalue and corresponding eigenfunction $y = b$, $b \neq 0$.

 II. $\lambda > 0$; let $\lambda = k^2$, $k > 0$, $y = a\cos(kx) + b\sin(kx)$,

$$y(-\pi) = y(\pi) \rightarrow a\cos(k\pi) - b\sin(k\pi) = a\cos(k\pi) + b\sin(k\pi),$$
$$2b\sin(k\pi) = 0, \tag{3.49}$$

For

$$y'(-\pi) = y'(\pi) = ak\sin(k\pi) + bk\cos(k\pi)$$
$$= -ak\sin(k\pi) + bk\cos(k\pi) \tag{3.50}$$
$$2ak\sin(k\pi) = 0$$

From Eqs. (3.49) and (3.50) and $k > 0$, $b\sin(k\pi) = a\sin(k\pi) = 0$
We will have a trivial solution if $\sin(k\pi) \neq 0$ and $a = b = 0$. But if $\sin(k\pi) = 0$, a and b can be nonzero values and $k > 0$, then, $\lambda = n^2$ $n = 1, 2, 3, \ldots$, is the eigenvalue.
Eigenfunctions are $f_n(x) = a_n \cos(nx) + b_n \sin(nx)$.

III. $\lambda < 0$; $\lambda = -k^2$, $k > 0$, $y(x) = ae^{kx} + be^{-kx}$,

$$y(-\pi) = y(\pi) \text{ implies } ae^{-kx} + be^{kx} = ae^{kx} + be^{-kx} \tag{3.51}$$

$$y'(-\pi) = y'(\pi) \text{ implies } kae^{-kx} - kbe^{kx} = kae^{kx} - kbe^{kx} \tag{3.52}$$

We can write (3.51) as $a(e^{k\pi} - e^{-k\pi}) = b(e^{k\pi} - e^{-k\pi})$, thus $a = b$. Then Eq. (3.52) becomes $2kae^{-k\pi} = 2kae^{k\pi}$, which implies that $ae^{-k\pi} = ae^{k\pi}$, hence $a = 0$, $b = 0$. There are no negative eigenvalues.

Problem 3.3.7 In order to construct Green's function from a BVP in Sturm-Liouville form, we can use following equations:
The general form of SL operator is

$$\mathcal{L} = \frac{1}{r}\left[\frac{d}{dx}\left(p\frac{d}{dx}\right) + q\right] \tag{3.53}$$

$$G(x,z) = \{c_1(z) \cdot u_1(x) \quad q \le x \le z \tag{3.54a}$$

$$G(x,z) = \{c_2(z) \cdot u_2(x) \quad z \le x \le b \tag{3.54b}$$

where

$$c_1(z) = \frac{r \cdot u_2}{p \cdot W}, c_2 = \frac{r \cdot u_1}{p \cdot W} \tag{3.55a, b}$$

The Wronskian,

$$W = \begin{vmatrix} u_1 & u_2 \\ u_1' & u_2' \end{vmatrix} \tag{3.56}$$

If a BVP is given as

$$\frac{d^2u}{dx^2} = f(x), u(0) = u'(1) = 0 \tag{3.57}$$

Construct its Green's function via the SL form.

Solution
The SL form of given BVP is $u(x) = c_1x + c_2$,

$$u_1(0) = c_2 = 0 \rightarrow u_1 = x$$
$$u_2'(1) = (c_1(1) + c_2)' = 0$$
$$c_1 = 1 \rightarrow u_2 = 1$$
$$W = \begin{vmatrix} u_1 & u_2 \\ u_1' & u_2' \end{vmatrix} = \begin{vmatrix} x & 1 \\ 1 & 0 \end{vmatrix} = -1$$

From (3.55a,b) *with r = p = 1, q = 0*;

$$c_1 = \frac{(1)(1)}{(1)(-1)} = -1, \quad c_2 = \frac{(1)(z)}{(1)(-1)} = -z$$

From (3.54a, b),

$$G = \begin{cases} (-1)(x) = -x & 0 \le x < z \\ (-z)(1) = -z & z < x < 1 \end{cases}$$

Problem 3.3.8 Natural frequencies of transverse vibration of a uniform beam fixed at $x = 0$ and simply supported at $x = 1$ is given by the Sturm-Liouville BVP,

$$y^4 - \lambda^4 y = 0 \tag{3.58a}$$

subject to boundary conditions,

$$y(0) = 0 \tag{3.58b}$$

$$y'(0) = 0 \tag{3.58c}$$

$$y(1) = 0 \tag{3.58d}$$

$$y'(1) = 0 \tag{3.58e}$$

Find the smallest eigenvalue, λ_1, of this BVP.

Solution

Assume solution of the form $y = ce^{sx}$.

Auxiliary equation is $s^4 - \lambda^4 = 0 \rightarrow s_{1,2} = \mp\lambda, \ s_{3,4} = \mp i\lambda$

Hence, the general solution becomes

$$y(x) = a_1 e^{\lambda x} + a_2 e^{-\lambda x} + a_3 e^{j\lambda x} + a_4 e^{-j\lambda x} \tag{3.59}$$

$$y(x) = c_1 \cos \lambda x + c_2 \sin \lambda x + c_3 \cos h\lambda x + c_4 \sin h\lambda x \tag{3.60}$$

Condition (3.58b) yields $c_1 = -c_3$

$$y'(0) = \lambda[-c_1 \sin \lambda x + c_2 \cos \lambda x + c_3 \sin h\lambda x + c_4 \cos h\lambda x] = 0$$

which gives $c_2 = -c_4$.

Then, the solution is reduced to following form,

$$y(x) = c_1[\cos(\lambda x) - \cosh(\lambda x)] + c_2[\sin(\lambda x) - \sin h(\lambda x)] \tag{3.61}$$

Applying boundary conditions (3.58d) and (3.58e) gives

$$c_1[\cos(\lambda) - \cosh(\lambda)] + c_2[\sin(\lambda) - \sin h(\lambda)] = 0 \qquad (3.62)$$

$$-c_1[\cos(\lambda) + \cos h(\lambda)] - c_2[\sin(\lambda) + \sin h(\lambda)] = 0 \qquad (3.63)$$

From (3.62) and (3.63), determinant of coefficients must be zero,

$$\begin{vmatrix} \cos(\lambda) - \cos h(\lambda) & \sin(\lambda) - \sin h(\lambda) \\ -[\cos(\lambda) + \cos h(\lambda)] & -[\sin(\lambda) + \sin h(\lambda)] \end{vmatrix} = 0$$

$$\cos(\lambda)\sin h(\lambda) - \sin(\lambda)\cos h(\lambda) = 0$$

$$\sin h(\lambda) - \tan(\lambda)\cos h(\lambda) = 0$$

$$\tanh(\lambda) - \tan(\lambda) = 0 \qquad (3.64)$$

The roots of (3.64) yields λ_n, $n = 1, 2, 3, \ldots$

Pilot curves to determine the smallest eigenvalue of this BVP are displayed in Fig. 3.14. Note that zero is not an eigenvalue. Another zero-crossing around 1.6 is not a root but merely asymptotic jump-trace on the figure. We use MATLAB `fzero` function to find the roots of this nonlinear equation after plotting the frequency function described by (3.64). (MATLAB function `solve` cannot find explicit solution).

The smallest eigenvalue is computed to be $\lambda = 3.926602312047919$.

MATLAB script used to compute the smallest root of Eq. (3.64) is listed below.

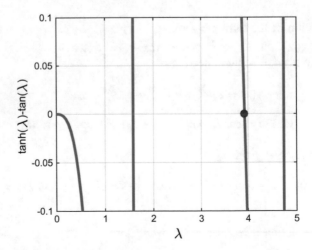

Fig. 3.14 Pilot curves to determine the smallest eigenvalue of BVP, $y^4 - \lambda^4 y = 0$, $y(0) = y'(0) = y(1) = y'(1) = 0$

```
%zerosx.m
clc;clear;format long
L=0:0.05:5;
plot (L,tanh(L)-tan(L) ,'linewidth',2);hold on;
% no root around 1.6, asymptote!
ylim([-0.1 0.1]);
r = fzero(@(L) tanh(L)-tan(L), 3)
plot(r,tanh(r)-tan(r),'.r','markersize',22);
grid; ylabel('tanh(\lambda)-tan(\lambda)'); xlabel('\lambda')
```

3.4 Exercises

1. Given the ODE, $y'' + \pi^2 y = 0$. Show that,

   (a) the IVP with initial conditions $y(0) = 0$, $y'(0) = 1$ has a unique solution over $0 \le x \le 1$,

   (b) the BVP with boundary conditions $y(0) = y(1) = 0$ has infinitely many solutions,

   (c) the BVP with boundary conditions $y(0) = y(1) = 1$ has no solution.

2. Which of the following boundary conditions are not "separated"?

   (a) $a_1 y(0) + a_2 y'(0) = 0$, $b_1 y(2) + b_2 y'(2) = 0$
   (b) $a_1 y(0) = 0$, $b_1 y(1) = 0$
   (c) $a_1 y(0) + a_2 y'(1) = 0$
   (d) $a_1 y(2) + a_2 y'(2) = 0$, $b_1 y(3) + b_2 y'(3) = 0$

   Solve the ODE, $y'' + y = 0$, subject to following boundary conditions:

3. $y(0) = 1, y(1) = 0$ Answer: $y(x) = -\sin(x-1)/\sin 1$

4. $y(0) = 1, y(\pi/2) = 1$ Answer: $y(x) = \cos(x) + \sin(x)$

5. $y(-\pi/4) = y(\pi/4) = 1$ Answer: $y(x) = \sqrt{2}\cos(x)$
   True or false?

6. A Sturm-Liouville BVP consists of an ODE,

$$[p(x)y']' - q(x)y + \lambda r(x)y = 0$$

on the interval $0 < x < 1$, subject to boundary conditions

$$\alpha_1 y(0) + \alpha_2 y'(0) = 0, \quad \beta_1(1) + \beta_2 y'(1) = 0$$

7. All the eigenvalues of the SL problem are real.

8. If ϕ_m, ϕ_n are two eigenfunctions of the SL problem with corresponding eigenvalues λ_m, λ_n respectively, and if $\lambda_m \ne \lambda_n$, then eigenfunctions are orthogonal with respect to the weight function r;

$$\int\limits_0^1 r(x)\phi_m\phi_n dx = 0$$

9. The eigenfunctions of the SL problem are all simple.
10. The eigenvalues of the SL problem form an infinite sequence and can be ordered according to the increasing magnitude, $\lambda_1 < \lambda_2 < \lambda_3 < \ldots$

References

1. Bailey PB, Shampine LF, Waltman PE (1968) Nonlinear two point boundary value problems. Academic Press, New York
2. Cole RH (1968) Theory of ordinary differential equations. Appleton-Century-Crofts, New York
3. Hille E (1969) Lectures on ordinary differential equations. Addison-Wesley, Reading, MA
4. Hochstadt H (1975) Differential equations: a modern approach. New York, Dover
5. Agarwal RP (1986) Boundary value problems for higher-order differential equations. World Scientific, Singapore
6. Kurzweil J (1986) Ordinary differential equations. Elsevier, Amsterdam
7. Rogers C, Ames WF (1989) Nonlinear boundary value problems in science and engineering. Academic Press, Cambridge
8. Birkhoff G, Rota G-C (1989) Ordinary differential equations, 4th edn. Wiley, Hoboken
9. Sagan H (1989) Boundary and eigenvalue problems in mathematical physics. Dover, New York
10. Agarwal RP, Gupta RC (1991) Essentials of ordinary differential equations. McGraw–Hill, Singapore
11. O'Regan D (1994) Theory of singular boundary value problems. World Scientific, Singapore
12. Miller RK, Michel AN (2007) Ordinary differential equations. Dover, New York
13. Jordan DW, Smith P (2007) Nonlinear ordinary differential equations: problems and solutions a sourcebook for scientists and engineers. Oxford University Press, Oxford
14. Edwards CH, Penney DE (2008) Differential equations, 6th edn. Pearson
15. Hairer E Nørsett SP, Wanner G (2008) Solving ordinary differential equations. I nonstiff problems, Second revised edition. Springer, New York
16. Agarwal RP, O'Regan D (2008) An introduction to ordinary differential equations. Springer, New York
17. Zill DG, Cullen MR (2009) Differential equations with boundary-value problems, 7th edn. Brooks/Cole Cengage Learning
18. Nagle RK, Saff EB, Snider D (2012) Fundamentals of differential equations. 8th edn. Pearson
19. Boyce WE, DiPrima RC (2012) Elementary differential equations and boundary value problems. 10th edn, Wiley, London
20. Simmons GF (2017) Differential equations with applications and historical notes, 3rd edn. Taylor & Francis Group, LLC, London
21. Ascher UM et al (1995) Numerical solution of boundary value problems for ordinary differential equations. SIAM, Philadelphia, p. 91
22. Mills RD (1992) Slope retention techniques for solving boundary-value problems in differential equations. J Symbolic Comput 1992(13):59–80

Chapter 4
Boundary Value Green's Functions

The linear superposition principle is one of the most elegant and effective methods to represent solutions of BVPs in terms of an auxiliary function known as Green's function which was first introduced by George Green[1] in 1828. Subsequently, the method of Green's functions became a useful analytical method [1]. If the solution of the Green's equation is known, then the general solution of the ODE is represented as a convolution of the Green's solution and the nonhomogeneous part of the ODE. Because it is based on the superposition principle, the Green's function method is applicable only to linear systems (see, however, [2, 3]).

A way of solving a BVP $\mathcal{L}[y] = f(x)$ with the boundary conditions of

$$A_1 y(a) + B_1 y'(a) = 0 \tag{4.1}$$

$$A_2 y(b) + B_2 y'(b) = 0 \tag{4.2}$$

is to proceed in the following steps:

(i) Decide on the fundamental solution $\{v_1, v_2\}$ of the homogeneous form of the given ODE, $\mathcal{L}[y] = 0$.
(ii) Find the solutions y_1, y_2 of $\mathcal{L}[y] = 0$ by taking linear combinations of v_1, v_2 that satisfy (4.1) and (4.2).
(iii) Find the Green's function,

$$G(x, z) = \begin{cases} y_1(z)y_2(x)/W(z) & 0 \le z \le x \\ y_1(x)y_2(z)/W(z) & x \le z \le b \end{cases}$$

where $W(z)$ is the Wronskian of two solutions.

---

[1]George Green (1793–1841), English mathematician.

© Springer Nature Switzerland AG 2019
A. Ü. Keskin, *Boundary Value Problems for Engineers*,
https://doi.org/10.1007/978-3-030-21080-9_4

(iv) Find the solution of $\mathcal{L}[y] = f(x)$ by integration,

$$y(x) = \int\limits_a^b G(x,z) \cdot f(z) \cdot dz$$

Green's functions[2] help to visually interpret the actions associated to a source (such as a force or a charge) concentrated at a point [4].

Problem 4.1 The solution of the BVP,

$$\frac{d^2 u}{dx^2} + P(x)\frac{du}{dx} + Q(x)u = f(x), \quad a < x < b \tag{4.3}$$

$$A_1 u(a) - B_1 u'(a) = 0 \tag{4.4}$$

$$A_2 u(b) + B_2 u'(b) = 0 \tag{4.5}$$

can be found by the integral equation

$$u(x) = \int\limits_a^b G(x,z) f(z) dz \tag{4.6}$$

where $G(x,z)$ is the Green's function for the BVP,

$$G(x,z) = \{ u_1(z)u_2(x)/W(z) \quad a < z \leq x \tag{4.7a}$$

$$G(x,z) = \{ u_1(x)u_2(z)/W(z) \quad x \leq z < b \tag{4.7b}$$

where u_1, u_2 are the independent solutions of the homogeneous ODE that satisfy the boundary conditions as required, while $W(z)$ is the Wronskian of the two solutions.

The advantage of the Green's function is that it shows how the inhomogeneity $f(x)$ influences the BVP.

(a) What are the possibilities that cause the solution using this method to fail?
(b) Find the Green's function for $y'' = f(x)$, $0 < x < 1$, $y(0) = y(1) = 0$.
(c) Use Green's method to solve the BVP, $y'' = 1$, $y(0) = y(1) = 0$

Solution
Besides the possibility that $P(x)$, $Q(x)$ in (4.3) might not be continuous, division by zero is a possible cause of a failure.

---

[2]For the proofs related to properties of boundary value Green's functions, see textbooks, for example, [5–10].

$$W(x) = \begin{vmatrix} u_1(x) & u_2(x) \\ u_1'(x) & u_2'(x) \end{vmatrix} = u_1(x)u_2'(x) - u_2(x)u_1'(x)$$

$$A_1 u_2(a) - B_1 u_2'(a) \text{ and } A_2 u_1(b) + B_2 u_1'(b)$$

All of these equations are null if any one of them is zero.

This makes the solutions $u_1(x)$, $u_2(x)$ dependent. Thus, the BVP (4.3)–(4.5) has a unique solution unless there is a nontrivial solution of the homogenous equation

$$\frac{d^2 u}{dx^2} + P(x)\frac{du}{dx} + Q(x)u = 0, \quad a < x < b$$

that satisfies (4.4), (4.5). The unique solution is given by (4.6), (4.7a, b).

(b) The Green's function for this BVP,

$$G(x, z) = \begin{cases} y_1(z)y_2(x)/W(z), & 0 < z \le x \\ y_1(x)y_2(z)/W(z), & x \le z < 1 \end{cases}$$

The general solution for the homogeneous ODE, $y'' = 0$ is $y(x) = c_1 x + c_2$.

$$y_1(0) = 0 \rightarrow c_1 = 1, c_2 = 0 \rightarrow y_1(x) = x$$

$$y_2(1) = 0 \rightarrow c_1.(1) + c_2 = 0 \rightarrow c_1 = -1, c_2 = 1 \rightarrow y_2(x) = -x + 1$$

The Wronskian and Green's function are

$$W(x) = \begin{vmatrix} u_1(x) & u_2(x) \\ u_1'(x) & u_2'(x) \end{vmatrix} = \begin{vmatrix} x & -x+1 \\ 1 & -1 \end{vmatrix} = -x - (-x+1) = -1$$

$$G(x, z) = \begin{cases} z(-x+1)/(-1) & 0 < z \le x \\ x(-z+1)/(-1) & x \le z < 1 \end{cases}$$

(c) Here, $f(x) = 1$,

$$y(x) = \int_0^1 G(x, z)f(x)dz = \int_0^x G_1(x, z)dz + \int_x^1 G_z(x, z)dz$$

$$G_1(x, z) = (x - 1)z, \quad 0 \le z \le x,$$

$$G_2(x,z) = (z-1)x, \quad x \le z \le 1,$$

$$y(x) = \int_0^x (xz - z)\,dz + \int_x^1 (xz - x)\,dz = \left(\frac{xz^2}{2} - \frac{z^2}{2}\right)\Big|_0^x + \left(\frac{xz^2}{2} - xz\right)\Big|_x^1 = \frac{x^2 - x}{2}$$

Problem 4.2 Use Green's function to solve the BVP,

$$y'' - y = 1, \quad 0 < x < 1, \quad y(0) = y(1) = 0$$

Solution
Two independent solutions of the homogenous ODE, $y'' - y = 0$ that satisfy the boundary conditions can be obtained from the general solution of the homogenous ODE,

$$y(x) = c_1 \cosh(x) + c_2 \sinh(x)$$

$$y_1(0) = 0 = c_1 \rightarrow y_1(x) = \sinh(x)$$

$$y_2(1) = 0 = c_1 \cosh(1) + c_2 \sinh(1) \rightarrow c_1 = \sinh(1), c_2 = -\cosh(1)$$

$$y_2(x) = \sinh(1)\cosh(x) - \cosh(1)\sinh(x) = \sinh(1 - x)$$

Note that $c_1 = -\sinh(1), c_2 = \cosh(1)$ is another possibility, however, this does not change the solution of BVP.
The Wronskian of these two basis solutions is

$$W(x) = \begin{vmatrix} \sinh(x) & \sinh(1-x) \\ \cosh(x) & -\cosh(1-x) \end{vmatrix}$$
$$= -\sinh(x)\cosh(1-x) - \sinh(1-x)\cosh(x)$$

Using the identities,

$$\sinh(1-x) = \sinh(1)\cosh(x) - \cosh(1)\sinh(x)$$
$$\cosh(1-x) = \cosh(1)\cosh(x) - \sinh(1)\sinh(x)$$

$$W(x) = -\sinh(x)[\cosh(1)\cosh(x) - \sinh(1)\sinh(x)]$$
$$\quad - \cosh(x)[\sinh(1)\cosh(x) - \cosh(1)\sinh(x)]$$
$$W(x) = -\sinh(x)\cosh(x)\cosh(1) + \sinh^2(x)\sinh(1) - \cosh^2(x)\sinh(1)$$
$$\quad + \cosh(x)\sinh(x)\cosh(1)$$
$$= \sinh(1)[\sinh^2(x) - \cosh^2(x)] = \sinh(1)[-(\cosh^2(x) - \sinh^2(x))]$$
$$= -\sinh(1)$$

Green's function for this BVP is

$$G(x,z) = \begin{cases} \sinh(z)\sinh(1-x)/W & 0<z\leq x \\ \sinh(x)\sinh(1-z)/W & x\leq z<1 \end{cases}$$

Since $f(x) = 1$, the solution is given by

$$y(x) = \int_0^1 G(x,z)dz = \int_0^x -\frac{\sinh(z)\sinh(1-x)}{\sinh(1)}dz + \int_x^1 -\frac{\sinh(x)\sinh(1-z)}{\sinh(1)}dz$$

$$= -\frac{\sinh(1-x)}{\sinh(1)}\cosh(z)\Big|_0^x - \frac{\sinh(x)}{\sinh(1)}[-\cosh(1-z)]\Big|_x^1$$

$$= -\left\{\frac{\sinh(1-x)}{\sinh(1)}[\cosh(x)-1] + \frac{\sinh(x)}{\sinh(1)}[\cosh(1-x)-1]\right\}$$

$$= -\frac{\sinh(1-x)\cosh(x)+\sinh(x)\cosh(1-x)}{\sinh(1)} + \frac{\sinh(1-x)+\sinh(x)}{\sinh(1)}$$

$$y(x) = \frac{\sinh(1-x)+\sinh(x)}{\sinh(1)} - 1$$

Problem 4.3 True (T), or False (F)?

(a) Unlike initial value problems, boundary value problems (BVPs) do not always have solutions.

(b) Green's functions are used for the solution of BVPs.

(c) BVPs have only one solution.

(d) In a second order BVP, a Green's function is constructed out of two independent solutions of the homogeneous equation.

(e) The Green's function can be thought as the response function to a unit impulse at $z = x$.

(f) Dirac delta function is the integral of the Heaviside (unit step) function.

(g) If the interval of interest is infinitely small, the BVP is singular.

(h) At a singular point, one can not specify a value for the solution of the BVP.

(i) At a singular point, one can specify a value for the derivative of solution of the BVP.

(j) At a singular point, we require that the solution and its derivative be finite, or bounded.

(k) A boundary point cannot be a mathematical boundary without being a physical boundary.

(l) The Green's function is related to the BVP in a similar same way that the inverse of a square matrix A is related to the linear algebraic system $y = Ax$.

(m) A BVP has a Green's function if the homogeneous BVP has only the non-trivial solution.

Answers

(a) T, (b) T, (c) F (they can have many solutions), (d) T, (e) T, (f) F (derivative of), (g) F (infinitely long), (h) T, (i) F, (j) T, (k) F (can bc), (l) T, (m) F (trivial).

Problem 4.4

(a) Determine the Green's function and the solution for the following BVP,

$$y'' - k^2 y = -1, \quad y(0) = y(b) = 0, \quad 0 < x < b$$

(b) Calculate $y(x)$ if $k = 1, b = 2$.

Solution

(a) General solution is of the form, $y(x) = c_1 \cosh(kx) + c_2 \sinh(kx)$

From given boundary conditions we find,

$$y(0) = c_1 \cdot 1 + c_2 \cdot 0 = 0 \rightarrow c_1 = 0 \rightarrow y_1(x) = \sinh(kx)$$

$$y_2(b) = c_1 \cosh(kb) + c_2 \sinh(kb) \rightarrow c_1 = -\sinh(kb), c_2 = \cosh(kb)$$

$$y_2(x) = -\sinh(kb)\cosh(kx) + \cosh(kb)\sinh(kx)$$

The Wronskian determinant is calculated as $W(z) = k \cdot \sinh(kb)$
The Green's function is

$$G(x,z) = \begin{cases} y_1(z)y_2(x)/W(z) & 0 < z \le x \\ y_1(x)y_2(z)/W(z) & x \le z < b \end{cases}$$

$$G(x,z) = \begin{cases} \sinh(kz)(-\sinh(kb)\cosh(kx) + \cosh(kb)\sinh(kx))/W(z) & 0 < z \le x \\ \sinh(kx)(-\sinh(kb)\cosh(kz) + \cosh(kb)\sinh(kz))/W(z) & x \le z < b \end{cases}$$

$$G(x,z) = \begin{cases} -\sinh(kz)(\sinh(k(b-x)))/W(z) & 0 < z \le x \\ -\sinh(kx)(\sinh(k(b-z)))/W(z) & x \le z < b \end{cases}$$

$$y(x) = \int_0^b G(x,z) \cdot f(z)dz$$

$$= \int_0^x \frac{-\sinh(kz)(\sinh(k(b-x)))}{k \cdot \sinh(kb)} \cdot (-1)dz + \int_x^b \frac{-\sinh(kx)(\sinh(k(b-z)))}{k \cdot \sinh(kb)} \cdot (-1)dz$$

$$= \frac{(\sinh(k(b-x)))}{k \cdot \sinh(kb)} \int_0^x \sinh(kz)dz + \frac{\sinh(kx)}{k \cdot \sinh(kb)} \int_x^b (\sinh(k(b-z)))dz$$

$$y(x) = \frac{\sinh(kb) - \sinh(kx) - \sinh(k(b-x))}{2k^2 \cdot \sinh\left(\frac{kb}{2}\right) \cdot \cosh\left(\frac{kb}{2}\right)}$$

The solution is

$$y(x) = K(\sinh(kb) - \sinh(kx) - \sinh(k(b-x)))$$

where

$$K = \frac{1}{2k^2 \cdot \sinh\left(\frac{kb}{2}\right) \cdot \cosh\left(\frac{kb}{2}\right)}$$

(b) For $k = 1, b = 2$,

$$y(x) = \frac{\sinh(x-2) + \sinh(2) - \sinh(x)}{\sinh(2)}$$

Problem 4.5 The Green's function associated with the nonhomogeneous equation $L[y] = y'' + p(x)y' + q(x)y = f(x)$ satisfies the differential equation:

$$L[G(x,z)] = \delta(x-z)$$

where $\delta(x-z)$ is Dirac delta function which is equal to 0 everywhere except $x = z$, and $p(x), q(x)$ are continuous functions.

Another algorithm of the BVP solution has the following steps:

1. Solve homogeneous problem first; $L[y] = 0$ which has a general solution $= C_1 y_1 + C_2 y_2$. If $x \neq z, L[G(x,z)] = 0$, this gives

$$G(x,z) = \begin{cases} ay_1 + by_2, & x < z \\ cy_1 + dy_2, & x > z \end{cases}$$

Four equations are needed to find four unknown parameters. The fundamental solutions y_1, y_2 should satisfy the following properties:

$$L[y_1] = 0, \quad B_0[y_1] = 0$$
$$L[y_2] = 0, \quad B_1[y_2] = 0$$

B_0, B_1 represent boundary condition operator.

Based on y_1, y_2 and their Wronskian, formulate first two equations to find the coefficients of Green's function.

2. Apply continuity and jump conditions to form two remaining equations.

$$ay_1(z) + by_2(z) = cy_1(z) + dy_2(z)$$
$$cy_1'(z) + dy_2'(z) - ay_1'(z) + by_2'(z) = 1$$

3. Solve all four equations to find the coefficients and construct the Green's function.
4. Multiply Greens function by forcing function and then integrate the result. Use this algorithm to determine the solution of BVP,

$$y'' = f(x) = x^3, \quad y(0) = y'(1) = 0$$

Solution

The Green's function for this problem satisfies,

$$L[G(x,z)] = G_{xx}(x,z) = \delta(x-z)$$
$$G(0,z) = G_x(1,z) = 0$$

If $x \neq z$, $L[G(x,z)] = G_{xx}(x,z) = 0$, Thus,

$$G(x,z) = \begin{cases} ax+b, & x<z \\ cx+d, & x>z \end{cases}$$

From boundary conditions,

$$G(0,z) = 0 \rightarrow b = 0 \rightarrow G(x,z) = ax, \quad x<z$$
$$G'(1,z) = 0 \rightarrow c = 0 \rightarrow G(x,z) = d, \quad x>z$$

From continuity condition, $az = d$
From jump condition, $0 - a = 1 \rightarrow a = -1$
Hence,

$$G(x,z) = \begin{cases} -x, & x<z \\ -z, & x>z \end{cases}$$

Since $f(x) = x^3$,

$$y(x) = \int_0^1 G(x,z)f(z)dz = \int_0^x (-z)z^3dz + \int_x^1 (-x)z^3dz = -\int_0^x z^4dz - \int_x^1 (x)z^3dz$$

$$y(x) = -\frac{x^5}{5} - \frac{x}{4} + \frac{x^5}{4} = \frac{x^5}{20} - \frac{x}{4}$$

Problem 4.6 Use Green's function method to solve the following nonhomogeneous BVP for the following forcing functions,

$$y'' = f(x), \quad y(0) = y(\pi/2) = 0, \quad x \in [0, \pi/2]$$

$$(a)f(x) = x; \quad (b)f(x) = x^2; \quad (c)f(x) = e^{-x}\sin(2x)$$

(i) Show all manual steps of calculations for part (a).

(ii) Write a MATLAB code and verify the result found above, then solve parts
(b) and (c) using this code and plot all solutions, separately (green1.m).

Solution

(i) The general solution of homogenous ODE $(y'' = 0)$ is $y = c_1 x + c_2$,

$$y(0) = 0 = c_2 \rightarrow y_1 = x$$

$$y(\pi/2) = 0 = c_1(\pi/2) + c_2 \rightarrow c_1 = 1, c_2 = -\pi/2 \rightarrow y_2(x) = x - \pi/2$$

Wronskian,

$$W = \begin{vmatrix} x & x - \pi/2 \\ 1 & 1 \end{vmatrix} = x - x + \pi/2 = \pi/2$$

Green's function,

$$G = \begin{cases} y_1(z)y_2(x)/W(2), & 0 < z \leq x \\ y_1(x)y_2(z)/W(2), & x \leq z < \pi/2 \end{cases}$$

$$G = \begin{cases} z(x - \pi/2)/(\pi/2) = 2z(x - \pi/2)/\pi, & 0 < z \leq x \\ x(z - \pi/2)/(\pi/2) = 2x(z - \pi/2)/\pi, & 0 < z \leq x \end{cases}$$

For the forcing function $f(x) = x$,

$$y(x) = \int\limits_0^{\pi/2} G(x,z) \cdot f(z) dz = \int\limits_0^x \frac{2z(x - \pi/2)}{\pi} \cdot z dz + \int\limits_x^{\pi/2} \frac{2x(z - \pi/2)}{\pi} \cdot z dz$$

$$= \frac{2}{\pi}\left(x - \frac{\pi}{2}\right) \int\limits_0^x z^2 dz + \frac{2x}{\pi} \int\limits_x^{\pi/2} z^2 dz - \frac{2x}{\pi} \cdot \frac{\pi}{2} \int\limits_x^{\pi/2} z dz$$

$$= \left(\frac{2x}{\pi} - 1\right)\left(\frac{z^3}{3}\Big|_0^x\right) + \frac{2x}{\pi}\left(\frac{z^3}{3}\Big|_x^{\pi/2}\right) - x\left(\frac{z^2}{2}\Big|_x^{\pi/2}\right)$$

$$= \left(\frac{2x}{\pi} - 1\right)\left(\frac{x^3}{3}\right) + \frac{2x}{\pi}\left(\frac{\pi^3}{24} - \frac{x^3}{3}\right) - x\left(\frac{\pi^2}{8} - \frac{x^2}{2}\right) = \frac{2x^4}{3\pi} - \frac{x^3}{3} + \frac{x\pi^2}{12} - \frac{2x^4}{3\pi} - \frac{x\pi^2}{8} + \frac{x^3}{2}$$

$$y(x) = x^3\left(-\frac{1}{3} + \frac{1}{2}\right) + x\left(\frac{\pi^2}{12} - \frac{\pi^2}{8}\right) = \frac{x^3}{6} - \frac{x\pi^2}{24} = \frac{x}{24}(4x^2 - \pi^2)$$

(ii) Wronskian of basis functions and Green's function for all cases are the same,

```
W = pi/2, G1 = -(2*z*(pi/2 - x))/pi, G2 = -(2*x*(pi/2 - z))/pi
```

(a) `f = x, y = (x*(4*x^2 - pi^2))/24`

(b) f = x^2, y = -(x*(- 8*x^3 + pi^3))/96

(c) f = exp(-x)*sin(2*x),

y = (8*x - 4*pi + 8*x*exp(-pi/2) + 4*pi*cos(2*x)*exp(-x) -
    3*pi*sin(2*x)*exp(-x))/(25*pi)

Figure 4.1 displays all three solutions.

```
%green1.m   y"=f(x),   y(a)=ya,     y(b)=yb
clc; clear; syms x z
a=0; b=pi/2; % x-boundaries
%f=x                     % f(x)
%f=x*x                   % f(x)
f=exp(-x)*sin(2*x)   % f(x)
% y=c1*x+c2  % solution of homogeneous part
y1=x;
y2=x-pi/2;
s = [ y1,   y2]; s2 = [ s; diff(s,x,1)];W = simplify(det(s2)) %Wronskian
f1=subs(f,x,z); y1z=subs(y1,x,z); y2z=subs(y2,x,z);
G1=y1z*y2/W % Green's upper
G2=y1*y2z/W % Green's lower
I1=int(f1*G1,z,a,x);   % Definite integral of f wrt z from 0 to x.
I2=int(f1*G2,z,x,b); % Definite integral of f wrt z from x to pi/2.
y=I1+I2 ; y=simplify(y)
pretty(y)
h=ezplot(y,[0 pi/2]); ylabel('y'),grid; set(h,'Linewidth',2.5);
```

Problem 4.7 Write a MATLAB code which finds the solution of a nonhomogeneous BVP using Green's function with homogeneous Dirichlet boundary conditions. Then, use the code to solve and plot the solutions in the same figure for $y'' = -1, y'' + y = -1, \ y'' + y' + y = -1$; all with the same boundary values of $y(0) = y'(1) = 0$ (bvp_Green1.m).

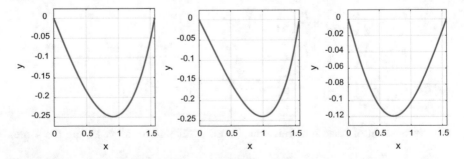

Fig. 4.1 Solutions for $y'' = f(x)$, $y(0) = y(\pi/2) = 0$, using Green's function when $f(x) = x$, (left), $f(x) = x^2$, (middle), $f(x) = e^{-x}\sin(2x)$ (right)

Solution

MATLAB code (bvp_Green1.m) is listed below. Figure 4.2 displays the solution curves of three different BVPs for $y'' = -1, y'' + y = -1, y'' + y' + y = -1$ having the same forcing functions, within the same interval and same boundary conditions, $y(0) = y'(1) = 0$.

```
% bvp_Green1.m
% Finds the solution of a Nonhomogeneous BVP
% using  Green's function with homogeneous Dirichlet boundary conditions.
% The general form of the BVP is : L[y] = f(x), y(a)=0, y(b)=0
% Consider first :    L[y] =  0,   y(a)=0, this gives solution  y1(x)
% Consider next  :    L[y] =  0,   y(b)=0, this gives solution  y2(x)
% Wronskian : W = y1 * y2'- y1' * y2
% Green's Function is defined by :
%      G(x,s)= y1(z)*y2(x)/W,    a <= z < x
%      G(x,s)= y1(x)*y2(z)/W,    x< z <= b
% The solution for non-homogeneous BVP is   y(x)=int(G(x,z)*f(z),z,a,b)
clc;clear; syms x z
%a=0; b=1;   eq='D2y=0'; bc1='y(0)=0'; bc2='y(1)=0';f=-1;
%a=0; b=1;   eq='D2y+y=0'; bc1='y(0)=0'; bc2='y(1)=0';f=-1;
 a=0; b=1;   eq='D2y+Dy+y=0'; bc1='y(0)=0'; bc2='y(1)=0';f=-1;
y1=simplify(dsolve(eq,bc1,x)); y2=simplify(dsolve(eq,bc2,x));
S1=setdiff(symvar(y1),x);
for k=1:numel(S1)
    y1=subs(y1,S1(k),1)
end
S1=setdiff(symvar(y2),x);
for k=1:numel(S1)
    y2=subs(y2,S1(k),1)
end
W=y1*diff(y2,x)-diff(y1,x)*y2; W=simplify(W); % Wronskian
W=subs(W,x,z)
if W~=0      % If W=0, the Green's f & solution to BVP does not exist
G1=simplify(subs(y1,x,z)*y2/W) % a<= x < z
G2=simplify(y1*subs(y2,x,z)/W) % z< x <= b
I1=int(G2*subs(f,x,z),z,a,x);I2=int(G1*subs(f,x,z),z,x,b);
y=I1+I2; y=simplify(y)
plot(linspace(a,b),subs(y,x,linspace(a,b)),'linewidth',2);hold on;
xlabel('x');ylabel( 'y(x)'); grid on;
legend('D2y=-1','D2y+y=-1','D2y+Dy+y=-1');
else
disp(' The solution to BVP does not exist, Wronskian = 0')
end
```

Problem 4.8 (*Piecewise continuous forcing function*)

(i) Use Green's function method to solve the following nonhomogeneous BVP,

$$y'' = f(x), \quad y(0) = y(b) = 0, \quad x \in [0, b],$$

Fig. 4.2 Solution curves of
three different BVPs having
the same forcing functions,
within the same interval and
same boundary conditions

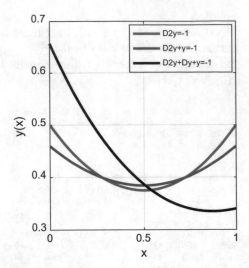

$$f(x) = \begin{cases} 0, & 0 < x < \frac{b}{2} \\ A, & \frac{b}{2} < x < b \end{cases}$$

where A is the magnitude of forcing function.

(ii) Write a MATLAB code and verify the result found above, use this code to plot
the solution and the forcing function in the same figure, for $A = -5, b = 2$
(green4.m).

Solution

(i) The general solution of homogenous ODE $(y'' = 0)$ is $y = c_1 x + c_2$,

$$y(0) = 0 = c_2 \rightarrow y_1 = x$$

$$y(b) = 0 = c_1 \cdot b + c_2 \rightarrow c_1 = -1, c_2 = b \rightarrow y_2(x) = b - x$$

Wronskian,

$$W = \begin{vmatrix} x & b - x \\ 1 & 1 \end{vmatrix} = x + x - b = -b$$

Green's function,

$$G = \begin{cases} y_1(z)y_2(x)/W(2), & 0 < z \leq x \\ y_1(x)y_2(z)/W(2), & x \leq z < b \end{cases}$$

$$G = \begin{cases} z(x - b)/b, & 0 < z \leq x \\ x(z - b)/b, & x \leq z \leq b \end{cases}$$

Taking forcing function $f(x)$ into consideration, we obtain

$$y(x) = \int_0^b G(x,z) \cdot f(z)dz = \int_0^x \frac{z(x-b)}{b} \cdot f(z)dz + \int_x^b \frac{x(z-b)}{b} \cdot f(z)dz$$

$$y(x) = \int_0^x \frac{z(x-b)}{b} \cdot 0 \cdot dz + \int_x^b \frac{x(z-b)}{b} \cdot f(z)dz = \int_x^b \frac{x(z-b)}{b} \cdot f(z)dz$$

$$x < \frac{b}{2}; \quad y(x) = \frac{A}{b} \int_{b/2}^b x(z-b)dz = -\frac{Abx}{8}$$

$$x > \frac{b}{2}; \quad y(x) = \frac{A}{b} \int_x^b -x(z-b)dz = -\frac{Ax(b-x)^2}{2b}$$

(ii) Figure 4.3 displays the solution with $A = -5$, $b = 2$.

Fig. 4.3 Nonhomogeneous BVP with a piecewise continuous forcing function. Magnitude of the forcing function is scaled by a factor of 10

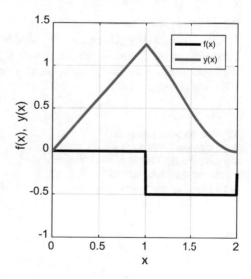

```
%green4.m    y"=f(x),   y(0)=0,     y(b)=0     0<x<b
clc;clear ;close;
syms x z b
f=-5 ;         % f(x) for (b/2)< x < b
% y=c1*x+c2  % homogeneous solution
y1=x; y2=-x+b;
s = [ y1,  y2]; s2 = [ s; diff(s,x,1)];W = simplify(det(s2)) %Wronskian
f1=subs(f,x,z);  y1z=subs(y1,x,z); y2z=subs(y2,x,z);
G1=y1z*y2/W ;G1=simplify(G1)     % Green's upper func.
G2=y1*y2z/W ;G2=simplify(G2)     % Green's lower func.
I1=int(f1*G2,z,b/2,b)   % Definite integral of f1 wrt z from b/2 to b.
I2=int(f1*G2,z,x,b)     % Definite integral of f1 wrt z from x to b.
%b=2;
I1=subs(I1,b,2);I2=subs(I2,b,2);b=2;
y=I1*(heaviside(x)-heaviside(x-b/2))+I2*(heaviside(x-b/2)-heaviside(x-b));
f1=(f/10)*(heaviside(x-b/2)- heaviside(x-b));
h1=ezplot(f1,[0 b]);grid;ylabel('y');
set(h1,'color','k','Linewidth',2.5); hold on;
h=ezplot(y);grid;ylabel('f(x),  y(x)');set(h,'Linewidth',2.5);grid;
axis([0,b,-1, 1.5]);title('');legend('f(x)','y(x)');
```

Problem 4.9 Write a MATLAB code which finds the solution of a nonhomogeneous BVP using Green's function with homogeneous Dirichlet boundary conditions. Then, use the code to solve and plot the solutions in the same figure for $y'' = -1, y'' + y = -1,\ y'' + y' + y = -1$; all with the same boundary values of $y(0) = y'(1) = 0$ (bvp_Green1.m).

Solution
MATLAB code (bvp_Green1.m) is listed below. Figure 4.4 displays the solution curves of three different BVPs for $y'' = -1, y'' + y = -1,\ y'' + y' + y = -1$ having the same forcing functions, within the same interval and same boundary conditions, $y(0) = y'(1) = 0$.

Fig. 4.4 Solution curves of three different BVPs having the same forcing functions, within the same interval and same boundary conditions

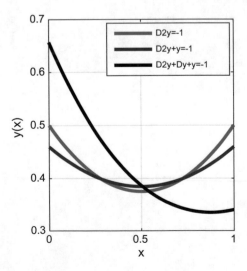

```
% bvp_Green1.m
% Finds the solution of a Nonhomogeneous BVP
% using  Green's function with homogeneous Dirichlet boundary conditions.
% The general form of the BVP is : L[y] = f(x), y(a)=0, y(b)=0
% Consider first :   L[y] =  0,  y(a)=0, this gives solution  y1(x)
% Consider next  :   L[y] =  0,  y(b)=0, this gives solution  y2(x)
% Wronskian : W = y1 * y2' - y1' * y2
% Green's Function is defined by :
%      G(x,s)= y1(z)*y2(x)/W,   a <= z < x
%      G(x,s)= y1(x)*y2(z)/W,   x< z <= b
% The solution for non-homogeneous BVP is  y(x)=int(G(x,z)*f(z),z,a,b)
clc;clear; syms x z
%a=0; b=1;  eq='D2y=0'; bc1='y(0)=0'; bc2='y(1)=0';f=-1;
%a=0; b=1;  eq='D2y+y=0'; bc1='y(0)=0'; bc2='y(1)=0';f=-1;
 a=0; b=1;  eq='D2y+Dy+y=0'; bc1='y(0)=0'; bc2='y(1)=0';f=-1;
y1=simplify(dsolve(eq,bc1,x)); y2=simplify(dsolve(eq,bc2,x));
S1=setdiff(symvar(y1),x);
for k=1:numel(S1)
    y1=subs(y1,S1(k),1)
end
S1=setdiff(symvar(y2),x);
for k=1:numel(S1)
    y2=subs(y2,S1(k),1)
end
W=y1*diff(y2,x)-diff(y1,x)*y2; W=simplify(W); % Wronskian
W=subs(W,x,z)
if W-=0     % If W-0, the Green's f & solution to BVP does not exist
G1=simplify(subs(y1,x,z)*y2/W) % a<= x < z
G2=simplify(y1*subs(y2,x,z)/W) % z< x <= b
I1=int(G2*subs(f,x,z),z,a,x);I2=int(G1*subs(f,x,z),z,x,b);
y=I1+I2; y=simplify(y)
plot(linspace(a,b),subs(y,x,linspace(a,b)),'linewidth',2.5);hold on;
xlabel('x');ylabel( 'y(x)'); grid on;
legend('D2y=-1','D2y+y=-1','D2y+Dy+y=-1');
else
disp(' The solution to BVP does not exist, Wronskian = 0')
end
```

Problem 4.10 Green's function $G(x,z)$ for any BVP satisfies the equation $\mathcal{L}_x G(x,z) = \delta(x-z)$ for some operator $\mathcal{L}_x$ with homogeneous boundary conditions. In other words, the Green's function is the response to a unit impulse forcing function at $x = z$ under homogeneous boundary conditions,

$$\mathcal{L}_x G(x,z) = \delta(x-z), \quad a < z < b, \quad G(a,z) = 0, \quad dG(b,z)/dx = 0$$

On the other hand, we can define the Sifting Property as

$$\int_{x-c}^{x+d} f(z)\,\delta(x-z)dz = f(x), \quad c,d > 0$$

(a) Use these information to find the solution of the BVP,

$$d^2u/dx^2 = f(x), \quad u(0) = u'(b) = 0, \quad b \in \mathbb{R}^+. \tag{4.8}$$

(b) Let $b = 2$. Find and plot solution of the BVP given in (4.8) for the following forcing functions:

$$(i) f(x) = 1, \quad (ii) f(x) = \sin(4x)$$

(green5.m).

Solution
The Green's function is the solution to the BVP, $d^2G(x,z)/dx^2 = \delta(x - z)$, with the boundary conditions of $G(0, z) = 0$, $dG(1, z)/dx = 0$.

Using the homogeneous case, $d^2G(x,z)/dx^2 = 0$, we consider this solution with "constants" as functions of z. Then, this equation has solutions $G(x, z) = A(z)x + B(z)$.

In the first interval $[0, z)$, $G_1(x, z) = A_1(z)x + B_1(z), G_1(0, z) = 0$,

$$B_1(z) = 0, \quad G_1(x, z) = A_1(z)x$$

in the second interval (z, b),

$$G_2(x, z) = A_2(z)x + B_2(z), \quad G_2'(b, z) = 0 \rightarrow A_2(z) = 0, \quad G_2(x, z) = B_2(z)$$

The Green s function becomes

$$G(x, z) = \begin{cases} A_1(z)u_1(x) = A_1(z)x & 0 < x \le z \\ B_2(z)u_2(x) = A_2(z) & z \le x < b \end{cases}$$

Since $G(x, z)$ is be continuous at $x = z$, this means $A_1(z)z = B_2(z)$

On the other hand, using the definition of Dirac delta function and integrating the equation $d^2G(x, z)/dx^2 = \delta(x - z)$,

$$\int_{z-\delta}^{z+\delta} d^2G(x, z)/dx^2 = \delta(x - z) = 1$$

This indicates that the derivative of the Green's function is discontinuous across the unit impulse. Then, we obtain

$$dG(x, z)/dx = \begin{cases} A_1(z) & 0 < x \le z \\ 0 & z \le x < b \end{cases}$$

Using these last two equations, we get $A_1(z) = -1, B_2(z) = -z$
The Green s function is

$$G(x, z) = \begin{cases} -x & 0 < x \le z \\ -z & z \le x < b \end{cases}$$

The Green's function may be expressed in terms of Heaviside[3] (unit step) function as

$$G(x, z) = -xH(z - x) - zH(x - z)$$

We know that the general solution of the BVP is given by

$$u(x) = \int_0^b G(x, z)f(z)dz = \int_0^b [-xH(z - x) - zH(x - z)]f(z)dz$$

$$u(x) = -\int_0^x zf(z)dz - x\int_x^b f(z)dz$$

(b) Here $b = 2$. We compute and plot the solution of the BVP given in (4.8) as follows:

(i) $f(x) = 1$,

$$u(x) = -\int_0^x z \cdot dz - x\int_x^2 dz = x(x - 2) - \frac{x^2}{2}$$

(ii) $f(x) = \sin(4x)$,

$$u(x) = -\int_0^x z \cdot \sin(4z) \cdot dz - x\int_x^2 \sin(4z) \cdot dz = \frac{x\cos(8)}{4} - \frac{\sin(4x)}{16}$$

Three curves arc displayed in Fig. 4.5. MATLAB m-file (green5.m) is listed below.

---

[3]The Heaviside (unit) step function, $H(x - z)$ is defined by

$$H(x - z) = \begin{cases} 0 & z < x \\ 1 & z > x \end{cases}$$

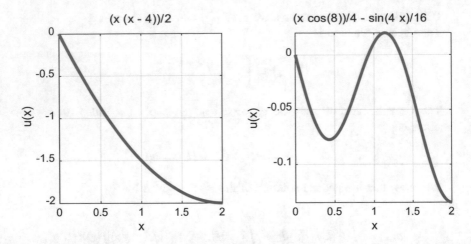

Fig. 4.5 The solution curves of $d^2u/dx^2 = f(x)$, $u(0) = u'(2) = 0$, for $(i)f(x) = 1$(left), $(ii)f(x) = \sin(4x)$, (right)

```
%green5.m    u"=f(x),   u(0)=0,    u'(2)=0 ,    0<x<b
clc;clear;close; syms x z b
b=2;
%f=1;                    % f(x) for  0 < x < b
f=sin(4*z);              % f(x) for  0 < x < b
I1=-int(f*z,z,0,x);      % Definite integral of f wrt z from b/2 to b.
I2=-x*int(f,z,x,b);      % Definite integral of f wrt z from x to b.
u=I1+I2;u=simplify(u)
h1=ezplot(u,[0 b]);ylabel('u(x)');grid on;set(h1,'Linewidth',2.5);axis
tight;
```

Problem 4.11

(a) Find the Green's function for the BVP,

$$y''(x) + y(x) = f(x), \quad y(0) = y'(b) = 0$$

(b) Solve the BVP and plot the solution curves for,

   (i) $b = 2$, $f(x) = -1$,
   (ii) $b = 2$, $f(x) = -x$ (green6.m).

Solution

General form of the solution $s^2 + 1 = 0 \rightarrow y = A \cos(x) + B \sin(x)$, fundamental system is $y_1 = \sin x, y_2 = \cos x$ with general boundary conditions,

$$A_1 y(a) + B_1 y'(a) = 0, \quad A_2 y(b) + B_2 y'(b) = 0$$

From given boundary conditions, $a = 0$, $b = 1$, $A_1 y(0) = 0$, $B_2 y'(b) = 0$
If we take $y = y_1 = \sin x \rightarrow y(0) = \sin(0) = 0$ is satisfied.
If we take $y = \cos(x - b) \rightarrow y'(b) = \sin(b - b) = 0$ is satisfied.
Therefore, $y_1 = \sin x$, $y_2 = \cos(x - b)$ are suitable functions.
The Wronskian is

$$W = \begin{vmatrix} y_1 & y_2 \\ y_1' & y_2' \end{vmatrix} = \begin{vmatrix} \sin x & \cos(x-b) \\ \cos x & -\sin(x-b) \end{vmatrix} = -\sin(x) \cdot \sin(x-b) - \cos x \cdot \cos(x-b)$$

$$= -\sin x(\sin x \cdot \cos b - \cos x \cdot \sin b) - \cos x(\cos x \cdot \cos b + \sin x \cdot \sin b)$$

$$= -\sin^2 x \cdot \cos b + \sin x \cdot \cos x \cdot \sin b - \cos^2 x \cdot \cos b - \sin x \cdot \cos x \cdot \sin b$$

$$W = -\cos(b)(\sin^2 x + \cos^2 x) = -\cos(b)$$

Green's function for the given BVP is

$$G(x,z) = \begin{cases} y_1(z)y_2(x)/W = -\sin(z) \cdot \cos(x-b)/\cos(b) & 0 \le z \le x \\ y_1(x)y_2(z)/W = -\sin(x) \cdot \cos(z-b)/\cos(b) & x \le z \le b \end{cases}$$

(b)

(i) $b = 2$, $f(x) = -1$,

$$y(x) = \int_0^b G(x,z) \cdot f(z)dz = \int_0^x G(x,z) \cdot f(z)dz + \int_x^b G(x,z) \cdot f(z)dz$$

$f(z) = -1$,

$$y(x) = \int_0^x (\sin(z) \cdot \cos(x-b)/\cos(b))dz + \int_x^b (\sin(x) \cdot \cos(z-b)/\cos(b))dz$$

$$y(x) = \frac{\cos(x-b)}{\cos(b)} \int_0^x \sin(z) \cdot dz + \frac{\sin(x)}{\cos(b)} \int_x^b \cos(z-b) \cdot dz$$

$b = 2$,

$$y(x) = \frac{\cos[2(x-1)]}{2\cos(2)} - \frac{\sin(1) \cdot \sin(x)}{2\sin^2(1) - 1} - \frac{1}{2}$$

(ii) $b = 2$, $f(x) = -x$,

$$y(x) = -\frac{\sin(x) \cdot [\sin(1) + 2\sin^2(1/2)]}{2\sin^2(1) - 1}$$
$$- \frac{\sin(x) \cdot [\cos(x-2) + x\sin(x-2) - 1]}{\cos(2)}$$

Fig. 4.6 Solutions of the
BVP, $y''(x) + y(x) =$
$f(x)$, $y(0) = y'(b) = 0$ for,
(i) $b = 2$, $f(x) = -1$,
and (ii) $b = 2$, $f(x) = -x$

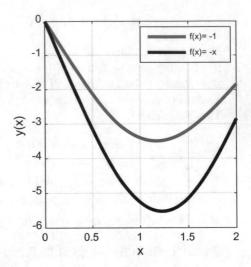

Both of these solutions are displayed in Fig. 4.6. MATLAB code (green6.m) is
listed below.

```
%green6.m    y"(x)+y(x)=f(x),    y(0)=0,    y'(b)=0    0<x<b
clc; clear; syms x z b
%f=-1 ;       % f(x) for (b/2)< x < b
f=-x;         % f(x) for (b/2)< x < b
% y=c1*cosx+c2*sinx  % homogeneous solution
y1=sin(x); y2=cos(x-b);
s =[ y1,   y2]; s2 =[ s; diff(s,x,1)]; W = simplify(det(s2)) %Wronskian
f1=subs(f,x,z); y1z=subs(y1,x,z); y2z=subs(y2,x,z);
G1=y1z*y2/W ;G1=simplify(G1)    % Green's upper
G2=y1*y2z/W ;G2=simplify(G2)    % Green's lower
I1=int(f1*G2,z,b/2,b);   % Definite integral of f wrt z from b/2 to b.
I2=int(f1*G2,z,x,b);   % Definite integral of f wrt z from x to b.
I1=subs(I1,b,2);I2=subs(I2,b,2) %b=2;
y=I1+I2
h=ezplot(y);grid;ylabel('y(x)');set(h,'Linewidth',2.5);grid on;hold on;
axis([0, 2, -6, 0]);title('');legend('f(x)= -1','f(x)= -x')
```

Problem 4.12

(a) Construct the Green's function for the BVP,

$$y'' + y = f(x), \quad y(0) + y'(0) = 0, \quad y(\pi) - y'(\pi) = 0$$

(b) Find the solution of this BVP if the forcing function is constant, $f(x) = 1$.
(c) Write a MATLAB code to compute the Green's function of this BVP and then find
    and plot the solution if the forcing function is constant, $f(x) = 1$ (green7.m).
(d) Use MATLAB dsolve function to find and plot the solution of this BVP
    when the forcing function is constant, $f(x) = 1$ (dsolve_bvp1.m).

Solution

(a) Fundamental solutions are $z_1 = \cos(x), z_2 = \sin x$.

$$\beta_1(y) = A_1 y(a) + B_1 y'(a), \quad \beta_2(y) = A_2 y(b) + B_2 y'(b)$$

In this case,

$$\beta_1(y) = y(0) + y'(0), \quad \beta_2(y) = y(\pi) - y'(\pi)$$

Then,

$$\beta_1(z_1) = \cos(0) + [-\sin(0)] = \cos(0) - \sin(0) = 1$$
$$\beta_2(z_1) = \cos(\pi) + [-\sin(\pi)] = -1$$
$$\beta_1(z_2) = \sin(0) + \cos(0) = 1$$
$$\beta_2(z_2) = \sin(\pi) - \cos(\pi) = 1$$
$$\beta_1(z_1)\beta_2(z_2) - \beta_1(z_2)\beta_2(z_1) = 2 \neq 0$$

Therefore, given BVP has unique solution. Let,

$$y_1 = \beta_1(z_2) \cdot z_1 - \beta_1(z_1) \cdot z_2 = \cos(x) - \sin(x)$$
$$y_2 = \beta_2(z_2) \cdot z_1 - \beta_2(z_1) \cdot z_2 = \cos(x) + \sin(x)$$

which makes $\beta_1(y_1) = \beta_2(y_2) = 0$.
The Wronskian of y_1, y_2 is

$$W(x) = \begin{vmatrix} \cos(x) - \sin(x) & \cos(x) + \sin(x) \\ -\sin(x) - \cos(x) & -\sin(x) + \cos(x) \end{vmatrix} = 2$$

The Green's function is

$$G(x,z) = \begin{cases} y_1(z) \cdot y_2(x)/W(z) = 0 \leq z < x \\ y_1(x) \cdot y_2(z)/W(z) = x \leq z \leq \pi \end{cases}$$

$$G_1 = \frac{y_1(z) \cdot y_2(x)}{W(z)} = \frac{[\cos(z) - \sin(z)][\cos(x) + \sin(x)]}{2}$$

$$G_2 = \frac{y_1(x) \cdot y_2(z)}{W(z)} = \frac{[\cos(x) - \sin(x)][\cos(z) + \sin(z)]}{2}$$

(b)
$$y(x) = \int\limits_0^\pi G(x,z) \cdot f(z) \cdot dz = \int\limits_0^x G_1 dz + \int\limits_x^\pi G_2 dz$$

$$y(x) = \frac{\cos(x) + \sin(x)}{2} \int\limits_0^x [\cos(z) - \sin(z)] dz$$

$$+ \frac{\cos(x) - \sin(x)}{2} \int\limits_x^\pi [\cos(z) + \sin(z)] dz$$

$$y(x) = 1 - \sin(x)$$

Note that this result checks with the given BVP, $y'' + y = 1, y(0) + y'(0) = 0,$ $y(\pi) - y'(\pi) = 0$.

(c) A plot of the solution of the BVP $y'' + y = 1, y(0) + y'(0) = 0, y(\pi) - y'(\pi) = 0$ is shown in Fig. 4.7.

```
% green7.m  D2y+y=f(x),    y(0)+y'(0)=0,    y(pi)-y'(pi)=0.
% the solution of a Nonhomogeneous  BVP using Green's function
% Green's Function is defined by   (W = Wronskian):
%     G(x,s)= y1(z)*y2(x)/W,    a <= z < x
%     G(x,s)= y1(x)*y2(z)/W,    x< z <= b
% The solution for non-homogeneous BVP is  y(x)=int(G(x,z)*f(z),z,a,b)
clc;clear; syms x z
f=1;
a=0; b=pi;    eq='D2y+y=0';
bc1='y(0)+Dy(0)=0'; bc2='y(pi)-Dy(pi)=0';
y1=simplify(dsolve(eq,bc1,x));y2=simplify(dsolve(eq,bc2,x));
S1=setdiff(symvar(y1),x);
for k=1:numel(S1)
    y1=subs(y1,S1(k),1)
end
S1=setdiff(symvar(y2),x);
for k=1:numel(S1)
    y2=subs(y2,S1(k),1)
end
W=y1*diff(y2,x)-diff(y1,x)*y2; W=simplify(W); % Wronskian
if W~=0      % If W=0, the Green's f & solution to BVP does not exist
W=subs(W,x,z);   G1=simplify(y2*subs(y1,x,z)/W) % a<= x < z
G2=simplify(y1*subs(y2,x,z)/W) % z< x <= b
I1=int(G1*subs(f,x,z),z,a,x);I2=int(G2*subs(f,x,z),z,x,b);
y=I1+I2; y=simplify(y)
plot(linspace(a,b),subs(y,x,linspace(a,b)),'linewidth',2.5);
xlabel('x');ylabel( 'y(x)'); grid on; xlim([a b]);
else
disp(' The solution to BVP does not exist, Wronskian = 0')
end
```

(d) Following MATLAB function can also be used to find and plot the solution of this BVP when the forcing function is constant, $f(x) = 1$ (dsolve_bvp1.m). The solution curve is the same as that one shown in Fig. 4.7.

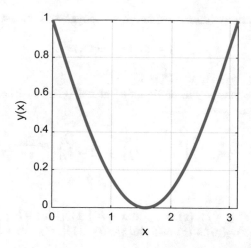

Fig. 4.7 The solution of the BVP, $y'' + y = 1$, $y(0) + y'(0) = 0$, $y(\pi) - y'(\pi) = 0$.

```
% dsolve_bvp1.m, Solve BVP using dsolve and plot the solution.
clc;clear; syms  y(x)
eq='D2y+y=1';                % ODE
bc1='y(0)+Dy(0)=0';          % boundary conditions1
bc2='y(pi)-Dy(pi)=0';        % boundary conditions2
S=dsolve(eq,bc1,bc2,'x');    % Solution
pretty(S)
x = linspace(0,pi,100); y=eval(vectorize(S));
plot(x,y,'linewidth',2.5);xlabel('x'),ylabel('y(x)');xlim([0  pi]);
grid on;
```

Problem 4.13

(a) (Bounded BVP) Construct the Green's function for the BVP,

$$\frac{1}{x}\frac{d}{dx}\left(x\frac{dy}{dx}\right) = f(x), \quad 0 < x < b, \quad y(0) \text{ bounded}, \quad y(b) = 0.$$

(b) Find the solution of this BVP if $f(x) = 1$ and $b = 1$, then plot the solution (BVP_GreenCE.m).

Solution

(a) By the chain rule,

$$\frac{1}{x}\frac{dy}{dx} + \frac{d^2y}{dx^2} = f(x)$$

The general solution of the corresponding homogeneous Euler-Cauchy equation is

$$y = c_1 + c_2 \ln(x).$$

We take $y_1 = 1$, $y_2 = \ln[x - (b - 1)]$.
This will assure that y_1 will be bounded at $x = 0$ and $y_2(b) = 0$.
Wronskian determinant:

$$W(z) = \begin{vmatrix} v_1 & v_2 \\ v_1' & v_2' \end{vmatrix} = \begin{vmatrix} 1 & \ln[z - (b-1)] \\ 0 & 1/[z - (b-1)] \end{vmatrix} = \frac{1}{z - (b-1)}$$

Green's function:

$$G(x,z) = \begin{cases} v_1(z)v_2(x)/W(z) = 1 \cdot \ln(x - b + 1)/[1/(z - b + 1)], & 0 < z \le x \\ v_1(x)v_2(z)/W(z) = 1 \cdot \ln(z - b + 1)/[1/(z - b + 1)], & x \le z < b \end{cases}$$

$$G(x,z) = \begin{cases} (z - b + 1)\ln(x - b + 1) & 0 < z \le x \\ (z - b + 1)\ln(z - b + 1) & x \le z < b \end{cases}$$

(b)
$$y = \int_{a=0}^{b=1} G(x,z) \cdot f(z) \cdot dz = \int_0^x z\ln(x)dz + \int_x^1 z\ln(z)dz$$

$$y = (I_1) + (I_2) = \left(\frac{\ln(x) \cdot x^2}{2} \right) + \left(\frac{x^2 - 1}{4} - \frac{x^2 \ln(x)}{2} \right)$$

$$y = \frac{1}{4}(x^2 - 1)$$

The solution curve of the BVP for $f(x) = 1$ and $b = 1$ is shown in Fig. 4.8.

Fig. 4.8 The solution curve of the BVP for $f(x) = 1$ and $b = 1$

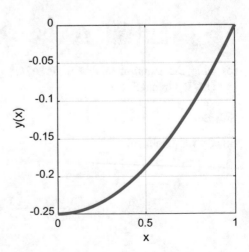

MATLAB m-file (BVP_GreenCE.m) is given below.

```
% BVP_GreenCE.m   D2y+Dy/x=f(x), with y(0) bounded and y(b)=0.
% the solution of a Nonhomogeneous Cauchy-Euler BVP using Green's function
% Wronskian : W = y1 * y2' - y1' * y2
% Green's Function is defined by :
%     G(x,s)= y1(z)*y2(x)/W,    a <= z < x
%     G(x,s)= y1(x)*y2(z)/W,    x< z <= b
% The solution for non-homogeneous BVP is   y(x)=int(G(x,z)*f(z),z,a,b)
clc; clear; syms x z k
f=1; %f(x)
y0_bounded=1  %If y(0) is bounded enter 1,otherwise 0
a=0; b=1;   eq='D2y+Dy/x=0'; bc1='y(0)=0';bc2='y(b)=0';
y1=simplify(dsolve(eq,bc1,x)), y2=simplify(dsolve(eq,bc2,x))
S1=setdiff(symvar(y1),x);
for k=1:numel(S1)
    y1=subs(y1,S1(k),1);
end
if y1==0 && (y0_bounded==1)
    y1=1
else
    end
S1=setdiff(symvar(y2),x);
for k=1:numel(S1)
    y2=subs(y2,S1(k),1)
end
W=y1*diff(y2,x)-diff(y1,x)*y2; W=simplify(W); % Wronskian
if W~=0     % If W=0, the Green's f & solution to BVP does not exist
W=subs(W,x,z);
G1=simplify(y2*subs(y1,x,z)/W) % a<= x < z
G2=simplify(y1*subs(y2,x,z)/W) % z< x <= b
I1=int(G1*subs(f,x,z),z,a,x);I2=int(G2*subs(f,x,z),z,x,b);
y=I1+I2; y=simplify(y)
y1=subs(y,x,1)% proof that y(b)=0
plot(linspace(a,b),subs(y,x,linspace(a,b)),'linewidth',2.5);hold on;
xlabel('x');ylabel( 'y(x)'); grid on;
else
disp(' The solution to BVP does not exist, Wronskian = 0'); end
```

Problem 4.14 Make a brief review on different (and relatively more recent) engineering applications of Green's Functions (GF) without delving into their mathematical descriptions.

Answer

GF of multifield materials (such as Piezoelectric and Ferroelectric Solids) are already well known topics in various areas of engineering. A Piezoelectric crystal is an anisotropic solid that produces dielectric polarization when deformed elastically, or vice versa. This interaction has many applications, such as in electromechanical sensors and transducers, crystal microphones, stabilized oscillators, and surface acoustic wave devices [11]. On the other hand, Ferroelectricity is defined by reversible polarization through application of an electric field. There are many ceramic, inorganic, polymeric and hybrid materials used for ferroelectricity.

Normally GF is understood to be a mathematical technique that is used for solving operator equations. An operator represents a process of measurement.

The GF corresponding to an operator is the inverse of that operator and gives the response of the material to the measurement process represented by that operator.

For example, A Green's function (GF) based method is the natural choice for modeling many experiments on 2D materials, such as Scanning tunneling microscopy (STM) which is a scanning probe microscopy (SPM) technique that uses raster scanning and tunneling electrons to map out the nanoscale topography and electronic properties of conductive surfaces [12]. The GF contains all the information about the material as modeled by the corresponding operator equation. At least for a 2D material, the GF can be measured directly by SPM. The measurability of the GF is particularly useful because, in principle, it can be used for modeling other related characteristics of that material. This is an interesting example of an apparently mathematical artefact becoming a tool for experimental characterization of 2D materials.

Cell migration plays a particular important role in the initiation and progression of many physical processes and pathological conditions such as tumor invasion and metastasis. Three-dimensional traction force microscopy (TFM) of high resolution and high accuracy is being developed in an effort to unveil the underlying mechanical process of cell migration in a vivo-like environment. Linear elasticity-based traction force microscopy (LETM) relies on the Green's function (that relates traction forces to matrix deformation), of which the inherent boundary conditions and geometry of the matrix can affect the result [13].

Distributed Point Source Method (DPSM) is a modeling technique which is based on the concept of Green's function [14]. First, a collection of source and target points are distributed over the solution domain. Then, the effects from all source points are superimposed at the location of every individual target point. Therefore, a successful implementation of DPSM entails an effective evaluation of Green's function between many pairs of source and target points. For homogeneous and isotropic media, the Green's function is available as a closed-form analytical expression. But for anisotropic solids, the evaluation of Green's function needs to be done numerically. Applications such as defect detection in composite materials require anisotropic analysis. The DPSM is used for ultrasonic field modeling in anisotropic materials.

Problem 4.15 Can one solve nonlinear BVPs via Green's function method?

Answer
The superposition principle is used in the derivation of Green's Function, thus it is applicable only to linear equations.

Nevertheless, the method can be extended to nonlinear equations via the backward and forward propagators and by using the short time expansion of the solution in terms of the nonlinear Green's function [15–18].

4.1 Exercises

Fill in the blanks.

1. The advantage of the Green's function is that it shows how the
   .... influences the BVP.
2. For the BVP,

$$u'' + P(x)u' + Q(x)u = f(x), \quad a < x < b$$
$$A_1 u(a) - B_1 u'(a) = 0$$
$$A_2 u(b) + B_2 u'(b) = 0$$

   besides the possibility that $P(x)$, $Q(x)$ might not be continuous,
   .... is a possible cause of a failure.
3. In a second order BVP, a Green's function is constructed out of
   .... of the homogeneous equation.
4. The Green's function can be thought as the .... to a unit impulse at
   $z = x$.
5. The Green's function is related to the BVP in a similar same way that the
   .... of a square matrix A is related to the linear algebraic system $y = Ax$.
   True, or False?
6. The Green's function method of solving BVPs is applicable only to linear
   systems?
7. The Green's function $G(x, z)$ associated with the inhomogeneous equation
   $L[y] = f(x)$ satisfies the differential equation: $L[G(x, z)] = \delta(x - z)$, where L is
   the differential operator, and the right hand side of the equation represents Dirac
   delta function.
8. Once the Green's function for a differential operator and certain homogeneous
   boundary conditions are known, one can write the solution for any inhomogeneity as a convolution of the Green's solution and the nonhomogeneous part
   of the ODE.
9. If we know the Green's function, we can solve the nonhomogeneous differential equation.
10. The Green's function satisfies the boundary conditions.
11. The Green's function is symmetric in its arguments, $G(x, z) = G(z, x)$.
12. The Green's function can be obtained from transform methods or as an
    eigenfunction expansion.
13. The Green's function of the BVP, $y'' + y - 0$, $y(0) = y(\pi/2) = 0$ is given by
    the expression,

$$G(x, z) = \begin{cases} -\cos(z)\sin(x), & 0 \le x \le z \\ -\sin(z)\cos(x), & z \le x \le \pi/2 \end{cases}$$

Hence, solve the nonhomogeneous BVP if the forcing function is,

$$(a)\, f(x) = x^2 + 1, \qquad (b)\, f(x) = 3x.$$

Answer:

(a) $x^2 - 1 + \cos(x) - (\pi^2/4 - 1)\sin(x)$
(b) $3x - (3\pi/2)\sin(x)$

14. Determine the Green's functions for the BVP

$$y'' - \frac{y'}{x} = f(x), y(0) = y(1) = 0$$

Answer:

$$G(x,z) = \left\{ \begin{array}{ll} (z^2 - 1)x^2/(2x), & x \le z \\ z(x^2 - 1)/2, & x \ge z \end{array} \right\}$$

15. Use the boundary value Green's function to solve the problem, $y'' - x^2 = 0$ subject to following boundary conditions.

$$(a)\, y(0) = y(1) = 0; \quad (b)\, y(0) = y'(1) = 0$$

Answer: $(a)\, y(x) = (x^4 - x)/12, \quad (b)\, y(x) = (x^4 - 4x)/12$
16. Consider the boundary value problem,

$$y'' - \cos(x) = 0, \quad y'(0) = y(\pi/2) = 0$$

(a) Solve by direct integration.
(b) Solve the BVP using the Green's function.
    Answer: $y(x) = -\cos(x)$

17. Why should one choose to solve a two point BVP via Green's Function method if a symbolic solver (such as dsolve) is available?

References

1. Myint-UT, Debnath L (2007) Green's functions and boundary-value problems. In: Linear partial differential equations for scientists and engineers. Birkhäuser Boston
2. Teterina OA (2013) The green's function method for solutions of fourth order nonlinear boundary value problem. Master's thesis, University of Tennessee
3. Morrison SM (2007) Application of the green's function for solutions of third order nonlinear boundary value problems. Master's thesis, University of Tennessee

4. Qin Q (2014) Green's functions of magneto-electro-elastic plate under thermal loading. Encyclopedia of thermal stresses. Springer Netherlands, Netherlands, pp 2096–2103
5. Stakgold I, Holst M (ed) (2011) Green's functions and boundary value problems, 3rd edn. Wiley, London
6. Duffy DG (2015) Green's functions with applications. CRC Press, Inc
7. Cole KD, Beck JV, Haji-Sheikh A, Litkouhi B (2011) Heat conduction using green's functions, 2nd edn. Taylor & Francis, Boca Raton, FL
8. Hartmann F (2013) Green's functions and finite elements. Springer-Verlag, Heidelberg, Germany
9. Bender CM, Orszag SA (1999) Advanced mathematical methods for scientists and engineers: asymptotic methods and perturbation theory. Springer, Berlin. Section 1.5
10. Roach GF (1982) Green's functions, 2nd edn. Cambridge University Press
11. Wang C-Y (1996) Green's functions and general formalism for 2d piezoelectricity. Appl Math Lett 9(4):1–7
12. Tewary VK, Quardokus RC, DelRio FW (2016) Green's function modeling of response of two-dimensional materials to point probes for scanning probe microscopy. Phys Lett A 380 (20):1750–1756
13. Du Y, Herath SCB, Wang QG, Asada H, Chen PCY (2018) Determination of green's function for three-dimensional traction force reconstruction based on geometry and boundary conditions of cell culture matrices. Acta Biomater 67:215–228
14. Fooladi S, Kundu T (2017) Ultrasonic field modeling in anisotropic materials by distributed point source method. Ultrasonics 78:115–124
15. Frasca M (2008) Green functions and nonlinear systems: short time expansion. Int J Mod Phys A 23(2):299–308
16. Frasca M (2007) Green functions and nonlinear systems. Mod Phys Lett A 22(18):1293–1299
17. Rey O (1990) The role of the green's function in a non-linear elliptic equation involving the critical sobolev exponent. J Funct Anal 89:1–52
18. Berger MS (1977) Nonlinearity and functional analysis. Lectures on nonlinear problems in mathematical analysis. Pure and applied mathematics. Academic Press [Harcourt Brace Jovanovich, Publishers], New York-London

Chapter 5
The Shooting Method for the Solution of One-Dimensional BVPs

The shooting method transforms the boundary-value ODE into a system of first order ODEs, which can be solved by the initial-value methods. The boundary conditions on one side of the given interval is used as initial conditions. The additional initial conditions needed are assumed, the initial-value problem is solved, and the solution at the other boundary is compared to the known boundary conditions on that boundary. The initial conditions guessed on one boundary is varied iteratively until the boundary conditions on the other boundary are satisfied.

The shooting method can be applied easily to the general nonlinear second-order boundary-value ODE with known function (Dirichlet or Neumann) boundary conditions, where the nonlinear terms pose no special problems. This is an advantage of using shooting method in contrast to Finite Difference (equilibrium or replacement) method, in which solution of finite difference equations is required [1].

Higher-order BVPs can be solved by the shooting method by replacing each higher-order ODE by a system of first-order ODEs.

Superposition and extrapolation techniques may improve the accuracy of shooting method.õ

Systems of coupled second-order BVPs can be solved by the shooting method by replacing each second-order ODE by two first-order ODEs and solving the coupled systems of first-order ODEs. Note, however that solving coupled systems of second-order BVPs can be difficult.

When the shooting method is used for the solution of a BVP, consistency, order, stability, and convergence of the initial-value ODE solution method must be considered.

Single Shooting method is simple to implement. It is more effective when the interval [a, b] is short. A large interval requires a large number of iterations. Single shooting method can be unstable for some problems, especially for those of highly nonlinear or unstable ODEs. Shooting method requires a good initial guess for the IVP.

Multiple shooting method can handle most of these problems by reducing the growth of the solutions of the IVPs that must be solved, using partitioning the

© Springer Nature Switzerland AG 2019
A. Ü. Keskin, *Boundary Value Problems for Engineers*,
https://doi.org/10.1007/978-3-030-21080-9_5

interval into a number of subintervals, and then simultaneously adjusting the "initial" data to satisfy the boundary conditions and appropriate continuity conditions [2].

BVPs constitute one of the most difficult classes of problems to solve on a computer. Convergence is by no means assured, good initial guesses must be available, and considerable trial and error, as well as large amounts of machine time, are usually required [3].

It should be kept in mind that such BVPs are considered in steady state for closed solution domains and they are not time dependent (transient) problems.

5.1 Linear Single Shooting Method

Problem 5.1.1 Consider two initial value problems (IVPs) with unique solutions:

$$y'' = p(x)y' + q(x)y + r(x), \quad x \in [a, b], \quad y(a) = \alpha, \quad y'(a) = 0 \qquad (5.1)$$

$$y'' = p(x)y' + q(x)y, \quad x \in [a, b], \quad y(a) = 0, \quad y'(a) = 1 \qquad (5.2)$$

Let $y_1(x)$ and $y_2(x)$ denote the solutions of each IVPs, respectively, and let $y_2(b) \neq 0$.
Show that,

$$w(x) \cong y(x) = y_1(x) + k \cdot y_2(x), \quad k = \frac{\beta - y_1(b)}{y_2(b)} \qquad (5.3)$$

is the unique solution to the (linear) BVP,

$$y'' = p(x)y' + q(x)y + r(x), \quad x \in [a, b], \quad y(a) = \alpha, \quad y(b) = \beta.$$

Solution
Since $y_1(x)$ and $y_2(x)$ are solutions to (5.1) and (5.2), on the other hand from (5.2) we have

$$y_2'' - p(x)y_2' - q(x)y_2 = 0$$

From (5.3),

$$y(a) = y_1(a) + \frac{\beta - y_1(b)}{y_2(b)} \cdot y_2(a)$$

But, $y_2(a) = 0, y_1(a) = \alpha$,

$$y(a) = y_1(a) + \frac{\beta - y_1(b)}{y_2(b)} \cdot (0) = \alpha$$

also we have

$$y(b) = y_1(b) + \frac{\beta - y_1(b)}{y_2(b)} \cdot y_2(b) = y_1(b) + \beta - y_1(b) = \beta$$

$$\therefore \text{Q.E.D.}$$

Problem 5.1.2 Consider the BVP of Euler type,

$$y'' + \frac{1}{x}y' - \frac{1}{x^2}y = 0, \quad y(1) = 2, \quad y(3) = 5$$

Solve the BVP using linear shooting method.

Solution
The solution will be of the form

$$y(x) = u(x) + \frac{5 - u(3)}{v(3)} \cdot v(x)$$

$$u(x): \quad \rightarrow \quad u'' + \frac{1}{x}u' - \frac{1}{x^2}u = 0, \quad u(1) = 2, \quad u'(1) = 0$$

$$v(x): \quad \rightarrow \quad v'' + \frac{1}{x}v' - \frac{1}{x^2}v = 0, \quad v(1) = 0, \quad v'(1) = 1$$

Since these are Euler type ODEs, their solutions are of the following form,

$$u(x) = Ax + \frac{1}{x}B, \quad v(x) = Cx + \frac{1}{x}D$$

Substituting these forms into the initial conditions, we obtain

$$u(1) = Ax + \frac{B}{x} = 2, \quad u'(1) = Ax + \frac{B}{x} = 0$$

$$v(1) = Cx + \frac{D}{x} = 0, \quad v'(1) = Cx + \frac{D}{x} = 1$$

$$u(x) = x + \frac{1}{x}, \quad v(x) = \frac{x^2 - 1}{2x}$$

$$u(3) = 3 + \frac{1}{3} = \frac{10}{3}, \quad v(3) = \frac{3^2 - 1}{8} = \frac{4}{3}$$

$$y(x) = \left(x + \frac{1}{x}\right) + \frac{5 - (10/3)}{(4/3)}\left(\frac{x^2 - 1}{2x}\right)$$

$$y(x) = \frac{1}{8x}(13x^2 + 3)$$

Problem 5.1.3

(a) The BVP $y'' = -y' - 2y + 10x^3$, for $0 \leq x \leq 1$ with $y(0) = y(1) = 0$ has a unique solution. Use linear shooting method to approximate the solution with a uniform step size of $h = 0.1$. (Use the Runge-Kutta[1] method of order four when differentiation is necessary). Tabulate x, y_1, y_2 and w (approximate solution), then plot them in a single figure.

Do not use any MATLAB external "function" functions, but only anonymous functions. (This will deliberately make the code seem lengthy, however, we will refine these in the following problem solutions) (shooting1.m).

(b) Compute the absolute error of approximation for conditions specified above.
(c) Run the code for different step size parameters and display the error of approximation as an Euclidean norm (= norm two) log-log plot. What would be the order of the approximation error if forward Euler's method or Heun's method were implemented in the code?
(d) Convert the m-file of part (b) to include external (function) calls, along with anonymous functions (shooting11.m), (rk4.m).
(e) Discuss disadvantages of linear shooting method.

Solution

(a) Boundary conditions indicate that $y(a) = \alpha = y(b) = \beta = 0$. Linear Shooting method will require approximation of the solutions to the following IVPs;

$$y_1'' = -y_1' - 2y_1 + 10x^3, \quad x \in [0 \quad 1], \quad y_1(0) = y_1'(0) = 0$$

$$y_2'' = -y_2' - 2y_2, \quad x \in [0 \quad 1], \quad y_2(0) = 0, \quad y_2'(0) = 1$$

---

[1]For refreshing information on the structure of RK4, Euler and Heun's methods and their MATLAB applications, readers may refer to textbooks on numerical methods, such as [4–7].

In the MATLAB m-file (shooting1.m) given below, we use RK4 method to solve both IVPs, then form the combination as given by (5.3),

$$y(x) = y_1(x) + ky_2(x)$$

where

$$k = \frac{\beta - y_1(b)}{y_2(b)}, \quad b = 1, \quad \beta = 0$$

A tabulation of x, y_1, y_2 and (approximate solution) w is listed below. These functions are displayed in Fig. 5.1 as functions of x.

x	y1	y2	y(comp.)
0	0	0	0
0.1000	0.0000	0.0948	-0.0873
0.2000	0.0002	0.1789	-0.1644
0.3000	0.0011	0.2515	-0.2303
0.4000	0.0048	0.3124	-0.2827
0.5000	0.0142	0.3616	-0.3185
0.6000	0.0347	0.3993	-0.3327
0.7000	0.0735	0.4257	-0.3182
0.8000	0.1402	0.4416	-0.2661
0.9000	0.2471	0.4476	-0.1647
1.0000	0.4090	0.4445	0

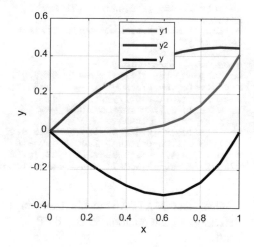

Fig. 5.1 The solution and sub-solutions for the BVP, $y'' = -y' - 2y + 10x^3$, for $0 \le x \le 1$ with $y(0) = y(1) = 0$, using linear shooting method

```
% shooting1.m y"=-y'-2y+10x^3, [a=0  b=1], y(a)=alpha=0, y(b)=beta=0
clc;clear;close;format short
span=[0 1];bv=[0 0]; alpha=bv(1); beta=bv(2);
h=0.1 %Step size
N=(1/h)+1 %Number of steps
x=linspace(0,1,N);
y=[alpha; 0];% = [y(0) y'(0)]= initial conditions for y1
yd=@(x,y) [y(2);
    -y(2)-2*y(1)+10*x^3];%y(2)=y',y(1)=y
for i=1:N-1
k1(:,i)=yd(x(i),y(:,i));
k2(:,i)=yd(x(i)+.5*h,y(:,i)+k1(:,i)*.5*h);
k3(:,i)=yd(x(i)+.5*h,y(:,i)+k2(:,i)*.5*h);
k4(:,i)=yd(x(i)+h,y(:,i)+k3(:,i)*h);
y(2,i+1)=y(2,i)+1/6*(k1(2,i)+2*k2(2,i)+2*k3(2,i)+k4(2,i))*h;
y(1,i+1)=y(1,i)+1/6*(k1(1,i)+2*k2(1,i)+2*k3(1,i)+k4(1,i))*h;
end
y1=y';
%A=[x' y1(:,1)]
y=[0; 1];% = [y(0) y'(0)]= initial conditions for y2
yd=@(x,y) [y(2);
    -y(2)-2*y(1)];%y(2)=y',y(1)=y
for i=1:N-1
k1(:,i)=yd(x(i),y(:,i));
k2(:,i)=yd(x(i)+.5*h,y(:,i)+k1(:,i)*.5*h);
k3(:,i)=yd(x(i)+.5*h,y(:,i)+k2(:,i)*.5*h);
k4(:,i)=yd(x(i)+h,y(:,i)+k3(:,i)*h);
y(2,i+1)=y(2,i)+1/6*(k1(2,i)+2*k2(2,i)+2*k3(2,i)+k4(2,i))*h;
y(1,i+1)=y(1,i)+1/6*(k1(1,i)+2*k2(1,i)+2*k3(1,i)+k4(1,i))*h;
end
y2=y'; Y1=y1(:,1); Y2=y2(:,1);
% A=[x' Y2]
k=(beta-Y1(end))/Y2(end); w=Y1+k*Y2;
A=[x' Y1 Y2 w]
plot(x,Y1,x,Y2,x,w,'linewidth',2);xlabel('x');ylabel('y');grid;
legend('y1','y2','y');
% Exact solution
syms y(x);format long
eq='D2y=-Dy-2*y+10*x^3'; bc='y(0)=0,y(1)=0';
S=dsolve(eq,bc,x); x = linspace(span(1),span(2),N);
y=eval(vectorize(S));y=y';
[y w abs(y-w)]; norme=norm(abs(y-w))
%error log-log plot
h=1./[0.1 0.05 0.025 0.0125 0.00625 0.003125];
e=[1.3e-5 1.2e-6 1.1e-7 1.0e-8 8.9e-10 7.97e-11];figure;
f=loglog(h,e,'.-','markersize',20);grid;
xlabel('steps');ylabel('norm2 error');toc
```

Note that MATLAB m-file (shooting1.m) provides approximations for the derivative of the solution to the BVP in addition to the solution of the problem itself.

(b) The absolute error of approximation for each step within the given interval of the BVP (excluding the boundary values) for conditions specified in part (b) of this problem are listed below in edited form, along with the exact and approximate solution values.

```
       y(exact)               y(computed)                    error

 -0.087268413110891   -0.087266497371131        0.000001915739760
 -0.164428711374031   -0.164425285633700        0.000003425740331
 -0.230256982478933   -0.230252433662851        0.000004548816082
 -0.282716299117864   -0.282711012966139        0.000005286151725
 -0.318502766226684   -0.318497143158035        0.000005623068649
 -0.332658840971110   -0.332653309618548        0.000005531352562
 -0.318254870868310   -0.318249898874533        0.000004971993778
 -0.266138409267443   -0.266134511071846        0.000003898195596
 -0.164749621532515   -0.164747363018844        0.000002258513671
```

(c) We input different step size (h) values in the code and compute norm2 error,
     $\|y - w\|$, corresponding to each step count which is (almost equal to) the
     inverse of the step size (at the end of the code) and then log-log plot these data.
     Figure 5.2 shows the resulting curve.

Overall order of approximation error in the linear shooting technique depends on
the method of integration. Since the order of approximation error for the forward
Euler's method is unity, $\mathcal{O}(h)$, and Heun's method has a second order approximation error performance, when they are implemented in the linear shooting algorithm they reduce the error performance of the linear shooting algorithm (relative
to RK4-based method) by a factor of 4, and 2, respectively.

(d) Main file (shooting11.m) calls fourth order Runge Kutta functions (rk4.m)
     twice, while each of these external functions call anonymous functions F1 and
     F2. Note that inputs to both function calls, as well as their outputs are different.

Fig. 5.2 Graph of Euclidian
norm of approximation error
for different number of step
counts

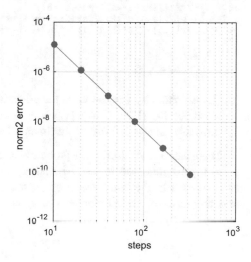

```
%shooting11.m BVP,  y"=-y'-2y+10x^3,  [a=0  b=1],  y(a)=alpha=0, y(b)=beta=0
clc;clear;close;format short;tic
span=[0 1];bv=[0 0];  a=span(1);  b=span(2);  alpha=bv(1); beta=bv(2);
h=0.1 %Step size
ic1=[alpha; 0];% = [y(0) y'(0)]= initial conditions for y1
ic2=[0; 1];    % = [y(0) y'(0)]= initial conditions for y2
F1=@(x,y) [y(2); -y(2)-2*y(1)+10*x^3];  %y(2)=y',y(1)=y
F2=@(x,y) [y(2); -y(2)-2*y(1)];         %y(2)=y',y(1)=y
[x,y1]=rk4(F1,a,b,h,ic1);
[x,y2]=rk4(F2,a,b,h,ic2);
Y1=y1(:,1); Y2=y2(:,1);
% A=[x Y2]
k=(beta-Y1(end))/Y2(end); w=Y1+k*Y2;
A=[x Y1 Y2 w]
plot(x,Y1,x,Y2,x,w,'linewidth',2);xlabel('x');ylabel('y');grid;
legend('y1','y2','y');
% Exact solution
syms  y(x);format long ; eq='D2y=-Dy-2*y+10*x^3'; bc='y(0)=0,y(1)=0';
S=dsolve(eq,bc,x); x = linspace(span(1),span(2),(1/h)+1 );
y=eval(vectorize(S)); y=y'; [y w abs(y-w)]; Euclid_norm=norm(abs(y-w))
%error log-log plot
h=1./[0.1 0.05 0.025 0.0125 0.00625 0.003125];
e=[1.3e-5 1.2e-6 1.1e-7 1.0e-8 8.9e-10 7.97e-11];figure;
f=loglog(h,e,'.-','markersize',20);grid;
xlabel('steps');ylabel('norm2 error');toc


function [x,y]=rk4(F,a,b,h,ic)
%F is string defining an IVP
%a,b are the boundaries of the interval
%h is step size
%ic is the initial conditions vector [y(0) y'(0)]
N=(1/h)+1; %Number of steps
x=linspace(a,b,N);
y=ic;% = [y(0) y'(0)]= initial conditions for y
for i=1:N-1
k1(:,i)=F(x(i),y(:,i));
k2(:,i)=F(x(i)+.5*h,y(:,i)+k1(:,i)*.5*h);
k3(:,i)=F(x(i)+.5*h,y(:,i)+k2(:,i)*.5*h);
k4(:,i)=F(x(i)+h,y(:,i)+k3(:,i)*h);
y(2,i+1)=y(2,i)+1/6*(k1(2,i)+2*k2(2,i)+2*k3(2,i)+k4(2,i))*h;
y(1,i+1)=y(1,i)+1/6*(k1(1,i)+2*k2(1,i)+2*k3(1,i)+k4(1,i))*h;
end
x=x';y=y';
```

(e) Linear shooting method takes the advantages of well-known numerical solution methods for IVPs. However, some IVPs with rapidly growing nature can be unstable although the BVP itself may be well posed and stable.

There can be round-off error problems in the background of this technique [8]. If $y_1(x)$ rapidly increases in $[a \quad b]$, then $y_1(b)$ will be large. If $|\beta|$ is relatively small as compared to the last element of y_1, the k term in Eq. (5.3) turns out to be the negative ratio of the last elements of y_1 and y_2,

$$w_i = y_{1,i} - \left(\frac{y_{1,N}}{y_{2,N}}\right) y_{2,i}$$

This will cause a loss of significant digits because of cancellation.

A remedy to this problem is to employ the linear shooting method in reverse direction,

$$y'' = p(x)y' + q(x)y + r(x), \quad x \in [a,b], \quad y(b) = \beta, \quad y'(b) = 0$$

$$y'' = p(x)y' + q(x)y, \quad x \in [a,b], \quad y(b) = 0, \quad y'(b) = 1$$

If the same problem persists with the revers shooting, then multiple shooting techniques (or others) can be applied.

Problem 5.1.4 (*Linear Single Shooting Method for an homogeneous ODE*)

(a) Use the linear shooting method for the BVP, $y'' = y$, $0 \le x \le 1$, $y(0) = 1$, $y(1) = e^{-1}$, to approximate the (exact) solution $y = e^{-x}$. Plot the approximate solution of this BVP along with its two modes.
(b) Determine the absolute error of approximation at each step.

Use fourth order Runge-Kutta (RK4) method for numerical differentiation of IVPs. Choose the uniform step size, $h = 0.1$ (shooting30.m), (rk4.m).

Solution

(a) Since $y'' = y$, $p = r = 0$, $q = 1$, $a = 0$, $b = 1$, $\alpha = 1$, $\beta = e^{-1}$.

Linear Shooting method will require approximation of the solutions to the following IVPs;

$$y'' = y, \quad 0 \le x \le 1, \quad y(0) = \alpha = 1, \quad y'(0) = 0,$$

$$y'' = y, \quad 0 \le x \le 1, \quad y(0) = 0, \quad y'(0) = 1.$$

In the MATLAB m-file (shooting3.m) given below, we use RK4 method to solve both IVPs, then form the combination as given by (5.3),

$$y(x) = y_1(x) + k y_2(x)$$

where

$$k = \frac{\beta - y_1(b)}{y_2(b)}, \quad b = 1, \quad \beta = e^{-1}$$

Fig. 5.3 The solution and
two sub-solutions (modes) for
the BVP, $y'' = -y$, for
$y(0) = 1, y(1) = e^{-1}$, using
linear shooting method

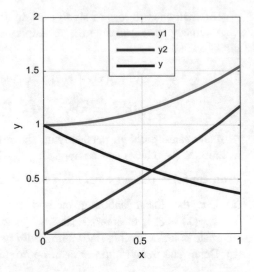

A tabulation of x, y_1, y_2 and (approximate solution) w is listed below. These
functions are displayed in Fig. 5.3 as functions of x. Note that MATLAB m-file
(shooting30.m) provides approximations for the derivative of the solution to the
BVP in addition to the solution of the problem itself.

x	y1	y2	w
0	1.0000	0	1.0000
0.1000	1.0050	0.1002	0.9048
0.2000	1.0201	0.2013	0.8187
0.3000	1.0453	0.3045	0.7408
0.4000	1.0811	0.4108	0.6703
0.5000	1.1276	0.5211	0.6065
0.6000	1.1855	0.6367	0.5488
0.7000	1.2552	0.7586	0.4966
0.8000	1.3374	0.8881	0.4493
0.9000	1.4331	1.0265	0.4066
1.0000	1.5431	1.1752	0.3679

(b) The absolute error of approximation for each step within the given interval of
the BVP, $0 \leq x \leq 1$, for conditions specified in part (a) of this problem are listed
below in edited form, along with the exact and approximate solution values.

y(exact)	w=y(approximate)	abs.error
1.000000000000000	1.000000000000000	0
0.904837418035960	0.904837471596625	0.000000053560666
0.818730753077982	0.818730844315230	0.000000091237248
0.740818220681718	0.740818335651127	0.000000114969409
0.670320046035639	0.670320172444189	0.000000126408550
0.606530659712633	0.606530786661126	0.000000126948493
0.548811636094027	0.548811753846256	0.000000117752230
0.496585303791410	0.496585403566562	0.000000099775153
0.449328964117222	0.449329037902317	0.000000073785095
0.406569659740599	0.406569700120103	0.000000040379504
0.367879441171442	0.367879441171442	0

Elapsed time is 0.332150 seconds.

```
%shooting30.m BVP,  y"=y,  [a=0  b=1],  y(a)=alpha=1, y(b)=beta=exp(-1)
clc;clear;close;format short;tic
h=0.1 %Step size
span=[0 1];bv=[1 exp(-1)]; a=span(1); b=span(2); alpha=bv(1); beta=bv(2);
ic1=[alpha; 0];% = [y(0) y'(0)]= initial conditions for y1
ic2=[0; 1];    % = [y(0) y'(0)]= initial conditions for y2
F1=@(x,y) [y(2); y(1)];   %y(2)=y',y(1)=y
F2=@(x,y) [y(2); y(1)];   %y(2)=y',y(1)=y
[x,y1]=rk4(F1,a,b,h,ic1);
[x,y2]=rk4(F2,a,b,h,ic2);
Y1=y1(:,1); Y2=y2(:,1);
% A=[x Y2]
k=(beta-Y1(end))/Y2(end); w=Y1+k*Y2; A=[x Y1 Y2 w]
plot(x,Y1,x,Y2,x,w,'linewidth',2);xlabel('x');ylabel('y');grid;
legend('y1','y2','y');
% absolute error
format long; [exp(-x)   w   abs(w-exp(-x))]
toc
```

Problem 5.1.5 Use linear shooting method to solve the BVP,

$$y'' = -y, \quad x \in [0, \pi], \quad y(0) - 0, \quad y(\pi) = 1$$

Use RK4 method for the solution of IVPs, with equal step size of h = 0.1 (shooting5.m), (rk4.m).

Solution
Before attempting to apply any numerical solution, it is necessary that we check the structure of a given BVP. In this case, general solution of $y'' = -y$ is known to be $y = A\cos(x) + B\sin(x)$. We calculate the coefficients of the solution from given boundary conditions,

$$y(0) = A\cos(0) + B\sin(0) = 0 \rightarrow A = 0$$
$$y(\pi) = A\cos(0) + B\sin(0) = 1 \rightarrow B = ?$$

We conclude that given BVP has no solution.

However, it is intuitive to see what comes out of an application of a shooting method, in such a case. Following m-file (shooting5.m) output list demonstrates that one of the modes of "approximate solution" turns out to be identically zero;

x	y1	y2	w
0	0	0	0
0.3142	0	0.0998	0.1186
0.6283	0	0.1987	0.2361
0.9425	0	0.2955	0.3512
1.2566	0	0.3894	0.4628
1.5708	0	0.4794	0.5697
1.8850	0	0.5646	0.6710
2.1991	0	0.6442	0.7656
2.5133	0	0.7174	0.8525
2.8274	0	0.7833	0.9309
3.1416	0	0.8415	1.0000

```
%shooting5.m BVP,   y" = -y ,    y(a)=alpha, y(b)=beta,
clc;clear;close;format short;tic
h=0.1 %Step size
a=0; b=1*pi; span=[a b];bv=[0 1];
alpha=bv(1); beta=bv(2);
ic1=[alpha; 0];            % = [y(0) y'(0)]= initial conditions for y1
ic2=[0; 1];                % = [y0)  y'(0)]= initial conditions for y2
F1=@(x,y) [y(2); -y(1)];   %y(2)=y',y(1)=y
F2=@(x,y) [y(2); -y(1)];   %y(2)=y',y(1)=y
[x,y1]=rk4(F1,a,b,h,ic1); [x,y2]=rk4(F2,a,b,h,ic2);
Y1=y1(:,1); Y2=y2(:,1); k=(beta-Y1(end))/Y2(end); w=Y1+k*Y2;
A=[x Y1 Y2 w]
```

Problem 5.1.6 Given the BVP,

$$y'' - 10^4 y = 0 \quad y(0) = 1, \quad y(10) = B$$

Apply shooting method and show that the solution of a BVP can be insensitive to the changes in the boundary conditions, but IVPs involved in the method can be sensitive to changes in the initial values.

Solution
Shooting from left to right involves the IVP with initial values $y(0) = 1$, $y'(0) = p$ with the analytical solution;

$$y = A \cosh(\omega x) + B \sinh(\omega x)$$

$$y(0) = 1 = A$$

$$y'(0) = p = -\omega \sinh(\omega x) + B\omega \cosh(\omega x)|_{x=0}$$

$$p = B\omega \rightarrow B = p/\omega = p/10^2 = 10^{-2}p$$

$$y(x,p) = \cosh(100x) + 10^{-2}p \sinh(100x)$$

The slope for which the boundary condition is satisfied at $x = 10$

$$B = \cosh(100 \times 10) + 10^{-2} \times p \times \sinh(100 \times 10)$$

$$p = \frac{B - \cosh(1000)}{10^{-2}\sinh(1000)} = \frac{100[B - \cosh(1000)]}{\sinh(1000)}$$

We substitute this value of p into the general solution of IVP,

$$y = \cosh(100x) + \frac{B - \cosh(1000)}{\sinh(1000)} \sinh(100x)$$

take the derivative with respect to B,

$$\frac{dy}{dB} = \frac{\sinh(100x)}{\sinh(1000)}$$

Note that absolute value of this expression is

$$\left|\frac{dy}{dB}\right| = \left|\frac{\sinh(100x)}{\sinh(1000)}\right| \leq 1$$

Therefore, the solution of the IVP is more sensitive to changes in the initial slope p than the sensitivity of the BVP solution to changes in the boundary value $y(10) = B$.

Hence, if the IVP has unstable solution, then a shooting method may fail.

Problem 5.1.7 A pressure vessel is generally considered to be thick walled if its thickness is about 1/10th of the inner radius of the vessel or greater. For a thick metallic pressure vessel of inner radius $r = r_i$ and outer radius $r = r_o$, the differential equation for the radial displacement $u(r)$ of a point along the thickness is given by

$$u'' + r^{-1}u' - r^{-2}u = 0 \qquad (5.4)$$

with the following boundary conditions (if only the internal pressure P_i can be taken into account),

$$u(r_i) = P_i \frac{r_i}{E}\left(\frac{r_i^2 + r_o^2}{r_o^2 - r_i^2} + \mu\right)$$

$$u(r_o) = P_i \frac{2}{E}\frac{r_i^2 r_o}{r_o^2 - r_i^2}$$

where E, μ are the Young's modulus of elasticity and Poisson's ratio, respectively. Use Linear Shooting method to solve the BVP and plot the normalized radial displacement at any point of the wall, assuming that the inner radius is 10 cm, the outer radius 11 cm, and $\mu = 0.3$. Compute the absolute error at each step. Consider 10 equal segments between the two boundaries and use RK4 method to solve IVPs (shooting4.m).

Solution
Exact result of solution for a thick walled open ended cylinder is given by [9],

$$u(r) = C_1 r + C_2 r^{-1} \tag{5.5}$$

$$C_1 = \left(\frac{1-\mu}{E}\right)\frac{r_i^2 P_i - r_o^2 P_o}{r_o^2 - r_i^2}, \quad C_2 = \left(\frac{1+\mu}{E}\right)\frac{r_i^2 r_o^2 (P_i - P_o)}{r_o^2 - r_i^2}$$

$$P_o = 0,$$

$$u(r) = \left(\frac{1-\mu}{E}\right)\frac{r_i^2 P_i}{r_o^2 - r_i^2} r + \left(\frac{1+\mu}{E}\right)\frac{P_i r_i^2 r_o^2}{r_o^2 - r_i^2} r^{-1}$$

We introduce the normalized radial displacement as

$$u_n(r) = \frac{u(r)}{K} = (c_1 r + c_2 r^{-1}), \quad K = \left(\frac{P_i}{E}\right) \tag{5.6}$$

$$c_1 = \frac{r_i^2(1-\mu)}{r_o^2 - r_i^2}, \quad c_2 = \frac{r_i^2 r_o^2 (1+\mu)}{r_o^2 - r_i^2} \tag{5.7}$$

With the normalized boundary conditions and given numerical values of the parameters,

$$u_n(r_i) = r_i\left(\frac{r_i^2 + r_o^2}{r_o^2 - r_i^2} + \mu\right) = 108.2380, \quad u_n(r_o) = 2\frac{r_i^2 r_o}{r_o^2 - r_i^2} = 104.7619 \tag{5.8}$$

Therefore Eqs. (5.4) and (5.8) constitute the BVP to be solved.

Fig. 5.4 Approximate solution and its larger mode for normalized displacement. Second mode, (u2) does not appear because of its much lower values. The units of the two axes are not indicated, since they are normalized values by a factor K given in Eq. (5.6)

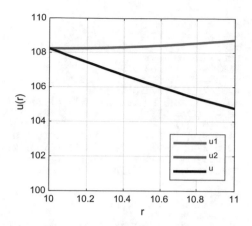

We re-write (5.4) as

$$u'' = -r^{-1}u' + r^{-2}u$$

Using linear shooting method we obtain an approximate solution of the BVP for normalized displacement in the vessel wall. This solution curve as well as one of its modes of the solution are displayed in Fig. 5.4. A resulting edited output tabulation and the listing of the m-file (shooting4.m) is also given below.

```
          c1 = 3.3333,     c2 =   749.0476

            u(exact)           w(approximate)          abs.error
  1.0e+02 *
         1.082380952380952   1.082380952380952                      0
         1.078297972654408   1.078297972657118   0.000000000002710
         1.074360410830999   1.074360410835689   0.000000000004690
         1.070564031437818   1.070564031443813   0.000000000005995
         1.066904761904762   1.066904761911438   0.000000000006676
         1.063378684807256   1.063378684814035   0.000000000006778
         1.059982030548068   1.059982030554414   0.000000000006345
         1.056711170449488   1.056711170454904   0.000000000005416
         1.053562610229277   1.053562610233303   0.000000000004026
         1.050532983835736   1.050532983837947   0.000000000002211
         1.047619047619048   1.047619047619048                      0
Elapsed time is 0.178798 seconds.
```

```
%shooting4.m BVP,   u" = -r^(-1)*u' + r^(-2)*u ,    y(a)=alpha, y(b)=beta
clc; clear; close; format short; tic
h=0.1 %Step size
ri=10; ro=11; mu=0.3;
urin=ri*((ri.^2+ro.^2)./(ro.^2-ri.^2)+mu) % normalized bv1
uron=2*ri^2*ro/(ro^2-ri^2)                 % normalized bv2
span=[ri ro]; bv=[urin uron]; alpha=bv(1); beta=bv(2);
ic1=[alpha; 0];           % = [u(0) u'(0)]= initial conditions for y1
ic2=[0; 1];               % = [u(0) u'(0)]= initial conditions for y2
F1=@(r,u) [u(2); -r^(-1)*u(2)+r^(-2)*u(1)];  %u(2)=u',u(1)=u
F2=@(r,u) [u(2); -r^(-1)*u(2)+r^(-2)*u(1)];  %u(2)=u',u(1)=u
[r,u1]=rk4(F1,ri,ro,h,ic1);   [r,u2]=rk4(F2,ri,ro,h,ic2);
U1=u1(:,1); U2=u2(:,1); k=(beta-U1(end))/U2(end); w=U1+k*U2;
%A=[r U1 U2 w]
plot(r,U1,r,U2,r,w,'linewidth',2);xlabel('r');ylabel('u(r)');grid;
legend('u1','u2','u');
% absolute error
c1=ri^2*(1-mu)/(ro.^2-ri.^2), c2=ri^2*ro^2*(1+mu)/(ro.^2-ri.^2)
format long
[c1*r+c2./r   w   abs(w-(c1*r+c2./r))]
toc
```

Problem 5.1.8

(a) Use shooting method along with MATLAB dsolve function[2] to solve the BVP,

$$y'' - 2y = 8x(3 - x); \quad y(0) = y(3) = 0.$$

Convert the BVP into IVP by introducing an initial derivative value at the point of initial value and then find a derivative value at $y'(0)$ so that solution of the IVP gives the value at $y(3) = 0$. Assume $y'(0) = [1 - 10]$ with unit increments (run in the positive direction). Observe pattern (slope) change in these figures. If there is no change, then repeat the same procedure for $y'(0) = [(-1) - (-10)]$ with unit increments (run in negative direction). If a change is observed, then concentrate on the interval in which the change occurs.

(b) A more efficient shooting strategy is possible for linear ODEs in which the trajectory of the perfect shot is linearly related to the results of our two erroneous shots. Consequently, linear interpolation can be employed to arrive at the required trajectory [10].

As a modification of above procedure, shoot at derivative values of $y'(0) = p_0$, and $y'(0) = p_1$ and determine the values of the function at the boundary,

---

[2]It is particularly advantageous to employ dsolve function here, since the analytical solution obtained using this function is useful in assessing the accuracy of the solutions obtained with the (approximate) numerical methods.

$y(3) = q_0, y(3) = q_1$. Then estimate the zero crossing point using the following equation,

$$p = p_0 + \frac{p_1 - p_0}{q_1 - q_0}(q - q_0)$$

Solution

(a) We convert the BVP into IVP by introducing an initial derivative value at the point of initial value. Here, we must find a derivative value at $y'(0)$ so that solution of the IVP gives the value at $y(3) = 0$.

Figure 5.5 displays the actual solution curve for the given BVP (We would not actually know this solution, but our objective here is to show how we can approach this solution using the shooting method).

Assume $y'(0) = [1 - 10]$ with unit increments (run in the positive direction). Observe pattern (slope) change in these figures.

If there is no change, then repeat the same procedure for $y'(0) = [(-1) - (-10)]$ with unit increments (run in negative direction).

If a change is observed, then concentrate on the interval in which the change occurs.

In this problem, no change can be observed in positive direction. However, a change of slope sign is apparent in moving between $y'(0) = -6$, and $y'(0) = -7$ (Fig. 5.6a, b, c). Therefore, we test the initial derivative values within this interval. A test value $y'(0) = -6.5$ shows that we shoot almost the right value for the initial derivative at time zero, so that $y(3) = 0.0840$ (Fig. 5.6d). A fine tuning estimate results in $y'(0) = -6.5034$ with $y(3) = 0.0004$.

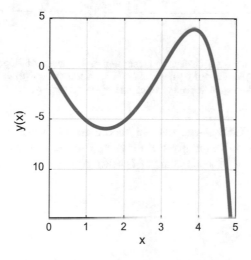

Fig. 5.5 Actual solution curve of the BVP, $y'' - 2y = 8x(9 - x)$; $y(0) = y(3) = 0$

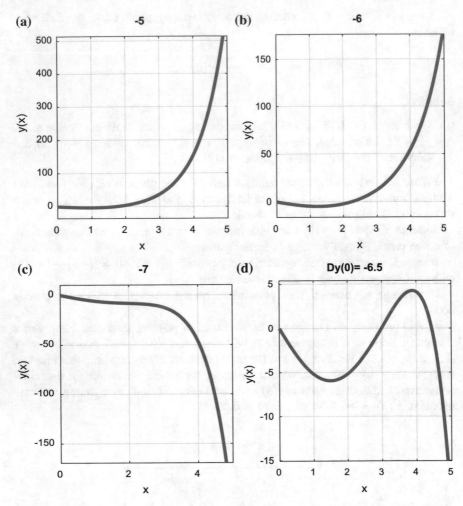

Fig. 5.6 Observing the slope change and fine tuning. A change of slope sign is apparent in moving between $y'(0) = -6$, and $y'(0) = -7$ (**a, b, c**). A test value $y'(0) = -6.5$ shows the right value for the initial derivative at time zero, so that $y(3) = 0.0840$ (**d**)

Note that MATLAB m-file (dsolve_shooting.m) consists of three stages. The first part is used to solve the given BVP using dsolve. The second part is a search stage for the interval of slope change. The last part is a fine tuning stage.

```
% dsolve_shooting1.m,   Shooting method BVP, y"-2y=8x(3-x), y(0)=y(3)=0.
 clc;clear; syms  y(x) k
% Actual BVP solution
% eq='D2y-2*y=8*x*(3-x)'; %ODE
% ic1='y(0)=0';          % ic1
% ic2='y(3)=0';          % ic2
% S=dsolve(eq,ic1,ic2,'x');    pretty(S)
% h=ezplot(S,[0 5]); set(h,'linewidth',2.5);
% title('');grid;ylabel('y(x)');

% search for proper IVP y'(0)  to make y(3)=0 --------------
% for j=-10:-1
% eq='D2y-2*y=8*x*(3-x)'; %ODE
% ic1='y(0)=0';          % ic1
% ic2='Dy(0)=k'; ic2=subs(ic2,k,j);       % ic2
% S=dsolve(eq,ic1,ic2,'x');       %Solution
% h=ezplot(S,[0 5]); set(h,'linewidth',2.5);
% title(num2str(j));grid; ylabel('y(x)');figure;
% end
%fine tuning-----------------------------------------------
eq='D2y-2*y=8*x*(3-x)'; ic1='y(0)=0';     ic2='Dy(0)=-6.5';
S=dsolve(eq,ic1,ic2,'x'); h=ezplot(S,[0 5]); set(h,'linewidth',2.5);
title('Dy(0)= -6.5'); grid; ylabel('y(x)');
```

(b) We shoot at derivative values of $y'(0) = p_0 = -1$, and $y'(0) = p_1 = -10$ and
     determine   the   values   of   the   function   at   the   boundary,
     $y(3) = q_0 = 135.3796, y(3) = q_1 = -86.0131$. Then we estimate the zero
     crossing point,

$$p = p_0 + \frac{p_1 - p_0}{q_1 - q_0}(q - q_0) = -1 + \frac{-10 - (-1)}{-86.0131 - 135.3796}(0 - 135.3796)$$
$$= -6.5034$$

Note that this value is identical with the result we found earlier, but takes less
effort to compute.

```
% dsolve_shooting2.m,    shooting method BVP , y"-2y=8x(3-x), y(0)=y(3)=0.
clc;clear; syms  y(x)
eq='D2y-2*y=8*x*(3-x)'; %ODE
ic1='y(0)=0';          % ic1
ic2='Dy(0)=-6.5034';       % ic2
S=dsolve(eq,ic1,ic2,'x')       %Solution
S=vpa(subs(S,x,3))
h=ezplot(S,[0 5]); set(h,'linewidth',2.5);
grid;ylabel('y(x)');
p0=-1
q0=135.3796
p1=-10
q1=-86.0131
q=0
p=p0+(q-q0)*(p1-p0)/(q1-q0)
```

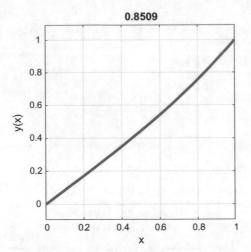

Fig. 5.7 Solution curve for the BVP, $y'' - y = 0$, $0 \leq x \leq 1$, $y(0) = 0, y(1) = 1$. Best initial slope value is found to be $y'(0) = 0.8509$

Problem 5.1.9 Find the value of initial slope, $y'(0)$, which solves the BVP,

$$y'' - y = 0, \quad 0 \leq x \leq 1, \quad y(0) = 0, \quad y(1) = 1$$

Plot the optimal solution. (Hint: It is given that $0.8505 \leq y'(0) \leq 0.8510$) (shootinit.m).

Solution
Application of MATLAB script (shootinit.m) to the problem reveals that best initial slope is $y'(0) = 0.8509$. Minimum absolute error computed at $y(1) = 1$ (corresponding this initial slope) is 0.0000213043. The solution curve is shown in Fig. 5.7.

```
% shootinit.m
% search for  y'(0)  to make y(1)=1, in y"-y=0, y(0)=0,y(1)=beta.
clc;clear; syms  y(x) k
beta=1; m=0;
for j=.8505:0.0001:0.8510
m=m+1;
eq='D2y-y=0';            % ODE
ic1='y(0)=0';           % ic1
ic2='Dy(0)=k'; ic2=subs(ic2,k,j); % ic2
S=dsolve(eq,ic1,ic2,'x');
h=ezplot(S,[0 1]); set(h,'linewidth',2.5);
title(num2str(j));grid; ylabel('y(x)');figure;
yb=subs(S,x,beta); err=abs(yb-beta); A(m,:)=[ic2 err];
end
%A
i=find(A(:,2)==min(A(:,2)));
dy0=vpa(A(i,:),6) % best shooting angle at x=0 and error value at x=beta.
```

Problem 5.1.10 (*Unknown boundary value, free boundary problem*) Use MATLAB dsolve function and the theory of single shooting to solve the following BVP,

$$y'' + y = 1, \quad 0 \le x \le b, \quad y(0) = 0, \quad y(b) = 1, \quad y'(b) = 2,$$

where the value of right-end boundary b is unknown [11] (shootinit2.m).

Solution
These problems may come up in various fields of engineering. An example of having an unknown parameter is when the final time is unknown in trajectory optimization.

It looks like the given BVP has an overdetermined structure with the value of right-end boundary b is unknown.

For second-order BVPs, the essence of the shooting method is to guess a value for the initial slope at left-end point a, integrate the ODE with this value of slope to the second (right-end) boundary point b and compare the computed value of $y(b)$ with the correct (given) value at b. The initial guess for s is then improved via shooting to produce the correct value at $y(b)$.

A so called 'slope retention' approach for the solution of such BVPs takes just one or two steps with many terms in the Taylor's expansion of the solution [12]. Retaining $s = y'(0)$ as the symbolic parameter, we find the following expansion up to x^7,

$$y(x, s) = sx + \frac{1}{2}x^2 - \frac{1}{6}sx^3 - \frac{1}{24}x^4 + \frac{1}{120}sx^5 + \frac{1}{720}x^6 - \frac{1}{5040}sx^7$$

Solving the equations $y(b, s) = 1$ and $y'(x, s) = 2$ for s and b of the free boundary gives the relevant values $s = 1.73204862$, $b = 0.523599258$.

Analytical expression for the solution is computed as $y(x) = s \cdot \sin(x) - \cos(x) + 1$.

Figure 5.8 displays this solution curve computed by forward shooting method, using MATLAB script (shootinit2.m).

Fig. 5.8 Solution curve for
the BVP, $y'' + y = 1$, $y(0) =$
0, $y(b) = 1$, $y'(b) = 2$ where
the value of b is unknown

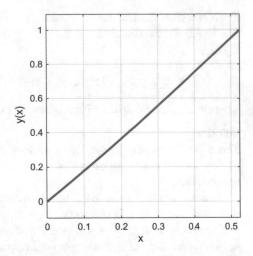

```
%shootinit2.m,    y"+y=1, y(0)=0,y(b)=1,y'(b)=2; Find y'(0) & b, y
clc;clear; syms  y(x,s) b u(x)
y=s*x + x.^2/2 -s*x.^3/6 - x.^4/24 + s*x.^5/120 + x.^6/720-s*x.^7/5040;
y1=subs(y,x,b);
y2=diff(y,x);y2=subs(y2,x,b);
[Sb,Ss] = solve(y1== 1,y2 == 2); % Ss:initial slope
Ss=vpa(Ss);Sb=vpa(Sb);
for i=1:length(Ss)
if(imag(Ss(i))~=0)
Ss(i)=0; Sb(i)=0;
end
end
zs=find(Ss==0);Ss(zs)=[]
rb=find(Sb==0);Sb(rb)=[]
y1=subs(y1,[b,s],[Sb(4),Ss(4)]), y2=subs(y2,[b,s],[Sb(4),Ss(4)])
% Solve and plot the solution using computed values of s,b
Du = diff(u);
u=dsolve(diff(u,2,x)+u == 1, u(0)==0, Du(0) ==Ss(4))
g=ezplot(u, [0    0.523599258585]);set(g,'linewidth',2); grid;
ylabel('y(x)');
```

Problem 5.1.11 (*Unknown boundary value, free boundary problem*) Use
MATLAB ode45 solver and single Reverse Shooting Method to solve the following
BVP,

$$y'' + y = 1, \quad 0 \le x \le b, \quad y(0) = 0, \quad y(b) = 1, \quad y'(b) = 2,$$

where the value of right-end boundary b is unknown (reverse_shoot1.m).

Solution

Since right-end boundary conditions contain both $y(b) = 1$, $y'(b) = 2$, while the
left-end boundary condition has only a single known value, $y(0) = 0$, it turns out to

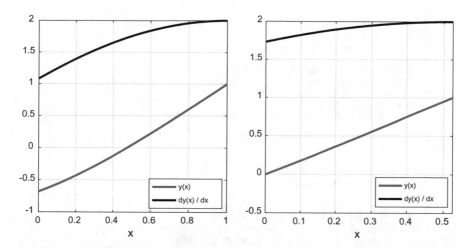

Fig. 5.9 *Left* Reversed shooting solution curve for the BVP, $y'' + y = 1$, $y(b) = 0$, $y(1) = 1$, $y'(1) = 2$ where the unknown value of b is determined by finding x-axis zero crossing point of the solution. This point also gives $y'(b) \approx \sqrt{3}$, the value of derivative function at point $x = b$. *Right* Actual solution curves are obtained by using the computed value of $b \approx 0.5237$ and then transforming x-axis values, $y(1) \rightarrow y(b)$, $y(b) \rightarrow y(0)$ by the second run of the code, and providing the conversion, $y(0) = 0, y'(0) \approx \sqrt{3}$

be simpler to use Reverse Shooting Method than to implement other types of forward shooting strategies (like slope retention method).

Reversed Shooting solution for the BVP $y'' + y = 1$ can be performed using new (reversed) set of boundary conditions, $y(1) = 1$, $y'(1) = 2$, $y(b) = 0$, where the unknown value of b is determined by finding x-axis zero crossing point of the solution and subtracting it from 1. This point also gives $y'(b) \approx \sqrt{3}$, the value of derivative function at point $x = b$. We obtain following edited list at the output of MATLAB script (reverse_shoot1.m), and corresponding solutions are displayed in Fig. 5.9.

```
       x                      y                    dy/dx
1.000000000000000    1.000000000000000    2.000000000000000
0.476262773473124   -0.000239794457555    1.731912340040196
               0    -0.682941969615793    1.080604611736281
```

Second pass:

```
0.523737226526876    1.000000000000000    2.000000000000000
               0    -0.000239794457555    1.731912340040196
```

```
% reverse_shoot1.m
clc;clear;close;format long
syms y(x)
[V] = odeToVectorField (diff(y, 2) == 1-y);
M = matlabFunction(V,'vars', {'x','Y'});
options = odeset('RelTol',1e-15,'AbsTol',1e-18);
[x,Y]= ode45(M, [1 0],[1 2],options); % reversed IVP:[1 0]
%[x,Y]= ode45(M,[1 0],[1 2]); % reversed IVP:[1 0]
U=[x  Y]
plot(x,Y,'linewidth',2);legend('y(x)','dy(x) / dx'); xlabel('x');grid;
xlim([0 1])

% reverse_shoot1.m  second run
clc;clear;close;format long
syms y(x)
b = 1-0.476262773473124
[V] = odeToVectorField (diff(y, 2) == 1-y);
M = matlabFunction(V,'vars', {'x','Y'});
options = odeset('RelTol',1e-15,'AbsTol',1e-18);
[x,Y]= ode45(M,b*[1 0],[1 2],options); % reversed IVP:[b 0]
U=[x  Y]
plot(x,Y,'linewidth',2);legend('y(x)','dy(x) / dx'); xlabel('x');grid;
xlim([0 b])
```

Problem 5.1.12 Use a reverse shooting strategy[3] to solve the BVP,

$$y'' + y = 0, \quad y(0) = 0, \quad y(b) = 1, \quad y'(b) = 0.$$

where b is an unknown boundary point (reverse_shoot2.m).

Solution
Since $y(0) = 0$, we search for the zero crossing point by letting (guessing) $b = 1$, first.

Running the code (reverse_shoot2.m) for this initial guess value does not provide a zero crossing point, therefore we set $b = 2$ and run the code again (without changing anything besides the entry for b). In this case $x = b \approx 0.426$ is found to be the zero point (from the list in the Command Window). Finally we obtain the value of $b = 2 - 0.426 = 1.574 \approx \pi/2$ as the value of right boundary. Figure 5.10 displays the results of these attempts.

---

[3]Apparently, analytical solution is easy to obtain and more feasible for that linear, homogeneous ODE.

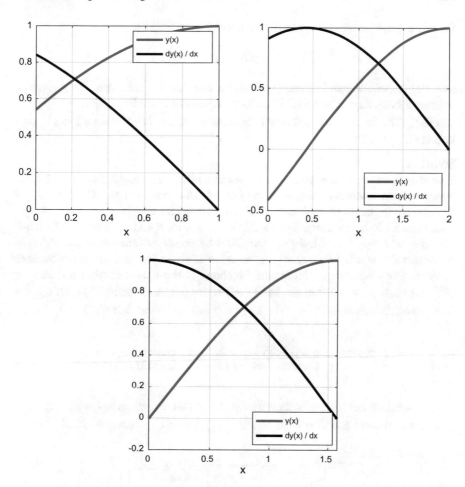

Fig. 5.10 Solution phases of the BVP, $y'' + y = 0$, $y(0) = 0$, $y(b) = 1$, $y'(b) = 0$ using a reverse shooting strategy. Above left; $b = 1$, above right; $b = 2$. Below; $b = 2 - 0.426 = 1.574 \approx \pi/2$

```
% reverse_shoot2.m
clc;clear;close;format long;syms y(x)
%b=1
%b=2
b=2-0.426
[V] = odeToVectorField (diff(y, 2) == -y);
M = matlabFunction(V,'vars', {'x','Y'});
options = odeset('RelTol',1e-9,'AbsTol',1e-9);
[x,Y]= ode45(M,b*[1 0],[1 0],options); % reversed IVP
U=[x  Y]  %size(U)
plot(x,Y,'linewidth',2);legend('y(x)','dy(x) / dx'); xlabel('x');grid;
xlim(b*[0 1])
```

Problem 5.1.13 Use ODE45 solver to solve the BVP [13],

$$y'' - y = 1, \quad y(0) = 0, \quad y(b) = e^b, \quad y'(b) = -1, \quad |b| \leq 1,$$

and plot the solution and its derivative. Note here that the unknown b also gives an unknown boundary value in addition to an unknown boundary point.

Hint: The value of unknown boundary point is $b = \ln(\sqrt{2} - 1)$ (unknownbndry1.m).

Solution
One should first compute the value of unknown boundary point b. Since $|b| \leq 1$, we apply reverse shooting starting at $x = b = 1$ and integrate the ODE by small decrements in the value of b towards $x = b = 0$, while checking the value of $y(0)$ each time for its equivalence to zero. Alternatively, we could have started (forward) shooting at $b = -1$ and integrate the ODE by small increments in the value of b towards $x = b = 0$, while checking the value of $y(0)$ each time for its equivalence to zero. Stopping criteria during the iterations is the approximation tolerance at $y(0)$, which is set at 10^{-4} in the MATLAB script (unknownbndry1.m). This yields $b = -0.881299999999825 \approx \ln(\sqrt{2} - 1)$. It is also found that $y'(0) = 0$.

x	y	dy/dx
-0.881299999999825	0.414244044236198	-1.000000000000000
0.0	0.000043108350438	-0.000043108330277

The solution curves are plotted in Fig. 5.11. Due to symmetry with respect to y-axis, we observe that $b = Re[\ln(-\sqrt{2} - 1)] \approx 0.8813$. Hence, $b \approx \pm 0.8813$.

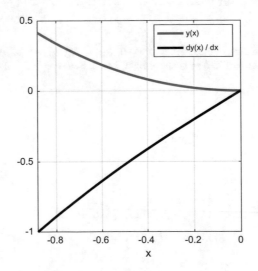

Fig. 5.11 Plots of the function and its derivative for the BVP, $y'' - y = 1$, $y(0) = 0$, $y(b) = e^b$, $y'(b) = -1$, $|b| \leq 1$

```
% unknownbndry1.m        y"-y=1, y(0)=0, y(b)=exp(b), y'(b)=-1
clc;clear;close;format long; syms y(x); tic
[V] = odeToVectorField (diff(y, 2) == 1+y);
M = matlabFunction(V,'vars', {'x','Y'});
b = 1; %b = -1;
f = 1;
while abs(f) > 0.0001
b = b - 0.0001;      % b = b + 0.0001;
options = odeset('RelTol',1e-9,'AbsTol',1e-12);
[x,Y]= ode45(M,[b 0],[exp(b) -1],options); f = Y(end,1); %U=[x  Y];
end
b = b %size(Y)
plot(x,Y,'linewidth',2); xlim([b  0]); legend('y(x)','dy(x) / dx');
xlabel('x');grid; toc
```

Problem 5.1.14

(a) Use MATLAB ode23 solver and the shooting method to solve the following linear BVP,

$$y'' - 0.5y = -0.1, \quad y(0) = 3, \quad y(5) = 7$$

(shooting2.m).

(b) Use ode45 solver for the solution of the same problem. Comment on the results.

Solution

(a) We first convert given second order ODE into a pair of first-order ODEs (order reduction),

$$y' = u, \quad u' = 0.5y - 0.1$$

We guess a value of $u_1(0) = -1$ along with the initial value for $y(0) = 3$. The solution is obtained by integrating the pair of ODEs from $x = 0$ to $x = 5$ by first setting up an inline (anonymous) function to hold the ODEs. Then, we use MATLAB ode23 function to generate the solution.

We obtain a value at $x = 5$, $y_1(5) = 23.991$, which differs from the desired boundary condition of $y(5) = 7$.

We make another guess $u_2(0) = -10$ and perform the computation again. This time, the result of $y_2(5) = -193.7397$ is obtained.

Since the original ODE is linear, we can use following equation to determine the correct trajectory to provide the best shot with $q = y(5) = 7$,

$$p - p_0 + \frac{p_1 - p_0}{q_1 - q_0}(q - q_0) = -1 + \frac{-10 - (-1)}{-193.7397 - 23.991}(7 - 23.991)$$
$$= -1.7023$$

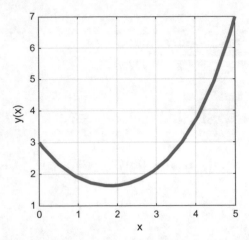

Fig. 5.12 The solution curve for the BVP, $y'' - 0.5y = -0.1$, $y(0) = 3$, $y(5) = 7$

This value of p can then be used to generate the correct solution, as shown in Fig. 5.12. The value of $y(5) = 6.9974$ is the best shot value obtained in this numerical process.

(b) Solution of this problem using ode45 solver (with default parameter settings) yields slightly better approximation to the solution as compared to the solution method using ode23 solver. Both results are given as follows (here, **p** is the derivative estimates vector, **q** is the corresponding solutions for the second boundary values vector):

Solution using ode23 solver;

```
p =  -1.000000000000000 -10.000000000000000  -1.702331174289034

q =   1.0e+02 *  0.239910050565791  -1.937396768599841   0.069966387179042
```

Solution using ode45 solver;

```
p =  -1.000000000000000 -10.000000000000000  -1.702767574028646

q =  1.0e+02 *   0.240369430321089  -1.941468391392779   0.069999993579911
```

Note that the use of different solvers affect the first and second solution values, as well.

MATLAB code (shooting2.m) is listed below.

```
%shooting2.m ,   solve BVP y"-0.5y=-0.1, y(0)=3,y(5)=7.
clc;clear;close;tic;format long
x1=0; x2=5; y1=3; y2=7; % boundary data
p0=[-1 -10  0];          % guess vector
%-----------------------------------------------------------------
x0=[x1, x2];             % x-span
q=[0 0 0];               % initialization
for i=1:3
odefun = @(x, y)([y(2); 0.5*y(1)-0.1]); % anonymous function
if i==3
p0(i)=p(1)+(y2-q(1))*(p(2)-p(1))/(q(2)-q(1));
end
y0=[y1, p0(i)];          % initial conditions
[x, y]=ode23(odefun, x0, y0);  p(i)=p0(i); q(i)=y(end,1);
end
p,q
plot(x, y(:,1),'linewidth', 2.5);grid; xlabel('x');ylabel('y(x)');toc
```

Problem 5.1.15 (*Boundary condition at infinity*) Use a Linear Shooting method (implementing a MATLAB solver from the ODE suite, which will be necessary for the solution of IVPs) to solve the BVP, $y'' = -y + 1 - 1.01e^{-0.1x}$, $y(0) = 0$, $y(\infty) = 1$ (shooting6.m).

Solution
Let $100 \rightarrow \infty$. We apply Linear Shooting method with formula (yielding the solution vector as a combination of its two linearly independent modes) to solve the given linear BVP with one of its boundary values is specified at infinity.

Graph of the solution and its two sinusoidal modes for given BVP are displayed in Fig. 5.13. Final part of the m-file (shooting6.m) uses dsolve function to check for the analytical solution.

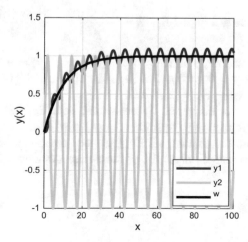

Fig. 5.13 Graph of the solution and its two sinusoidal modes for given BVP, $y'' = -y + 1 - 1.01e^{-0.1x}$, $y(0) = 0$, $y(b) = 1$, $b = 100 \rightarrow \infty$, $w(x) \cong y(x)$

```
%shooting6.m     y" = 1-y-1.01exp(-0.1x),   y(0)=0, y(inf)=1
clc;clear;close;format long
span=[0 100]; bv=[0  1]; h=0.01; %Step size
a=span(1);b=span(2); alpha=bv(1);beta=bv(2);
icl=[alpha 0.0];              % = [y(0) y'(0)]= initial conditions for y1
ic2=[0.0  1.0];               % = [y0)  y'(0)]= initial conditions for y2
F1=@(x,y) [y(2);  1-y(1)-1.01*exp(-0.1*x)]; %y(2)=y',y(1)=y
F2=@(x,y) [y(2);  -y(1)];                   %y(2)=y',y(1)=y
[x,y1]=ode45(F1,linspace(a,b,1000),ic1);
[x,y2]=ode45(F2,linspace(a,b,1000),ic2);
Y1=y1(:,1);length(Y1);Y2=y2(:,1);length(Y2);k=(beta-Y1(end))/Y2(end);
w=Y1+k*Y2;
plot(x,Y1,x,Y2,x,w,'linewidth',2);xlabel('x');ylabel('y(x)');grid;
legend('y1','y2', 'w');
% %Check
% clc;clear; syms y(x)
% eq='D2y+y=1-1.01*exp(-0.1*x)';bc='y(0)=0,y(100)=1';
% S=dsolve(eq,bc,x);simplify(S)
% x = linspace(0,100,1000); y=eval(vectorize(S));y=y'
% plot(x,y,'linewidth',2);grid
```

Problem 5.1.16 (*Mixed boundary conditions*) Dirichlet boundary condition discussed to this point is one of several types of boundary conditions. An alternative for Dirichlet (= fixed) boundary condition is the Neumann boundary condition involving derivatives at both boundaries. Here we deal with a mixed case of boundary conditions.

Use MATLAB Symbolic Toolbox commands to demonstrate the shooting process for the solution of the BVP, $y'' + y' + y = 1$, $y(0) = 0$, $y'(2) = 3$ (shooting_mixed1.m).

Solution

$$y(a) = \alpha, \quad y'(b) = \beta$$

$$y \cong w = y_1 + ky_2, \quad k = \frac{\beta - y_1'(b)}{y_2'(b)} = \frac{3 - y_1'(2)}{y_2'(2)} \tag{5.9}$$

We solve the first IVP, $y'' + y' + y = 1$, $y(0) = 0$, $y'(0) = 0$, and then the second IVP, $y'' + y' + y = 0$, $y(0) = 0$, $y'(0) = 1$. These will give us $y_1(x)$, $y_2(x)$. Taking derivatives will give $y_1'(x)$, $y_2'(x)$.

The coefficient k is computed by evaluating both derivatives at the second boundary, b, and subtracting $y_1'(b)$ from $y'(b) = \beta$ then dividing the result by $y_2'(b)$.

Finally we obtain the solution via Eq. (5.9) as a combination of y_1, k, y_2.

Following is the edited output list of these computations. The results are displayed in Fig. 5.14.

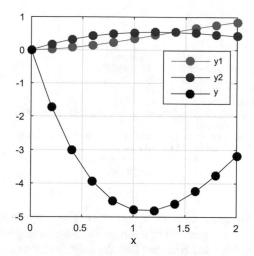

Fig. 5.14 Solution plot and its modes for the BVP, $y'' + y' + y = 1$, $y(0) = 0$, $y'(2) = 3$

```
eq1= D2y+1*Dy+1*y=1,   ic1 = y(0)=0,Dy(0)=0

eq2=D2y+1*Dy+1*y=0,    ic2 = y(0)=0,Dy(0)=1

y1=1-(3^(1/2)*exp(-x/2)*sin((3^(1/2)*x)/2))/3-exp(-x/2)*cos((3^(1/2)*x)/2)

y2=(2*3^(1/2)*exp(-x/2)*sin((3^(1/2)*x)/2))/3

dy1=(2*3^(1/2)*exp(-x/2)*sin((3^(1/2)*x)/2))/3

dy2=exp(-x/2)*cos((3^(1/2)*x)/2)-(3^(1/2)*exp(-x/2)*sin((3^(1/2)*x)/2))/3

k=-9.60428
```

x	y1	y2	y
[0,	0,	0,	0]
[0.2,	0.018669,	0.180064,	-1.71072]
[0.4,	0.069413,	0.320982,	-3.01338]
[0.6,	0.144584,	0.424757,	-3.93490]
[0.8,	0.237037,	0.494373,	-4.51106]
[1.0,	0.340300,	0.533507,	-4.78365]
[1.2,	0.448681,	0.546270,	-4.79785]
[1.4,	0.557338,	0.536980,	-4.59997]
[1.6,	0.662293,	0.509972,	-4.23562]
[1.8,	0.760425,	0.469433,	-3.74814]
[2.0,	0.849426,	0.419280,	-3.17745]

```
%shooting_mixed1.m
clc; clear; syms y(x)
eq1='D2y+1*Dy+1*y=1',ic1='y(0)=0,Dy(0)=0'
eq2='D2y+1*Dy+1*y=0',ic2='y(0)=0,Dy(0)=1'
y1=dsolve(eq1,ic1,x);y1=simplify(y1)
y2=dsolve(eq2,ic2,x);y2=simplify(y2)
dy1=diff(y1),dy2=diff(y2)
dy1end=vpa(subs(dy1,x,2),6);dy2end=vpa(subs(dy2,x,2),6);
k=vpa((3-dy1end)/dy2end,6)
x = linspace(0,2,11);y1=eval(vectorize(y1));y2=eval(vectorize(y2));
y=y1+k*y2;  A=[x' y1' y2' y'];A=vpa(A,6)
plot(x,y1,'.-',x,y2,'.-',x,y,'.-','markersize',24);
xlabel('x'); legend('y1','y2','y'); grid;
```

Problem 5.1.17 An application of a derivative boundary condition is the heat transfer problem in which one end of a rod (rather than having a fixed temperature) is subject to convection, which means that the heat flux is zero at this end.

Given the heated rod equation,

$$\frac{d^2T}{dx^2} = -k(T_\infty - T), \quad dT(0)/dx = T_a', \quad T(L) = T_b \tag{5.10}$$

where L is the length of the rod, T_a and T_b are the temperature values at both ends of the rod, T_∞ is the temperature of surrounding medium. We have a mixed boundary condition, since T_a' is a given boundary condition, along with the fixed (Dirichlet) boundary condition of $T(L) = T_b$.

(a) Use Linear Shooting method to determine the temperature distribution in the rod if $L = 5\,\text{m}$, $K = 0.05\,\text{m}^{-2}$, $T_\infty = 300\,\text{K}$, $T_a' = 0\,\text{K} \cdot \text{m}^{-1}$, $T_b = 385\,\text{K}$.

(b) Repeat (a) if $T_a' = -5\,\text{K} \cdot \text{m}^{-1}$. Plot axial temperature pattern in the rod for both cases.

Use ode23 and ode45 ODE solvers to compute IVPs (shooting_mixed2.m).

Solution
Equation (5.10) is first expressed as a system of two first order equations,

$$T'' = p$$

$$p' = -K(T_\infty - T)$$

$$dT(0)/dx = T_a', \quad T(L) = T_b$$

(a) MATLAB m-file (shooting_mixed2.m) employs single shooting with direct
    guess method and gives following results for insulated-end case:

ODE23:

```
p =1.0e+02 *   3.000000000000000   4.000000000000000   3.502131124840167
q =1.0e+02 *   3.000000000000000   4.692784927981837   3.849999124071677
Elapsed time is 0.266665 seconds.
```

ODE45:

```
p =1.0e+02 *   3.000000000000000   4.000000000000000   3.502103445063416
q =1.0e+02 *   3.000000000000000   4.692878247215860   3.849999999964503
Elapsed time is 0.368487 seconds.
```

(b) For the case $dT(0)/dx = T'_a = -5\,\mathrm{K}\cdot\mathrm{m}^{-1}$,

ODE23:

```
p =1.0e+02 *   3.000000000000000   4.000000000000000   3.682557679876033
q =1.0e+02 *   2.694580636037131   4.387359719810267   3.850000974921936
Elapsed time is 0.272233 seconds.
```

ODE45:

```
p =1.0e+02 *   3.000000000000000   4.000000000000000   3.682528189627893
q =1.0e+02 *   2.694562874614802   4.387441121907605   3.850000000037066
Elapsed time is 0.339853 seconds.
```

Both steady state temperature profiles in the rod as described by the given BVP
are depicted in Fig. 5.15.

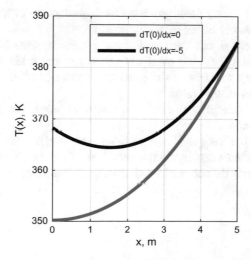

Fig. 5.15 Steady state temperature profiles in a rod, described by the BVP, $d^2T/dx^2 = -0.05(300 - T)$, $dT(0)/dx = T'_a$, $T(L = 5\,\mathrm{m}) = 385\,\mathrm{K}$ for $T'_a = 0$ K/m, and $T'_a = -5$ K/m

```
%shooting_mixed2.m, shooting by direct guessing boundary value(1)
clc;clear;tic;format long
x1=0; x2=5; y2=385; % boundary data
%dy1=0;    % Mixed boundary condition (run first, hold on, then %)
dy1=-5;    % mixed boundary condition (activate after the above line)
%guess left-end temperature where dy1=alpha=0
p0=[300 400  0];  % guess vector=[(1st guess)  (2nd guess)   0]
%-------------------------------------------------------------
x0=[x1, x2];              % x-span
q=[0 0 0];                % Initialization of "results vector"
for i=1:3
odefun = @(x,y)([y(2); 0.05*y(1)-15]); % Anonymous function
if i==3
p0(i)=p(1)+(y2-q(1))*(p(2)-p(1))/(q(2)-q(1)); % Linear Interpolation
end
y0=[ p0(i) dy1];          % Initial conditions
[x, y]=ode45(odefun,x0,y0);
p(i)=p0(i); q(i)=y(end,1);
end
p,q
plot(x,y(:,1),'linewidth',2.5); xlabel('x, m');ylabel('T(x), K');hold on;
legend('dT(0)/dx=0','dT(0)/dx=-5');grid on;
toc
% %True solution
% clc;clear;
% syms T(x)
% eq='D2T=0.05*T-15'
% bc='DT(0)=0,  T(5)=385'
% T=dsolve(eq,bc,x);T=simplify(T)
% x = linspace(0,5,11);T=eval(vectorize(T));
% plot(x,T,'.-','markersize',24);xlabel('x'),grid
```

5.2 Nonlinear Shooting Methods

Problem 5.2.1 True or False?

(a) In two-point boundary value problems the boundary conditions are specified at two different values of independent function.

(b) In shooting method, one takes a shot using a guess value of the derivative of function at initial (left-end) point and observe where it hits the target (second boundary point). Then, he or she corrects the aim and continue shooting again, until a good hit is succeeded.

(c) Shooting Methods give rise to linear sets of equations if the differential equation is nonlinear.

(d) All methods of solving nonlinear equations are iterative procedures.

(e) Iterative procedures usually do not need lots of computational resources.

(f) Iterative methods do not need good starting values in order to converge.

Answers

(a) T, (b) T, (c) F (nonlinear), (d) T, (e) F (usually need), (f) F (need).

Problem 5.2.2 Describe the principles of Shooting Method combined with Bisection Method for the solution of nonlinear BVPs.

Solution

When the second-order ODE is transformed into a system of first-order ODEs, it requires two initial conditions. The boundary condition at the "left end" of the interval also serves as an initial condition at that point, while the second initial condition must be guessed. The system is then solved by an ODE solver (such as a fourth order Runge Kutta method) and the value of this computed solution at the "right end" point is compared with the given boundary condition there. If the accuracy of this computed solution is not acceptable, the initial values for the IVP are guessed again and the system of first-order ODEs is solved once more. The same procedure is repeated until the computed value at the "right end" boundary agrees with the given boundary condition there. So far, this is the common procedure for all shooting methods for the solution of second order BVPs [14].

Because the data generated by two initial-value estimates can be linearly interpolated, a linear BVP is easier to solve using this method. On the other hand, since linear interpolation is not a choice for nonlinear BVPs, a remedy is to use root finding methods for obtaining the new initial value estimate.

Shooting Method combined with the Bisection Method requires a preprocessing stage to find out two estimates of the first derivative of the function for the boundary condition at left-end of the interval. These initial values should yield respective shooting values for the second boundary condition given in the problem. If this is less (greater) than the target value, shooting process continues until we get a number greater (less) than targeted value. Once we have two numbers below and above the target value, we then apply bisection method and stop when a prescribed tolerance level is met.

Problem 5.2.3 Motion of an oscillating mass attached to two springs is given by [15],

$$y'' + y^3 = 0, \quad y(0) = 0.2000, \quad y(2) = 0.1846$$

Use shooting Method combined with Bisection Method to solve the BVP. Specify the value of $y(1)$, and the slope values at two boundaries in four digit accuracy (shooting9.m).

Solution

Explicit solution could not be found by dsolve command of Symbolic Math Toolbox. We apply numerical method of Shooting combined with the Bisection Method.

The m-file (shooting9.m) we use for the solution of this problem consists of three sections. The first one has data, the second one (between the two dashed lines) computes bracketing values for bisection, and the final part performs computions for bisection algoritm. In the first run, only the first two parts are activated to compute the two initial derivative estimates which yield upper and lower values of the boundary at the right end (target). We obtain the following results at the end of this stage;

```
%betaL=0.176413361468470 ;    % lower  guessed value for beta
%betaU=0.215626787178892 ;    % upper  guessed value for beta
xL = -2e-2;                   % lower  guess dy
xU =  2e-2;                   % upper  guess dy
```

We transfer this information to last section and activate first and the third parts, then run the code again. We observe that at the end of 15th iteration we reach the (targeted) second boundary value of $y(2) = 0.184599$. The solution at $y(1) = 0.1960$, and the slope values at two boundaries are $y'(0) = 0$, $y'(2) = -0.0148$.

A graph of the solution is displayed in Fig. 5.16.

```
%shooting9.m   y"+y^3=0,   y(0)=1/5,  y(2)=0.1846
clc; clear; close; format long
a=0;b=2; alpha=1/5;
beta=0.1846;  % boundary condition  on the rightend (target)
tol = 1e-6; % prescribed tolerance
F =@(x,y) [y(2); -y(1)^3];
%-------------------------------------------------
%select bracketing values of beta
% for i= 1:5
% dy0(i)=-i*.01;
% ic1 = [alpha; dy0(i)];  % initial condition (guess, 5 elements)
% [x,y]=rk4(F,a,b,h,ic1);
% betaest(i) = y(end,1);  % estimate second boundary value, bv(b)
% end
% [dy0' betaest']
%-------------------------------------------------
% bisection stage
%betaL=0.176413361468470 ;    % lower  guessed value for beta
%betaU=0.215626787178892 ;    % upper  guessed value for beta
xL = -2e-2;                   % lower  guess dy(0) bracketing value
xU =  2e-2;                   % upper  guess dy(0) bracketing value
imax = 25; % Maximum number of iterations
for i = 1:imax
xr = xL +0.5*(xU-xL)       % Bisection
ic = [alpha; xr]           % Set initial state vector
[x,y] = ode45(F,[a,b],ic)  % Solve the IVP
Y(i) = y(end,1); err = Y(i) - beta; % Compare with target
if abs(err) < tol, break, end
if err > 0; xU = xr; else, xL = xr; end % Adjust xU and/or xL
end
i,Y(i),plot(x,y(:,1),'linewidth',2);grid;xlabel('x'),ylabel('y(x)');
```

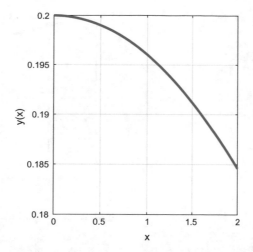

Fig. 5.16 Solution curve for the BVP, $y'' + y^3 = 0$, $y(0) = 0.2000$, $y(2) = 0.1846$, using shooting with bisection method

Problem 5.2.4 Use Shooting Method combined with Bisection Method to compute and plot one of the two solutions [16] of the BVP,

$$y'' = -|y| = f(x, y, y'), \quad y(a = 0) = 0, \quad y(b = 4) = \beta = -2$$

(shooting99.m).

Solution
Here, dsolve command can not be used, since the function $f(x, y, y')$ is not continuous everywhere and the $|y|$ term is not differentiable as y passes through the origin. We apply numerical method of Shooting combined with the Bisection Method. Guessing upper and lower initial slope values as $(0.2, -0.2)$ corresponds estimated values of $\beta = -0.194$ and $\beta = -5.458$, respectively, which bracket given β. The method converges to the solution and stops at the end of 20th iteration for an error tolerance value of 10^{-6}, with $y'(0) = -0.0733$.

Figure 5.17 displays resulting solution curve.

```
%shooting99.m  Shooting with Bisection,  y"+|y|=0,  y(0)=0, y(4)=-2
clc;clear;
a=0;b=4; alpha=0;beta=-2; tol = 1e-6;F =@(x,y) [y(2); -abs(y(1))];
xL = -0.2;    % lower  guess dy,  betaL=-5.458 ;
xU =  0.2;    % upper  guess dy,  betaU=-0.194 ;
imax = 25; % Maximum number of iterations
for i = 1:imax
xr = xL +0.5*(xU-xL);ic = [alpha; xr];[x,y] = ode45(F,[a,b],ic);
Y(i) = y(end,1); err = Y(i) - beta % Compare with target
if abs(err) < tol, break, end
if err > 0; xU = xr; else, xL = xr; end; end;
i,Y(i),plot(x,y(:,1),'linewidth',2);grid;xlabel('x'),ylabel('y(x)');
```

Fig. 5.17 One solution curve for the BVP, $y'' + |y| = 0$; $y(0) = 0$, $y(4) = -2$ using shooting method combined with the bisection method

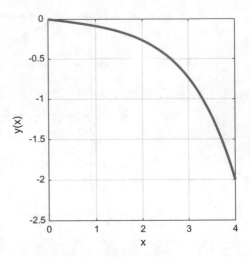

Problem 5.2.5 (*Shooting Method combined with Secant Method*) Given a BVP,

$$y'' = f(x, y, y'), \quad y(a) = \alpha, \quad y(b) = \beta,$$

remember that in the shooting method, a BVP is considered as an IVP and attempt is made to determine the value $y'(a)$ which results in $y(b) = \beta$.

When shooting using secant method, we let

$$m(z) = f(b, z) - \beta$$

where $f(b, z)$ is the solution to $y'' = f(x, y, y')$, using the value $y'(a) = z$.

In this method, one determines a root of $m(z)$ to solve the BVP using the secant method.

In secant method, one makes successive approximations using the following iteration formula,

$$z_{n+1} = z_n - \frac{z_n - z_{n-1}}{f(z_n) - f(z_{n-1})} f(z_n) \tag{5.11}$$

which requires two initial approximations (guess values) to initiate the process. The iteration in (5.11) is continuously applied until either one or both of the following two convergence criteria are satisfied [17]:

$$|z_{n+1} - z_n| \leq \epsilon_1, \quad |f(z_{n+1})| \leq \epsilon_2 \tag{5.12}$$

Use shooting with secant method to solve the following nonlinear autonomous BVP,

$$y'' = -(1+y)y' + y^2, \quad y(0) = 1, \quad y(1) = e^{-1}$$

Plot the solution and its first derivative, as well as the convergence of initial value of the first derivative versus the number of iterations. Use MATLAB ode45 solver, when it is necessary to solve IVPs (shooting8.m).

Solution

The analytical solution of given BVP is $y(x) = e^{-x}$.

The convergence for the initial value of the first derivative versus the number of iterations and the graph of solution and its first derivative are displayed in Fig. 5.18. It is seen here that convergence occurs at the end of fifth iteration (with respect to defined accuracy) and it is relatively fast.

MATLAB m-file (shooting8.m) is also listed below.

```
%shooting8.m    y" =-(1+y)y'-y^2,   y(0)=1, y(1)=exp(-1)
clc;clear;close;
span=[0 1];bv=[1  exp(-1)]; alpha=bv(1);beta=bv(2);
F=zeros(2,1);%initialization
F=@(t,y) [y(2);   -(1+y(1))*y(2)-y(1)^2];
[x,y1]=ode45(F,span,[1,alpha]);[x,y2]=ode45(F,span,[1,beta]);
i=1;
m1=y1(end,1)-bv(2);m2=y2(end,1)-bv(2);
while(abs(beta-alpha)>1e-6)
xp=beta;
beta=alpha-(alpha-beta)/(m1-m2)*m1; %secant
alpha=xp;
[x,y1]=ode45(F,span,[1,alpha]);[x,y2]=ode45(F,span,[1,beta]);
m1=y1(end,1)-bv(2);m2=y2(end,1)-bv(2);
A(i)=alpha; %for convergence study
i=i+1;
end
i=1:length(A);% convergence analysis
plot(i,A,'o-','linewidth',2);grid;xlabel('iterations');ylabel('dy(0) / dx')
figure;
y=y2; plot(x,y,'linewidth',2);xlabel('x');grid;legend('y(x)','dy(x) / dx');
```

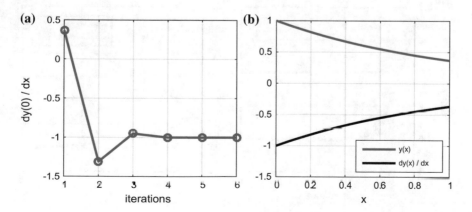

Fig. 5.18 a The convergence of initial value of the first derivative versus the number of iterations. **b** Graph of solution and its first derivative for the BVP, $y'' = -(1+y)y' + y^2$, $y(0) = 1$, $y(1) = e^{-1}$

Problem 5.2.6 Use shooting with secant method to solve the following nonlinear BVP,

$$y'' = -(y')^2/y, \quad y(0) = 1, \quad y(3) = \sqrt{10}$$

Plot the solution and its first derivative, as well as the convergence of initial value of the first derivative versus the number of iterations. Use MATLAB ode45 solver, when it is necessary to solve IVPs (shooting7.m).

Solution

The convergence for the initial value of the first derivative versus the number of iterations and the graph of solution and its first derivative are displayed in Fig. 5.19. It is seen here that convergence occurs at the end of fifth iteration (with respect to defined accuracy) and it is relatively fast. MATLAB m-file (shooting7.m) is listed below.

```
% shooting7.m nonlinear shooting y" = -(y')^2/y,  y(0)=1, y(3)=sqrt(10)
% using secant method
clc;clear;close;
span=[0 3];bv=[1 sqrt(10)];alpha=bv(1);beta=bv(2);
F=@(x,y) [y(2);  -y(2)^2/y(1)];
[x,y1]=ode45(F,span,[1,alpha]);[x,y2]=ode45(F,span,[1,beta]);
i=1;
m1=y1(end,1)-bv(2); m2=y2(end,1)-bv(2);
while(abs(beta-alpha)>1e-6)
xp=beta; beta=alpha-(alpha-beta)/(m1-m2)*m1; %secant
alpha=xp; [x,y1]=ode45(F,span,[1,alpha]);[x,y2]=ode45(F,span,[1,beta]);
m1=y1(end,1)-bv(2); m2=y2(end,1)-bv(2);
A(i)=alpha; %for convergence study
i=i+1;
end
i=1:length(A);% convergence analysis
plot(i,A,'o-','linewidth',2);grid;
xlabel('iterations');ylabel('dy(0) / dx');figure; y=y2;
plot(x,y,'linewidth',2);grid; legend('y(x)','dy(x) / dx');
```

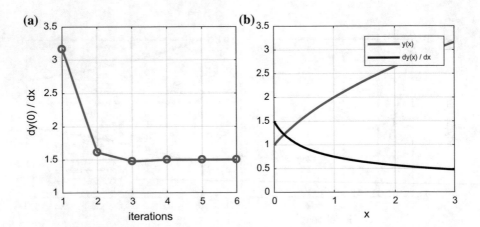

Fig. 5.19 a The convergence of initial value of the first derivative versus the number of iterations. **b** Graph of solution and its first derivative for $y'' = -(y')^2/y$, $y(0) = 1$, $y(3) = \sqrt{10}$

Problem 5.2.7 Use Shooting Method combined with the Secant Method to solve the nonlinear BVP, $2yy'' - (y')^2 + 4y^2 = 0$, $y(\pi/6) = 1/4$, $y(\pi/2) = 1$ [18].

Use MATLAB ode45 solver when it is necessary to solve an IVP (shooting777.m), (shoot_secant.m).

Solution

Explicit solution of this BVP could not been found using dsolve function. Application of Shooting Method combined with the Secant Method rapidly converges to the solution, as shown in Fig. 5.20. Exact solution is $y = \sin^2(x)$, and the slope value at two boundaries are $y'(\pi/6) = \sqrt{3}/2$, $y'(\pi/2) = 0$.

```
% shooting777.m % shooting using secant method
% y" = (y')^2/(2*y)-2*y,  y(pi/6)=1/4, y(pi/2)=1
clc;clear;close;
span=[pi/6 pi/2]; bv=[1/4 1];
F=@(x,y) [y(2);  (y(2))^2/(2*y(1))-2*y(1)];
[x,y]=shoot_secant(F,span,bv)
plot(x,y,'linewidth',2); xlabel('x'); grid; legend('y(x)','dy(x) / dx');
xlim([x(1) x(end)]);

function [x,y]=shoot_secant(F,span,bv)
alpha=bv(1);beta=bv(2);
[~,y1]=ode45(F,span,[bv(1),alpha]);[x,y2]=ode45(F,span,[bv(1),beta]);
i=1; m1=y1(end,1)-bv(2); %f(x_i)
m2=y2(end,1)-bv(2); %f(x_i-1)
while(abs(beta-alpha)>1e-6), xp=beta;
beta=alpha-(alpha-beta)/(m1-m2)*m1; %secant
alpha=xp;
[x,y1]=ode45(F,span,[bv(1),alpha]); [x,y2]=ode45(F,span,[bv(1),beta]);
m1=y1(end,1)-bv(2); m2=y2(end,1)-bv(2);
A(i)=alpha; %for convergence study
i=i+1;
end
i=1:length(A);% convergence analysis
plot(i,A,'.-','markersize',20);grid;
xlabel('iterations');ylabel('dy(0) / dt'); figure; y=y2;
```

Problem 5.2.8 Motion of an oscillating mass attached to two springs is given by [15],

$$y'' + y^3 = 0, \quad y(0) = 0.2000, \quad y(2) = 0.1846$$

Use Single Shooting Method combined with Secant Method to solve the BVP. Specify the value of $y(1)$, and the slope values at two boundaries (shooting7777.m), (shoot_secant.m).

Solution

Explicit solution of the given BVP could not been found using dsolve function. Application of Shooting Method combined with the Secant Method rapidly converges to the solution, as shown in Fig. 5.21. The solution at $y(1) = 0.196$, and the slope values at two boundaries are $y'(0) = 0$, $y'(2) = -0.0148$.

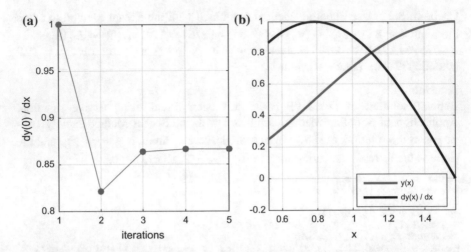

Fig. 5.20 a The convergence of initial value of the first derivative versus the number of iterations.
b Graph of solution and its first derivative for the BVP, $2yy'' - (y')^2 + 4y^2 = 0$, $y(\pi/6) = 1/4$, $y(\pi/2) = 1$. In the solution of this problem, we use a MATLAB function file (secant.m) and call it from the main program (shooting777.m). The short main file includes data required for input to function file, as well as the anonymous function which defines the ODE

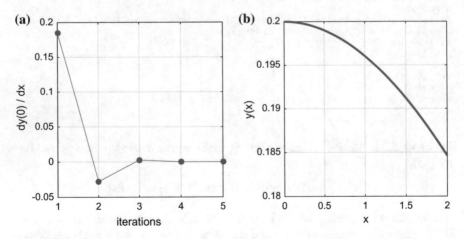

Fig. 5.21 a The convergence of initial value of the first derivative versus the number of iterations.
b Graph of solution for the BVP, $y'' + y^3 = 0$, $y(0) = 0.2$, $y(2) = 0.1846$

In the solution of this problem, we use a MATLAB function file (secant.m) and call it from the main program (shooting7777.m). The short main file includes data required for input to function file, as well as the anonymous function which defines the ODE.

```
%shooting7777.m %Shooting with Secant M., y"=-y^3, y(0)=1/5, y(2)=0.1846
clc;clear;close;
span=[0 2]; bv=[1/5 0.1846];
F=@(x,y) [y(2);  -(y(1))^3];
[x,y]=shoot_secant(F,span,bv)
plot(x,y(:,1),'linewidth',2); xlabel('x'); ylabel('y(x)'); grid;
```

Problem 5.2.9 The unidirectional thin film flow of an incompressible fluid down an inclined plane is described by the following nonlinear second-order BVP [19],

$$y'' + \beta y^2 y'' + \lambda = 0, \quad y(0) = y(1) = 0, \quad \beta = 0.3, \quad \lambda = 2.5$$

Use Shooting Method combined with the Secant Method to solve the BVP and plot y, y' as functions of independent variable (time) (shooting_secant2.m), (shoot_secant.m).

Solution

$$y''\left(1 + \beta y^2\right) + \lambda = 0 \quad \rightarrow \quad y'' = -\frac{\lambda}{1 + \beta y^2}$$

The results are shown in Fig. 5.22.

(a)   **(b)**

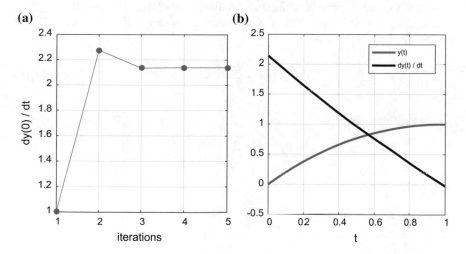

Fig. 5.22 a The convergence of initial slope value versus the number of iterations. **b** Graph of solution for the BVP, $y'' + \beta y^2 y'' + \lambda = 0$, $y(0) - y(1) = 0$, $\beta = 0.3$, $\lambda = 2.5$. Computed values of initial and final derivatives are $y'(0) = 2.1389, y'(1) = -0.0254$

In the solution of this problem, we use a MATLAB function file (secant.m) and call it from the main program (shooting_secant2.m). The short main file includes data required for input to function file, as well as the anonymous function which defines the ODE.

```
%shooting_secant2.m % nonlinear shooting using secant method
clc;clear;close;
span=[0 1]; bv=[0 1]; b=0.3; La=2.5;
F=@(x,y) [y(2);  -La/(1+b*y(1)*y(1))];
[x,y]=shoot_secant(F,span,bv)
plot(x,y,'linewidth',2);xlabel('t');grid;legend('y(t)','dy(t) /dt');
```

Problem 5.2.10 (*Shooting Method combined with Newton's method*) In the solution of a nonlinear BVP of the form

$$y'' = f(x, y, y'), \quad y(a) = \alpha, \quad y(b) = \beta \tag{5.13}$$

using the shooting method, we consider the BVP as an IVP and try to compute the value of $y'(a)$ that yields $y(b) = \beta$.

We let

$$y'(a) = p, \quad m(p) = f(b; p) - \beta \tag{5.14}$$

where $f(b; p)$ is the solution to the IVP using the value of p. Therefore, we compute a root of $m(p)$ to solve the BVP.

When we use Newton's method in nonlinear shooting, we must know $m'(p)$, which is the derivative of $m(p)$. This function is defined as the solution of IVP at the rightend of interval, b.

$$m'(p) = \left. \frac{\partial f(x; p)}{\partial t} \right|_{x=b} \tag{5.15}$$

On the other hand,

$$\frac{\partial y''}{\partial p} = \frac{\partial f}{\partial y} \frac{\partial y}{\partial p} + \frac{\partial f}{\partial y'} \frac{\partial y'}{\partial p} \tag{5.16}$$

Let

$$z(x, p) = \frac{\partial f(x, p)}{\partial p}$$

Then,

$$z'' = \frac{\partial f}{\partial y} z + \frac{\partial f}{\partial y'} z', \quad z(0) = 0, \quad z'(0) = 1 \tag{5.17}$$

with

$$m'(p) = z(b)$$

Set up defining system of first order ODEs to be used in shooting method combined with Newton's method for the following BVPs:

(a) $\qquad y'' - 10^{-8}y^4 - 20y = -5000, \quad y(0) = 500, \quad y(0.15) = 300.$

(b) $\qquad\qquad\qquad y'' = -\dfrac{(y')^2}{y}, \quad y(0) = 1, \quad y(1) = 2.$

(c) $\qquad\qquad\qquad y'' = y - y^2, \quad y(0) = 1, \quad y(1) = 4.6.$

Solution

(a) $a = 0, \quad b = 0.15, \quad \alpha = 500, \quad \beta = 300$

$$y_1' = y_2, \quad y_2' = f = 10^{-8}y_1^4 + 20y_1 - 5 \times 10^3, \quad y_3' = y_4,$$

$$y_4' = \frac{\partial f}{\partial y_1}y_3 + \frac{\partial f}{\partial y_2}y_4 = \left(4 \times 10^{-8}y_1^3 + 20\right)y_3$$

$$y_1(0) = \alpha = 500, \quad y_2(0) = p, \quad y_3(0) = 0, \quad y_4(0) = 1$$

Here, p is guessed. These are defied in MATLAB by an anonymous or external function.

(b) $a = 0, \quad b = 1, \quad \alpha = 1, \quad \beta = 2$

$$y_1' = y_2, \quad y_2' = f = -\frac{y_2^2}{y_1}, \quad y_3' = y_4, \quad y_4' = \frac{\partial f}{\partial y_1}y_3 + \frac{\partial f}{\partial y_2}y_4 = \frac{y_2^2}{y_1^2}y_3 - \frac{2y_2}{y_1}y_4$$

$$y_1(0) = \alpha, \quad y_2(0) = p, \quad y_3(0) = 0, \quad y_4(0) = 1$$

(c) $a = 0, \quad b = 1, \quad \alpha = 1, \quad \beta = 4.6$

$$y_1' = y_2, \quad y_2' = f = y_1 - y_1^2, \quad y_3' = y_4, \quad y_4' = \frac{\partial f}{\partial y_1}y_3 + \frac{\partial f}{\partial y_2}y_4 = (1 - 2y_1)y_3$$

$$y_1(0) = \alpha = 1, \quad y_2(0) = p, \quad y_3(0) = 0, \quad y_4(0) = 1$$

Problem 5.2.11 Solve the Van der Pol equation,

$$y'' - \mu(y^2 - 1)y' + y = 0, \quad y(0) = 0, \quad y(2) = 1, \quad \mu = 1.3$$

using Shooting Method combined with the Newton's Method (shooting10.m).

Solution

$$a = 0, \quad b = 2, \quad \alpha = 0, \quad \beta = 1$$

$$y_1' = y_2,$$

$$y_2' = f = \mu(y_1^2 - 1)y_2 - y_1 = \mu y_1^2 y_2 - \mu y_2 - y_1,$$

$$y_3' = y_4,$$

$$y_4' = \frac{\partial f}{\partial y_1} y_3 + \frac{\partial f}{\partial y_2} y_4 = (2\mu y_1 y_2 - 1)y_3 + (\mu y_1^2 - \mu)y_4$$

$$y_1(0) = \alpha = 0, \quad y_2(0) = p, \quad y_3(0) = 0, \quad y_4(0) = 1$$

The initial value is chosen as $y_2(0) = p = p_1 = -1$.

We input these data into m-file (shooting10.m). The output of the code shows that iterations converge to expected solution with prescribed tolerance at the end of eighth iteration. Solution curve for the Van der Pol equation and its derivative (for $\mu = 1.3$) are displayed in Fig. 5.23.

```
%shooting10.m  shooting combined with Newton's (Newton-Raphson) method
clc;clear;close;
p1=-1;                   % guess for initial derivative value
tol=1e-6;                % error tolerance
span=[0,2]; alpha=0; beta=1; mu=1.3; %span=[a,b], alpha=y(a), beta=y(b)
F=@(x,y) [y(2);
          mu*y(1)^2*y(2)-mu*y(2)-y(1) ;
          y(4);
          (2*mu*y(1)*y(2)-1)*y(3)+(mu*y(1)^2-mu)*y(4)];
[x,y]=ode45(F,span,[alpha,p1,0,1]); i=1; m=y(end,1)-beta;
p2=p1-m/y(end,3);%Newton's formula
while(abs(p2-p1)>tol)
p1=p2; [x,y]=ode45(F,span,[alpha,p1,0,1]); m=y(end,1)-beta;
p2=p1-m/y(end,3);%Newton's formula
i=i+1; end
%y
i, plot(x,y(:,1),'linewidth',2);xlabel('x'),ylabel('y(x)');
hold on;plot(x,y(:,2),'k-','linewidth',2);grid;legend('y(x)','dy(x)/dx');
```

Problem 5.2.12 Use Shooting Method combined with Newton's Method to find the solution of the nonlinear BVP, $y'' = e^y, y(0) = y(1) = 0$. Determine the range of initial guess value of $y_{est}'(0)$, and actual values of $y'(0), y'(1)$ (shooting101.m).

Fig. 5.23 Solution curves for
Van der Pol ODE and its
derivative ($\mu = 1.3$)

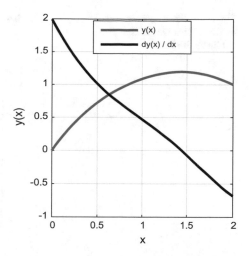

Solution

Explicit solution could not be found using dsolve built in MATLAB function.

We run the MATLAB m-file (shooting101.m) for a starting estimate of $y'_{est}(0) = 0$, then gradually increase its value in positive direction. It is found that the range of (guessed value of) this parameter is $y'_{est}(0) \leq 3.30116$. Computations stop at the end of 10th iteration, for $tol = 10^{-6}$.

Actual values at the boundaries are $y'(0) = -0.4636$, $y'(1) = 0.4636$.

The solution curve for the nonlinear BVP and its first derivative are displayed in Fig. 5.24.

```
%shooting101.m  Shooting Method combined with Newton's method
clc;clear;close;
p1=1;                  % guess for initial derivative value
tol=1e-6;              % error tolerance
span=[0,1]; alpha=0; beta=0;  %span=[a,b], alpha=y(a), beta=y(b)
F=@(x,y) [y(2);    exp(y(1))  ;   y(4);   exp(y(1))*y(3)];
%----------------------------------------------------------------------
[x,y]=ode45(F,span,[alpha,p1,0,1]);i=1;m=y(end,1)-beta;
p2=p1-m/y(end,3); while(abs(p2-p1)>tol)
p1=p2;[x,y]=ode45(F,span,[alpha,p1,0,1]);m=y(end,1)-beta;
p2=p1-m/y(end,3); i=i+1; end
i, plot(x,y(:,1),'linewidth',2);xlabel('x');hold on;
plot(x,y(:,2),'k-','linewidth',2);grid;legend('y(x)','dy(x)/dx');
```

Fig. 5.24 Solution curve and its first derivative for the nonlinear BVP, $y'' = e^y$, $y(0) = y'(1) = 0$

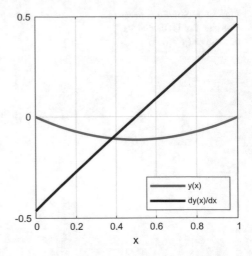

Problem 5.2.13 The classical one dimensional Bratu problem with homogeneous Dirichlet boundary conditions is given by the BVP, $y'' + Ce^y = 0$, $C \in \mathbb{R}^+$, $x \in [0, 1]$, $y(0) = y(1) = 0$. If $C < C_C$, the BVP has two solutions. There is only one solution for $C = C_C$ and no solution for $C > C_C$ where C_C is a critical bifurcation point (known as Frank-Kamenetskii parameter in chemistry). The numerical value of critical point is $C_C \cong 3.51383$.

This problem appears in a variety of application areas such as the fuel ignition model of thermal combustion, chemical reactor theory, radiative heat transfer, thermal reaction, and the Chandrasekhar model of the expansion of the universe and appeals much interest for its solutions using different methods [20–24].

Use Shooting Method combined with the Newton's Method to solve one dimensional Bratu's problem, plot the solution and its first derivative for the critical bifurcation point (shooting102.m).

Solution
The BVP is $y'' + 3.51383e^y = 0$, $y(0) = y(1) = 0$. Application of Shooting Method combined with the Newton's Method converges to the solution, at the end of 14th iteration for the error tolerance value set at 10^{-6}. We use initial guess value for the slope as $y'_{est}(0) = 0$. The slope values at two boundaries are computed as $y'(0) = 3.9975$, $y'(1) = -3.9975$. Graph of solution and its first derivative for this BVP are depicted in Fig. 5.25.

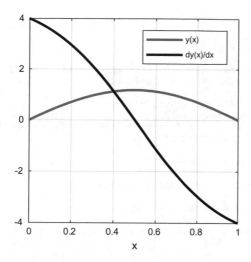

Fig. 5.25 Graph of solution and its first derivative for the BVP, $y'' + C_C e^y = 0$, $y(0) = y(1) = 0$, $C_C = 3.51383$

```
%shooting102.m  shooting combined with Newton's (Newton-Raphson) method
clc;clear;close;
p1=0; tol=1e-6;span=[0,1]; alpha=0; beta=0;
F=@(x,y) [y(2);  -3.51383*exp(y(1)) ; y(4); -3.51383*exp(y(1))*y(3)];
[x,y]=ode45(F,span,[alpha,p1,0,1]); i=1; m=y(end,1)-beta;
p2=p1-m/y(end,3); while(abs(p2-p1)>tol); p1=p2;
[x,y]=ode45(F,span,[alpha,p1,0,1]); m=y(end,1)-beta;p2=p1-m/y(end,3);
i=i+1; end; i, plot(x,y(:,1),'linewidth',2); xlabel('x'); hold on;
plot(x,y(:,2),'k-','linewidth',2); grid; legend('y(x)','dy(x)/dx');
```

Problem 5.2.14 (*Single Shooting with Brent's Method*) Solve the Van der Pol equation,

$$y'' - \mu(y^2 - 1)y' + y = 0, \quad y(0) = 0, \quad y(2) = 1, \quad \mu = 1.3$$

using Single Shooting combined with Brent's Method (shooting_brent1.m).

Solution

Solution curve for the Van der Pol equation and its derivative (for $\mu = 1.3$) using Shooting Method combined with Brent's method are displayed in Fig. 5.23, which is exactly the same figure obtained when using Shooting Method combined with the Newton's Method (hence, we do not duplicate the figure here).

```
%shooting_brent1.m
clc; clear; close; tic;
global a b alpha mu
a = 0; b = 2; span=[a  b];bv=[0  1]; alpha=bv(1); mu=1.3;
u1 = 1; u2 = 2; % Trial values for initial condition u.
x = a;
u = brent(@res,u1,u2); ic= [alpha u];
[x,y] = ode45(@odes, span, ic);
plot(x,y,'linewidth',2); grid; legend('y(x)','dy(x) / dx');
xlabel('x');ylabel('y(x)');toc;

function F = odes(~,y) % System of first-order ODEs
global mu
F= [y(2);mu*(y(1)^2-1)*y(2)-y(1)];%Van der Pol ODE

function r = res(u) % Boundary residual
global a b alpha
[~,y] = ode45(@odes,[a  b],[alpha u]);
r = y(size(y,1),1) - 1;

function root = brent(fun,a,b)% Finds a root of f(x)=0 by Brent's method
% fun = function handle; a, b =  interval bracketing the root
% root = one root of f(x)
x1 = a; f1 = feval(fun,x1); if f1 == 0; root = x1; return; end
x2 = b; f2 = feval(fun,x2); if f2 == 0; root = x2; return; end
if f1*f2 > 0.0, error('Root is not bracketed in (a,b)'), end
x3 = 0.5*(a + b);
for i = 1:30
f3 = feval(fun,x3);
if abs(f3) < 1e-10; root = x3; return, end
if f1*f3 < 0.0; b = x3; else a = x3; end
if (b - a) < 1e-10*max(abs(b),1.0);root = 0.5*(a + b); return,end
%   set up rational function for quadratic interpolation
den = (f2-f1)*(f3-f1)*(f2-f3);
num = x3*(f1-f2)*(f2-f3+f1)+ f2*x1*(f2-f3) + f1*x2*(f3-f1);
% For zero division, get x out of bracket
if den == 0; dx = b - a; else dx=f3*num/den; end
x = x3 + dx;
% use bisection, when interpolation is out of bracket
if (b - x)*(x - a) <0, dx =(b - a)/2;  x=a+dx; end
if x < x3; x2 = x3; f2 = f3; else x1 = x3; f1 = f3; end
x3 = x;
end
root = []; % root does not converge.
```

Problem 5.2.15 Use Shooting Method combined with the Brent's Method to solve the nonlinear BVP, $y'' + (y')^2 = x$; $y(0) = 3$, $y(5) = 1$. Plot the solution curve and its first derivative in the same figure (shooting_brent2.m).

Solution

Here, the selection of trial values for initial conditions is significant. It is observed that selecting both of them with positive numerical values does not give a solution. A possible choice that yields a solution is to keep lower initial condition at more

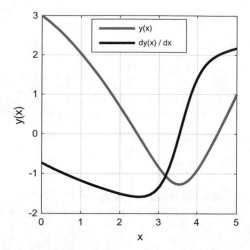

Fig. 5.26 Computed solution and its first derivative for the nonlinear BVP, $y'' + (y')^2 = x$, $y(0) = 3$, $y(5) = 1$, using shooting method combined with the Brent's method

negative level than the other one, which can have a zero value. When the latter is selected at zero, lower maximum initial value (with four digit accuracy) is computed to be $u_1 = -0.7289$. A list of some critical values of the solution is given below. The solution curve and its first derivative are displayed in Fig. 5.26.

x	y	y'
0	3.000000000000000	-0.728831017968814
0.068929185766818	2.948512669816100	-0.764921895382415
2.400716743067273	0.069145476751016	-1.574610885353950
2.525716743067273	-0.128064193150330	-1.577816928650846
3.522821593508172	-1.266849854539965	-0.058387812812241
4.513678586084559	-0.016813279169279	1.986573711012347
4.591040058236281	0.138425017267501	2.027230274135664
4.955781118672862	0.904202995149078	2.160082066931861
5.000000000000000	0.999999999999373	2.172660938843943

The list of MATLAB functions given below do not include (brent1.m) and (res.m), since they are the same functions which remain intact as given in the previous problems.

```
%shooting_brent2.m
clc;clear;close;tic;
global a b alpha
a = 0 ; b = 5; span=[a  b]; bv=[3 1];
u1 = -.7289 ;   u2 = 0; % Trial values for initial condition u.
alpha=bv(1); x = a; u = brent(@res,u1,u2); ic= [alpha u];
[x,y] = ode45(@odes,span,ic); plot(x,y,'linewidth',2); grid;
legend('y(x)','dy(x) / dx'); xlabel('x');ylabel('y(x)');toc;

function F = odes(x,y) % System of first-order ODEs.
F = [y(2); x-y(2)^2];
```

Problem 5.2.16 Solve the nonlinear autonomous BVP, $y'' = -y^2$; $y(0) = y(1) = 2$ by Shooting Method using MATLAB's fzero built-in function. Plot the solution (shootfzero1.m, F.m).

Solution

No closed form solution is found for this BVP using MATLAB dsolve function.

The nonlinear second-order ODE is first expressed as two first-order ODEs as

$$y_1' = y_2,$$

$$y_2' = -y_1^2$$

A MATLAB anonymous function can compute the right-hand sides of last two equations. Additionally, a MATLAB function is developed to hold the residual that is driven to zero. Then, the zero is computed with the fzero function. Thus, we see that if we set the initial slope properly, the residual function will be driven to zero and the boundary condition at the right-end of the interval will be satisfied within the specified tolerance limits, which can be verified by printing a list of the solution and plotting the solution curve.

Solution curve of the BVP using Shooting Method combined with MATLAB's fzero built-in function is shown in Fig. 5.27. The slope values at two boundaries are $y'(0) = 3.5607$, $y'(2) = -3.5607$.

```
%shootfzero1.m
clc;clear;close;
global ya yb span
ya=2; yb=2; span=[0, 1]; root=fzero(@F,-1)
dy=@(x,y) [y(2); -y(1)^2];[x,y]=ode45(dy, span, [ya, root])
plot(x,y(:,1),'linewidth',2);grid;xlabel('x'),ylabel('y(x)');

function z=F(guess_yb)
global ya yb span
dy=@(x,y) [y(2); -y(1)^2];[x,y]=ode45(dy, span, [ya, guess_yb]);
z=y(end,1)-yb; %residual
```

Fig. 5.27 Solution curve of
the nonlinear BVP,
$y'' = -y^2$; $y(0) = y(1) = 2$
obtained by shooting method,
using MATLAB's fzero built
in function

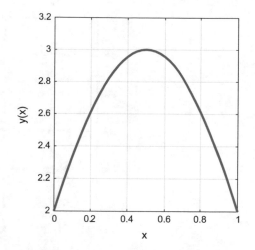

Problem 5.2.17 Use Shooting Method including MATLAB's built-in fzero function to solve the following second order nonlinear BVP, $y'' = Ay^4 + By - C$,
$y(0) = 277$, $y(4) = 352$,

$$A = 2.7 \times 10^{-9}, \quad B = 5 \times 10^{-2}, \quad C = 14.32.$$

(shootfzero2.m), (Rsd.m).

Solution
The nonlinear second-order ODE is first expressed as two first-order ODEs as

$$y'_1 = y_2,$$

$$y'_2 = Ay_1^4 + By_1 - C$$

A MATLAB anonymous function computes the right-hand sides of these
equations.

The initial slope (the root of residual function) and the slope at right-end
boundary are computed as $y'(0) = -20.7793$, $y'(4) = 63.9573$, respectively. The
solution is displayed in Fig. 5.28.

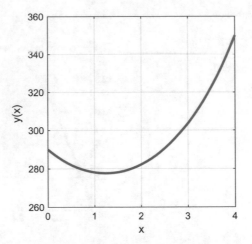

Fig. 5.28 The graph of the BVP, $y'' = 2.7 \times 10^{-9}y^4 + 0.05y - 14.32$, $y(0) = 277$, $y(4) = 352$, using shooting method combined with MATLAB's fzero built-in function

```
%shootfzero2.m     y"=Ay^4+By-C      y(0)=277, y(4)=350
clc;clear;close;
global alpha beta span    A B C
alpha=290; beta=350; span=[0 4]; A=2.7e-9; B=0.05; C=14.32;
root=fzero(@Rsd, -40)% guess  a zero for residual
fun=@(x,y) [y(2); A*y(1)^4+B*y(1)-C];
[x,y]=ode45(fun, span, [alpha, root])
plot(x,y(:,1),'linewidth',2);grid;xlabel('x'),ylabel('y(x)');

function z=Rsd(dy0est)
global alpha beta span A  B  C
fun=@(x,y) [y(2);  A*y(1)^4+B*y(1)-C];
[x,y]=ode45(fun,span,[alpha, dy0est]); z=y(end,1)-beta; %residual
```

Problem 5.2.18 If the ODE is linear,[4] any root-computing technique needs only a single interpolation to find initial estimate value, u. Therefore a linear interpolation may replace any of the root-finding method whenever the ODE is linear.

Use Shooting Method to solve the linear BVP,

$$y'' - (1+x^2)y = 0, \quad y(0) = 1, \quad y(2) = 0$$

Solution
We can use linear interpolation method for the solution of this BVP.

The zero of a linear function by straight line interpolation is

---

[4]The study of "nonlinear methods of shooting" brings more insight to the study of linear Shooting techniques, helping to devise more general algorithms for BVP solvers.

$$r = x_2 - f(x_2)\frac{x_2 - x_1}{f(x_2) - f(x_1)}$$

$$y(0) = \alpha = 1, \quad y'(0) = u, \quad y(2) = \beta = 0, \quad y'(2) = p$$

Computed boundary values are $y'(0) = -1.1337$, $y'(2) = -0.15343$.

Solution curve of this linear BVP using linear interpolation for computing the zero of boundary residual function is given in Fig. 5.29.

```
% shooting_lint1.m, y''-(1+x^2)y=0 , y(0)=1,y(2)=0, y(0)=alpha, y(2)=beta
clc;clear;close;format shortE; format compact
global span alpha beta
span=[0;2];% Interval
alpha=1;beta=0;
u=[-2 2]    % initial guess values
odes=@(x,y) [y(2); (1+x^2)*y(1)];
%w = fzero(@res,u)
w = linInterp(@res,u);
[X,Y] = ode45(odes,span,[alpha  w]);
X;Y, plot(X,Y(:,1),'linewidth',2);xlabel('x');ylabel('y(x)');grid;

function r = linInterp(fx,x)
% Computes r for linear  f(r)=0 by lin. interpolation between x1 and x2.
% fx function handle
x1=x(1);x2=x(2); f1 = feval(fx,x1); f2 = feval(fx,x2);
r = x2 - f2*(x2 - x1)/(f2 - f1);
end

function r = res(u) % Boundary residuals.
global  span alpha  beta
odes=@(x,y) [y(2); (1+x^2)*y(1)]; [~,Y] = ode45(odes,span,[alpha u]);
r= Y(end,1)-beta;
end
```

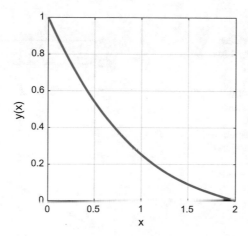

Fig. 5.29 Solution curve of the linear BVP, $y'' - (1+x^2)y = 0$, $y(0) = 1$, $y(2) = 0$

Problem 5.2.19 (*Bifurcation point*) Consider the nonlinear autonomous BVP,

$$y'' = -e^y, \quad y(0) = 0, \quad y(1) = 1.4.$$

Determine the value of initial slope (bifurcation point) of the function (with an accuracy better than 10^{-4}) beyond which the BVP will have two solutions (shooting_fzero1.m).

Solution

Two out of four boundary conditions are known,

$$y(0) = 0, \quad y'(0) = u, \quad y(1) = 1.4, \quad y'(1) = p$$

Let the initial slope (bifurcation point) of the function be $y'(0) = u$.

By using Shooting Method (shooting_fzero1.m), it is observed that the values of $u \geq 5.73458$ yields two solutions as depicted in Fig. 5.30.

```
%shooting_fzero1.m, y'' = -exp(y), y(0)=alpha, y(1)=beta
clc;clear;close;format shortE; format compact
global span alpha beta
span=[0;1];% Interval
alpha=0; beta=1.4
for i=1:2
if i==1, u = 5.73458; else, u=0 ;end  % Trial value
odes=@(x,y) [y(2); -exp(y(1))];
w = fzero(@res_fzero1,u); [X,Y] = ode45(odes,span,[alpha  w]);
X;Y, plot(X,Y(:,1),'linewidth',2);hold on;
end
xlabel('x');ylabel('y(x)'); grid; legend('u= 5.73458','u=0');

function r = res_fzero1(u) % Boundary residuals.
global  span alpha  beta
odes=@(x,y) [y(2); -exp(y(1))];[~,Y] = ode45(odes,span,[alpha  u]);
r= Y(end,1)-beta; end
```

Fig. 5.30 The slope values of $y'(0) \geq 5.73458$ yields two solutions for the BVP, $y'' = -e^y$, $y(0) = 0$, $y(1) = 1.4$. There exists single solution curve when $y'(0) < 5.73458$, however, the second solution curve (in blue color) appears as (additional) second solution for the BVP above this (bifurcation) slope threshold

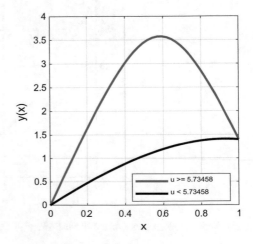

Problem 5.2.20 (*Boundary condition at infinity*) Solve the nonlinear autonomous BVP,

$$y'' + 3yy' = 0, \quad y(0) = 0, \quad y(\infty) = 1,$$

using single shooting with secant method (shooting77.m).

Solution

This BVP could not be solved analytically via Symbolic commands.

Re-writing given ODE as a system of two first order ODEs gives

$$y_1' = y_2$$

$$y_2' = -3yy', \quad y_1(0) = 0, \quad y_1(\infty) = 1$$

Let us arbitrarily assign $5 \to \infty$. Applying a nonlinear BVP solution algorithm (single shooting with secant method), we get the solution curves of Fig. 5.31 along with a plot showing the convergence of initial value of the first derivative versus the number of iterations.

A MATLAB m-file listed below (shooting77.m) applies nonlinear single shooting using secant method for the solution of given BVP which includes an infinite boundary condition.

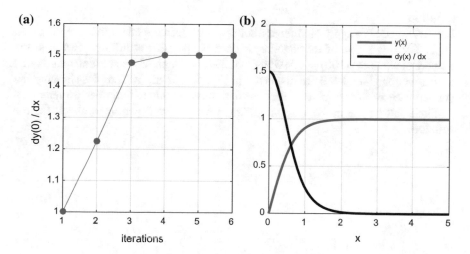

Fig. 5.31 **a** The convergence of initial value of the first derivative versus the number of iterations. **b** Graph of solution and its first derivative for the BVP, $y'' = -3yy'$, $y(0) = 0$, $y(\infty) = 1$

```
%shooting77.m % nonlinear shooting using secant method, infinite bc
%            y" = -3y'y,   y(0)=0, y(inf)=1
clc;clear;close;
% Let 5--> inf
span=[0 5]; bv=[0 1];
alpha=bv(1);beta=bv(2);
F=@(x,y) [y(2);   -3*y(2)*y(1)];
[x,y1]=ode45(F,span,[bv(1),alpha]);[x,y2]=ode45(F,span,[bv(1),beta]);
i=1;
m1=y1(end,1)-bv(2);m2=y2(end,1)-bv(2);
while(abs(beta-alpha)>1e-6)
xp=beta;
beta=alpha-(alpha-beta)/(m1-m2)*m1; %secant
alpha=xp;
[x,y1]=ode45(F,span,[bv(1),alpha]);[x,y2]=ode45(F,span,[bv(1),beta]);
m1=y1(end,1)-bv(2);m2=y2(end,1)-bv(2);
A(i)=alpha; %for convergence study
i=i+1;
end
i=1:length(A);% convergence analysis
plot(i,A,'.-','markersize',20);grid;
xlabel('iterations');ylabel('dy(0) / dx');figure;
y=y2;
plot(x,y,'linewidth',2);xlabel('x');grid;legend('y(x)','dy(x) / dx');
```

Proõblem 5.2.21 (*Boundary condition at infinity*) Use a Nonlinear Shooting method to solve the linear BVP, $y'' = -y + 1 - 1.01e^{-0.1x}$, $y(0) = 0$, $y(\infty) = 1$ (shooting88.m).

Solution

Let $100 \to \infty$. We apply Nonlinear Shooting with secant method to solve the given linear BVP with one of its boundary values is specified at infinity, and note the numerical value at y(100) = 0.999999935229229. Graph of the solution and its first derivative for given BVP are displayed in Fig. 5.32 along with a plot showing the convergence of initial value of the first derivative versus the number of iterations.

Final part of the m-file (shooting88.m) uses dsolve function to check for the solution.

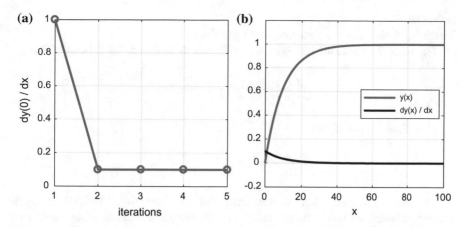

Fig. 5.32 a The convergence of initial value of the first derivative versus the number of iterations.
b Graph of solution and its first derivative for $y'' = -y + 1 - 1.01e^{-0.1x}$, $y(0) = 0$, $y(\infty) = 1$

```
%shooting88.m    y" = 1-y-1.01exp(-0.1x),  y(0)=0, y(inf)=1
clc;clear;close;
span=[0 100];bv=[0  1]; alpha=bv(1);beta=bv(2);
F=@(x,y) [y(2);  1-y(1)-1.01*exp(-0.1*x)];
[x,y1]=ode45(F,span,[0,alpha]);[x,y2]=ode45(F,span,[0,beta]);
i=1;
m1=y1(end,1)-bv(2);m2=y2(end,1)-bv(2);
while(abs(beta-alpha)>1e-6)
xp=beta; beta=alpha-(alpha-beta)/(m1-m2)*m1; %secant
alpha=xp;
[x,y1]=ode45(F,span,[0,alpha]);[x,y2]=ode45(F,span,[0,beta]);
m1=y1(end,1)-bv(2);m2=y2(end,1)-bv(2); A(i)=alpha; i=i+1;
end
i=1:length(A);% convergence analysis
plot(i,A,'o-','linewidth',2);grid;xlabel('iterations');ylabel('dy(0) / dx')
figure; y=y2;
plot(x,y,'linewidth',2);xlabel('x');grid;legend('y(x)','dy(x) / dx');
%%Check
% clc;clear; syms  y(x)
% eq='D2y+y=1-1.01*exp(-0.1*x)';bc='y(0)=0,y(100)=1';
% S=dsolve(eq,bc,x);simplify(S)
% x = linspace(0,100,1000); y=eval(vectorize(S));
% plot(x,y,'.-','markersize',24);grid
```

Problem 5.2.22 (*Shooting Method for a Singular BVP with a Neumann boundary condition*) Solve the BVP,

$$y'' + \frac{2}{x}y' = \frac{y}{1+y}, \quad y'(0) = 0, \quad y(1) = 1$$

Using Shooting Method combined with the Bisection Method (shooting999.m).

Solution

There is no closed form solution of the problem.

The second term in the left side of equation is evaluated using L'hospital's rule,

$$\lim_{x \to 0}\left(\frac{y'}{x}\right) = \lim_{x \to 0}\left(\frac{y''}{x'}\right) = \frac{y''}{1} = y''$$

Therefore, the ODE becomes

$$y'' + 2y'' = 3y'' = \frac{y}{1+y}$$

For using Shooting Method combined with the Bisection Method for given Neumann boundary condition at the left-end of interval of definition, we guess two initial values of the boundary at the left-end point (rather than its derivative which is used for Dirichlet boundary condition), so that shooting at these values gives neighbouring (upper and lower) values for the right-end boundary condition,

$$y_{est}(0) = 0.7 \to \beta = 0.7693, \quad y_{est}(0) = 1.3 \to \beta = 1.3947$$

Using these initial estimates in MATLAB m-file (shooting999.m), computed initial value is found to be $y(0) = 0.9196$, while $y'(1) = 0.1620$. For the error tolerance value of 10^{-6}, convergence to solution occurs at the end of 18th iteration. Figure 5.33 displays both of these curves, $y(x)$ and $y'(x)$.

```
%shooting999.m  Shooting with Bisection, 3*y"=y/(1+y), y'(0)=0, y(1)=1
clc;clear;close;
a=0;b=1; dy0=0; % Neumann boundary condition at leftend boundary
beta=1;  % Given boundary condition  on the rightend (target)
tol = 1e-6; % prescribed tolerance
F =@(x,y) [y(2); y(1)/(3*(1+y(1)))];
%-----------------------------------------------------
% %select bracketing values of beta
% for i= 1:15; y0(i)=i*0.1;
% ic1 = [y0(i); dy0];  % initial condition (guess, 15 elements)
% [x,y]=ode45(F,[a,b],ic1); betaest(i) = y(end,1); % estimate, bv(b)
% end
% [y0' betaest']
%-----------------------------------------------------
%betaL=0.7693 ;    % lower  guessed value for beta
%betaU=1.3947 ;    % upper  guessed value for beta
xL = 0.7; xU =  1.3; imax = 25; for i = 1:imax
xr = xL +0.5*(xU-xL);ic = [xr; dy0]; % Set initial state vector
[x,y] = ode45(F,[a,b],ic); Y(i) = y(end,1); err = Y(i) - beta ;
if abs(err) < tol, break, end
if err > 0; xU = xr; else, xL = xr; end % Adjust xU and/or xL
end
y, i, Y(i), plotyy(x,y(:,1),x,y(:,2));legend('y(x)','dy(x)/dx');
grid;xlabel('x');
```

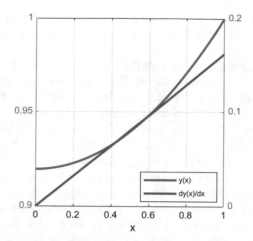

Fig. 5.33 Graph of the function and its first derivative for the solution of BVP, $y'' + (2/x)y' = y/(1+y)$, $y'(0) = 0$, $y(1) = 1$

Problem 5.2.23 True or False?

(a) The essence in a shooting method is the replacement of boundary value problem by an initial value problem.

(b) Almost all numerical packages containing BVP solvers in the market operate on shooting method.

(c) If an induced stability is observed in a forward shooting method, most probably the reverse shooting will also show the similar instability.

(d) Guessing and systematic adjustment becomes necessary in the treatment of nonlinear problems.

(e) In an iterative solution of a nonlinear BVP, nearer the starting approximation, the more likely is convergence to that solution.

Answers
(a) T, (b) F, (c) F, (d) T, (e) T.

5.3 Shooting Method for the Solution of High Order and the Systems of BVPs

Problem 5.3.1 Briefly describe some engineering applications of higher order BVPs.

Answer
BVPs with orders equal to or higher than three arise in modeling of various engineering processes. For example, Falkner-Skan's treatment of the laminar boundary layer of an incompressible fluid and the Blasius equation are third order

nonlinear BVPs. Vibrational analysis of beam structures is governed by fourth order BVPs. Another fourth order BVP is the Orr-Sommerfeld equation in the field of hydrodynamic stability.

In electrical engineering, the induction motor behavior is represented by a fifth-order differential equation model [25]. Vibrations of ring structures are described by a sixth order differential equation. An eighth-order differential equation occurring in torsional vibration of uniform beams was investigated in [26].

When an infinite horizontal layer of fluid is heated from below and is subject to rotation, the instability sets in. When this instability sets in as overstability, it is represented by an eighth-order ordinary differential equation. If an infinite horizontal layer of fluid is heated from below, and a uniform magnetic field is applied as well across the fluid in the same direction as gravity and the fluid is subject to the action of rotation, the instability sets in. When this instability sets in as ordinary convection, it is modeled by tenth-order boundary value problem. A class of characteristic-value problems of high order (as high as 24) are known to arise in hydrodynamic and hydromagnetic stability [27, 28].

An ordinary convection yields a tenth order BVP, an ordinary overstability yields a 12th order BVP [29].

Various other examples as well as the existence and uniqueness of the solution for higher order BVPs are given in Agarwal's book [30].

Numerical approximations for these problems are challenging because of the higher order derivatives and boundary conditions involving higher order derivatives of the unknown function [31]. The first step in applying Shooting Method for the solution of a higher order BVP is to convert the ODE into a first order system of ODEs with the same boundary conditions.

Problem 5.3.2 The Blasius equation in fluid mechanics describes the boundary layer that forms near a flat plate with fluid moving by it. The nonlinear BVP is described by,

$$y''' + (1/2)yy'' = 0, \quad y(0) = y'(0) = 0, \quad y'(\infty) = 1$$

Use Newton's method for systems to solve Blasius equation. The aim is to evaluate the shear stress at the plate, thus determine the value of $y''(0)$, in particular. Plot the solution and its derivatives.

Is the use of Newton's Method (for systems) computationally an efficient choice for this particular problem? If not, propose alternative methods (shooting_Blasius.m).

Solution
We convert this BVP to a system of first-order differential equations,

$$y_1' = y_2$$

$$y_2' = y_3$$

$$y_3' = -(1/2)y_1 y_3$$

$$y_1(0) = y_2(0) = 0, \quad y_2(\infty) = 1$$

We need three initial conditions at $x = 0$ to solve this BVP using the shooting method, but only two of them are given. Let the unknown initial value be u_1, then we have the set of initial conditions,

$$y(0) = 0, \quad y'(0) = 0, \quad y''(0) = u_1$$

If the slope of the function is constant at $x = \infty$, this means the values of higher derivatives will tend to zero at $x = \infty$, thus the terminal conditions are

$$y(\infty) = \beta_1, \quad y'(\infty) = 1, \quad y''(\infty) = 0$$

Results: Computed initial and terminal values are $y''(0) = 0.33257$, $y(\infty) = 4.2849, y''(\infty) = 0.0023953$. Solution curves of $y(x), y'(x), y''(x)$ are displayed in Fig. 5.34.

Using Newton's Method for systems to solve Blasius equation is not computationally efficient choice. Since only a single initial estimate of $y''(0) = u_1$ is to be computed, a better choice is to employ Newton's Method for a single variable (as described earlier) or MATLAB's `fsolve` function, which is also included as a single line (in commented form) in the main m-file (shooting_Blasius.m). Hence, we can use either one of these functions for the solution of Blasius problem in computationally more efficient manner.

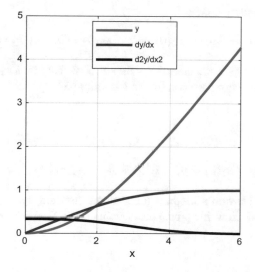

Fig. 5.34 Solution curves of the BVP, $y''' + (1/2)yy'' = 0$, $y(0) = y'(0) = 0$, $y'(\infty) = 1$

```
% shooting_Blasius.m,   y''' - 0.5*y*y" = 0, y(0)=y'(0)=0,y(Inf)=1
clc;clear;format shortE; format compact
global span
span=[0,6];   % Interval (Inf <-- 6)
u = 1;   % Initial trial value
odes=@(x,y) [y(2); y(3); -0.5*y(1)*y(3)];
w = newton_sys(@res_Blasius,u);
%w = fzero(@res_Blasius,u);
[X,Y] = ode45(odes,span,[0   0   w(1)]); X,Y
plot(X,Y,'linewidth',2);xlabel('x');grid;legend('y','dy/dx','d2y/dx2');

function r = res_Blasius(u) % Boundary residuals.
global   span
r = zeros(length(u),1);
odes=@(x,y) [y(2); y(3); -0.5*y(1)*y(3)];
[~,Y] = ode45(odes,span,[0   0   u(1)]);
r(1)= Y(end,2)-1;
end
```

Problem 5.3.3 Given is the following system of ODEs [32],

$$x_1' = x_2$$

$$x_2' = x_3$$

$$x_3' = -1.55x_1x_3 + 0.1x_2^2 + 1 - x_4^2 + 0.2x_2$$

$$x_4' = x_5$$

$$x_5' = -1.55x_1x_5 + 1.1x_2x_4 + 0.2x_4 - 0.2$$

subject to the boundary conditions,

$$x_1(0) = x_2(0) = x_4(0) = 0, \quad x_2(b) = 0, \quad x_4(b) = 1$$

Use Shooting Method to determine value of b so that given system of ODEs is solved for specified boundary conditions (shootingh51.m).

Solution

$$x_1(0) = x_2(0) = x_4(0) = 0, \quad x_3(0) = u_1 \quad x_5(0) = u_2$$

$$x_1(b) = p_1 \quad x_2(b) = 0, \quad x_3(b) = p_2, \quad x_4(b) = 1, \quad x_5(b) = p_3$$

When we use Newton's Method for systems (newton_sys.m), it is noted that iterations exceeded allowable preset count value (= 250), and following message is printed:

Warning: Failure at t=1.227708e+00. Unable to meet integration tolerances without reducing the step size below the smallest value allowed
(3.552714e-15) at time t.

Increasing error tolerance in Newton's Method by a factor of $\times 1000$, does not
improve the result.

When we use MATLAB's fsolve function, we obtain following results with
$b = \pi$:

$$x_3(0) = -0.9883, \quad x_5(0) = 0.6327$$
$$x_1(b) = -1.3966, \quad x_3(b) = 1.3064, \quad x_5(b) = -0.0563$$

The solution curves are plotted in Fig. 5.35.

```
% shootingh51.m   Shooting method, system of 5 ODEs
clc;clear;tic
global span
b=pi;
span=[0, b];  % Interval
u = [1 -1];   % Trial values
odes=@(x,y) [y(2); y(3);-1.55*y(1)*y(3)+0.1*y(2)*y(2)+1-y(4)*y(4)+0.2*y(2);
            y(5);-1.55*y(1)*y(5)+1.1*y(2)*y(4)+0.2*y(4)-0.2 ];
%w = fsolve(@res,u);
w=newton_sys(@res,u);
[X,Y] = ode45(odes,span,[0   0   w(1)  0   w(2)]);
X;Y, plot(X,Y(:,1),X,Y(:,2),X,Y(:,3),X,Y(:,4),X,Y(:,5),'linewidth',2);
xlabel('t');ylabel('x(t)');xlim([0,b]);legend('x1','x2','x3','x4','x5');
grid on;toc

function r = res(u) % Boundary residuals.
global  span
r = zeros(length(u),1);
odes=@(x,y) [y(2); y(3);-1.55*y(1)*y(3)+0.1*y(2)*y(2)+1-y(4)*y(4)+0.2*y(2);
            y(5);-1.55*y(1)*y(5)+1.1*y(2)*y(4)+0.2*y(4)-0.2 ];
[~,Y] = ode45(odes,span,[0   0   u(1)  0   u(2)]);
r(1)= Y(end,2);  r(2)  = Y(end,4)-1;
end
```

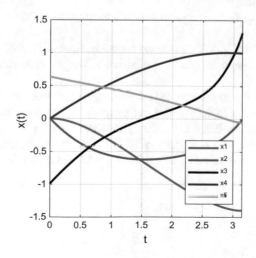

Fig. 5.35 The solution curves of given system of ODEs with the specified boundary conditions,
$x_1(0) = x_2(0) = x_4(0) = 0$, $x_2(b) = 0$, $x_4(b) = 1$, $b = \pi$

Problem 5.3.4 Given the following system of ODEs [33],

$$x_1' = -2x_2$$

$$x_2' = x_3$$

$$x_3' = x_1 x_3 + x_2^2 - x_4^2 + k$$

$$x_4' = x_5$$

$$x_5' = 2x_2 x_4 + x_1 x_5$$

and the boundary conditions,

$$x_1(0) = x_2(0) = 0, \quad x_4(0) = 1$$
$$x_1(b) = x_2(b) = 0, \quad x_4(b) = 0.5$$

Use Shooting Method to solve this eigenvalue problem (as k is to be determined so that all six boundary conditions are satisfied). Let the interval b as large as possible (shootingh52.m).

Solution

We replace k by a new dependent variable, x_6, and add the extra equation, $x_6' = 0$.
   Boundary conditions become,

$$x_1(0) = x_2(0) = 0, \quad x_3(0) = u_1, \quad x_4(0) = 1, \quad x_5(0) = u_2, \quad x_6(0) = u_3$$
$$x_1(b) = x_2(b) = 0, \quad x_3(b) = p_1, \quad x_4(b) = 0.5, \quad x_5(b) = p_2, \quad x_6(b) = p_3$$

Application of Newton's Method for systems (newton_sys.m) is found to be unsuccessful.
   A warning message is:
`Matrix is singular, close to singular or badly scaled. Results may be` inaccurate.
   When we use MATLAB's `fsolve` function, following results are obtained.
   Trial initial vector selection is a decisive factor on extending the interval of definition.
   For example, if the trial vector is $[1\ 1\ -1]$, maximum value of $b = 3.5467$, while $k = 0.5338$. Selection of initial trial vector as $[1\ -1\ 0]$ gives maximum value of $b = 8.5830$, while $k = 0.5264$. On the other hand, when the initial vector is $[1\ 0\ -1]$, maximum value of $b = 9.3688$, while $k = 0.5258$. When the initial trial vector is $[0\ 1\ -1]$, maximum value of $b = 9.9699$, and $k = 0.5253$.

Solution curves for the last case are displayed in Fig. 5.36.

```
% shootingh52.m  Shooting method, system of 5 ODEs (Osborne,1969)
clc;clear;tic
global span
b=3.5467; u = [1  1  -1];    % Trial values   k=0.5338
%b=9.9699; u = [0  1  -1];   % Trial values   k=0.5253
%b=9.3688; u = [1  0  -1]    % Trial values   k=0.5258
%b=8.5830; u = [1 -1  0];    % Trial values   k=0.5264
span=[0, b];        % Interval
odes=@(x,y) [-2*y(2); y(3); y(1)*y(3)+y(2)*y(2)-y(4)*y(4)+y(6);
             y(5); y(1)*y(5)+2*y(2)*y(4);0];
%w = fsolve(@resz,u);
w=newton_sys(@resz,u);
[X,Y] = ode45(odes,span,[0  0  w(1)  1  w(2) w(3)]);
X;Y
plot(X,Y(:,1),X,Y(:,2),X,Y(:,3),X,Y(:,4),X,Y(:,5),X,Y(:,6),'linewidth',2);
xlabel('t');ylabel('x(t)');xlim([0,b]);
legend('x1','x2','x3','x4','x5','x6');grid on;toc

function r = resz(u) % Boundary residuals.
global  span
r = zeros(length(u),1);
odes=@(x,y) [-2*y(2); y(3); y(1)*y(3)+y(2)*y(2)-y(4)*y(4)+y(6);
             y(5); y(1)*y(5)+2*y(2)*y(4);0];
[~,Y] = ode45(odes,span,[0  0  u(1)  1  u(2) u(3)]);
r(1)= Y(end,1); r(2)= Y(end,2); r(3) = Y(end,4)-0.5;
end
```

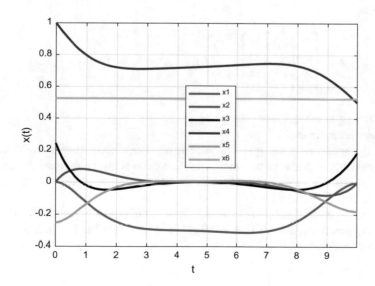

Fig. 5.36 Solution curves for b = 9.9699; u = [0 1 −1]; k = 0.5253

Problem 5.3.5 Use Shooting Method combined with built-in function `fsolve` to solve the following fourth order linear BVP,

$$\frac{d^{(4)}y}{dx^{(4)}} = x, \quad y(0) = y''(0) = y(1) = y''(1) = 0 \tag{5.18}$$

Solution
We need four initial conditions at $x = 0$ to solve this BVP using the shooting method, but only two of them are given. Let the two unknown initial values be u_1, u_2, then we have the set of initial conditions,

$$y(0) = 0, \quad y'(0) = u_1, \quad y''(0) = 0, \quad y'''(0) = u_2 \tag{5.19}$$

If Eq. (5.18) is solved with the shooting method using the initial conditions in Eq. (5.19), the computed boundary values at $x = 1$ depend on the choice of u_1, u_2. We express this dependence as

$$y(1) = \varphi_1(u_1, u_2), \quad y''(1) = \varphi_2(u_1, u_2) \tag{5.20}$$

The correct choice of u_1, u_2 satisfies the equations

$$\varphi_1(u_1, u_2) = 0, \quad \varphi_2(u_1, u_2) = 0 \tag{5.21}$$

The equivalent first-order equations and the boundary conditions can be written as

$$\mathbf{y}' = \begin{bmatrix} y_1' \\ y_2' \\ y_3' \\ y_4' \end{bmatrix} = \begin{bmatrix} y_2 \\ y_3 \\ y_4 \\ x \end{bmatrix}, \quad y_1(0) = y_3(0) = y_1(1) = y_3(1) = 0$$

Following is the output of the code:

`Equation solved.`

`fsolve completed because the vector of function values is near zero as measured by the default value of the function tolerance, and the problem appears regular as measured by the gradient.`

`w = 0.0194 -0.1667`

which indicates that $y'(0) = u_1 = 0.0194$, $y'''(0) = u_2 = -0.1667$. The code also computes the values $y'(1) = -0.0222$, $y'''(1) = 0.3333$. The solution curve is displayed in Fig. 5.37.

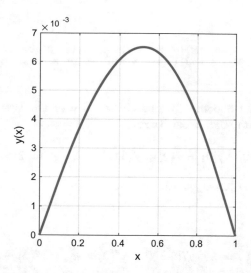

Fig. 5.37 The solution curve for the BVP, $y^{(4)} = x$, $y(0) = y''(0) = y(1) = y''(1)$

```
% shooting_4th_order.m  Shooting method for 4th-order BVP
clc; clear;
span=[0,1]; % Interval
u = [0 1];  % Trial values
odes=@(x,y) [y(2); y(3); y(4); x]; w = fsolve(@res2,u)
[X,Y] = ode45(odes,span,[0 w(1) 0 w(2)]);
plot(X,Y(:,1),'linewidth',2); xlabel('x'); ylabel('y(x)'); grid;

function r = res2(u) % residuals
global  span
r = zeros(length(u),1); odes=@(x,y) [y(2); y(3); y(4); x];
[~,Y] = ode45(odes, span,[0 u(1) 0 u(2)]);
r(1)- Y(end,1); r(2) - Y(end,3);
```

Note The closed form solution is obtained by direct integration of the ODE, or by MATLAB dsolve function as

$$y(x) = \frac{3x^5 - 10x^3 + 7x}{360}$$

In this case, MATLAB dsolve function has a simpler script for the solution of given BVP,

```
clc;clear;syms  y(x)
eq='D4y=x';  bc='y(0)=0,  D2y(0)=0,  y(1)=0,  D2y(1)=0';S=dsolve(eq,bc,x);
x = linspace(0,1,20);  y=eval(vectorize(S));plot(x,y);grid;
```

Problem 5.3.6 Use Shooting Method combined with `fsolve` function to solve the following fourth order linear BVP,

$$y^{(4)} = 0, \quad y(0) = y'(0) = y'(1/2) = 0, \quad y'''(1/2) = -1/2$$

(shootingh4.m), (dsolve_y4.m).

Solution

$$y(0) = 0, \quad y'(0) = 0, \quad y''(0) = u_1, \quad y'''(0) = u_2$$
$$y(1/2) = p_1, \quad y'(1/2) = 0, \quad y''(1/2) = p_2, \quad y'''(1/2) = -1/2$$

There are two unknown boundary conditions at each end of given interval. We choose shooting combined with `fsolve` function. These computations yield $y''(0) = 0.125$, $y'''(0) = -0.5$. Solution curve is displayed in Fig. 5.38.

The closed form solution which is obtained by direct integration of the ODE, or by MATLAB (dsolve_y4.m) verifies the result, $y(x) = (-1/12)x^3 + (1/16)x^2$.

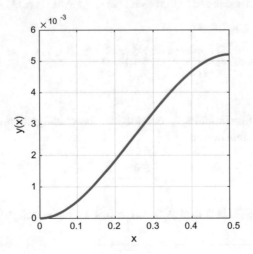

Fig. 5.38 Solution curve for the BVP, $y^{(4)} = 0$, $y(0) = y'(0) = y'(1/2) = 0$, $y'''(1/2) = -1/2$, using shooting method

```
% shootingh4.m  Shooting method for 4th-order BVP, y''''=0
clc;clear;close;
global span
span=[0, 1/2] ;%Interval
u = [0 1]; % Trials
odes=@(x,y) [y(2); y(3); y(4); 0];
z = fsolve(@res4,u)
[X,Y] = ode45(odes,span,[ 0  0  z(1)  z(2) ]);
X,Y, plot(X,Y(:,1),'linewidth',2);xlabel('x');ylabel('y(x)');grid;

function r = res4(u) % Boundary residuals.
global  span
r = zeros(length(u),1);
odes=@(x,y) [y(2); y(3); y(4); 0];
[~,Y] = ode45(odes,span,[0  0  u(1)  u(2)]);
r(1)= Y(end,2);r(2)  = Y(end,4)-(-1/2);
end

%dsolve_y4.m  d^4y/dx^4=0, y(0)=y'(0)=y'(1/2)=0, y'''((1/2)=-1/2
clc; clear; syms  y(x)
eq='D4y=0';bc='y(0)=0,Dy(0)=0,Dy(1/2)=0,D3y(1/2)=-1/2';
S=dsolve(eq,bc,x),  y2=diff(diff(S));y3=diff(y2);
y20=subs(y2,x,0) %D2y(0)
y30=subs(y3,x,0) %D3y(0)
x = linspace(0,1/2);y=eval(vectorize(S));
plot(x,y,'linewidth',2);grid;xlabel('x');ylabel('y(x)');
```

Problem 5.3.7 Solve the following third order linear BVP using shooting method,

$$y''' - y'' - 6y + 6x = 0, \quad y(0) = 0, \quad y(2) = 1.8899, \quad y'(2) = 0 \qquad (5.22)$$

Solution
We need three initial conditions at $x = 0$ to solve this BVP using the shooting method, but only one of them are given. Let the two unknown initial values be u_1, u_2, then we have the set of initial conditions,

$$y(0) = 0, \quad y'(0) = u_1, \quad y''(0) = u_2 \qquad (5.23)$$

If Eq. (5.22) is solved with the shooting method using the initial conditions in Eq. (5.23), the computed boundary values at $x = 2$ depend on the choice of u_1, u_2. The set of first-order equations and the boundary conditions are

$$y' - \begin{bmatrix} y_2 \\ y_3 \\ y_3 + 6y_1 - 6x \end{bmatrix}$$

$$y(0) = 0, \quad y(2) = 1.8899, \quad y'(2) = 0$$

Following is the output of the code: $y'(0) = 3.2088$, $y''(0) = -2.7100$, $y''(2) = 0$. The solution curve is displayed in Fig. 5.39.

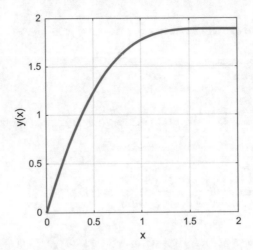

Fig. 5.39 The solution curve for the third order linear BVP using shooting method, $y''' - y'' - 6y + 6x = 0$, $y(0) = 0$, $y(2) = 1.8899$, $y'(2) = 0$

```
% shooting_3th_order.m  Shooting method for 4th-order linear BVP
clc;clear;close;
global span
span=[0,2] ;%Interval
u = [1 -1]; % Trials
odes=@(x,y) [y(2); y(3); y(3)+6*y(1)-6*x];
z = fsolve(@res3,u);
[X,Y] = ode45(odes,span,[0 z(1) z(2)]);
X,Y
plot(X,Y(:,1),'linewidth',2);xlabel('x');ylabel('y(x)');grid;

function r = res3(u) % Boundary residuals.
global  span
r = zeros(length(u),1);
odes=@(x,y) [y(2); y(3); y(3)+6*y(1)-6*x];
[X,Y] = ode45(odes,span,[0 u(1) u(2)])
r(1)= Y(end,2); r(2) = Y(end,3);
```

Alternative solution of the problem using dsolve function verifies the results;

```
clc;clear;
eq='D3y=D2y+6*y-6*x'; bc='y(0)=0, y(2)=1.8899, Dy(2)=0';
y=dsolve(eq,bc,x); h=ezplot(y,[0,2]);
```

Problem 5.3.8 Use Shooting method to solve the fourth order homogeneous linear BVP,

$$y^{(4)} - y = 0, \quad y(0) = y''(0) = y(\pi) = y''(\pi) = 0$$

(shootingh5.m).

Solution

We find that the initial (guess) values of $[1\ -1]$ yields following values at initial and terminal boundaries,

$(1.0e-07*)$	y	y'	y''	y'''
	0	-0.3714	0	0.3714
	0.0000	0.3714	0.0000	-0.3714

when the initial (guess) values of $[-1\ 1]$ are used, values at boundaries are,

$(1.0e-07*)$				
	0	0.3714	0	-0.3714
	0.0000	-0.3714	0.0000	0.3714

The solution curves for the BVP is shown in Fig. 5.40.

```
% shootingh5.m  Shooting method, y''''-ky=0
clc;clear;
global span
span=[0,pi];  % Interval
%u = [-1 1]; % Trial values
u = [1 -1];  % Trial values
odes=@(x,y) [y(2); y(3); y(4); y(1)];
w = fsolve(@res5,u); [X,Y] = ode45(odes,span,[0 w(1) 0  w(2)]);
X,Y,plot(X,Y(:,1),'linewidth',2);xlabel('x');ylabel('y(x)');hold on;
xlim([0 pi]);legend('u=[1 -1]','u=[-1 1]');grid on;

function r = res5(u) % Boundary residuals.
global  span
r = zeros(length(u),1);
odes=@(x,y) [y(2); y(3); y(4); y(1)];
[~,Y] = ode45(odes,span,[0 u(1) 0 u(2)]);
r(1)= Y(end,1); r(2) = Y(end,3);
end
```

Problem 5.3.9 Use Shooting Method to solve the fifth order BVP [31],

$$y^{(5)} - y = -15e^x - 10xe^x,$$
$$y(0) = 0, \quad y'(0) = 1, \quad y''(0) = 0, \quad y(1) = 0, \quad y'(1) = -e$$

(shooting6.m), (res6.m).

Solution

In order to apply the Shooting Method to given BVP, let $y'''(0) = u_1$, $y^{(4)}(0) = u_2$.

Five specified and five computed initial (left-end, first line) and terminal (right-end, second line) boundary values are as follows;

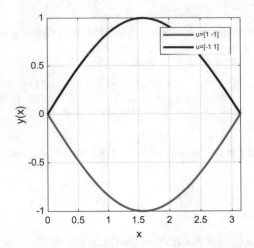

Fig. 5.40 Two solution curves for the BVP, $y^{(4)} - y = 0$, $y(0) = y''(0) = y(\pi) = y''(\pi) = 0$

y1	y2	y3	y4	y5
0	1.0000e+00	0	-3.0000e+00	-8.0000e+00
6.0373e-09	-2.7183e+00	-1.0873e+01	-2.4465e+01	-4.3493e+01

The solution curve $y(x)$ displayed in Fig. 5.41, which is in agreement with the exact solution, $y(x) = x(x - 1)e^x$.

MATLAB m-file (shooting6.m) and function file (res6.m) are given below.

```
% shooting6.m  Shooting method, y""-y=-15*exp(x)-10*x*exp(x)
clc;clear;format shortE; format compact
global span
span=[0,1];  % Interval
u = [-1 1];  % Trial values
odes=@(x,y) [y(2); y(3); y(4);y(5); y(1)-15*exp(x)-10*x*exp(x)];
w = fsolve(@res6,u);[X,Y] = ode45(odes,span,[0   1   0  w(1)   w(2)]);
X,Y, plot(X,Y(:,1),'linewidth',2); xlabel('x'); ylabel('y(x)'); grid;


function r = res6(u) % Boundary residuals.
global  span
r = zeros(length(u),1);
odes=@(x,y) [y(2); y(3); y(4);y(5); y(1)-15*exp(x)-10*x*exp(x)];
[~,Y] = ode45(odes,span,[0   1   0  u(1)   u(2)]);
r(1)= Y(end,1); r(2) = Y(end,2)+exp(1);
```

Problem 5.3.10 Use Shooting Method to solve the sixth order BVP [31],

$$y^{(6)} - y = -6e^x,$$
$$y(0) = 1, \quad y'(0) = 0, \quad y''(0) = -1, \quad y(1) = 0, \quad y'(1) = -e, \quad y''(1) = -2e$$

(shooting7.m), (res7.m).

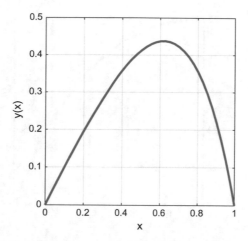

Fig. 5.41 Solution curve obtained by the shooting method for the fifth order BVP, $y^{(5)} - y = -15e^x - 10xe^x$, $y(0) = 0$, $y'(0) = 1$, $y''(0) = 0$, $y(1) = 0$, $y'(1) = -e$

Solution

In order to apply the Shooting Method to given sixth order BVP, let $y'''(0) = u_1$, $y^{(4)}(0) = u_2$, $y^{(5)}(0) = u_3$. Six specified and six computed initial (left-end, first line) and terminal (right-end, second line) boundary values are as follows;

y1	y2	y3	y4	y5	y6
1.00e+00	0	-1.00e+00	-2.00e+00	-3.00e+00	-4.00e+00
-5.33e-08	-2.71e+00	-5.43e+00	-8.15e+00	-1.08e+01	-1.35e+01

Actual solution is $y(x) = -(x-1)e^x$. Infinite norm of approximation error is computed as $err(\infty) = 9.0675 \times 10^{-8}$. The solution curve $y(x)$ displayed in Fig. 5.42.

MATLAB m-file (shooting7.m) and function file (res7.m) are given below.

```
% shooting7.m  Shooting,  6th order BVP,  y""" - y = - 6*exp(x)
clc;clear;format shortE; format compact
global span
span=[0,1];   u = [-1 1 0]; % Trial values
odes=@(x,y) [y(2); y(3); y(4);y(5);y(6); y(1)-6*exp(x) ];
w = fsolve(@res7,u)
[X,Y] = ode45(odes,span,[1  0  -1  w(1)  w(2) w(3)]);
X,Y, plot(X,Y(:,1),'linewidth',2);xlabel('x');ylabel('y(x)');grid;
yact=(1-X).*exp(X); err=(Y(:,1)-yact); norminf=norm(err,Inf)

function r = res7(u) % Boundary residuals.
global  span
r = zeros(length(u),1);
odes=@(x,y) [y(2); y(3); y(4);y(5);y(6); y(1)-6*exp(x) ];
[~,Y] = ode45(odes,span,[1  0  1  u(1)  u(2) u(3)]);
r(1)= Y(end,1); r(2) = Y(end,2)+exp(1); r(3) = Y(end,3)+2*exp(1);
```

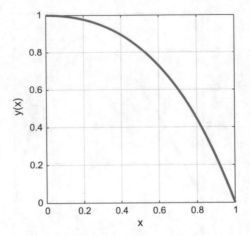

Fig. 5.42 Solution curve for the sixth order BVP, $y^{(6)} - y = -6e^x$, $y(0) = 1$, $y'(0) = 0$, $y''(0) = -1$, $y(1) = 0$, $y'(1) = -e$, $y''(1) = -2e$

5.4   Multiple Shooting Method

Problem 5.4.1

(a) Explain the need for the use of Multiple (Parallel) Shooting Method (MSM).
(b) Describe the method of Direct (Standard) Multiple Shooting.

Solution

(a) A major disadvantage of the single shooting method is the round-off error
accumulation, particularly occurring when an unstable IVP has to be integrated.
This results in an exponential error bound, with the length of the interval being
part of the exponent. There can be some problems that even the computation of
approximate initial value to full machine accuracy does not guarantee that
additional values of solution can be determined accurately [34]. In order to
decrease this bound, it is natural to restrict the size of domains over which IVPs
are integrated. Thus, the interval of integration is subdivided and initial value
integrations are performed on each subinterval. The resulting solution segments
are patched up to form a continuous solution over the entire interval [a, b],
leading to the method of multiple shooting. In order to obtain a solution to the
given differential equation on complete interval $[a, b]$, the solutions on the
subintervals must also coincide at the grid points together with their first
derivatives [35, 36].

(b) Multiple (Parallel) Shooting Method:

In this method, interval of definition $[a, b]$ is subdivided into m subintervals

$$a = t_0 < t_1 < \cdots < t_{m-1} < t_m = b \qquad (5.24)$$

where t_k are the shooting points.[5]

On each sub-interval $[t_j, t_{j+1}]$, $0 \leq j \leq m - 1$, an IVP is defined as

$$u'_j(x) = f\left(x, u_j(x)\right), \quad t_j \leq x \leq t_{j+1}$$
$$u_j(t_j) = s_j \qquad (5.25)$$

where s_j are unknown vectors. The solutions $u_j(x; s_j)$ on the subintervals $[t_j, t_{j+1}], j = 0, 1, \ldots, m - 1$ are so organized that a continuous function results on the interval $[a, b]$, satisfying both the ODE and its boundary conditions;

Continuity (matching) conditions:

$$u_j(t_{j+1}; s_j) - s_{j+1} = 0, \quad j = 0, 1, \ldots, m - 2 \qquad (5.26)$$

The solution satisfies the boundary conditions,

$$g(s_0, u_{m-1}(b; s_{m-1})) = 0 \qquad (5.27)$$

A nm-dimensional system of nonlinear algebreic equations for the unknown initial vectors s_j is obtained by

$$F^{(m)}\left(s^{(m)}\right) = 0 \qquad (5.28)$$

$$s^{(m)} = [s_0, s_1, s_2, \ldots, s_{m-1}]^T,$$

$$F^{(m)}\left(s^{(m)}\right) = \begin{bmatrix} g(s_0, u_{m-1}(b; s_{m-1})) \\ u_0(t_1; s_0) - s_1 \\ u_1(t_2; s_1) - s_2 \\ \vdots \\ u_{m-2}(t_{m-1}; s_{m-2}) - s_{m-1} \end{bmatrix}$$

If a multidimensional Newton-like method is used for the solution of (5.28), the k th iteration step involves the solution of linear algebraic equations,

$$M_k^{(m)} c_k^{(m)} = q_k^{(m)} = q^{(m)}\left(s_k^{(m)}\right) \qquad (5.29)$$

where $M_k^{(m)}$ is the Jacobian matrix of size $(mn \times mn)$,

---

[5]This method is not thoroughly understood as is evidenced by the fact that there is no generally accepted rule for picking the nodes [37]. However, some theoretical results and available strong heuristic motivations are useful in selecting the shooting points.

$$M_k^{(m)} = M^{(m)}\left(s_k^{(m)}\right) \tag{5.30}$$

and $q_k^{(m)}$ is a vector of size (mn).

There are different numerical techniques for solving (5.28).

One method is based on the compactification. Another method uses LU factorization of $M_k^{(m)}$ with partial pivoting, scaling and iterative refinement, which has more stable nature as compared to compatification solution. Sparse matrix methods can solve the linear shooting equations, as well [38].

Problem 5.4.2 Consider the linear BVP,

$$y' = Ay + q, \quad a \le x \le b$$
$$B_a y(a) + B_b y(b) = \beta$$

with $a = 0$, $b = 6$, $B_a = B_b = I$,

$$A = \begin{bmatrix} 1 - 2\cos 2x & 0 & 1 + 2\sin 2x \\ 0 & 2 & 0 \\ -1 + 2\sin 2x & 0 & 1 + 2\cos 2x \end{bmatrix},$$

$$q = \begin{bmatrix} (-1 + 2\cos 2x - 2\sin 2x)e^x \\ -e^x \\ (1 - 2\cos 2x - 2\sin 2x)e^x \end{bmatrix}, \quad \beta = \begin{bmatrix} 1 + e^6 \\ 1 + e^6 \\ 1 + e^6 \end{bmatrix}$$

Solve this BVP using,

(a) Single Shooting Method combined with any nonlinear system solvers.
(b) Multiple Shooting Method.

Solution

It is known, a priori, that the exact solution is $y = \begin{bmatrix} e^x & e^x & e^x \end{bmatrix}^T$.

An explicit solution could not be found using dsolve function.

(a) Solution using Single Shooting Method;

Initial (first row) and final (second row) values of three solutions of given BVP are as follows;

```
        0.9840      1.0000      1.0000
      403.4448    403.4288    403.4288
Elapsed time is 0.155757 seconds.
```

Three solution curves for the given system of BVP are displayed in Fig. 5.43. (Individual curves can not be distinguished due to slight differences).

Fig. 5.43 Three solution
curves for the given system of
BVP. First solution is slightly
different than the others.
However, they can not be
distinguished from each other
in this figure

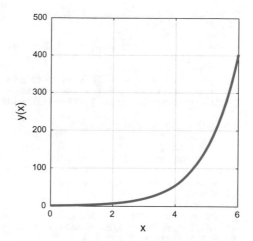

MATLAB code (shooting_system3.m) works for both MATLAB's fsolve and
(newton_sys.m) functions. This main file calls functions (ressys1.m) and an
equation solver (fsolve) or (newton_sys.m). The latter one calls a Jacobian matrix
computing function (jac.m). IVPs can be solved by employing an IVP solver from
MATLAB ODE Suite. In place of using a separate function file, the same anonymous function (odes) is used for the definition of the first order ODE system in the
main file as well as in the external function file (ressys1.m).

It has been observed that both nonlinear system solvers provide similar fast
execution time for the program.

```
% shooting_system3.m % BVP systems solution
% Nonlinear Shooting Method combined with Newton's Method
clc;clear;close;tic
global span beta
span=[0,6]; beta=[(1+exp(6)) (1+exp(6)) (1+exp(6))];
u = [1 1 1]; % Trial values
odes=@(x,y) [y(3)*(2*sin(2*x) + 1) - y(1)*(2*cos(2*x) - 1)...
    - exp(x)*(2*sin(2*x) - 2*cos(2*x) + 1);
    2*y(2) - exp(x);
    y(3)*(2*cos(2*x) + 1) + y(1)*(2*sin(2*x) - 1)...
    - exp(x)*(2*cos(2*x) + 2*sin(2*x) - 1)]
%w = fsolve(@ressys1,u)
w = newton_sys(@ressys1,u)
[X,Y] = ode45(odes,span,[ w(1)  w(2) w(3)]);
X,Y
plot(X,Y(:,1),'linewidth',2);xlabel('x');ylabel('y(x)');grid;toc
```

```
function r = ressys1(u) % Boundary residuals.
global  span beta
r = zeros(length(u),1);
odes=@(x,y) [y(3)*(2*sin(2*x) + 1) - y(1)*(2*cos(2*x) - 1)...
    - exp(x)*(2*sin(2*x) - 2*cos(2*x) + 1);
    2*y(2) - exp(x);
    y(3)*(2*cos(2*x) + 1) + y(1)*(2*sin(2*x) - 1)...
    - exp(x)*(2*cos(2*x) + 2*sin(2*x) - 1)];
[~,Y] = ode45(odes,span,[u(1)  u(2) u(3)]);  r=Y(1,:)+Y(end,:)-beta;


function r = newton_sys(f,u)
% Newton's method of finding  solution set of simultaneous
% equations fi(x1,x2,...,xn) = 0, i = 1,2,...,n.
% f = function handle  returning [f1,f2,...,fn].
% u = starting solution vector [x1,x2,...,xn].
if size(u,1) == 1; u = u'; end
tol=1e-6;
for i = 1:50
[J,fx] = jac(f,u); DP=dot(fx,fx);
if (DP/length(u))^(1/2) < tol, r = u; return, end
dx = J \(-fx); u = u + dx;
if sqrt(dot(dx,dx)/length(u)) < tol*norm(u,Inf), r = u; return, end
end, error(' iterations exceeded allowable count')


function [J,fx] = jac(f,x)
% J= Jacobian matrix
J = zeros(length(x)); fx = feval(f,x); fx=fx';
for k =1:length(x)
g = x(k); x(k) = g + 1.0e-5; fa = feval(f,x); fa=fa'; x(k) = g;
J(:,k) = (fa - fx)/1.0e-5;
end
```

(b) Solution using Multiple Shooting Method: MSM4

First, we use the MSM for two subintervals. Edited results are shown below;

n = 3, m = 2

Iteration	Func-count	f(x)	Norm of step	First-order optimality	Trust-region radius
0	1	1.92192e+10		1.11e+09	1
1	2	1.7042e+10	1	1.04e+09	1
2	3	1.21735e+10	2.5	8.81e+08	2.5
3	4	3.59292e+09	6.25	4.76e+08	6.25
4	5	1.94397e+07	15.625	1.78e+06	15.6
5	6	7.39735e-08	16.5362	2.18	39.1
6	7	6.89315e-21	6.11641e-08	6.66e-07	41.3
7	8	4.43693e-21	8.03578e-13	5.34e-07	41.3

Equation solved, inaccuracy possible.
The vector of function values is near zero, as measured by the selected value of the function tolerance. However, the last step was ineffective.

Y =

1.0000	20.0855	**403.6430**
1.0000	20.0855	**403.4561**
1.0000	20.0855	**402.6954**

It is observed that each of the last highlighted right-end boundary values are different.

We increase the number of partitions gradually, and observe that each right-end boundary values become identical (with better than 10^{-4} absolute accuracy), when the number of subintervals is 7 (or higher). Multiple Shooting points and solution curves for given BVP are displayed in Fig. 5.44.

n = 3, m = 7

Iteration	Func-count	f(x)	Norm of step	First-order optimality	Trust-region radius
0	1	3.78877e+06		1.36e+04	1
1	2	3.74617e+06	1	1.35e+04	1
2	3	3.64096e+06	2.5	1.33e+04	2.5
3	4	3.38602e+06	6.25	1.27e+04	6.25
4	5	2.79902e+06	15.625	1.11e+04	15.6
5	6	1.64225e+06	39.0625	7.29e+03	39.1
6	7	458547	97.6562	3.11e+03	97.7
7	8	8.56832e-12	240.162	3.24e-05	244
8	9	9.79966e-24	1.12488e-06	1.54e-11	600

Equation solved.

fsolve completed because the vector of function values is near zero as measured by the selected value of the function tolerance, and the problem appears regular as measured by the gradient.

Fig. 5.44 Multiple shooting points and three solution curves for the given system of BVP (number of subintervals is 7)

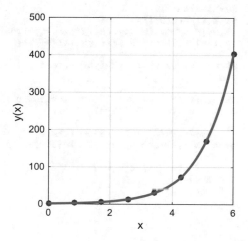

```
y =
    1.0000    2.3564    5.5527    13.0845    30.8326    72.6544    171.2042    403.4288
    1.0000    2.3564    5.5527    13.0045    30.8326    72.6544    171.2042    403.4288
    1.0000    2.3564    5.5527    13.0845    30.8326    72.6544    171.2042    403.4288
```

```
%MSM4.m    Multiple Shooting Method
clc;clear;close;
global  beta
beta=[1+exp(6);  1+exp(6);  1+exp(6)];
n=3 % order of ODE (or number of 1st order ODEs)
m=2 % number of shooting points (m>=2)
a=0; % left bc
b=6; % right bc
[t,tk,yk,y]=msh(n,m,a,b);
% [tk yk]
plot(t,y(1,:),'r.','markersize',18);hold on;
plot(tk,yk(:,1)','linewidth',2);
grid;xlabel('x');ylabel('y(x)');hold off;
figure;
plot(t,y(2,:),'k.','markersize',18);hold on;
plot(tk,yk(:,2)','linewidth',2);
grid;xlabel('x');ylabel('dy / dx');

function [t,tk,yk,y] = msh(n,m,a,b)
options=optimset('Display ','iter ','Jacobian ','on','TolFun ', ...
1e-8,'TolX ',1e-8,'MaxIter ' ,500,'MaxFunEvals ' ,500*n*m);
s=zeros(n*m ,1);% starting solution guess values
Y=fsolve(@MSN ,s(:),options,a,b,n,m);
% Find solution matrix y, each column of it corresponds to the solution
% at a shooting point.This is an (n,m) matrix, elements are taken
% columnwise from Y.
y=reshape(Y,n,m);% shooting points (without right-end point)
h=(b-a)/(m); t=a:h:b;
[~,yend ]= ode45(@ode4,[t(end -1),t(end)],y(:,m));
Ye=yend(end,:)' ;
y=[y,Ye]% shooting points including right-end
yk=zeros(m*20,n); % initialization
h=h/20;            % step length in a sub-interval
for i=1:m
tt=t(i):h:t(i+1);
[~,yy]= ode45(@ode4 ,tt ,y(:,i),options );
yk((i-1)*20+1:i*20 ,:)= yy (1:20 ,:);
end
tk=a:h:b; tk=tk';  yk=[yk; yy(end , :) ];
end

function f=ode4(x,y) % Function defining the ODEs
A=[1-2*cos(2*x) 0 1+2*sin(2*x);
    0 2 0;
    -1+2*sin(2*x) 0 1+2*cos(2*x);];
q=exp(x)*[-1+2*cos(2*x)-2*sin(2*x);
    -1;
    1-2*cos(2*x)-2*sin(2*x)];
f = A*y+q;
end
```

```
function [Q,Jq]= MSN(p,a,b,n,m) % nonlinear algebraic equations
%  computes the vector Q(p) and the Jacobian Jq(p)
% p : nxm vector,  Q : nxm  vector , Jq : (nxm) x (nxm) Jacobian
options = odeset('RelTol ',1e-8 ,'AbsTol ',1e-6);
s=(b-a)/m;   d=a:s:b+s/10;   q=zeros(n,m); Q=q; %(n x m) Matrix
p=reshape(p,n,m);
for k=1:m % Compute Q
[~,qj]= ode45(@ode4 ,[d(k),d(k+1)] ,p(:,k),options );
q(:,k)=qj(end ,:);
if k ~= m, % continuity
Q(:,k+1)=q(:,k)-p(:,k+1);
else
Q(: ,1)=bc4(p(:,1),q(:,m));
end
end
X=zeros(n,n,m+1);%(n x n) x (m+1)  matrix
for k=1:m
ph=p(:,k);
for j=1:n
h=(1e-6 + 1e-3*abs(p(j))); ph(j)=ph(j)+h;
[~,yh]= ode45(@ode4 ,[d(k),d(k+1)] ,ph ,options ); yh=yh(end ,:)';
if k==1
Fh=bc4(ph,q(:,m)); X(:,j,m+1)=( Fh(:)-Q(: ,1))/h; end
if k<m
X(:,j,k)=(yh-p(:,k+1)-Q(:,k+1))/h;
else
Fh=bc4(p(:,1),yh);  X(:,j,m)=(Fh(:)-Q(: ,1))/h;   end
ph(j)=p(j,k); end
end
Q = Q(:);
Jq=spdiags(-ones(n*m ,1), 0, n*m, n*m);   Jq(1:n,1:n)=X(:,:,m+1);
for j=1:m-1
Jq(j*n+1:n*(j+1),(j-1)*n+1:n*j)=X(:,:,j);%sparse indexing
end
Jq(1:n,(m -1)*n+1:n*m)=X(:,:,m);

function r = bc4(ya,yb) % Function defining the boundary conditions
global beta
r = eye(3)*ya+eye(3)*yb-beta;
```

Problem 5.4.3 Briefly report topics of recent literature on the industrial applications of MSM.

Solution

Large scale optimization problems such as the control of distillation columns can be efficiently treated by the direct multiple shooting method [39].

A paper [40] presents the application of multiple shooting technique to minimax optimal control problems. A standard transformation is used to convert the minimax problem into an equivalent optimal control problem with state variable inequality constraints. Using this technique the highly developed theory on the necessary conditions for state-restricted optimal control problems can be applied advantageously. It is shown that, in general, these necessary conditions lead to a boundary value problem with switching conditions, which can be treated numerically by a special version of a multiple shooting algorithm.

The work in [41] present the generalized multiple shooting method (GMSM) to analyze the dynamics of elastic mechanisms. The GMSM solves a boundary value problem by treating it as an initial value problem. Its accuracy depends on the order of space marching schemes rather than size of discretization. Dynamic equations with joint boundary conditions are derived by using Hamilton's principle to be systematically solved by the GMSM. Comparing with existing solutions and experiments, the GMSM is shown to be efficient.

In [42], a time-domain approach for computing the steady-state response of nonlinear oscillatory circuits using multiple shooting techniques is proposed. For nonautonomous circuits, the algorithm is easily adapted from the multiple shooting techniques currently used to solve boundary value problems. The nonautonomous multiple shooting algorithms must be modified to work in the autonomous case as the period is unknown. The two algorithms developed in this paper are compared with previously proposed single shooting methods on several example circuits.

In silico investigations by simulating dynamical models of biochemical processes play an important role in systems biology. If the parameters of a model are unknown, results from simulation studies can be misleading. Such a scenario can be avoided by estimating the parameters before analysing the system. Almost all approaches for estimating parameters in ordinary differential equations have either a small convergence region or suffer from an immense computational cost. The method of multiple shooting can be situated in between of these extremes. In spite of its good convergence and stability properties, the literature regarding the practical implementation and providing some theoretical background is rarely available. Necessary information for a successful implementation is supplied in [43] and the basic facts of the involved numerics are discussed. To show the performance of the method, two illustrative examples are discussed.

Kirches et al. [44] address direct MSM algorithm for nonlinear model predictive control, with a focus on problems with long prediction horizons. Proposed techniques comprise structure exploiting linear algebra on the one hand, and approximation of derivative information in an adjoint Sequential Quadratic Programming method on the other hand. They discuss the applicability of the proposed methods for three benchmark problems.

A heuristics model to reduce the computing time of multiple shooting based SQP methods for the numerical solution of optimal control problems with inherent homogeneities in the state variables are presented in [45] by a transitional model reduction approach. An approximate model is solved and its solution is used as starting point for the optimization of the detailed model. The method results in a speedup as it reduces the number of expensive SQP iterations.

Hybrid zero dynamics (HZD) has emerged as a popular framework for the stable control of bipedal robotic gaits, but typically designing a gait's virtual constraints is a slow and undependable optimization process. To expedite and boost the reliability of HZD gait generation, Hereid et al. [46] use methods of trajectory optimization to formulate a smoother and more linear optimization problem, and present a multiple-shooting formulation for the optimization of virtual constraints, combining the stability-friendly properties of HZD with an optimization-conducive problem formulation. They use the same process to generate periodic planar walking gaits on two different robot models, and in one case, demonstrate stable walking on the hardware prototype.

Another approach to nonlinear model predictive control (NMPC) is proposed in [47]. The multiple shooting method is used for discretizing the dynamic system, through which the optimal control problem is transformed to a nonlinear program (NLP) by employing the method of collocation on finite elements. Due to its high numerical accuracy, the computation efficiency for the integration of model equations can be enhanced, in comparison to the existing multiple shooting method where an ODE solver is applied for the integration and the chain-rule for the gradient computation.

Nonlinear model predictive control (NMPC) has become an appealing control concept for chemical processes because it can directly take into account the multivariable character, nonlinearities and constraints on inputs and states. While linear MPC is frequently applied in the process industries, practical applications of NMPC are rare. One reason is the relatively large computation time still needed to solve the nonlinear constrained optimization problems inherent to NMPC schemes. For faster controller response, the use of an extended partially reduced SQP method within the direct multiple shooting framework is proposed in [48]. This method is implemented in the code MSOPT which allows for an accelerated calculation of the directional derivatives and thereby saves computation time. Furthermore, this method is adapted to the real-time iteration scheme, which leads to a tremendous reduction of the time needed to provide new controls. The new NMPC variant is compared with a previously introduced scheme based on the optimization code MUSCOD-II, where the extended reduction is not in use. Numerical results are presented for both NMPC schemes and a decentralized PI controller in closed loop with a simulation model of a highly nonlinear thermally coupled distillation column which separates a ternary mixture.

For extreme maneuvers, feasible flight trajectories can be difficult to compute. If the aircraft is controlled based on linear approximations of the system dynamics that are computed along infeasible trajectories, poor control actions and violations of the flight envelope constraints can result. Gros et al. [49] propose a Non-linear Model Predictive Control (NMPC) and multiple shooting approach to handle extreme maneuvers, respect the flight envelope, handle actuator failure, and perform emergency obstacle avoidance with a single, non-hierarchical controller that can be implemented in real time.

A nonlinear model predictive control (NMPC) approach for steering an autonomous mobile robot is presented in [50]. The vehicle dynamics with a counter steering system is described by a nonlinear bicycle model. The NMPC problem is formulated taking into account the obstacles description as inequality constraints which will be updated at each sampling time based on a laser scanner detection. The nonlinear optimal control problem (NOCP) is efficiently solved by a combined multiple-shooting and collocation method. Experimentation results illustrate the viability of our approach for active autonomous steering in avoiding spontaneous obstacles.

An article [51] introduces a new method for the solution of optimal control problems for which the system is composed by many subsystems whose dynamics are coupled through input–output connections. The proposed approach can be regarded as a generalization of the direct multiple shooting method and exploits the structure of the problem to achieve a highly parallelizable algorithm. The method is applied to the control of a hydro power plant composed of several connected reaches.

MSM can be used to transform an optimal control problem into a nonlinear programming problem. The constraints (differential equations) can be turned into a system of ordinary non-linear algebraic equations. Then, the constraints and the objective function are combined using Lagrange Multiplier. Finally, a Lagrangian function that represents the original optimal control problem is obtained, which is an optimization problem without any constraints and its optimum values can be determined through finding its first derivative in the form of system of non-linear equations and the solution of this system are found using Newton method [52].

Capolei and Jørgensen [53] describe a novel numerical algorithm for solution of constrained optimal control problems of the Bolza type for stiff and/or unstable systems. The numerical algorithm combines explicit singly diagonally implicit Runge-Kutta (ESDIRK) integration methods with a multiple shooting algorithm. As they consider stiff systems, implicit solvers with sensitivity computation capabilities for initial value problems must be used in the multiple shooting algorithm. Traditionally, multi-step methods based on the BDF algorithm have been used for such problems. The main novel contribution of this paper is the use of ESDIRK integration methods for solution of the initial value problems and the corresponding sensitivity equations arising in the multiple shooting algorithm. Compared to BDF-methods, ESDIRK-methods are advantageous in multiple shooting algorithms in which restarts and frequent discontinuities on each shooting interval are present.

The optimal solutions of air-to-surface missile using a direct multiple shooting are presented in [54]. The missile must attack a fixed target by a vertical dive. The missile guidance problem is reinterpreted using optimal control theory resulting in the formulation of minimum integrated altitude problem. The formulation entails nonlinear, 2-dimensional missile flight dynamics, boundary conditions and path constraints. The influence of the boundary conditions on the structure of the optimal solution and the performance index are investigated. The results are then interpreted from the operational and computational perspective.

An et al. [55] introduce an algorithm for solving approximate shortest path problems in motion planning which is based on a direct multiple shooting dis-cretization that includes a collinear condition, multiple shooting structure, and approximation conditions. It is claimed that (in the case of monotone polygons), their proposed algorithm significantly reduces the running time and memory usage of the system.

Usman and Yorai [56] present a shooting method for a class of optimal control problems where the input is expressed in terms of its associated state and costate (adjoint) along the optimal trajectory. The problem is formulated in terms of computing the initial costate that drives the final costate to a prescribed value, using a combination of gradient descent and multiple shooting. The proposed technique is applied to a power-aware problem of co-optimizing motion and transmission power in networks of mobile agents. A novel parameter estimation and state identification algorithm for nonlinear dynamical systems from data by embedding a multiple shooting methodology in the framework of a genetic algorithm (EMSGA) is described in [57]. For chaotic dynamics with extremely high sensitivity to parameter values and initial conditions, EMSGA accurately estimates all process parameters while at the same time recovering the true dynamics of monitored and unmonitored variables from limited extents of noisy dynamic data. The superiority in accuracy and computational time of EMSGA over standalone genetic or multiple shooting algorithms for parameter estimation is shown for comparison purposes. They show that from noisy in vivo dynamic data it becomes possible to completely recover the noise-free dynamical behavior of unmonitored species concentrations using EMSGA.

Carbonell et al. [58] study the feasibility of the Local Linearization (LL) approach for the simultaneous numerical integration of the ODE and the evaluation of such derivatives. This integration approach results in a stable method for the accurate approximation of the derivatives with no more computational cost than that involved in the integration of the ODE. Their numerical simulations show that the proposed Multiple Shooting-Local Linearization method recovers the true parameters value under different scenarios of noisy data.

Assassa and Marquardt [59] present a wavelet-based grid refinement approach for direct multiple shooting applied to dynamic optimization problems. The algorithm, named adaptive multiple shooting, automatically generates a problem-dependent parameterization of the control profiles. Additional grid points are only inserted where required and redundant grid points are eliminated. Hence, the algorithm minimizes the number of grid points required to obtain accurate optimal control trajectories and demonstrates the efficiency of adaptive grid refinement compared to an equidistant discretization employing a multiple shooting method for a chemical reactor simulation example.

A predictor corrector two-point block method is proposed by Zanariah and Phang [60] to solve the well-known Blasius and Sakiadis flow numerically as an example to provide a new method that can solve the higher order BVP directly without reducing it to a system of first order equation. The proposed method is adapted with multiple shooting techniques via a three-step iterative method. The advantage of the proposed code is that the multiple shooting converges faster than the shooting method that has been implemented in other software.

5.5 Exercises

1. Comment on Fig. 5.2.
2a. In Problem 5.1.7, vary thickness value and observe its influence on the absolute error of approximation.
2b. Keeping the thickness of vessel wall the same, vary internal radius value and then note the change of the absolute error of approximation.
3. Solve the Problem 5.1.16 by using Runge-Kutta 4 (RK4) algorithm with the step size of 0.1.
4. Use RK4 method in Problem 5.2.5 (instead of ode45 solver) for h = 0.1. Compare both shooting results.
5. Use Shooting Method combined with the Secant Method to solve the problem, $2yy'' - (y')^2 + 4y^2 = 0$, $y(0) = 0$, $y(\pi/2) = 1$. Use MATLAB ode45 solver when it is necessary to solve an IVP.
6. Compare and contrast performances of the Shooting methods combined with Bisection and Secant methods for the solution of the BVP given in Problem 5.2.9.
7. Run the code given in Problem 5.2.11 using two other initial values of $p = p_1 = 0, 1$. What do you observe?

   a. Check the range of values for μ which makes the solution possible.
   b. Make a brief literature survey on the industrial applications of this equation.

8. Discuss the advantages and disadvantages of using linear interpolation method for numerically solving linear one dimensional BVPs.
9. Use Shooting Method combined separately with each of the Secant Method and Brent's Method and Newton's Method to solve the BVP, $y'' + y^3 = 0$, $y(0) = 0.2000$, $y(2) = 0.1846$. Specify the value of $y(1)$, and the slope values at two boundaries in four digit accuracy.
10. Use Shooting Method combined separately with each of the Secant Method, Brent's Method and Newton's Method to compute and plot one of the two solutions of the BVP,

$$y'' = -|y| = f(x, y, y'), \quad y(a = 0) = 0, \quad y(b = 4) = \beta = -2$$

11. Use Shooting Method combined separately with each of the Bisection Method, Secant Method and Brent's Method to find the solution of the nonlinear BVP, $y'' = e^y$, $y(0) = y(1) = 0$. Determine the range of initial guess value of $y'_{est}(0)$, and actual values of $y'(0)$, $y'(1)$, in each case.
12. Solve the Van der Pol equation, $y'' - \mu(y^2 - 1)y' + y = 0$, $y(0) = 0$, $y(2) = 1$, $\mu = 1.3$ using Single Shooting combined separately with each of the Bisection Method, Secant Method and Newton's Method.

13. Use Shooting Method combined separately with each of the Bisection Method, Secant Method and Newton's Method to solve the nonlinear BVP, $y'' + (y')^2 = x$; $y(0) = 3$, $y(5) = 1$. Plot the solution curve and its first derivative in the same figure.

14. Solve the nonlinear autonomous BVP, $y'' + 3yy' = 0$, $y(0) = 0$, $y(\infty) = 1$, using single Shooting Method combined separately with each of the Bisection Method, Brent's Method and Newton's Method.

15. Use Shooting Method combined separately with each of the Bisection Method, Brent's Method and Newton's Method to solve the linear BVP, $y'' = -y + 1 - 1.01e^{-0.1x}$, $y(0) = 0$, $y(\infty) = 1$.

16. Use Shooting Method combined separately with each of the Bisection Method, Brent's Method and Newton's Method to solve the BVP, $y'' = -(1+y)y' + y^2$, $y(0) = 1$, $y(1) = e^{-1}$. Plot the solution and its first derivative, as well as the convergence of initial value of the first derivative versus the number of iterations. Use MATLAB ode45 solver whenever it is necessary to solve IVPs.

17. Use shooting method combined separately with each of the Brent's Method and Newton's Method to solve the BVP [61],

$$y'' - 1.5y^2 = 0, \quad 0 \le x \le 1, \quad y(0) = 4, \quad y(1) = 1$$

18. Use Shooting Method combined separately with each of the Bisection Method, Brent's Method and Newton's Method to solve the BVP, $y'' = -(y')^2/y$, $y(0) = 1$, $y(3) = \sqrt{10}$.

19. Use Shooting Method combined separately with each of the Bisection Method, Brent's Method and Newton's Method to solve the nonlinear BVP, $2yy'' - (y')^2 + 4y^2 = 0$, $y(\pi/6) = 1/4$, $y(\pi/2) = 1$. Use MATLAB ode45 solver whenever it is necessary to solve an IVP.

20. Use Shooting Method combined separately with each of the Bisection Method, Brent's Method and Secant Method to solve one dimensional Bratu's problem, $y'' + 3.51383 e^y = 0$, $y(0) = y(1) = 0$. Plot the solution and its first derivative.

21. Solve the following singular BVP,

$$y'' + \frac{2}{x}y' = \frac{y}{1+y}, \quad y'(0) = 0, y(1) = 1.$$

Use Shooting Method combined separately with each of the Secant Method, Brent's Method and Newton's Method.

22. Solve the nonlinear autonomous BVP, $y'' = -y^2$; $y(0) = y(1) = 2$ by Shooting Method combined separately with each of the Secant Method, Brent's Method and Newton's Method.

23. Solve all problems on nonlinear BVPs of this chapter using Shooting Method emplying fzero built-in function of MATLAB.

24. Use Shooting Method to solve the BVP, $y'' = 2yy'$, $y(0) = -1$, $y(\pi/2) = 1$.

   Exact solution: $y = \tan(x - \pi/4)$.

25. Since two out of three boundary conditions are given at the right-end of the interval in Problem 5.3.7, whereas the third condition $y''(2)$ is unknown, try to use reverse shooting to solve the problem.

26. What happens if the interval of definition of the BVP is partitioned into infinite subdivisions when using multiple shooting method? What if the interval is not partitioned, but the MSM algorithm is still run?

References

1. Hoffman JD (1992) Numerical methods for engineers and scientists, 2nd edn. (Chap. 8). Marcel Dekker, Inc.
2. Davis ME (1984) Numerical methods and modeling for chemical engineers. Wiley, pp 63–65
3. Conte SD, De Boor C (1980) Elementary numerical analysis, an algorithmic approach, 3rd edn. McGraw-Hill, p 415
4. Boyce WE, diPrima RC, Hamilton EP (2012) Elementary differential equations and boundary value problems, 10th edn. Wiley, p 468
5. Keskin AU (2019) Ordinary differential equations for engineers, problems with MATLAB solutions. Springer
6. Faires JD, Burden RL (2003) Numerical methods (Chap. 5). Thomson/Brooks/Cole
7. Chapra SC (2012) Applied numerical methods with MATLAB for engineers and scientists, 3rd edn. (Chap. 22). McGraw-Hill
8. Keller HB (1968) Numerical methods for two-point boundary-value problems. Blaisdell, New York
9. Fryer DM, Harvey JF (2012) High pressure vessels. Springer Science & Business Media, p 42
10. Chapra SC (2012) Applied numerical methods with MATLAB for engineers and scientists, 3rd edn. McGraw-Hill, p 622
11. Fox L, Mayers DF (1987) Numerical solution of ordinary differential equations. Chapman & Hall, London, p 107
12. Mills RD (1992) Slope retention techniques for solving boundary-value problems in differential equations. J Symb Comput 13:59–80
13. Fox L, Mayers DF (1987) Numerical solution of ordinary differential equations. Chapman & Hall, London, p 125
14. Esfandiari RS (2017) Numerical methods for engineers and scientists using MATLAB, 2nd edn. (Chap. 8.4). CRC Press, Inc.
15. Kubicek M, Hlavacek V (1983) Numerical solution of nonlinear boundary value problems with applications. Prentice Hall, p 228
16. Bailey PB, Shampine LF, Wattman PF (1968) Nonlinear two point BVPs. Academic Press, pp 7–9
17. Hoffman, JD (1992) Numerical methods for engineers and scientists, 2nd edn. Marcel Dekker, Inc., pp 150–152
18. Conte SD, De Boor C (1980) Elementary numerical analysis: an algorithmic approach, 3rd edn. McGraw-Hill, p 416
19. Sajid M, Hayat T (2008) The application of homotopy analysis method to thin film flows of a third order fluid. Chaos, Solitons Fractals 38:506–515
20. Bratu G (1914) Sur les equation integrals non-lineaires. Bull Math Soc France 42:113–142

21. Ascher UM et al (1995) Numerical solution of boundary value problems for ordinary differential equations. SIAM, Philadelphia
22. Boyd JP (2011) One-point pseudo spectral collocation for the one-dimensional Bratu equation. Appl Math Comput 217:5553–5565
23. Caglar H, Caglar N, Özerc M, Valaristosd A, Anagnostopoulos AN (2010) B-spline method for solving Bratu's problem. Int J Comput Math 87:1885–1891
24. Romero N (2015) Solving the one dimensional Bratu problem with efficient fourth order iterative methods. SeMA, Soc Esp Mat Apl 71:1–14
25. Richards G, Sarma PRR (1994) Reduced order models for induction motors with two rotor circuits. IEEE Trans Energy Convers 9(4):673–678
26. Bishop RED, Cannon SM, Miao S (1989) On coupled bending and torsional vibration of uniform beams. J Sound Vib 131:457–464
27. Chandrasekhar S (1961) Hydrodynamic and hydromagnetic stability. The international series of monographs on physics. Clarendon Press, Oxford, UK
28. Wazwaz AM (2000) Approximate solutions to boundary value problems of higher order by the modified decomposition method. Comput Math Appl 40:679–691
29. Wanga Y, Zhaoa YB, Weia GW (2003) A note on the numerical solution of high-order differential equations. J Comput Appl Math 159:387–398
30. Agarwal RP (1986) Boundary value problems from higher order differential equations. World Scientific, Singapore
31. Bhowmik SK (2015) Tchebychev polynomial approximations for mth order boundary value problems. Int J Pure Appl Math 98(1):45–63
32. Holt JF (1964) Numerical solution of two point boundary value problems by finite difference methods. Comm ACM 7:366–373
33. Osborne MR (1969) On shooting methods for boundary value problems. J Math Anal Appl 27:417–433
34. Stoer J Bulirsch R (2002) Introduction to numerical analysis, 3rd edn. Springer-Verlag, Berlin, New York, see Sections 7.3.5 and further
35. Ascher UM, Mattheij RMM, Russell RD (1995) Numerical solution of boundary value problems for ordinary differential equations. SIAM, p 145
36. Kress R (1998) Numerical analysis (graduate texts in mathematics). Springer, pp 261–262
37. Keller HB (1968) Numerical methods for two-point boundary-value problems. Blaisdell, New York, p 5
38. Hermann M, Saravi M (2016) Nonlinear ordinary differential equations: analytical approximation and numerical methods (Chap. 4.4). Springer
39. Bock HG, Plitt KJ (1984) A multiple shooting algorithm for direct solution optimal control problems. In: Proceedings of the 9th IFAC world congress. Pergamon Press, Budapest, pp 243–247
40. Oberle HJ (1985) Numerical treatment of minimax control problems by multiple shooting. IFAC Proc 18(2):131–138
41. Lan C-C, Lee K-M, Liou J-H (2009) Dynamics of highly elastic mechanisms using the generalized multiple shooting method: simulations and experiments. Mech Mach Theory 44 (12):2164–2178
42. Parkhurst JR, Ogborn LL (1995) Determining the steady-state output of nonlinear oscillatory circuits using multiple shooting. IEEE Trans Comput Aided Des Integr Circuits Syst 14 (7):882–889
43. Peifer M, Timmer J (2007) Parameter estimation in ordinary differential equations for biochemical processes using the method of multiple shooting. IET Syst Biol 1(2):78–88
44. Kirches C, Wirsching L, Bock HG, Schloeder JP (2012) Efficient direct multiple shooting for nonlinear model predictive control on long horizons. J Process Control 22(3):540–550
45. Sager S, Pollmann UB, Diehl M, Lebiedz D, Bock HG (2007) Exploiting system homogeneities in large scale optimal control problems for speedup of multiple shooting based SQP methods. Comput Chem Eng 31:1181–1186

46. Hereid A, Hubicki CM, Cousineau EA, Hurst JW, Ames AD (2015) Hybrid zero dynamics based multiple shooting optimization with applications to robotic walking. In: IEEE international conference on robotics and automation (ICRA), pp 5734–5740

47. Tamimi J, Li P (2009) Nonlinear model predictive control using multiple shooting combined with collocation on finite elements. IFAC Proc 42(11):703–708

48. Schäfer A, Kühl P, Diehl M, Schlöder J, Bock HG (2007) Fast reduced multiple shooting methods for nonlinear model predictive control. Chem Eng Process 46(11):1200–1214

49. Gros S, Quirynen R, Diehl M (2012) Aircraft control based on fast non-linear MPC & multiple-shooting. In: IEEE 51st IEEE conference on decision and control (CDC), pp 1142–1147

50. Drozdova E, Hopfgarten S, Lazutkin E, Li P (2016) Autonomous driving of a mobile robot using a combined multiple-shooting and collocation method. IFAC-Papers OnLine 49 (15):193–198

51. Savorgnan C, Romani C, Kozma A, Diehl M (2011) Multiple shooting for distributed systems with applications in hydro electricity production. J Process Control 21(5):738–745

52. Munzir S, Halfiani V, Marwan (2012) An optimal control solution using multiple shooting method. Bull Math 4(2):143–160

53. Capolei A, Jørgensen JB (2012) Solution of constrained optimal control problems using multiple shooting and ESDIRK methods. In: American control conference (ACC), pp 295–300

54. Dubcan S (2008) A direct multiple shooting for the optimal trajectory of missile guidance. In: IEEE international conference on control applications, 3–5 Sept 2008

55. An PT, Hai NN, Hoai TV (2013) Direct multiple shooting method for solving approximate shortest path problems. J Comput Appl Math 244:67–76

56. Usman A, Yorai W (2015) Multiple shooting technique for optimal control problems with application to power aware networks. IFAC-PapersOnLine 48(27):286–290

57. Sarode KD, Kumar VR, Kulkarni BD (2015) Embedded multiple shooting methodology in a genetic algorithm framework for parameter estimation and state identification of complex systems. Chem Eng Sci 134(29):605–618

58. Carbonell F, Iturria-Medina Y, Jimenez JC (2016) Multiple shooting local linearization method for the identification of dynamical systems. Commun Nonlinear Sci Numer Simul 37:292–304

59. Assassa F, Marquardt W (2014) Dynamic optimization using adaptive direct multiple shooting. Comput Chem Eng 60(10):242–259

60. Zanariah AM, Phang P-S (2017) Study of predictor corrector block method via multiple shooting to Blasius and Sakiadis flow. Appl Math Comput 314(1):469–483

61. Lesnic DA (2007) Nonlinear reaction diffusion process using the Adomian decomposition method. Int Comm Heat Mass Transfer 34:129–135

Chapter 6
Finite Difference Methods for the Solution of BVPs

The solution of BVPs by a one dimensional finite difference (FD) method can be accomplished by the following steps (the equilibrium or replacement method):

a. Discretize the solution domain into a one dimensional finite difference grid,
b. Approximate the derivatives in the ODE by rational algebraic approximations,
c. Substitute the approximations into the ODE,
d. Solve the resulting system of algebraic FD Equations.

In this process, a system of coupled FD equations must be solved simultaneously.

The problems concerning the FD method is presented first for the linear, variable coefficient, second-order BVPs with Dirichlet or Neumann boundary conditions. A mixed boundary condition is implemented in the FD method in the same manner as a derivative boundary condition is implemented.

Later in this chapter, nonlinear boundary value problems are studied. The solution of nonlinear BVPs by the FD method results in a system of nonlinear FD equations. Two methods (Iteration and Newton's method) are presented here for solving nonlinear boundary-value problems by the FD method.

When solving boundary-value problems using the FD method, convergence of the solution must be considered. FD methods have better rounding characteristics with respect to shooting methods, but they may require more computation to obtain a specified accuracy.

6.1 Basics and Linear Finite Difference Methods

Problem 6.1.1 Finite differences (replacement, substitution, equilibrium [1]) method is a numerical technique to solve ODEs and BVPs. It is based on replacing derivatives in a differential equation with approximations in the form of algebraic

© Springer Nature Switzerland AG 2019
A. Ü. Keskin, *Boundary Value Problems for Engineers*,
https://doi.org/10.1007/978-3-030-21080-9_6

rational functions. Solution of these equations ends up with simultaneous algebraic equations.

There are three schemes to use for the first derivative: forward, backward and central differences with $x_i = i \cdot h$,

Forward difference:

$$f_i' = (f_{i+1} - f_i)/h$$

Backward difference:

$$f_i' = (f_i - f_{i-1})/h$$

Central difference:

$$f_i' = (f_{i+1} - f_{i-1})/2h$$

Boundary condition, starting point derivative for central difference:

$$\frac{dy(0)}{dx} \rightarrow \frac{y_1 - y_{-1}}{2h}$$

Boundary condition, ending point derivative for central difference:

$$\frac{dy}{dx}(n) \rightarrow \frac{y_{n+1} - y_{n-1}}{2h}$$

where $h = \Delta x$ is the step size for equal spacing, and x_i is the independent point of interest.

The difference approximation for a second derivative is

$$f_i'' = \frac{f_{i+1} - 2f_i + f_{i-1}}{h^2} \tag{6.1}$$

Prove that difference approximation for a second derivative is given by (6.1).

Solution
Consider the Taylor series for third order approximation around x_i,

$$f(x_i + \Delta x) = f(x_i) - \Delta x f'(x_i) + \frac{\Delta x^2}{2!} f''(x_i) - \frac{\Delta x^3}{3!} f'''(x_i) + R_3$$

where $R_3 = (\Delta x^4 / 4!) f^{(4)}(\varepsilon_4)$.

Now, adding two series

$$f(x_i + \Delta x) + f(x_i - \Delta x) = 2f(x_i) + (\Delta x)^2 f''(x_i) + 2R_3$$

let $x_{i+1} \rightarrow x_i + \Delta x$, and ignoring R_3,

$$f_{i+1} + f_{i-1} = 2f_i + h^2 f_i''$$

Solving this equation for f_i'' yields Eq. (6.1).

Problem 6.1.2

(a) Consider the linear BVP,

$$y'' = p(x)y' + q(x)y + r(x), x \in [a, b], \quad x(a) = A, \quad x(b) = B.$$

Use central difference formulas for the first and second derivative terms to obtain a general form of the solution for this linear BVP in matrix form.

(b) Use this formulation to approximate the solution for the BVP,

$$y'' - 16y = 0, \quad x(0) = 0, \quad x(1) = 10.$$

with the number of internal nodes $n = 3$.

(c) Repeat part (b) for $n = 7$. Make a list of exact and computed values for the solution, as well as the absolute error. Plot the solution (fin_diff_bvp4_1.m).

Solution

(a) Central difference formulas for the derivatives are

$$y'(x_j) = \frac{y(x_{j+1}) - y(x_{j-1})}{2h} + O(h^2)$$

$$y''(x_j) = \frac{y(x_{j+1}) - 2y(x_j) + y(x_{j-1})}{h^2} + O(h^2)$$

where h is step size.

Substituting these equations into the given ODE, then dropping the terms $O(h^2)$ and using the notation $p_j = p(x_j)$, $q_j = q(x_j)$ for simplicity, we obtain

$$\frac{y_{j+1} - 2y_j + y_j}{h^2} \, 1 = p_j \frac{y_{j+1} - y_{j-1}}{2h} + q_j y_j + r_j$$

which can be used to solve the given linear BVP.

Arranging this last equation in a system of linear equations gives

$$\left(\frac{-h}{2}p_j - 1\right)x_{j-1} + (2 + h^2 q_j)x_j + \left(\frac{h}{2}p_j - 1\right)x_{j+1} = -h^2 r_j$$

Let $j = 1, 2, \ldots, N - 1$ with $y_0 = A$, $y_N = B$. This leads to a matrix equation with α having a tridiagonal form with

$$\alpha y = \beta$$
$$\alpha(i, i) = 2 + h^2 q_i$$
$$\alpha(i, i+1) = (hp_i/2) - 1$$
$$\alpha(i+1, i) = (-hp_i/2) - 1$$

$$\beta^T = \left[-h^2 r_1 + c \quad -h^2 r_2 \quad \cdots \quad -h^2 r_j \quad \cdots \quad -h^2 r_{N-2} \quad -h^2 r_{N-1} + d \right]$$

where

$$c = (hp_i/2 + 1)A,$$
$$d = (-hp_{N-1}/2 + 1)B$$

$$y^T = \left[y_1 \quad y_2 \quad \cdots \quad y_j \quad \cdots \quad y_{N-2} \quad y_{N-1} \right]$$

(b) $p(x) = 0$, $q(x) = 16$, $r(x) = 0$, $x \in [a = 0, b = 1]$, $x(a) = A = 0$, $x(b) = B = 10$.

Since the number of internal nodes $n = 3, i = 1, 2, 3$;

$$h = \frac{(b - a)}{n + 1} = \frac{1}{4}$$
$$\alpha(i, i) = 2 + h^2 q_i = 2 + (1/16)16 = 3$$
$$\alpha(i, i+1) = \left(\frac{hp_i}{2}\right) - 1 = 0 - 1 = -1$$
$$\alpha(i+1, i) = (-hp_i/2) - 1 = 0 - 1 = -1$$

$$\beta^T = [c \ 0 \ d] = [A \ 0 \ B] = [0 \ 0 \ 10]$$
$$\begin{bmatrix} 3 & -1 & 0 \\ -1 & 3 & -1 \\ 0 & -1 & 3 \end{bmatrix} \begin{bmatrix} y_1 \\ y_2 \\ y_3 \end{bmatrix} = \begin{bmatrix} 0 \\ 0 \\ 10 \end{bmatrix}$$

Final solution vector including boundary values is

$$y^T = [0 \quad 0.4762 \quad 1.4286 \quad 3.8095 \quad 10]$$

(c) $p(x) = 0, q(x) = 16, r(x) = 0, \qquad x \in [a = 0, b = 1], \qquad x(a) = A = 0,$
    $x(b) = B = 10.$

Since the number of internal nodes is $n = 7, i = 1, 2, \ldots, 6, 7;$

$$h = \frac{(b-a)}{n+1} = \frac{1}{8}$$
$$\alpha(i, i) = 2 + h^2 q_i = 2 + (1/64)16 = 2.25$$
$$\alpha(i, i+1) = \left(\frac{hp_i}{2}\right) - 1 = 0 - 1 = -1$$
$$\alpha(i+1, i) = (-hp_i/2) - 1 = 0 - 1 = -1$$

$$\beta^T = [c \ 0 \ d] = [A \ 0 \ B] = [0 \ 0 \ 10]$$
$$\alpha y = \beta$$

Following is a list of exact and computed values for the solution (at the internal nodes), as well as the absolute error. Computed values are displayed in Fig. 6.1.

Fig. 6.1 Graph of the solution of the BVP, $y'' - 16y = 0$; $x(0) = 0, x(1) = 10$ (including boundary conditions)

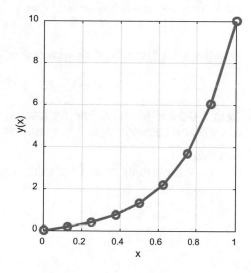

Exact	Computed	Abs.Error
0.1909	0.1967	0.0057
0.4306	0.4425	0.0119
0.7802	0.7990	0.0187
1.3290	1.3552	0.0262
2.2170	2.2502	0.0332
3.6709	3.7078	0.0369
6.0618	6.0924	0.0306

MATLAB m-file (fin_diff_bvp4_1.m) is given below.

```
clc;clear;close;
g=2.25
A=[g    -1    0   0    0  0    0;
   -1    g   -1   0    0  0    0;
    0   -1    g  -1    0  0    0;
    0    0   -1   g   -1  0    0;
    0    0    0  -1    g  -1   0;
    0    0    0   0   -1  g   -1;
    0    0    0   0    0  -1   g];
b=[ 0    0    0   0    0  0   10 ]';
Y=A\b
%Analytical (exact) solution
syms   y(x)
bc='y(0)=0,y(1)=10';eq='D2y-16*y=0';S=dsolve(eq,bc,x);
x = linspace(0,1,9); z = eval(vectorize(S));
plot(x,z,'kO-','linewidth',2);grid;xlabel('x'); ylabel('T(x)');
%e=absolute error
z(1)=[]; z(8)=[]; e=z'-Y; z=[z',Y, e]
```

Problem 6.1.3 Use finite difference method to solve the linear, autonomous BVP,

$$y'' + 1 = 0, \quad y(0) = y(1) = 0$$

for $h = 0.2$ and $h = 0.02$. Plot the solution curve for $h = 0.02$. Check the result using dsolve function (fin_diff_bvp2.m), (dsolve_fin_diff2.m).

Solution

For $h = 0.2$, rewriting given ODE using equivalent algebraic form,

$$\frac{y_{i+1} + y_{i-1} - 2y_i}{h^2} + 1 = 0$$

$$y_{i+1} - 2y_i + y_{i-1} = -h^2$$

$$i = 1, \quad y_0 - 2y_1 + y_2 = -h^2$$

$$i = 2, \quad y_1 - 2y_2 + y_3 = -h^2$$

$$i = 3, \quad y_2 - 2y_3 + y_4 = -h^2$$

$$i = 4, \quad y_3 - 2y_4 + y_5 = -h^2$$

$$y_0 = y_5 = 0$$

Let,

$$\mathbf{Ay = b}$$

$$\begin{bmatrix} -2 & 1 & 0 & 0 \\ 1 & -2 & 1 & 0 \\ 0 & 1 & -2 & 1 \\ 0 & 0 & 1 & -2 \end{bmatrix} \begin{bmatrix} y_1 \\ y_2 \\ y_3 \\ y_4 \end{bmatrix} = \begin{bmatrix} -h^2 \\ -h^2 \\ -h^2 \\ -h^2 \end{bmatrix}$$

Solution of this equation gives

$$y_1 = 0.08, \quad y_2 = 0.12, \quad y_3 = 0.12, \quad y_4 = 0.08$$

which means that the solution vector is

$$\mathbf{y} = \begin{bmatrix} 0 & 0.08 & 0.12 & 0.12 & 0.08 & 0 \end{bmatrix}.$$

When the (equidistant) step size is reduced by a factor of 10 ($h = 0.02$), the solution curve shown in Fig. 6.2 is obtained using the MATLAB code (fin_diff_bvp2.m).

Fig. 6.2 Solution curve for the BVP, $y'' + 1 = 0$, $y(0) = y(1) = 0$

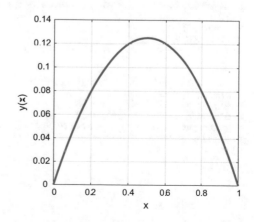

Analytical solution:

$$y = y_h + y_p$$

$$y_h = c_1 + c_2 x, y_p = -\frac{x^2}{2}$$

$$y = c_1 + c_2 x - \frac{x^2}{2}$$

From boundary conditions, $c_1 = 0, c_2 = 1/2$

$$y = \frac{1}{2}\left(x - x^2\right) = \frac{1}{2}x(1 - x)$$

This analytical solution can be verified by using MATLAB dsolve function.
First part of the code given below (commented lines) solves the BVP for
h = 0.2.

```
%fin_diff_bvp2.m
clc;clear;close;
% h=0.2;
% H=-h*h;
% A=[-2  1  0  0;
%     1 -2  1  0;
%     0  1 -2  1;
%     0  0  1 -2];
% b=[ H H H H ]';;
% y=A\b

%general case
h=0.02; H=-h*h; N=1/h;
A=zeros(N-1);
for i=1:N-2
    A(i,i+1)=1; B=A';
end
A=A+B;
for i=1:N-1
    A(i,i)=-2;
end
A; b=H*ones(N-1,1); y=A\b ; Y=zeros(N+1,1);
i=2:N;
Y(i)=y(i-1)
x=linspace(0,1,N+1);
plot(x,Y,'linewidth',2);grid;xlabel('x');ylabel('y(x)');


% dsolve_fin_diff2.m    Solves the  BVP y"+1=0 y(0)=0,y(1)=0
clc;clear; syms  y(x)
bc='y(0) ==0,y(1)==0';
S=dsolve(diff(diff(y))+1 == 0,bc,x)
x = linspace(0,1,50); z = eval(vectorize(S));
plot(x,z,'linewidth',2);grid;xlabel('x'); ylabel('y(x)');
```

Problem 6.1.4

(a) Solve the autonomous BVP, $y'' + y = 0$, $y(0) = 0$, $y(1) = 1$, $h = 0.02$, using finite differences method.

(b) Repeat (a) for $h = 0.02$ using MATLAB. Plot the solution curve (fin_diff_bvp1.m), (dsolve_fin_diff1.m).

Solution

(a) Substituting equivalent difference form of y'' results in

$$\frac{y_{i+1} - 2y_i + y_{i-1}}{h^2} + y_i = 0$$

$$y_{i-1} + \left(h^2 - 2\right)y_i + y_{i+1} = 0$$

$$h^2 - 2 = (0.2)^2 - 2 = -1.96$$

$$i = 1 \quad y_0 - 1.96y_1 + y_2 = 0$$

$$i = 2 \quad y_1 - 1.96y_2 + y_3 = 0$$

$$i = 3 \quad y_2 - 1.96y_3 + y_4 = 0$$

$$i = 4 \quad y_3 - 1.96y_4 + y_5 = 0$$

$$y(0) = y_0 = 0, \quad y(5) = y_5 = 1$$

$$- 1.96y_1 + y_2 = 0$$

$$y_1 - 1.96y_2 + y_3 = 0$$

$$y_2 - 1.96y_3 + y_4 = 0$$

$$y_3 - 1.96y_4 + 1 = 0$$

Put this last system of equations in matrix form,

$$Ay = b$$

$$\begin{bmatrix} -1.96 & 1 & 0 & 0 \\ 1 & -1.96 & 1 & 0 \\ 0 & 1 & -1.96 & 1 \\ 0 & 0 & 1 & -1.96 \end{bmatrix} \begin{bmatrix} y_1 \\ y_2 \\ y_3 \\ y_4 \end{bmatrix} = \begin{bmatrix} 0 \\ 0 \\ 0 \\ -1 \end{bmatrix}$$

Solution of this equation yields

$$y_1 = 0.2362, \quad y_2 = 0.463, \quad y_3 = 0.6713, \quad y_4 = 0.8527.$$

which means that the solution vector is

$$y = [0 \quad 0.2362 \quad 0.463 \quad 0.6713 \quad 0.8527 \quad 1].$$

(b) Note that A is a symetric matrix with

$$A(i,i) = h^2 - 2 = -1.96$$
$$A(i,i+1) = A(i+1,i)$$
$$b(i) = \begin{cases} 0 & i \neq N \\ -1 & i = N \end{cases}$$
$$N = 1/h$$

Figure 6.3 displays the solution curve, and MATLAB code is given below.
Analytical check of the solution is made by using MATLAB dsolve function, which yields

$$y(x) = \sin(x)/\sin(1).$$

A plot of this function verifies the results found, above.

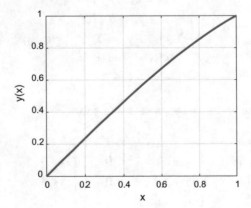

Fig. 6.3 Solution curve for the BVP, $y'' + y = 0, y(0) = 0, y(1) = 1$

```
%fin_diff_bvp1.m
clc;clear;close;

% A=[-1.96 1 0 0;
% 1 -1.96 1 0;
% 0 1 -1.96 1;
% 0 0 1 -1.96];
% b=[0 0 0 -1]';
% y=A\b

%general case
h=0.02; N=1/h;
A=zeros(N-1);
for i=1:N-2
    A(i,i+1)=1; B=A';
end
A=A+B;
for i=1:N-1;A(i,i)=h^2-2;end
A;
b=zeros(N-1,1); b(N-1,1)=-1;
y=A\b;
Y=zeros(N+1,1); Y(1)=0;Y(N+1)=1;
i=2:N;
Y(i)=y(i-1);
x=linspace(0,1,N+1);
plot(x,Y,'linewidth',2);grid;xlabel('x');ylabel('y(x)');

% dsolve_fin_diff1.m    SOLVE  BVP y"+y=0 y(0)=0,y(1)=1
clc;clear; syms y(x)
bc='y(0) --0,y(1)==1';
S=dsolve(diff(diff(y))+y == 0, bc)
x = linspace(0,1,50);z = eval(vectorize(S));
plot(x,z,'k','linewidth',2);grid; xlabel('x'); ylabel('y(x)');
```

Problem 6.1.5 Solve the linear nonhomogeneous BVP, $y'' + y = 1, y(0) = 2$, $y(3) = 0$, using FD approximation. Let the number of internal nodes be $n = 3$ (fin_diff_bvp_mat1.m).

Solution

Explicit solution could not be found using MATLAB dsolve function. We use replacement (equilibrium) method. The MATLAB code is based on generating tridiagonal matrix in a different way than the other codes presented in this chapter.

$$y'' = \frac{y_{i+1} + 2y_i + y_{i-1}}{h^2},$$

$$\frac{y_{i+1} + 2y_i + y_{i-1}}{h^2} + y_i = 1$$

$$y_{i-1} - (2 - h^2)y_i + y_{i+1} = h^2$$

For $n = 3$, $h = (3 - 0)/(n + 1) = 3/4$,

$$y_0 - (2 - h^2)y_1 + y_2 = h^2$$
$$y_1 - (2 - h^2)y_2 + y_3 = h^2$$
$$y_2 - (2 - h^2)y_3 + y_4 = h^2$$
$$y_0 = y(0) = 1, \quad y_4 = y(1) = 0$$

$$\begin{bmatrix} -(2 - h^2) & 1 & 0 \\ 1 & -(2 - h^2) & 1 \\ 0 & 1 & -(2 - h^2) \end{bmatrix} \begin{bmatrix} y_1 \\ y_2 \\ y_3 \end{bmatrix} = \begin{bmatrix} h^2 \\ h^2 \\ h^2 \end{bmatrix}$$

Internal node values are computed and then boundary values are added to the solution vector. Figure 6.4 displays the solution. MATLAB m-file is (fin_diff_bvp_mat1.m) is also given below.

```
%fin_diff_bvp_mat1.m
clc;clear;close;
span=[0 3]; %interval
bv=[2 0]; % boundary values
n=20; % number of internal nodes
v=ones(n+1,1); % initialization
h=(span(2)-span(1))/(n+1); % step size
A=-(2-h^2)*eye(n+2) % main diagonal matrix
B=diag(v,1)+ diag(v,-1) % diag(V,K), K tells which diagonal to place V in
C=A+B;
C(1,2)=0; C(n+2,n+1)=0; C(1,1)=1; C(n+2,n+2)=1;
b=h^2*ones(n+2,1);b(1)=bv(1);b(n+2)=bv(2);
y=C\b
x=linspace(span(1),span(2),n+2);
plot(x,y,'linewidth',2.5);xlim(span),xlabel('x'),ylabel('y(x)');grid
```

Problem 6.1.6 Use FDM to solve the two point BVP, $y'' = 6x, y(0) = 0, y(1) = 2$ (fin_diff_bvp_mat6.m).

Solution
Here, we demonstrate the use of sparse matrix concept in MATLAB, which can speed up computations for problems that involve large number of nodes, in particular.

$$\frac{y_{i+1} + y_{i-1} - 2y_i}{h^2} = 6x_i$$
$$y_{i+1} - 2y_i + y_{i-1} = h^2 6x_i$$

the boundary conditions are $y_0 = 0, y_{N+1} = 2$, N = number of internal nodes. Let N = 4,

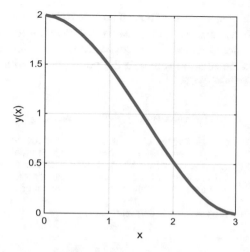

Fig. 6.4 Solution curve for the BVP, $y'' + y = 1, y(0) = 2, y(3) = 0$

$$y_{i+1} - 2y_i + y_{i-1} = h^2 \cdot 6x_i$$
$$i = 1, \quad y_0 - 2y_1 + y_2 = h^2 \cdot 6x_1$$
$$i = 2, \quad y_1 - 2y_2 + y_3 = h^2 \cdot 6x_2$$
$$i = 3, \quad y_2 - 2y_3 + y_4 = h^2 \cdot 6x_3$$
$$i = 4, \quad y_3 - 2y_4 + y_5 = h^2 \cdot 6x_4$$
$$y_0 = 0, y_5 = 2 \text{ (boundary conditions)}.$$

This can be put in matrix form,

$$ay = b$$

$$\begin{bmatrix} -2 & 1 & 0 & 0 \\ 1 & -2 & 1 & 0 \\ 0 & 1 & -2 & 1 \\ 0 & 0 & 1 & -2 \end{bmatrix} \begin{bmatrix} y_1 \\ y_2 \\ y_3 \\ y_4 \end{bmatrix} = \begin{bmatrix} 6h^2x_1 \\ 6h^2x_2 \\ 6h^2x_3 \\ 6h^2x_4 \end{bmatrix}$$

the matrix a is always tridiagonal when finite difference methods is applied for two-point BVPs for second order ODEs.

Alternatively,

$$AY = B$$

with A is an 6×6 element matrix with $A(1, 1) = A(6, 6) = 1$ all other first and last row elements are zero, Y is a 6×1 column vector of approximate solutions and $B = \begin{bmatrix} y_0 & 6h^2x_1 & 6h^2x_2 & 6h^2x_3 & 6h^2x_4 & y_5 \end{bmatrix}^T$.

Here, it is seen that **A** contains many zero elements. MATLAB code (fin_diff_bvp_mat6.m) employs sparse matrix computation to solve the system of equations. Below, some lines of the code results are illustrated for N = 9. Figure 6.5 displays the solution values at the nodes, along with the exact solution $(y = x + x^3)$ of the given BVP.

x	Y(computed)	y(exact)
0	0	0
0.100000000000000	0.101000000000000	0.101000000000000
0.200000000000000	0.208000000000000	0.208000000000000
0.800000000000000	1.312000000000000	1.312000000000000
0.900000000000000	1.629000000000000	1.629000000000000
1.000000000000000	2.000000000000000	2.000000000000000

Elapsed time is 0.075048 seconds.

```
%fin_diff_bvp_mat6.m
clc;clear;tic
format long
span=[0 1];bv=[0 2];
N=9; h=(span(2)- span(1))/(N+1)
x=linspace(span(1), span(2), N+2);x=x';
b = zeros(N+2,1);b(N+2)=bv(2);
b(2:N+1) = 6 * h^2 * x(2:N+1);
A = sparse(N+2,N+2);
A(1,1) = 1.0; A(N+2,N+2) = 1.0;
for k=2:N+1
A(k,[k-1, k, k+1]) = [1, -2, 1];
end
Y = A \ b;
yexact = (x.^3 +x);
[x Y yexact]
plot(x,Y,'.','markersize',20);hold on;
plot(x,yexact,'linewidth',1);grid;xlabel('x');ylabel('y(x)');
toc
```

Fig. 6.5 Solution curve for the BVP,
$y'' = 6x, y(0) = 0, y(1) = 2$

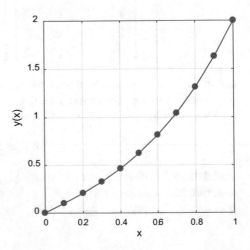

Problem 6.1.7 The main sources of error in the Finite Difference Method are the truncation error and the error made in solving the system of equations. The truncation error is made by using the centered-difference formulas which dominates for $h > \sqrt{\epsilon}$ where h, ϵ are the step size and machine epsilon, respectively. Machine epsilon for single precision is $2^{-23} \approx 1.19 \times 10^{-7}$ while for double precision (64 bits) $2^{-52} \approx 2.22 \times 10^{-16}$. (See, MATLAB command eps, floating point relative accuracy). This error is O(h2), so the error is inversely proportional to the squared number of subintervals n + 1. Alternatively, the error is proportional to the squared value of linear spacing of nodes (step size), h.

Test this expectation by solving the following BVP using the Finite Difference Method,

$$y'' - y = 0; \quad y(0) = 0, \quad y(1) = 1$$

and drawing the magnitude of the error E of the solution at t = 0.5 on a log-log plot, for various step size values, *h*.

Solution
Analytical solution gives $y(0.5) = 0.443409441985037$.

```
clc;clear;format long
syms  y(t)
eq='D2y-y=0'; bc='y(0)=0,y(1)=1';
S=dsolve(eq,bc); t = linspace(0,1,5);y=eval(vectorize(S))
```

Applying finite difference scheme to the ODE and using the centered difference form for the second derivative, one obtains,

$$\frac{y_{i-1} - 2y_i + y_{i+1}}{h^2} - y_i = 0$$
$$y_{i-1} - 2y_i + y_{i+1} - h^2 y_i = 0$$
$$y_{i-1} - (2 + h^2)y_i + y_{i+1} = 0$$

For $n = 3, h = 1/(n+1) = 0.25$, $y(0) = y_0 = 0, y(1) = y_4 = 1$,

$$i = 1: \quad y_0 - (2+h^2)y_1 + y_2 = 0$$
$$i = 2: \quad y_1 - (2+h^2)y_2 + y_3 = 0$$
$$i = 3: \quad y_2 - (2+h^2)y_3 + y_4 = 0$$

$$- (2+h^2)y_1 + y_2 = 0$$
$$y_1 - (2+h^2)y_2 + y_3 = 0$$
$$y_2 - (2+h^2)y_3 = -1$$

This is a tridiagonal matrix equation, $Ay = b$, with $2 + h^2 = 2.0625$

$$\begin{bmatrix} -2.0625 & 1 & 0 \\ 1 & -2.0625 & 1 \\ 0 & 1 & -2.0625 \end{bmatrix} \begin{bmatrix} y_1 \\ y_2 \\ y_3 \end{bmatrix} = \begin{bmatrix} 0 \\ 0 \\ -1 \end{bmatrix}$$

which yields $y_1 = 0.2151$, $y_2 = 0.4437$, $y_3 = 0.7000$.

Following MATLAB script is used for this solution:

```
h=0.25; a=-(2+h.^2);
A=[a   1   0 ; 1   a   1 ; 0   1   a ]; b=[ 0   0   -1 ]';
y=A\b
```

The absolute error at $x = 0.5$ is computed as

$$|E(0.5)| = |0.443409441985037 - 0.4437| = 2.6473 \times 10^{-4}$$

Repeating these computations for increasing number of n (decreasing values of h) and plotting the resulting values of h versus $|E|$ at $x = 0.5$ yields the straight line in the log-log plot shown in Fig. 6.6. MATLAB m-file (fin_diff_bvp0.m) is listed below.

```
%fin_diff_bvp0.m
clc;clear;close;
%n=3;    % number of equal spacing within given interval
%n=15;
%n=63;
%n=255;
n=1023
U=-1;   % Forcing function of given nonhomogeneous ODE
y(1)=0; y(n)=1;  % boundary values
h=1/(n+1)
for i=1:n
x=i*h;
a(i,i)= -(2+h.^2); % diagonal elements
end
for i=1:n-1 %upper & lower diagonal elements
a(i,i+1)=1;a(i+1,i)=1;
end
b=zeros(n,1);
b(n,1)=U; % f vector
Y=a\b;    % solution vector
y=[y(1) Y' y(n)]' % solution vector including boundary values
R=floor(length(y)/2); eps= abs(0.443409441985037-y(R+1))

h=[1/4 1/16 1/64 1/256 1/1024];
E=[2.6473e-4 1.6667e-5 1.0422e-6 6.5138e-8  4.0712e-9]
loglog(h,E,'linewidth',2);grid;
xlabel('step size,h');ylabel('Error magnitude');
```

Fig. 6.6 The magnitude of the error E of the solution at x = 0.5, for various step size values. The error as a function of h is a straight line with slope 2, or $\log(E) \approx a + b \cdot \log(h)$, where $b = 2$; this means that $E \propto h^2$, in other words, the error will change quadratically with the step size values

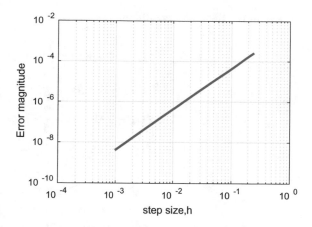

Problem 6.1.8

(a) Solve the non-homogeneous and non-autonomous two point BVP with a variable coefficient, $y'' - xy = -1$, $0 < x < 1, y(0) = y(1) = 1$, by setting up replacement equations. Let the number of nodes be $n = 5$.

(b) Write a MATLAB m-file to compute the solution of this BVP for any n and $f(x) \in \mathbb{R}$.

Plot the solution for $n = 60, f(x) = -1$ (fin_diff_bvp3.m).

Solution

(a) Given the BVP, $y'' - xy = -1$, $0 < x < 1$, $y(0) = y(1) = 1, n = 5$.

It is noted that MATLAB dsolve command does not explicitly solve this ODE. Replacement equation with boundary conditions can be written as

$$\frac{y_{i+1} - 2y_i + y_{i-1}}{(\Delta x)^2} - x_i \cdot y_i = -1, \quad i = 1, 2, 3, \ldots, n - 1 \tag{6.2}$$

$$y_0 = 1, \quad y_n = 1$$

Since $n = 5, \Delta x = 1/5, i = 1, 2, 3, 4$;

$$25(y_2 - 2y_1 + y_0) - \frac{1}{5}y_1 = -1$$

$$25(y_3 - 2y_2 + y_1) - \frac{2}{5}y_2 = -1$$

$$25(y_4 - 2y_3 + y_2) - \frac{3}{5}y_3 = -1$$

$$25(y_5 - 2y_4 + y_3) - \frac{4}{5}y_4 = -1$$

Using the boundary conditions $y_0 = 1, y_5 = 1$

$$25(y_2 - 2y_1 + 1) - \frac{1}{5}y_1 = -1$$

$$25(y_3 - 2y_2 + y_1) - \frac{2}{5}y_2 = -1$$

$$25(y_4 - 2y_3 + y_2) - \frac{3}{5}y_3 = -1$$

$$25(1 - 2y_4 + y_3) - \frac{4}{5}y_4 = -1$$

$$-(50.2)y_1 + 25y_2 = -26$$
$$25y_1 - (50.4)y_2 + 25y_3 = -1$$
$$25y_2 - (50.6)y_3 + 25y_4 = -1$$
$$25y_3 - (50.8)y_4 = -26$$

in matrix form, $ay = f$,

$$\begin{bmatrix} -50.2 & 25 & 0 & 0 \\ 25 & -50.4 & 25 & 0 \\ 0 & 25 & -50.6 & 25 \\ 0 & 0 & 25 & -50.8 \end{bmatrix} \begin{bmatrix} y_1 \\ y_2 \\ y_3 \\ y_4 \end{bmatrix} = \begin{bmatrix} -26 \\ -1 \\ -1 \\ -26 \end{bmatrix}$$

The solution of this matrix equation gives the following result:

$$x: \quad 0 \quad 0.2 \quad 0.4 \quad 0.6 \quad 0.8 \quad 1.0$$
$$y: \quad 1 \quad 1.0464 \quad 1.0612 \quad 1.0530 \quad 1.03 \quad 1.0$$

(b) First we write down the general form of Eq. (6.2), $ay = f$. Note that,

$$a_{i,i} = -(2d + x_i), \quad a_{i,i+1} = a_{i+1,i} = d, \quad f = [U - dy_0 \quad Y \quad U - dy_n]$$

where

$$Y_{1,1} = Y_{1,n-1} = U, \quad d = 1/(\Delta x)^2, \quad U = f(x) \in \mathbb{R}.$$

MATLAB m-file (fin_diff_bvp3.m) computes the solution of the BVP, $y'' - xy = U$ for any n and $f(x) = U \in \mathbb{R}$, as well as for any given boundary conditions. Figure 6.7 displays the solution curve for $n = 60$, $f(x) = U = -1$, $y_0 = y_{60} = 1$.

Fig. 6.7 The solution curve
of the BVP, $y'' - xy = U$,
$0 < x < 1$, for $n = 60$,
$f(x) = U = -1$, $y_0 = y_{60} = 1$

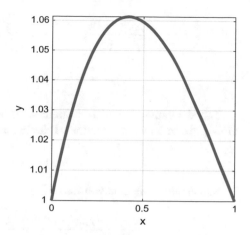

```
%fin_diff_bvp3.m  Solves linear system of simultaneous equations, a.y=f
%y=unknowns vector, while a and f are set up from given BVP.
clc;clear;
n=5;    % number of equal spacing within given interval
n=60;   % number of equal spacing within given interval
U=-1;   % Forcing function of given nonhomogeneous ODE
dx=1/n; d=1/(dx)^2;
y(1)=1; y(n)=1;  % boundary values
for i=1:n-1
x=i*dx; a(i,i)= -(2*d+x); % diagonal elements
end
for i=1:n-2
a(i,i+1)=d; a(i+1,i)=d; f(i+1,1)=U;% upper & lower diagonal elements
end
f(1,1)=U-d*y(1); f(n-1,1)=U-d*y(n); % f vector
%a,f
Y=a\f;     % solution vector
y=[y(1) Y' y(n)]' % solution vector including boundary values
x=0:dx:1; % interval
%plot(x,y,'ko','markersize',10,'linewidth',2.5); hold on;
plot(x,y,'linewidth',2.5);axis tight;
grid on;xlabel('x');ylabel('y');
```

Problem 6.1.9 Use finite differences method to solve the following BVP;

$$y'' - y = f(x) \quad 0 < x < 1, \quad y(0) = 1, \quad y'(1) = -1$$

$$f(x) = \begin{cases} 0 & 0 < x < 1/2 \\ -5 & x = 1/2 \\ -10 & (1/2) < x < 1 \end{cases}$$

Solution

It is noted that MATLAB dsolve command does not explicitly solve this ODE.
   Replacement equations yield,

$$\frac{y_{i+1} - 2y_i + y_{i-1}}{(\Delta x)^2} - y_i = f(x_i), \quad i = 1, 2, \ldots, n \tag{6.3}$$

$$y_0 = 1, \quad y'(1) = \frac{y_{n+1} - y_{n-1}}{2\,\Delta x} = -1$$

The derivative boundary condition at $x = 1$ requires the inclusion of y_{n+1} among the unknowns. Solution of the boundary condition replacement for y_{n+1}

$$y_{n+1} = y_{n-1} - 2\,\Delta x$$

Substituting this equation into (6.3),

$$\frac{2y_{n-1} - 2\Delta x - 2y_n}{(\Delta x)^2} - y_n = f(x_n) \tag{6.4}$$

This gives n equations for $y_1, y_2, \ldots y_n$.
Let $n = 4$, $\Delta x = 1/4$,

$$
\begin{aligned}
i &= 1, \quad 16(y_2 - 2y_1 + y_0) - y_1 = 0 \\
i &= 2, \quad 16(y_3 - 2y_2 + y_1) - y_2 = -5 \\
i &= 3, \quad 16(y_4 - 2y_3 + y_2) - y_3 = -10
\end{aligned}
$$

From (6.4), for $n = 4$,

$$16(2y_3 - 0.5 - 2y_4) - y_4 = -10$$

Applying boundary condition $y_0 = 1$ and cleaning up these equations, we get

$$
\begin{array}{ll}
16(y_2 - 2y_1 + 1) - y_1 = 0 & \rightarrow \quad 16y_2 - 32y_1 + 16 - y_1 = 0 \\
16(y_3 - 2y_2 + y_1) - y_2 = -5 & \rightarrow \quad 16y_3 - 32y_2 + 16y_1 - y_2 = -5 \\
16(y_4 - 2y_3 + y_2) - y_3 = -10 & \rightarrow \quad 16y_4 - 32y_3 + 16y_2 - y_3 = -10 \\
16(2y_3 - 0.5 - 2y_4) - y_4 = -10 & \rightarrow \quad 32y_3 - 8 - 32y_4 - y_4 = -10
\end{array}
$$

$$
\begin{bmatrix}
-33 & 16 & 0 & 0 \\
16 & -33 & 16 & 0 \\
0 & 16 & -33 & 16 \\
0 & 0 & 32 & -33
\end{bmatrix}
\begin{bmatrix}
y_1 \\ y_2 \\ y_3 \\ y_4
\end{bmatrix}
=
\begin{bmatrix}
-16 \\ -5 \\ -10 \\ -2
\end{bmatrix}
$$

$$
\begin{array}{lccccc}
x: & 0 & 0.25 & 0.50 & 0.75 & 1.00 \\
y: & 1 & 1.5274 & 2.1503 & 2.5951 & 2.5770
\end{array}
$$

Let $n = 16$, $\Delta x = 1/256$,

$$i = 1, \quad 256(y_2 - 2y_1 + y_0) - y_1 = 0$$
$$i = 2, \quad 256(y_3 - 2y_2 + y_1) - y_2 = -5$$
$$i = 3, \quad 256(y_4 - 2y_3 + y_2) - y_3 = -10$$

From (6.4), for $n = 4$,

$256(2y_3 - 1/8 - 2y_4) - y_4 = -10$
$256(y_2 - 2y_1 + 1) - y_1 = 0 \qquad \rightarrow \quad 256y_2 - 512y_1 + 256 - y_1 = 0$
$256(y_3 - 2y_2 + y_1) - y_2 = -5 \qquad \rightarrow \quad 256y_3 - 512y_2 + 256y_1 - y_2 = -5$
$256(y_4 - 2y_3 + y_2) - y_3 = -10 \qquad \rightarrow \quad 256y_4 - 512y_3 + 256y_2 - y_3 = -10$
$256(2y_3 - 1/8 - 2y_4) - y_4 = -10 \qquad \rightarrow \quad 512y_3 - 32 - 512y_4 - y_4 = -10$

$$
\begin{bmatrix}
-513 & 256 & 0 & 0 \\
256 & -513 & 256 & 0 \\
0 & 256 & -513 & 256 \\
0 & 0 & 512 & -513
\end{bmatrix}
\begin{bmatrix}
y_1 \\ y_2 \\ y_3 \\ y_4
\end{bmatrix}
=
\begin{bmatrix}
-256 \\ -5 \\ -10 \\ 22
\end{bmatrix}
$$

$x:$ 0   0.25   0.50   0.75   1.00
$y:$ 1   1.5274   2.1503   2.5951   2.5770

Problem 6.1.10 An example of a derivative boundary condition is the heat transfer problem in which the end of a rod is insulated, which means that the heat flux is zero at this end.

Given the heated rod equation,

$$\frac{d^2T}{dx^2} = -k(T_\infty - T) \tag{6.5}$$

with the boundary conditions, $dT(0)/dx = T_a', T(L) = T_b$ where L is the length of the rod, T_a and T_b are the temperature values at both ends of the rod, T_∞ is the temperature of surrounding medium. We have a derivative boundary condition, since T_a' is a given boundary condition, along with the fixed boundary condition of $T(L) = T_b$.

We assume that the rod is axially divided into a series of nodes with equidistant spacing (h) between each node. However, we also locate an imaginary node at a distance (h) to the left of the rod's left end (at a).

(a) Determine the temperature distribution in the rod if $L = 5\,\text{m}, h = 1\,\text{m}$, $k = 0.05\,\text{m}^{-2}$, $T_\infty = 300\,\text{K}, T_a' = 0\,\text{K}, T_h = 385\,\text{K}$.

(b) Repeat (a) if $T_a' = -10\,\text{K}$. Plot axial temperature pattern in the rod for both cases (fin_diff_bvp4.m).

Solution

(a) Replacing the second derivative term in (6.5) by the algebraic difference equation, we get

$$\frac{T_{i-1} - 2T_i + T_{i+1}}{h^2} + k(T_\infty - T_i) = 0$$

Collecting terms gives

$$-T_{i-1} + (2 + kh^2)T_i - T_{i+1} = kh^2 T_\infty \tag{6.6}$$

In this equation, there are $n + 1$ nodes, the first and the last node temperatures are denoted by T_0 and T_n. However, we included an extra node to the left of the left (insulated) end of the rod. Writing (6.6) for this node gives

$$-T_{-1} + (2 + kh^2)T_0 - T_{-1} = kh^2 T_\infty \tag{6.7}$$

If we represent the first derivative at (0) by the centered difference,

$$\frac{dT}{dx} = \frac{T_1 - T_{-1}}{2h} \tag{6.8}$$

and solve (6.8) for T_{-1},

$$T_{-1} = T_1 - 2h \cdot \frac{dT}{dx} \tag{6.9}$$

substituting this value into (6.7) yields

$$-T_1 + 2h \cdot \frac{dT}{dx} + (2 + kh^2)T_0 - T_1 = kh^2 T_\infty$$

$$(2 + kh^2)T_0 - 2T_1 + 2h \cdot \frac{dT}{dx} = kh^2 T_\infty$$

$$(2 + kh^2)T_0 - 2T_1 = kh^2 T_\infty - 2h \cdot \frac{dT}{dx} \tag{6.10}$$

At node 0, Eq. (6.10) takes the following form,

$$(2 + 0.05 \times 1)T_0 - T_1 = 0.05 \times 1 \times 300 \rightarrow 2.05T_0 - 2T_1 = 15$$

At node (1) we use Eq. (6.6), with $i = 1$,

$$-T_0 + 2.05T_1 - T_2 = 15$$

Repeating the similar procedure for the other nodes yields the final system of equations in the following matrix form,

$$AT = b$$

$$\begin{bmatrix} 2.05 & -2 & 0 & 0 & 0 \\ -1 & 2.05 & -1 & 0 & 0 \\ 0 & -1 & 2.05 & -1 & 0 \\ 0 & 0 & -1 & 2.05 & -1 \\ 0 & 0 & 0 & -1 & 2.05 \end{bmatrix} \begin{bmatrix} T_0 \\ T_1 \\ T_2 \\ T_3 \\ T_4 \end{bmatrix} = \begin{bmatrix} 15 \\ 15 \\ 15 \\ 15 \\ 400 \end{bmatrix}$$

Note that at node (4) we use Eq. (6.6) with $i = 4$,

$$-T_3 + 2.05T_4 - 385 = 15 \rightarrow -T_3 + 2.05T_4 = 400$$

The matrix equation can be solved for node temperature values as

$$T_0 = 350.3\,\text{K}, \quad T_1 = 351.6\,\text{K}, \quad T_2 = 355.4\,\text{K}, \quad T_3 = 362.0\,\text{K}, \quad T_4 = 371.7\,\text{K}$$

Following is the list of results obtained through MATLAB m-file fin_diff_bvp4. m when $dT(0)/dx = 0$.

```
T = 350.3042  351.5618  355.3975  362.0031  371.7088
S = Tinf - (exp(k^(1/2)*x)*(Tinf - 385))/(exp(-5*k^(1/2)) + exp(5*k^(1/2)))
  - (exp(-k^(1/2)*x)*(Tinf - 385))/(exp(-5*k^(1/2)) + exp(5*k^(1/2)))
z = 350.2103  351.4708  355.3156  361.9377  371.6696  385.0000
```

(b) For the case where $T'(0) = -10$,

$$2h\frac{dT}{dx} = 2 \times 1 \times (-10) = -20$$

Equation (6.10) can be used to present node 0 as

$$2.05T_0 - 2T_1 = 15 - (-20) = 35$$

The matrix form of the system is

$$
\begin{bmatrix}
2.05 & -2 & 0 & 0 & 0 \\
-1 & 2.05 & -1 & 0 & 0 \\
0 & -1 & 2.05 & -1 & 0 \\
0 & 0 & -1 & 2.05 & -1 \\
0 & 0 & 0 & -1 & 2.05
\end{bmatrix}
\begin{bmatrix}
T_0 \\ T_1 \\ T_2 \\ T_3 \\ T_4
\end{bmatrix}
=
\begin{bmatrix}
35 \\ 15 \\ 15 \\ 15 \\ 400
\end{bmatrix}
$$

which is solved for

$$
T_0 = 386.1\,\text{K}, \quad T_1 = 378.3\,\text{K}, \quad T_2 = 374.4\,\text{K}, \quad T_3 = 374.1\,\text{K}, \quad T_4 = 377.6\,\text{K}
$$

Following is the list of results obtained through MATLAB m-file fin_diff_bvp4. m when $dT(0)/dx = -10$.

```
T = 386.1297 378.2830 374.3504 374.1353 377.6270
S = Tinf - (exp(k^(1/2)*x)*(10*exp(-5*k^(1/2)) + Tinf*k^(1/2) -
385*k^(1/2)))/(k^(1/2)*(exp(-5*k^(1/2)) + exp(5*k^(1/2)))) + (exp(-
k^(1/2)*x)*(10*exp(5*k^(1/2)) - Tinf*k^(1/2) +
385*k^(1/2)))/(k^(1/2)*(exp(-5*k^(1/2)) + exp(5*k^(1/2)))))
z = 386.2953 378.3781 374.3962 374.1497 377.6261 385.0000
```

The two cases that reflect different derivative values at $x = 0$ are displayed in Fig. 6.8.

Note that incorporating a derivative boundary condition into the ODE increases the size of the square coefficient matrix by 1.

MATLAB m-file fin_diff_bvp4.m is listed below which performs analytical solution of the problem, as well (the z vector is the resulting temperature values that are obtained through the analytical solution).

```
%fin_diff_bvp4.m
clc;clear;close;
A=[2.05   -2       0       0       0;
   -1    2.05     -1       0       0;
    0     -1     2.05     -1       0;
    0      0      -1     2.05     -1;
    0      0       0      -1     2.05];
b=[ 15 15 15 15 400 ]';
T=A\b
%Analytical solution
syms  y(x)
k=0.05;Tinf=300;
bc='Dy(0)=0,y(5)=385';
S=dsolve('D2y+k*(Tinf-y)=0',bc,x)
x = linspace(0,5,6); z = eval(vectorize(S))
plot(x,z,'kO-','linewidth',2); grid; xlabel('x'); ylabel('y(x)');
```

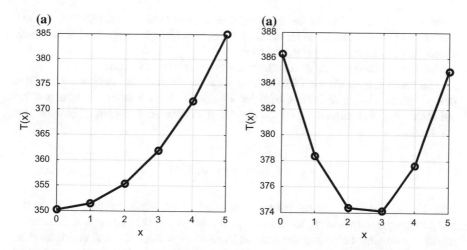

Fig. 6.8   a Axial temperature profile in the rod when $dT(0)/dx = 0$. **b** Axial temperature profile in the rod when $dT(0)/dx = -10$. $L = 5\,\text{m}, h = 1\,\text{m}, k = 0.05\,\text{m}^{-2}$, $T_\infty = 300\,\text{K}, T_b = 385\,\text{K}$

Problem 6.1.11 How can one measure the error of finite difference approximation (with uniform step size) when there is no analytical solution of a BVP?

Answer
One method is to take a numerical solution on a "very fine step size" as the 'exact' solution. Caution! When step size is "too small" cancellation may occur.

Problem 6.1.12 How can the order of accuracy of finite difference scheme with nonuniform step lengths be calculated when analytical solution is not available?

Answer
Choosing grid points N, 2N and 4N and assuming solutions for 4N points as the most accurate, one may compute maximum absolute errors in N and 2N grid points. Finally, using formula $\log(E_N/E_{2N})/\log(2)$, one can get an idea about the convergence rate towards this function. However, it is not a way to decide on the accuracy. The reason is that an inconsistent method may converge to some function which may not be the actual solution.

   Another method to get an idea about the error estimate is to use the same formula for the uniform mesh with mesh size h. In the nonuniform case, h will be the maximum length of the step sizes.

6.2   Solution of Nonlinear BVPs Using FD Approximations

Problem 6.2.1 (*Finite differences method for* nonlinear *BVPs with Dirichlet boundary conditions*) For nonlinear BVPs, linear interpolation or extrapolation may not provide a good estimate of the required boundary condition to attain an exact solution. One method involves recasting it as a roots finding problem, i.e., to find the value of x that makes the function equal to zero (Newton's method is the usual choice in performing this solution process).

Another method which we use here is the replacement method.

The following nonlinear ODE can be used to simulate the temperature of the heated rod as a more realistic model for a heated rod, as it takes the heat loss by radiation into the consideration (which is a nonlinear parameter) besides the convection;

$$\frac{d^2T}{dx^2} = -k(T_\infty - T) - \sigma(T_\infty^4 - T^4) \tag{6.11}$$

where $T_a(0)$ and $T_b(L)$ are the temperature values at both ends of the rod, T_∞ is the temperature of surrounding medium, L is the length of the rod, k is a bulk heat-transfer parameter reflecting the relative impacts of convection and conduction in units of m^{-2}, σ is a bulk heat-transfer parameter reflecting the relative impacts of radiation and conduction, in units of $K^{-3} m^{-2}$.

Determine the temperature distribution in the rod if $L = 5\,m, h = 1\,m$, $k = 0.05\,m^{-2}$, $T_\infty = 300\,K, T(0) = 350\,K, T_b = 385\,K$, $\sigma = 2.5 \times 10^{-9}\,K^{-3}\,m^{-2}$ (fin_diff_bvp5.m).

Solution
Explicit solution could not be found by MATLAB dsolve function.

Replacing the second derivative term in (6.11) by the algebraic difference equation, we get

$$\frac{T_{i-1} - 2T_i + T_{i+1}}{h^2} = -k(T_\infty - T_i) - \sigma(T_\infty^4 - T_i^4)$$

Collecting terms and solving for T_i,

$$-T_{i-1} + (2 + kh^2)T_i - T_{i+1} = kh^2 T_\infty + \sigma h^2(T_\infty^4 - T_i^4)$$

$$T_i = \frac{kh^2 T_\infty + \sigma h^2(T_\infty^4 - T_i^4) + T_{i-1} + T_{i+1}}{2 + kh^2} \tag{6.12}$$

The equation can be solved iteratively, (if the term T_i on the right is equal to its value from the previous iteration) to calculate the temperature of each node. Iterations continue until the difference between the previous and present temperature values become less than an acceptable tolerance. In the MATLAB code given below, the infinite norm of present and past temperature vectors is taken as a measure of tolerance.

Following is the list of solution set (temperature values in K, at each node) for four internal equidistant nodes (besides the endpoints, which are the boundary values) obtained after each iteration. Setting the tolerance at e = 0.1 stops the program at the end of eight iteration.

350.0000	187.9268	108.8667	70.3009	239.2931	385.0000
350.0000	239.5115	168.1518	215.9189	306.3277	385.0000
350.0000	265.9389	251.2728	286.5447	334.0397	385.0000
350.0000	304.3991	300.5989	318.5534	345.2082	385.0000
350.0000	324.0901	320.7226	329.4815	348.4042	385.0000
350.0000	330.9229	326.4402	332.0158	348.9901	385.0000
350.0000	332.5410	327.5207	332.3813	349.0472	385.0000
350.0000	332.7800	327.6313	332.3978	349.0434	385.0000

A plot of the temperature profile along the rod is displayed in Fig. 6.9. MATLAB m-file (fin_diff_bvp5.m) is also listed below.

```
%fin_diff_bvp5.m
clc;clear;close;
k=0.05;h=1;s=2.5e-9;
Tinf=300;T0=350;T6=385;
T = zeros(6,1);T(1)-T0;T(6)=T6;
e=0.1;%tolerance
N=10; % number of iterations
n=4;% number of internal nodes
y=zeros(N,n+2);
K=2+k*h.^2;
u0=k*h.^2*Tinf;
    for k = 1:N
        Tprev = T;% previous temperature vector
        for i = 2:n+1
 u1=s*h.^2*(Tinf.^4-Tprev(i).^4); u2=T(i-1); u3=T(i+1);
 T(i) =(u0+u1+u2+u3)/K;%Temperature at ith node
        end
        r=norm( T - Tprev, inf ); % measure of tolerance
        if r < e
        return;
        end
        y(k,:)=T
        I=y(k,:);plot(0:n+1,I,'o-','linewidth',2);
        xlabel('x, m');ylabel('T, K');grid;
    end
        error 'the method did not converge';
```

Problem 6.2.2 Solve the nonlinear autonomous BVP, $y'' = -y^4$; $y(0) = y(1) = 1$ by FDM.

Let the number of internal nodes be $n = 20$. Plot the solution which will be reached either at the end of 50 iterations or when the infinite norm of the difference between the ith solution vector and the previous solution vector becomes less than $\epsilon < 0.01$ (fin_diff_bvp3.m).

Fig. 6.9 The solution of a
second-order nonlinear BVP
with a Dirichlet boundary
condition

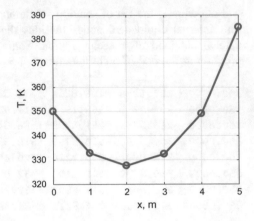

Solution

Analytical solution could not be found using dsolve function. We use replacement method.

$$\frac{y_{i-1} - 2y_i + y_{i+1}}{h^2} = -y_i^4$$

$$y_{i-1} - 2y_i + y_{i+1} = -h^2 y_i^4$$

$$y_{i-1} - 2y_i + h^2 y_i^4 + y_{i+1} = 0$$

Collecting terms and solving for y_i,

$$y_i = \frac{y_{i-1} - h^2 y_i^4 + y_{i+1}}{2} \tag{6.13}$$

where $y_0 = y(0) = 1$, $y_b = y(1) = 1$.

Equation (6.13) can be solved iteratively, (if the term y_i on the right is equal to its value from the previous iteration) to calculate the present value of function at each node. The solution curve of this nonlinear BVP is shown in Fig. 6.10. MATLAB m-file (fin_diff_bvp3.m) is listed below.

The minimum number of iterations for the given epsilon value is seen to be 50.

```
%fin_diff_bvp3.m
clc;clear;close;
e=0.01;        %tolerance
N=50;          % number of iterations
n=20;          % number of internal nodes
span=[0 1];a=span(1);b=span(2);
bv=[1 1];   % Boundary values
y=zeros(n+2,1); % Initialization
y(1)=bv(1);y(n+2)=bv(2);
h=(b-a)/(n+1);   % Step size
    for k = 1:N
        yprev = y; % previous temperature vector
        for i = 2:n+1
 u1=h.^2*yprev(i).^4; u2=y(i-1); u3=y(i+1);
 y(i) =(u1+u2+u3)/2;%Temperature at ith node
        end
        r=norm( y - yprev, inf ); % measure of tolerance
        if r < e
        return;
        end
        plot(a:h:b,y,'linewidth',2);
        xlabel('x');ylabel('y(x)');grid;
    end
 error 'the method did not converge';
```

Problem 6.2.3 Solve the nonlinear BVP, $y'' + 2yy' = x^3 + 1$; $y(1) = 1/2$, $y(2) = 2$, using a nonlinear FD equation with iteration method and centered difference approximations for both derivative terms, assuming three interior points ($n = 3$) of the uniform grid.

Solution

$h = (2-1)/(3+1) = 1/4$,

$$\left(\frac{y_{i+1} - 2y_i + y_{i+1}}{h^2}\right)^{(k+1)} + 2y^{(k)}\left(\frac{y_{i+1} - y_{i-1}}{2h}\right)^{(k+1)} = x_i^3 + 1 \qquad (6.14)$$

Multiply (6.14) by h^2 and collect the terms,

$$\left(y_{i+1} - 2y_i + y_{i+1}\right)^{(k+1)} + hy^{(k)}\left(y_{i+1} - y_{i-1}\right)^{(k+1)} = h^2\left(x_i^3 + 1\right)$$

$$y_{i-1}^{(k+1)}\left(1 - hy_i^{(k)}\right) - 2y_i^{(k+1)} + \left(1 + hy_i^{(k)}\right)y_{i+1}^{(k+1)} = h^2\left(x_i^3 + 1\right) \qquad (6.15)$$

At the three interior grids, Eq. (6.15) becomes
$i = 2$, $x = 1 + 1/4 = 1.25$,

$$y_1^{(k+1)}\left(1 - hy_2^{(k)}\right) - 2y_2^{(k+1)} + \left(1 + hy_2^{(k)}\right)y_3^{(k+1)} = (h)^2\left(1.25^3 + 1\right) \qquad (6.16a)$$

$i = 3$, $x = 1 + 1/2 = 1.50$,

Fig. 6.10 Solution curve of
the nonlinear BVP,
$y'' = -y^4$; $y(0) = y(1) = 1$

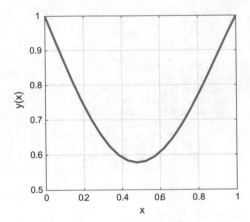

$$\left(1 - hy_3^{(k)}\right)y_2^{(k+1)} - 2y_3^{(k+1)} + \left(1 + hy_3^{(k)}\right)y_4^{(k+1)} = (h)^2\left(1.5^3 + 1\right) \qquad (6.16b)$$

$i = 4, \quad x = 1 + 3/4 = 1.75,$

$$\left(1 - hy_4^{(k)}\right)y_3^{(k+1)} - 2y_4^{(k+1)} + \left(1 + hy_4^{(k)}\right)y_5^{(k+1)} = (h)^2\left(1.75^2 + 1\right) \qquad (6.16c)$$

$$y(1) = y_1 = 1/2, \quad y(2) = y_5 = 2$$

First term of (6.16a) becomes $(1/2)\left(1 - hy_2^{(k)}\right) = (1/2) - (1/2)hy_2^{(k)}$.

Last term to the left of equal sign in Eq. (6.16c) becomes
$\left(1 + hy_4^{(k)}\right) = 2 = 2 + 2hy_4^{(k)}$.

Transferring constants to the right side of both Eqs. (6.16a), (6.16c) yields a
tridiagonal system of FD equations,

$$(-1/2)hy_2^{(k)} - 2y_2^{(k+1)} + \left(1 + hy_2^{(k)}\right)y_3^{(k+1)} = (-1/2) + h^2\left(1.25^3 + 1\right) \qquad (6.17a)$$

$$\left(1 - hy_3^{(k)}\right)y_2^{(k+1)} - 2y_3^{(k+1)} + \left(1 + hy_3^{(k)}\right)y_4^{(k+1)} = h^2\left(1.5^3 + 1\right) \qquad (6.17b)$$

$$\left(1 - hy_4^{(k)}\right)y_3^{(k+1)} - 2y_4^{(k+1)} + 2hy_4^{(k)} = -2 + h^2\left(1.75^3 + 1\right) \qquad (6.17c)$$

For the initial guess of $y^{(0)}$, assuming that all elements are zero at the internal
nodes

$$k = 0, \quad y^{(0)} = [1/2 \ 0 \ 0 \ 0 \ 2]$$

Substituting these values into (6.17) gives the following system of linear algebraic FD equations,

$$-(1/2)h \cdot (0) - 2y_2^{(1)} + [1 + h \cdot (0)]y_3^{(1)} = -1/2 + h^2(1.25^3 + 1)$$
$$[1 - h \cdot (0)]y_2^{(1)} - 2y_3^{(1)} + [1 + h \cdot (0)]y_4^{(1)} = h^2(1.5^3 + 1)$$
$$[1 - h \cdot (0)]y_3^{(1)} - 2y_4^{(1)} + 2h \cdot (0) = -2 + h^2(1.75^3 + 1)$$

or,

$$-2y_2^{(1)} + y_3^{(1)} = -0.315429$$
$$y_2^{(1)} - 2y_3^{(1)} + y_4^{(1)} = 0.273438$$
$$y_3^{(1)} - 2y_4^{(1)} = -1.602539$$

$$\begin{bmatrix} -2 & 1 & 0 \\ 1 & -2 & 1 \\ 0 & 1 & -2 \end{bmatrix} \begin{bmatrix} y_2^{(1)} \\ y_3^{(1)} \\ y_4^{(1)} \end{bmatrix} = \begin{bmatrix} -0.315429 \\ 0.273438 \\ -1.602539 \end{bmatrix}$$

$$y^{(1)} = [1/2 \quad 0.500488 \quad 0.685546 \quad 1.14043 \quad 2].$$

The solution $y^{(1)}$ (including the boundary conditions) is then substituted with $y^{(1)}$ and a new tridiagonal system of linear FD equations is set up and solved for $y^{(2)}$.

This process is repeated a total of N times until the solution satisfies a given convergence criterion. (Note: While the explicit solution is $y = x^2/2$, this analytical solution could not be found using dsolve function).

Problem 6.2.4 Solve the nonlinear BVP, $y'' + (y')^2 = 4x^2 - 2$; $y(0) = 0$, $y(2) = -4$, using matrical iteration method, employing centered difference approximations for both derivative terms, assuming three interior points (N = 3) of the uniform grid. Test different initial (guess) solution arrays with stopping criteria being solely the number of iterations (Let N = 40,000) or the infinite norm of the error of approximation to the solution array. Record elapsed time of computation for each case (fin_diff_bvp_mat5.m).

Solution
Analytical solution of this BVP is known to be $y = -x^2$. Figure 6.11 displays exact (analytical) and computed solution values for the given BVP when N = 3, M = 40,000, zero initial (guessed) solution array.

Below, computed results are presented along with exact solution values at the internal node points for different initial (guessed) solution arrays (zero, unity and linear).

Fig. 6.11 Exact and
computed solution values for
the BVP when N = 3,
M = 40,000

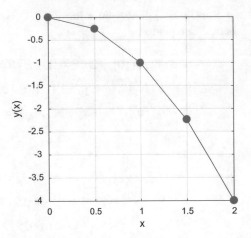

```
                            Yo=zeros (ones)
        x                   y (exact)              y(computed)
        0                       0                       0
0.500000000000000   -0.250000000000000   -0.256597673892484
1.000000000000000   -1.000000000000000   -1.005339204998119
1.500000000000000   -2.250000000000000   -2.240759971916852
2.000000000000000   -4.000000000000000   -4.000000000000000
          Elapsed time is 3.081339 seconds.(M=40000)


                            Yo= -(ones)
        0                       0                       0
0.500000000000000   -0.250000000000000   -0.252325792270791
1.000000000000000   -1.000000000000000   -1.012755966054557
1.500000000000000   -2.250000000000000   -2.268405280328639
2.000000000000000   -4.000000000000000   -4.000000000000000
          Elapsed time is 3.101966 seconds.  (M=40000)


                            Yo= linear
        0                       0                       0
0.500000000000000   -0.250000000000000   -0.241887258561555
1.000000000000000   -1.000000000000000   -0.990193178817884
1.500000000000000   -2.250000000000000   -2.254643224616111
2.000000000000000   -4.000000000000000   -4.000000000000000
          Elapsed time is 3.162605 seconds.
```

First part of MATLAB m-file (fin_diff_bvp_mat5.m) computes the analytical
solution using dsolve command, then plots the solution curve and holds it on. In the
second part of the code, matrix based iterations are performed to compute the FD
approximate solution of the given BVP. These are listed in Command Window
along with the exact solution values at the internal nodes (as shown in the edited
list, above) and also marked on the previously drawn (exact) solution figure.

When infinite norm of the error of approximation to the solution array becomes
less than the threshold value (defined by parameter eps), iterations are set to stop
and print the recent iteration count as well as approximate solution array on
Command Window;

```
         i = 239774
              w
      -0.247007937971106
      -0.997015208015728
      -2.252988717257702
Elapsed time is 63.801021 seconds.
```

```
%fin_diff_bvp_mat5.m  nonlinear finite difference method, Dirichlet bc.
clc;clear;close;tic;
N=3; % number of internal nodes
%  y"+(y')^2=-2+4x^2,   y(0)=0, y(2)=-4, using dsolve
syms  y(x)
eq='D2y+(Dy)^2=4*x^2-2';bc='y(0)=0, y(2)=-4';
S=dsolve(eq,bc,x);
x = linspace(0,2,N+2);y=eval(vectorize(S));
plot(x,y,'k');hold on;
%---------------------------------------------------------------------
%  FDM, using iterations
M=250000 % total number of iterations
%format rat
format long
span=[0 2]; % interval
bv=[0 -4];alpha=bv(1);beta=bv(2);% boundary values
h=(span(2)-span(1))/(N+1);% step size
x=linspace(span(1)+h,span(2)-h,N);% internal node positions
w=zeros(N,1); % preallocation,all zeros initial approximation
%w=-1*ones(N,1); % preallocation,all ones initial approximation
%linear initial approximations
% for i=1:N
% w(i)=alpha+i'*(beta-alpha)/(span(2)-span(1))*h;
% end
% w=w'; this is required for linear initial approximation
eps=3e-3;  y(1)=[];y(N+1)=[];y=y';
Y=zeros(N,1);Y(1)=alpha;Y(N)=beta
v=ones(N-1,1);
C=2*eye(N);
B=-diag(v,1)-diag(v,-1); % diag(V,K), K tells which diagonal to place V in
A=C+B   % tridiagonal  matrix
F=zeros(N,1); % preallocation
for i=1:M
    f = @(x,y,dy) (-2+4*x^2-(dy)^2); % y"=f(x,y,y')
    for j=2:N-1
    F(j,1)=f(x(j),w(j),(w(j+1)-w(j-1))/(2*h));
    F(1,1)=f(x(1),w(1),(w(2)-alpha)/(2*h));
    F(N,1)=f(x(N),w(N),(beta-w(N-1))/(2*h));
    end
    % F
D=Y-h.^2*F;
w=A\D;
%[w y abs(w-y)]
r=norm(w-y,inf) % measure of tolerance
        if r < eps
        i,w,y
 X=linspace(span(1),span(2),N+2);X=X';
  Y=[alpha; w; beta];
%  [X y Y]
plot(X,Y,'.','markersize',20),xlabel('x'),ylabel('y(x)'),grid on;toc
        return;
        end
end
```

Problem 6.2.5

(a) Derive a solution equation in matrix form for nonlinear second order BVPs using equilibrium method (nonlinear implicit FDE using iteration).
(b) Find the numerical values of elements of the solution vector at the end of first iteration for the BVP, $y'' + 2yy' = x^3 + 1$, $y(1) = 1/2, y(2) = 2$ using matrix notation, assuming three interior nodes and uniform grid. Let $\mathbf{y_0} = \begin{bmatrix} 0 & 0 & 0 \end{bmatrix}^T$.
(c) Find the solution of the BVP after 10 iterations. List the values of the elements of solution vector. Plot iteration versus solution values in a figure to demonstrate the convergence rate (fin_diff_bvp_mat2.m).
(d) Extend the MATLAB code to solve nonlinear BVP iteratively for larger number of nodes, ≥ 3. Plot the solution curve (fin_diff_bvp_mat3.m).

Solution

(a) A nonlinear BVP for a second order ODE is of the form

$$y'' = f(x, y, y'), \ x \in [a, b], \ y(a) = \alpha, \ y(b) = \beta \tag{6.18}$$

Considering central FD representation of first and second derivatives,

$$y'(x) = \frac{y(x+h) - y(x-h)}{2h}, \quad y''(x) = \frac{y(x+h) - 2y(x) + y(x-h)}{h^2} \tag{6.19}$$

where $h = b - a/(N+1)$, N = internal number of nodes,

$$x_0 = a, \quad x_1 = a + h, \ldots, \quad x_N = a + Nh, \quad x_{N+1} = a + (N+1)h = b$$

with the boundary values $Y_0 = y(a) = \alpha, Y_{N+1} = y(b) = \beta$.

We find $y_1, y_2, \ldots, y_N$ by solving the following system of equations which is obtained when Eq. (6.19) is placed in Eq. (6.18),

$$\frac{y_{i+1} - 2y_i + Y_{i-1}}{h^2} = f\left(x_i, y_i, \frac{y_{i+1} - y_{i-1}}{2h}\right) \tag{6.20}$$

$$-y_{i-1} + 2y_i - y_{i+1} + h^2 \cdot f\left(x_i, y_i, \frac{y_{i+1} - y_{i-1}}{2h}\right) = 0 \tag{6.21}$$

where $i = 1, 2, \ldots, N$.

Rewriting (6.21) using matrix notation,

$$diag\{-1, 2, 1\}\mathbf{y} + h^2 F(\mathbf{x}, \mathbf{y}) = A\mathbf{y} + h^2 F(\mathbf{x}, \mathbf{y}) = \mathbf{Y} \tag{6.22}$$

Where the first term means a tridiagonal square matrix,

$$y = [y_1, y_2, \ldots, y_N]^T, \, y = [\alpha, 0, 0, \ldots, 0, \beta]^T, \, A = diag\{-1, 2, 1\},$$
$$F(x, y) = [F_1(x, y), \ldots, F_N(x, y)]^T, \, F_i \cdot (x, y) = f(x_i, y_i, (y_{i+1} - y_{i-1})/2h)$$

Equation (6.22) can be solved for y, starting at the guess solution vector $y_0 = \left[y_1^{(0)}, \ldots, y_N^{(0)}\right]$,

$$Ay_{k+1} = Y - h^2 F(x, y_k) \tag{6.23}$$

(b) Given: $y'' + 2yy' = x^3 + 1$, $y(1) = 1/2$, $y(2) = 2$, $N = 3$, $y_0 = [0 \quad 0 \quad 0]^T$.

$$y'' = 1 + x^3 - 2yy' = f(x, y, y')$$

$$h = \frac{(b-a)}{(N+1)} = \frac{1}{4}, \quad \alpha = \frac{1}{2}, \quad \beta = 2$$
$$x = [a \quad x_1 \quad x_2 \quad x_3 \quad b] = [1 \quad 5/4 \quad 6/4 \quad 7/4 \quad 2]$$
$$Y = [\alpha \quad 0 \quad \beta]^T = [1/2 \quad 0 \quad 2]^T$$

We will compute $y_1 = \left[y_1^{(1)}, y_2^{(1)}, y_3^{(1)}\right]$, using (6.23);

$$Ay_{k+1} = Y - h^2 F(x, y_k) \rightarrow y_{k+1} = A^{-1} \cdot (Y - h^2 F(x, y_k))$$

We first compute $F(x, y)$,

$$F(x, y) = \begin{bmatrix} F_1(x, y) \\ F_2(x, y) \\ F_3(x, y) \end{bmatrix} = \begin{bmatrix} f(x_1, y_1, (y_2 - \alpha)/2h) \\ f(x_2, y_2, (y_3 - y_1)/2h) \\ f(x_3, y_3, (\beta - y_2)/2h) \end{bmatrix}$$
$$= \begin{bmatrix} 1 + x_1^3 - 2y_1(y_2 - \alpha)/2h \\ 1 + x_2^3 - 2y_2(y_3 - y_1)/2h \\ 1 + x_3^3 - 2y_3(\beta - y_2)/2h \end{bmatrix}$$

Since the initial (guessed) solution is given as $y_1 = y_2 = y_3 = 0$, substituting numerical values for h and x variables in the last equation along with guessed solution values, we get

$$F(x,y) = \begin{bmatrix} 1+(5/4)^3 \\ 1+(6/4)^3 \\ 1+(7/4)^3 \end{bmatrix} = \begin{bmatrix} 189/64 \\ 280/64 \\ 407/64 \end{bmatrix}$$

$$y_{k+1}^{(1)} = \begin{bmatrix} y_1^{(1)} \\ y_2^{(1)} \\ y_3^{(1)} \end{bmatrix} = \begin{bmatrix} 2 & -1 & 0 \\ -1 & 2 & -1 \\ 0 & -1 & 2 \end{bmatrix}^{-1} \cdot \left(\begin{bmatrix} 1/2 \\ 0 \\ 2 \end{bmatrix} - \left(\frac{1}{4}\right)^2 \begin{bmatrix} 189/64 \\ 280/64 \\ 407/64 \end{bmatrix} \right)$$

$$= \begin{bmatrix} 2 & -1 & 0 \\ -1 & 2 & -1 \\ 0 & -1 & 2 \end{bmatrix}^{-1} \cdot \begin{bmatrix} 323/1024 \\ -35/128 \\ 1641/1024 \end{bmatrix} = \begin{bmatrix} 512/1024 \\ 351/512 \\ 2343/2048 \end{bmatrix} = \begin{bmatrix} 0.500488 \\ 0.685547 \\ 1.144043 \end{bmatrix}$$

We insert boundary values to obtain the solution vector at the end of first iteration,

$$y^{(1)} = \begin{bmatrix} \frac{1}{2} & 0.500488 & 0.685547 & 1.144043 & 2 \end{bmatrix}^T$$

where T stands for the transpose operation.

(c) The values of the three solution vector elements (values at the internal nodes) at the end of each iteration are listed below. Figure 6.12 displays a plot of iteration versus solution values to demonstrate the convergence rate (10 iterations).

```
M = 10
      y =0.500488281250000    0.685546875000000    1.144042968750000
      y =0.667035579681396    0.995425462722779    1.486956119537354
      y =0.757831847090529    1.117617404002005    1.546798015217974
      y =0.783772200231180    1.135102178444836    1.539429076491491
      y =0.784255198925621    1.128636852408412    1.532019064291227
      y =0.781856270246606    1.125029923058829    1.530652612033305
      y =0.781123438037558    1.124646297466291    1.531002084198269
      y =0.781153432450002    1.124895711541430    1.531238429894438
      y =0.781234931279277    1.125005317560086    1.531271354611692
      y =0.781255292551433    1.125011901024558    1.531257268344503
```

Fig. 6.12 Convergence for the solution of BVP, $y'' + 2yy' = x^3 + 1$, $y(1) = 1/2, y(2) = 2$ using FD Method (number of internal nodes, N = 3)

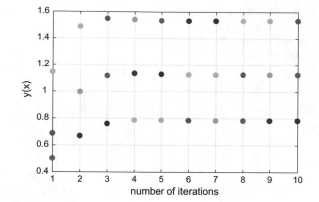

```
%fin_diff_bvp_mat2.m    y"+2yy'=1+x^3,    y(1)=1/2, y(2)=2,   iterations
clc;clear;close;
M=10 % total number of iterations
%format rat
format long
span=[1 2];
bv=[1/2 2];alpha=bv(1);beta=bv(2);
N=3;
h=(span(2)-span(1))/(N+1);
x=linspace(span(1)+h,span(2)-h,N);
y=zeros(N,1);
Y=zeros(N,1);Y(1)=alpha;Y(N)=beta;
v=ones(N-1,1);
C=2*eye(N);
B=-diag(v,1)-diag(v,-1); % diag(V,K), K tells which diagonal to place V in
A=C+B
for i=1:M
F=[1+x(1).^3-2*y(1)*(y(2)-alpha)/2/h;
    1+x(2).^3-2*y(2)*(y(3)-y(1))/2/h;
    1+x(3).^3-2*y(3)*(-y(2)+beta)/2/h];
D=Y-h.^2*F;
y=A\D
plot(i,y,'.','markersize',20);hold on;
end
xlabel('number of iterations'),ylabel('y(x)');grid;
```

(d) MATLAB m-file (fin_diff_bvp_mat3.m) is an extension of the previous code. Solution curve for the nonlinear BVP, $y'' + 2yy' = x^3 + 1$, $y(1) = 1/2, y(2) = 2$ is displayed in Fig. 6.13 using five internal nodes within the specified interval.

Fig. 6.13 Solution curve for
the nonlinear BVP,
$y'' + 2yy' = x^3 + 1,$
$y(1) = 1/2, y(2) = 2$

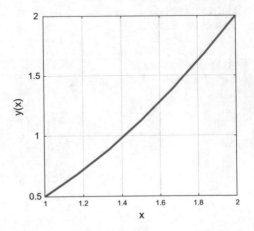

```
%fin_diff_bvp_mat3.m    y"+2yy'=1+x^3,    y(1)=1/2, y(2)=2,   iterations
clc;clear;close;
N=5; % number of internal nodes
M=10 % total number of iterations
%format rat
format long
span=[1 2];
bv=[1/2 2];alpha=bv(1);beta=bv(2);
h=(span(2)-span(1))/(N+1);
x=linspace(span(1)+h,span(2)-h,N)
w=zeros(N,1); % preallocation
Y=zeros(N,1);Y(1)=alpha;Y(N)=beta
v=ones(N-1,1);
C=2*eye(N);
B=-diag(v,1)-diag(v,-1); % diag(V,K), K tells which diagonal to place V in
A=C+B
F=zeros(N,1); % preallocation
for i=1:M
    f = @(x,y,dy) (1+x^3-2*y*dy); % dy=y'
    for j=2:N-1
  F(j,1)=f(x(j),w(j),(w(j+1)-w(j-1))/(2*h));
  F(1,1)=f(x(1),w(1),(w(2)-alpha)/(2*h));
  F(N,1)=f(x(N),w(N),(beta-w(N-1))/(2*h));
    end
D=Y-h.^2*F;
w=A\D
plot(i,w,'.','markersize',20);hold on;
end
xlabel('number of iterations'),ylabel('y(x)');grid;figure;
X=linspace(span(1),span(2),N+2);X=X';
Y=[alpha; w; beta];
plot(X,Y,'linewidth',2);xlabel('x'),ylabel('y(x)');grid;
```

Problem 6.2.6 MATLAB has a built-in nonlinear equation solver, `fsolve`. Solve the nonlinear BVP, $y'' + 2yy' = x^3 + 1$; $y(1) = 1/2, y(2) = 2$, using this function with iteration method and employing centered difference approximations for both derivative terms, assuming three interior points (N = 3) of the uniform grid (fin_diff_bvp_fsolve1.m).

Solution

While MATLAB's fzero command solves single nonlinear algebraic equation of one independent variable, `fsolve` (an Optimization Toolbox command[1]) can find the roots of systems of nonlinear continuous functions provided that a set of functions are supplied such that all the functions are zero or almost zero,

$$F(x) = 0$$

where $F(x)$ is a function that returns a vector value, and (x) is a vector (or a matrix). Because this function works iteratively, an initial guess must be provided.

The objective function we use to minimize here is

$$P(w) = Aw - Y + h^2 F$$

where A, w, are tridiagonal square matrix, vector of unknowns, respectively, and F is a vector the elements of which are the functions $f(x, y, y')$,

$$y'' = f(x, y, y')$$

Specifically, we define an anonymous function in MATLAB code,

```
f = @(x,y,dy) (1+x^3-2*y*dy); % dy=y'
```

and call this function from F

```
for j=2:N-1
F(j,1)=f(x(j),w(j),(w(j+1)-w(j-1))/(2*h));
F(1,1)=f(x(1),w(1),(w(2)-alpha)/(2*h));
F(N,1)=f(x(N),w(N),(beta-w(N-1))/(2*h));
```

where h is the step size, N is the number of internal nodes, and $Y = [\alpha \ 0 \ \beta]^T$.
We assume zero initial solution vector.

---

[1]More information on this command regarding syntax, input and output arguments, examples, algorithms and limitations can be found in MATLAB Documentations.

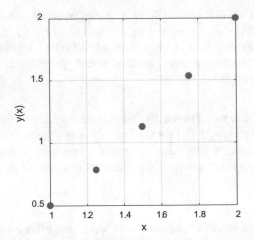

Fig. 6.14 The solution plot of the nonlinear BVP, $y'' + 2yy' = x^3 + 1$; $y(1) = 1/2, y(2) = 2$, using MATLAB dsolve function with iteration method and centered difference approximations for both derivative terms, assuming three interior points $(n = 3)$ of the linear first order grid

Following is the list of approximate and exact solution values for the given BVP. The approximate solution is also displayed in Fig. 6.14.

```
        x                    y                    exact

1.000000000000000    0.500000000000000    0.500000000000000
1.250000000000000    0.781250000577238    0.781250000000000
1.500000000000000    1.125000002270362    1.125000000000000
1.750000000000000    1.531249998700435    1.531250000000000
2.000000000000000    2.000000000000000    2.000000000000000
```

Equation solved.

fsolve completed because the vector of function values is near zero as measured by the default value of the function tolerance, and the problem appears regular as measured by the gradient.

Elapsed time is 0.080093 seconds.

MATLAB m-file (fin_diff_bvp_fsolve1.m) is listed below.

```
%fin_diff_bvp_fsolve1.m    y"+2yy'=1+x^3,    y(1)=1/2, y(2)=2,   iterations
% solves a nonlinear bvp of   y"=f(x,y,y'),  a<x<b, y(a)=alpha , y(b)=beta
clc;clear;close;tic;
N=3; % number of internal nodes
%format rat
format long
span=[1 2];
bv=[1/2 2];alpha=bv(1);beta=bv(2);
h=(span(2)-span(1))/(N+1);
x=linspace(span(1)+h,span(2)-h,N)
w=zeros(N,1); %   zero initial solution approximation
% for i=1:N
% w(i)=alpha+i'*(beta-alpha)/(span(2)-span(1))*h;%linear initial
approximations
% end
% w=w';
f = @(x,y,dy) (1+x^3-2*y*dy); % dy=y'
P=@(w) [2*w(1)-w(2)+(h^2)*f(x(1),w(1),(w(2)-alpha)/(2*h))-alpha;
-w(1)+2*w(2)-w(3)+(h^2)*f(x(2),w(2),(w(3)-w(1))/(2*h));
-w(2)+2*w(3)+(h^2)*f(x(3),w(3),(beta-w(2))/(2*h))-beta];
[w,fval]= fsolve(P,w)
X=linspace(span(1),span(2),N+2);X=X';
Y=[alpha; w; beta];
U=X.^2/2; %exact solution
F = [X Y U]
plot(X,Y,'.','MarkerSize',20); grid on; xlabel('x');ylabel('y(x)');toc
```

Problem 6.2.7 Solve the nonlinear BVP, $y'' + (y')^2 = x$; $y(0) = 3, y(5) = 1$, using
fsolve function with iteration method and employing centered difference approximations for both derivative terms, assuming 19 interior points (N = 19) of the
uniform grid (fin_diff_bvp_fsolve2.m).

Solution
Analytical solution is obtained by means of dsolve function, consisting of airy's and
gamma functions,

```
y(x) = log(3^(1/3) - 3*exp(2)*gamma(2/3)*airy(0, 5))
 - log(3^(1/3)*airy(2, 5) - 3^(5/6)*airy(0, 5)) + log(airy(2, x)
 - ((3^(5/6) - 3*exp(2)*gamma(2/3)*airy(2, 5))*airy(0, x))/(3^(1/3)
 - 3*exp(2)*gamma(2/3)*airy(0, 5))) + 1
```

This function is plotted in the first part of the code (fin_diff_bvp_fsolve2.m) as
the exact solution. Second part of the code computes the (solution) function values
at the 19 internal nodes using fsolve function and applying default tolerances and
employing centered difference approximations for both derivative terms and they
are overlaid on the same figure, as shown in Fig. 6.15.

Computed and exact function values at the nodes along with their corresponding
x-axis values are printed on Command Window, as listed (in edited form) below.
Here, not all "fval" values are listed, but the maximum of the absolute value of this
parameter is given, which is quite acceptable.

Fig. 6.15 Computed and
exact solution of the nonlinear
BVP, $y'' + (y')^2 = x$,
$y(0) = 3, y(5) = 1$, using
fsolve function with
iteration method

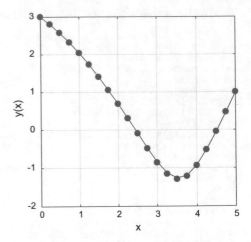

Equation solved.fsolve completed because the vector of function
values is near zero as measured by the default value of the function
tolerance, and the problem appears regular as measured by the gradient.

Max(abs((fval)) = 1.0e-07 *0.574740177394784

x	y(computed)	y(exact)
0	3.000000000000000	3.000000000000000
0.250000000000000	2.801291927228123	2.801745889737312
0.500000000000000	2.572525175457599	2.573361498239985
0.750000000000000	2.316173436080795	2.317348963009132
1.000000000000000	2.034265947614452	2.035765718841745
1.250000000000000	1.728526115929598	1.730371471330281
1.500000000000000	1.400494825423935	1.402764746586347
1.750000000000000	1.051685092056093	1.054561442103791
2.000000000000000	0.683858426864998	0.687714682243510
2.250000000000000	0.299638087870961	0.305199983464228
2.500000000000000	-0.096033346653869	-0.087412888244947
2.750000000000000	-0.492111716237055	-0.478369828647295
3.000000000000000	-0.863675030646041	-0.842294904356797
3.250000000000000	-1.158884978091369	-1.129187721837601
3.500000000000000	-1.298165375791154	-1.264192287318421
3.750000000000000	-1.219617885112230	-1.188959232785196
4.000000000000000	-0.938953653752069	-0.917444690581716
4.250000000000000	-0.527905775547510	-0.516013848390915
4.500000000000000	-0.049164183444495	-0.043895974544154
4.750000000000000	0.464573593175036	0.466260333533032
5.000000000000000	1.000000000000000	1.000000000000000

Elapsed time is 0.306470 seconds.

```
%fin_diff_bvp_fsolve2.m    y"+(y')^2=x,    y(0)=3, y(5)=1
% solves a nonlinear bvp of  y"=f(x,y,y'), a<x<b, y(a)=alpha , y(b)=beta
clc;clear;close;tic;
N=19; % number of internal nodes
%solution using "dsolve"
syms   y(x)
eq='D2y+(Dy)^2=x'; bc='y(0)=3,y(5)=1';
S=dsolve(eq,bc,x);simplify(S)
x = linspace(0,5,N+2);y=eval(vectorize(S));
plot(x,y,'k');grid; hold on;
%----------------------------------------------------------------
%  FDM, using iterations
%format rat
format long
span=[0 5];
bv=[3 1];alpha=bv(1);beta=bv(2);
h=(span(2)-span(1))/(N+1);
x=linspace(span(1)+h,span(2)-h,N)
w=zeros(N,1); %   zero initial solution approximation
% for i=1:N
% w(i)=alpha+i'*(beta-alpha)/(span(2)-span(1))*h;%linear initial
approximations
% end
% w=w';
%w=[2   0   12];w=w';%   arbitrary initial solution approximation
f = @(x,y,dy) x-(dy)^2; % dy=y'
 P=@(w) [2*w(1)-w(2)+(h^2)*f(x(1),w(1),(w(2)-alpha)/(2*h))-alpha;
-w(1)+2*w(2)-w(3)+(h^2)*f(x(2),w(2),(w(3)-w(1))/(2*h));
-w(2)+2*w(3)-w(4)+(h^2)*f(x(3),w(3),(w(4)-w(2))/(2*h));
-w(3)+2*w(4)-w(5)+(h^2)*f(x(4),w(4),(w(5)-w(3))/(2*h));
-w(4)+2*w(5)-w(6)+(h^2)*f(x(5),w(5),(w(6)-w(4))/(2*h));
-w(5)+2*w(6)-w(7)+(h^2)*f(x(6),w(6),(w(7)-w(5))/(2*h));
-w(6)+2*w(7)-w(8)+(h^2)*f(x(7),w(7),(w(8)-w(6))/(2*h));
-w(7)+2*w(8)-w(9)+(h^2)*f(x(8),w(8),(w(9)-w(7))/(2*h));
-w(8)+2*w(9)-w(10)+(h^2)*f(x(9),w(9),(w(10)-w(8))/(2*h));
-w(9)+2*w(10)-w(11)+(h^2)*f(x(10),w(10),(w(11)-w(9))/(2*h));
-w(10)+2*w(11)-w(12)+(h^2)*f(x(11),w(11),(w(12)-w(10))/(2*h));
-w(11)+2*w(12)-w(13)+(h^2)*f(x(12),w(12),(w(13)-w(11))/(2*h));
-w(12)+2*w(13)-w(14)+(h^2)*f(x(13),w(13),(w(14)-w(12))/(2*h));
-w(13)+2*w(14)-w(15)+(h^2)*f(x(14),w(14),(w(15)-w(13))/(2*h));
-w(14)+2*w(15)-w(16)+(h^2)*f(x(15),w(15),(w(16)-w(14))/(2*h));
-w(15)+2*w(16)-w(17)+(h^2)*f(x(16),w(16),(w(17)-w(15))/(2*h));
-w(16)+2*w(17)-w(18)+(h^2)*f(x(17),w(17),(w(18)-w(16))/(2*h));
-w(17)+2*w(18)-w(19)+(h^2)*f(x(18),w(18),(w(19)-w(17))/(2*h));
-w(18)+2*w(19)+(h^2)*f(x(19),w(19),(beta-w(18))/(2*h))-beta];
[w,fval]= fsolve(P,w)
X=linspace(span(1),span(2),N+2);X=X';
Y=[alpha; w; beta];
F = [X Y y']
plot(X,Y,'.','MarkerSize',20); grid on; xlabel('x');ylabel('y(x)');toc
```

Problem 6.2.8 True, or False?

(a) The FD equilibrium (replacement) method is based on relaxing a system of FDEs simultaneously, excluding the boundary conditions.

(b) An advantage of the FD equilibrium method is that the boundary conditions are applied directly and they are automatically satisfied.

(c) It is very easy to achieve higher than second-order accuracy using FD equilibrium method.

(d) Application of FD equilibrium method requires that a system of FDEs to be solved.

(e) Nonlinear ODEs result in a system of linear FDEs.

(f) Equilibrium methods work well for smoothly varying problems.

(g) Equilibrium methods work well for problems with delicate boundary conditions.

(h) Non-iterative methods are used when solving nonlinear FDEs.

Answers

(a) F (including), (b) T, (c) F (difficult), (d) T, (e) F (nonlinear FDEs), (f) T, (g) T, (h) F (iterative).

Problem 6.2.9 Solve the nonlinear autonomous BVP, $y'' = -y^2$; $y(0) = y(1) = 2$ by FDM, via Multivariate Newton's method. Let the number of internal nodes be $n = 20$. Plot the solution by making five iterations in Newton's method. Observe the convergence for the value of y_{max} within four digits of decimal accuracy (fin_diff_bvp6.m).

Solution

No analytical explicit solution is found for this BVP using MATLAB dsolve function.

The discretized form of the ODE at x_i is

$$\frac{y_{i-1} - 2y_i + y_{i+1}}{h^2} = -y_i^2$$

$$y_{i-1} - 2y_i + y_{i+1} = -h^2 y_i^2$$

$$y_{i-1} - 2y_i + h^2 y_i^2 + y_{i+1} = 0 \tag{6.24}$$

$$i = 1: \quad y_0 - 2y_1 + h^2 y_1^2 + y_2 = 0 \tag{6.25}$$

$$i = n: \quad y_{n-1} - 2y_n + h^2 y_n^2 + y_b = 0 \tag{6.26}$$

where $y_0 = y(0) = 2$, $y_b = y(1) = 2$.

(Multivariate) Newton's Method uses the iteration

$$y^{k+1} = y^k - Df(y^k)^{-1} f(y^k)$$
$$\Delta y = y^{k-1} - y^k,$$
$$\Delta y = -Df(y^k)^{-1} f(y^k)$$
$$Df(y^k)\Delta y = -f(y^k)$$

The function $f(y)$ is obtained from (6.24) to (6.26),

$$f \begin{bmatrix} y_1 \\ y_2 \\ \vdots \\ y_{n-1} \\ y_n \end{bmatrix} = \begin{bmatrix} y_0 - 2y_1 + h^2 y_1^2 + y_2 \\ y_1 - 2y_2 + h^2 y_2^2 + y_3 \\ \vdots \\ y_{n-2} - 2y_{n-1} + h^2 y_{n-1}^2 + y_n \\ y_{n-1} - 2y_n + h^2 y_n^2 + y_b \end{bmatrix}$$

The Jacobian $Df(y)$ of f is determined by taking the partial derivative of the ith row of f with respect to each y_m;

$$J = \begin{bmatrix} -2+2h^2 y_1 & 1 & 0 & 0 & \cdots & 0 \\ 1 & -2+2h^2 y_2 & 1 & 0 & \cdots & 0 \\ 0 & 1 & -2+2h^2 y_3 & 1 & \cdots & 0 \\ \cdots & \cdots & \cdots & \cdots & \cdots & \cdots \\ 0 & 0 & 0 & 0 & -2+2h^2 y_{n-1} & 1 \\ 0 & 0 & 0 & 0 & 1 & -2+2h^2 y_n \end{bmatrix}$$

Solution curve of the BVP is shown in Fig. 6.16. MATLAB m-file (fin_-diff_bvp6.m) is listed below.

Peak point data converges to $y_{max} = 3.0004$ at the end of fourth iteration of the Newton's formula:

$$P = 2.0000 \quad 2.8498 \quad 2.9960 \quad 3.0004 \quad 3.0004$$

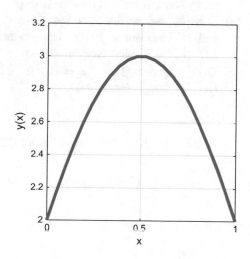

Fig. 6.16 Solution curve of the BVP, $y'' = -y^2, y(0) = y(1) = 2$ by FD method, via multivariate Newton's method with five iterations. The number of internal nodes is $n = 20$

```
% fin_diff_bvp6.m  Nonlinear FDM for BVP (Dirichlet b. conditions)
% This code employs Multivariate Newton's Method
clc;clear;%close;
n=20; % Number of nodes within the interval
span=[0 1];% x-interval
bv=[2 2];  % boundary conditions
% y=output array (solution)
A=span(1); B=span(2); ya=bv(1); yb=bv(2);
h=(span(2)-span(1))/(n+1); % h = step size
a=zeros(n,n); %initialization
y=zeros(n,1); % initialize array y
  for i=1:5  % loop for Newton's formula
f=zeros(n,1);
%System equations
f(1)=bv(1)-2*y(1)+h^2*y(1)^2+y(2);
f(n)=y(n-1)-2*y(n)+h^2*y(n)^2+bv(2);
for j=2:n-1
 f(j)=y(j-1)-2*y(j)+h^2*y(j)^2+y(j+1);
end
%Jacobien
for k=1:n
 a(k,k)=2*h^2*y(k)-2;
end
for k=1:n-1
 a(k,k+1)=1;a(k+1,k)=1;
end
y=y-a\f ; %Multivariate Newton's formula
P=max(y)
end
plot([A A+(1:n)*h  B],[ya  y'  yb],'linewidth',2.5);
xlabel('x');ylabel('y(x)');grid on;
%y=[ya  y'  yb] % solution vector including boundary values
```

Problem 6.2.10

(a) Derive a solution equation in matrix form for nonlinear second order BVPs using Newton method (nonlinear implicit FDE using iteration).

(b) Compute and list the numerical values of elements of the solution vector at the end of each iteration for the nonlinear BVP, $y'' + 2yy' = x^3 + 1$, $y(1) = 1/2, y(2) = 2$ using Newton method, assuming three interior nodes and uniform grid. Let $y_0 = \begin{bmatrix} 0 & 0 & 0 \end{bmatrix}^T$. Perform $M = 5$ iterations (fin_diff_bvp_Newton1.m).

Solution

(a) A nonlinear BVP for a second order ODE is of the form

$$y'' = f(x, y, y'), \ x \in [a, b], \ y(a) = \alpha, \ y(b) = \beta \tag{6.27}$$

Considering central FD representation of first and second derivatives,

$$y'(x) = \frac{y(x+h) - y(x-h)}{2h}, \quad y''(x) = \frac{y(x+h) - 2y(x) + y(x-h)}{h^2} \quad (6.28)$$

where $h = (b-a)/(N+1)$, N = internal number of nodes,

$$x_0 = a, \quad x_1 = a+h, \ldots, x_N = a+Nh, \quad x_{N+1} = a+(N+1)h = b$$

with the boundary values $Y_0 = \alpha, Y_{N+1} = \beta$.

We find $y_1, y_2, \ldots, y_N$ by solving the following system of equations which is obtained when Eq. (6.28) is replaced in Eq. (6.27),

$$\frac{y_{i+1} - 2y_i + y_{i-1}}{h^2} = f\left(x_i, y_i, \frac{y_{i+1} - y_{i-1}}{2h}\right) \quad (6.29)$$

$$-y_{i-1} + 2y_i - y_{i+1} + h^2 \cdot f\left(x_i, y_i, \frac{y_{i+1} - y_{i-1}}{2h}\right) = 0 \quad (6.30)$$

where $i = 1, 2, \ldots, N$.

Rewriting (6.30) using matrix notation,

$$diag\{-1, 2, 1\}y + h^2 F(x, y) = Ay + h^2 F(x, y) = Y \quad (6.31)$$

Where the first term means a tridiagonal square matrix,

$$y = [y_1, y_2, \ldots, y_N]^T, \quad Y = [\alpha, 0, 0, \ldots, 0, \beta]^T, \quad A = diag\{-1, 2, 1\},$$
$$F(x, y) = [F_1(x, y), \ldots, F_N(x, y)]^T, \quad F_i \cdot (x, y) = f(x_i, y_i, (y_{i+1} - y_{i-1})/2h)$$

Let $D = Ay + h^2 F - Y = [d_1 \quad d_2 \quad \ldots \quad d_N]^T$,
By linearization,

$$D(x, y) \cong D(x, y_k) + J(x, y_k) \cdot (y_{k+1} - y_k)$$

where J is the Jacobi Matrix of D.

Solution of linearized equations for y_{k+1} is performed as follows:

$$D(x, y_k) + J(x, y_k)(y_{k+1} - y_k)$$
$$y_{k+1} = y_k - [J(x, y_k)]^{-1} D(x, y_k)$$

Since $J(x,y) = A + h^2 J_F(x,y)$
where

$$J_F(x,y) = \left[\frac{F_i(x,y)}{y_j}\right] = diag\left\{\frac{F_i(x,y)}{y_{j-1}}, \frac{F_i(x,y)}{y_j}, \frac{F_i(x,y)}{y_{j+1}}\right\}$$

by applying Newton Method, we obtain

$$y_{k+1} = y_k - \left[A + h^2 J_F(x,y_k)\right]^{-1}\left[Ay_k + h^2 F(x,y_k) - Y\right] \qquad (6.32)$$

Following steps can be summarized as a procedure of finding the solution of a nonlinear BVP using Newton method (implicit FD method):

1. Form x, A, Y, J_F
2. Compute $B = A + h^2 J_F$ (this is a square matrix)
3. Compute $b = Ay_k + h^2 F(x,y_k) - Y$ (this is a vector)
4. Compute $y_{k+1} = y_k - B^{-1}b$

(b) Given:

$$y'' + 2yy' = x^3 + 1, \quad y(1) = 1/2, y(2) = 2, \quad N = 3, \quad y_0 = \begin{bmatrix} 0 & 0 & 0 \end{bmatrix}^T.$$

$$y'' = 1 + x^3 - 2yy' = f(x,y,y')$$

$$h = \frac{(b-a)}{(N+1)} = \frac{1}{4}, \quad \alpha = \frac{1}{2}, \quad \beta = 2$$

$$x = \begin{bmatrix} a & x_1 & x_2 & x_3 & b \end{bmatrix} = \begin{bmatrix} 1 & 5/4 & 6/4 & 7/4 & 2 \end{bmatrix}$$

$$Y = \begin{bmatrix} \alpha & 0 & \beta \end{bmatrix}^T = \begin{bmatrix} 1/2 & 0 & 2 \end{bmatrix}^T$$

Although MATLAB dsolve function does not provide an explicit solution of this BVP, we know that the analytical solution of this BVP is $y = x^2/2$.

Following is the output list of the MATLAB code (fin_diff_bvp_newton1.m). It is noted that values of components of the approximate solution vector converges into true solution (nodal values) at the end of the fourth iteration (u values).

```
A =
      2    -1     0
     -1     2    -1
      0    -1     2

Y = (alpha)    0   (beta)
```

```
F =
 (y1*(alpha - y2))/h + 189/64
     (y2*(y1 - y3))/h + 35/8
  407/64 - (y3*(beta - y2))/h

J =
[ (alpha - y2)/h,          -y1/h,                0]
[              y2/h, (y1 - y3)/h,          -y2/h]
[              0,          y3/h, -(beta - y2)/h]

B =
[ h*(alpha - y2) + 2,       - h*y1 - 1,                    0]
[            h*y2 - 1, h*(y1 - y3) + 2,          - h*y2 - 1]
[                   0,      h*y3 - 1, 2 - h*(beta - y2)]

b =
 2*y1 - y2 - conj(alpha) + h^2*((y1*(alpha - y2))/h + 189/64)
             2*y2 - y1 - y3 + h^2*((y2*(y1 - y3))/h + 35/8)
   2*y3 - y2 - conj(beta) - h^2*((y3*(beta - y2))/h - 407/64)

u = 0.66299716  1.0934393   1.7973189
u = 0.78159094  1.1263419   1.5304829
u = 0.78125004  1.1250002   1.5312499
u = 0.78125      1.125        1.53125
u = 0.78125      1.125        1.53125
```

MATLAB code (fin_diff_bvp_newton1.m) is listed below. This code employs Symbolic Toolbox commands. The code is written as ready to plot nodal solution values (last line), resulting plot is shown in Fig. 6.17.

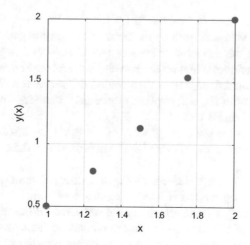

Fig. 6.17 The solution plot of the nonlinear BVP, $y'' + 2yy' = x^3 + 1$; $y(1) = 1/2, y(2) = 2$ using Newton method (see, also Fig. 6.14 for the equivalence of results)

```
%fin_diff_bvp_Newton1.m
%solves nonlinear BVP using implicit FDEs via Newton method.
clc; clear; close;
x=[1 5/4 6/4 7/4 2];
x1=x(2);x2=x(3);x3=x(4);     % internal nodes
M=5;                          % Number of iterations
u=[0 0 0]';                   % solution vector initialization
syms h alpha beta y1 y2 y3
A=[2 -1 0;-1 2 -1;0 -1 2]    % tridiagonal matrix
Y=[alpha 0 beta]';
F = [1+x1^3-2*y1*(y2-alpha)/(2*h);
     1+x2^3-2*y2*(y3-y1)/(2*h);
     1+x3^3-2*y3*(beta-y2)/(2*h)]
w=[y1;y2;y3];
J = jacobian(F, w)           % Jacobi matrix
B=A+h^2*J , b=A*w+h^2*F-Y
y=w-B\b ;                     % Multivariate Newton's formula
for i=1:M
y=subs(y,[alpha,beta,h],[1/2, 2, 1/4 ]);
u=vpa(subs(y,w,u),8)
end
v=[1/2; u; 2];plot(x,v,'o-','linewidth',2);grid;xlabel('x');ylabel('y(x)');
```

6.3 Exercises

1. Use the formulas described in Problem 6.1.2 and write a MATLAB m-file to compute the solution of a general second order linear nonhomogenous ODE discretized for any given number of internal nodes, with Dirichlet boundary conditions.

2. Run the code given in Problem 6.1.6 for N = 90, N = 900 and observe the elapsed time values as well as the error of approximation.

3. What will be the rate of convergence as a function of step size, if a second order ODE includes a first order term and it is replaced by the forward Euler's algebraic rational form, rather than central difference form?

4. Extend the m-file fin_diff_bvp0.m to present "Step size versus Error magnitude" data through a loop.

5. Use FDM to solve the BVP $y'' = y, y(0) = 0, y(1) = 1$ with $h = 0.05$ [2].

6. Use FDM to solve the BVP, $y'' + y = x, y(0) = 1, y(0.5\pi) = 0.5\pi - 1$, with $h = 0.1$ [3].

7. In Problem 6.2.1, all else remaining the same, determine the temperature distribution along the rod if $h = 0.1$ m, and $h = 1$ cm.

8. In Problem 6.2.2, all else remaining the same, determine epsilon value for the convergence of the solution, when the number of iterations is 20.

9. Considering the nature of nonlinear algebraic equations, what can be concluded from the results obtained in Problem 6.2.4?

10a. In Problem 6.2.7, Show that approximation to solution array becomes better as the number of internal nodes is increased from N = 3 to higher values. In what sense the term "better" could be used here?

10b. The function P is computed outside the code presented in this problem, and then copied into it. Write a general MATLAB code to compute this function for any number of internal nodes, N.

10c. Note that for large values of N, external computations of the function P, its check and transfer to the main code becomes tedious. Re-write the code presented in this problem by computing the function P within the code, for any number of internal nodes. Observe the computation time for the overall code.

11. Consider the solution of Problem 6.2.9. Make a more detailed convergence analysis by computing the maximum value of the solution array in seven digits accuracy (all else remaining the same),

   (a) by varying the number of internal nodes (n),
   (b) by varying the number of iterations in Newton's formula (i).

12. Extend the code presented in Problem 6.2.10 to service for larger number of internal nodes, N.

13. Convert the code given in Problem 6.2.10 to perform "double" operations (i.e., without recourse to symbolic toolbox commands) for a large number of internal nodes, N.

14. Use FDM to solve the BVP [4],

$$y'' - 1.5y^2 = 0, \quad 0 \leq x \leq 1, \quad y(0) = 4, \quad y(1) = 1$$

15. Use FDM to solve the BVP, $y'' = 2yy', y(0) = -1, y(\pi/2) = 1$. Exact solution: $y = \tan(x - \pi/4)$.

References

1. Hoffman JD (1992) Numerical methods for engineers and scientists, 2nd edn. (Chap. 8). Marcel Dekker, Inc.
2. Conte SD, de Boor C (1980) Elementary numerical analysis: an algorithmic approach, 3rd edn. (Chap. 9). McGraw-Hill Book Company
3. Dahlquist G, Bjorck A, Anderson N (1974) Numerical methods (Chap. 8). Prentice Hall Inc.
4. Lesnic DA (2007) Nonlinear reaction—diffusion process using the Adomian decomposition method. Int Commun Heat Mass Transf 34:129–135

Chapter 7
Adomian Decomposition Method (ADM)

The Adomian decomposition method (ADM) is a method for the solution of both linear and nonlinear differential equations and BVPs seen in different fields of science and engineering. However, the implementation of this method mainly depends upon the calculation of Adomian polynomials for nonlinear operators. The computation of Adomian polynomials for various forms of nonlinearity is the first step required to solve nonlinear problems.

MATLAB codes given in this chapter are written to show all steps of computations and most of them use symbolic math.

7.1 Adomian Polynomials and Solving Univariate Equations Using ADM

Problem 7.1.1

(a) What is the Adomian[1] decomposition method (ADM)?
(b) Use Symbolic MATLAB to solve the following equation [1] via ADM.

$$x^3 + 4x^2 + 8x + 8 = 0$$

(Adomian2.m).

---

[1]George Adomian (1922–1996); American mathematician and aerospace engineer. He was a faculty member at the University of Georgia.

© Springer Nature Switzerland AG 2019
A. Ü. Keskin, *Boundary Value Problems for Engineers*,
https://doi.org/10.1007/978-3-030-21080-9_7

Solution

(a) The Adomian decomposition method (ADM) [2] is a method mainly developed
    for solving nonlinear differential equations. The technique uses a decomposi-
    tion of the nonlinear term into a series of functions. Each term of this series is a
    polynomial called Adomian's polynomial which generate an infinite set of
    functions whose sum determines the actual solution. Note that Adomian
    polynomials are not unique.

Consider the nonlinear univariate equation of the form $f(x) = 0$, which can be
transformed into

$$x = F(x) + c \tag{7.1}$$

where F is a nonlinear function and c is a constant. It can be written as

$$\sum_{i=0}^{\infty} x_i = \sum_{i=0}^{\infty} A_i + c \tag{7.2}$$

where the A_i's are Adomian polynomials. Each term in the left side of (7.2) can be
calculated as

$$
\begin{aligned}
x_0 &= c \\
x_1 &= A_0 \\
&\vdots \\
x_{n+1} &= A_n
\end{aligned} \tag{7.3}
$$

An approximation to the solution of (7.1) can be

$$\varphi_n = \sum_{i=0}^{n-1} x_i, \quad \lim_{n \to \infty} \phi_n = x \tag{7.4}$$

Adomian polynomials are obtained from

$$A_n(u_0, u_1, \ldots, u_n) = \frac{1}{n!} \left[\frac{d^n}{d\lambda^n} \sum_{i=0}^{\infty} \lambda^i u_i \right]_{\lambda=0} \tag{7.5}$$

First few of these polynomials can be generated as follows:

$$A_0 = f(u_0)$$

$$A_1 = u_1 f'(u_0)$$

$$A_2 = u_2 f'(u_0) + \frac{u_1^2}{2!} f''(u_0)$$

$$A_3 = u_3 f'(u_0) + u_1 u_2 f''(u_0) + \frac{u_1^3}{3!} f'''(u_0)$$

$$A_4 = u_4 f'(u_0) + \left(\frac{1}{2!}u_2^2 + u_1 u_3\right) f''(u_0) + \frac{1}{2!}u_1^2 u_2 f'''(u_0) + \frac{u_1^4}{4!} f^{(4)}(u_0)$$

$$A_5 = u_5 f'(u_0) + (u_1 u_4 + u_2 u_3) f''(u_0) + \frac{1}{2}\left(u_3 u_1^2 + u_1 u_2^2\right) f'''(u_0)$$

$$+ \frac{u_1^3 u_2}{6} f^{(4)}(u_0) + \frac{u_1^5}{5!} f^{(5)}(u_0)$$

For example, if $u = (x^3)$,

$$x_0 = c$$

$$x_1 = A_0 = x_0^3$$

$$x_2 = A_1 = 3x_0^2 x_1$$

$$x_3 = A_2 = 3x_0^2 x_2 + 3x_0 x_1^2$$

$$x_4 = A_3 = 3x_0^2 x_3 + 6x_0 x_1 x_2 + x_0^3$$

$$x_5 = A_4 = 3x_0^2 x_4 + 3x_1^2 x_2 + 3x_2^2 x_0 + 6x_0 x_1 x_3$$

$$x_6 = A_5 = 3x_0^2 x_5 + 3x_1 x_2^2 + 3x_1^2 x_3 + 6x_0 x_1 x_4 + 6x_0 x_2 x_3$$

A characteristic of these polynomials is that the sum of subscripts of the components of u in each term of the polynomial is equal to n.

There are numerous articles related to the computation of these polynomials, and several Adomian polynomials of most frequently met functions are given in Refs. [3–12].

Recently, there are increasing number of engineering applications of ADM [13–22].

This method provides a rapidly convergent series, but gives only locally convergent results. In other words, although the series rapidly converges in a small region, it has slow convergence rate in the wider region where the truncated series solution is inaccurate.

ADM does not assure the existence and uniqueness of solutions, but it can be safely applied in cases when a fixed point theorem holds.

(b) Isolating the unknown variable results in

$$x = -1 - \frac{1}{2}x^2 - \frac{1}{8}x^3 = -1 - b\sum_{i=0}^{\infty} A_i(x^2) - a\sum_{i=0}^{\infty} A_i(x^3)$$

we compute the Adomian polynomials $A_i(x^2), A_i(x^3)$ using MATLAB script (Adomian2.m). Following list is an edited output of this code.

```
b = -0.5000,  a = -0.1250
A1 =[ x0^2, 2*x0*x1, x1^2 + 2*x0*x2, 2*x0*x3 + 2*x1*x2, x2^2 + 2*
x0*x4 + 2*x1*x3]

A2 =[ x0^3, 3*x0^2*x1, 3*x2*x0^2 + 3*x0*x1^2, 3*x3*x0^2 + 6*x2*x0*x1 +
x1^3, 3*x1^2*x2 + 3*x0^2*x4 + 6*x0*(x2^2/2 + x1*x3)]

X =[ -0.375, -0.234375, -0.1640625, -0.117919921875,
 -0.083587646484375]

zero =-1.9749
```

MATLAB code given below is written to show all steps of computations. First part computes the first Adomian polynomial, while the second part computes the second Adomian polynomial, both using symbolic Math. Final section demonstrates how these two parts are combined to yield the solution. Note that this result is obtained by the sum of first six terms of Adomian polynomials, while the exact solution of the given equation is $x = -2$.

```
%Adomian2.m
clc;syms x0 x1 x2 x3 x4
f0=x0^(2);
b=-1/2,a=-1/8
A(1)=(f0);
A(2)=x1*diff(f0,1,'x0');
A(3)=x2*diff(f0,1,'x0')+(x1.^2/2)*diff(f0,2,'x0');
A(4)=x3*diff(f0,1,'x0')+x1*x2*diff(f0,2,'x0')+(x1.^3/6)*diff(f0,3,'x0');
A(5)=x4*diff(f0,1,'x0')+((1/2)*x2.^2+x1*x3)*diff(f0,2,'x0')...
     +(1/2)*x1.^2*x2*diff(f0,3,'x0')+(1/120)*x1.^4*diff(f0,4,'x0');
A1=A
f0=x0^(3);
A(1)=(f0);
A(2)=x1*diff(f0,1,'x0');
A(3)=x2*diff(f0,1,'x0')+(x1.^2/2)*diff(f0,2,'x0');
A(4)=x3*diff(f0,1,'x0')+x1*x2*diff(f0,2,'x0')+(x1.^3/6)*diff(f0,3,'x0');
A(5)=x4*diff(f0,1,'x0')+((1/2)*x2.^2+x1*x3)*diff(f0,2,'x0')...
     +(1/2)*x1.^2*x2*diff(f0,3,'x0')+(1/120)*x1.^4*diff(f0,4,'x0');
A2=A
X(1)=b*A1(1)+a*A2(1);X(1)=vpa(subs(X(1),x0,-1));
X(2)=b*A1(2)+a*A2(2);X(2)=subs(X(2),[x0,x1],[-1,X(1)]);
X(3)=b*A1(3)+a*A2(3);X(3)=subs(X(3),[x0,x1,x2],[-1,X(1),X(2)]);
X(4)=b*A1(4)+a*A2(4);X(4)=subs(X(4),[x0,x1,x2,x3],[-1,X(1),X(2),X(3)]);
X(5)=b*A1(5)+a*A2(5);
X(5)=vpa(subs(X(5),[x0,x1,x2,x3,x4],[-1,X(1),X(2),X(3),X(4)]))
zero=vpa((sum(X))-1,5)
```

Problem 7.1.2 Derive first four Adomian polynomials for $f(y) = (y')^2$.

Solution

Let $y' = u$. We derive four Adomian Polynomials for u^2, first. Then replace $u_i = (y_i)'$.

This yields,

$$A_0 = (y'_0)^2$$
$$A_1 = 2y'_0 y'_1$$
$$A_2 = 2y'_0 y'_2 + (y'_1)^2$$
$$A_3 = 2y'_0 y'_3 + 2y'_1 y'_2$$

Problem 7.1.3 Calculate the first three Adomian polynomials for $f(y) = e^{-y^2}$, manually.

Solution

$$A_0 = f(y_0) = e^{-y_0^2}$$

$$A_1 = y_1 \left[\frac{df(y_0)}{dy_0} \right] = -2y_0 y_1 e^{-y_0^2}$$

$$A_2 = y_2 \left[\frac{df(y_0)}{dy_0} \right] + \frac{y_1^2}{2!} \left[\frac{d^2 f(y_0)}{dy_0^2} \right] = y_2 \left(-2y_0 e^{-y_0^2} \right) + \frac{y_1^2}{2} \left(-2y_0 e^{-y_0^2} \right)'$$

$$= -2y_0 y_2 e^{-y_0^2} - y_1^2 \left(y_0 e^{-y_0^2} \right)'$$

$$= -2y_0 y_2 e^{-y_0^2} - y_1^2 \left(e^{-y_0^2} - 2y_0^2 e^{-y_0^2} \right)$$

$$A_2 = e^{-y_0^2} \left(-2y_0 y_2 - y_1^2 + 2y_0^2 y_1^2 \right)$$

Problem 7.1.4 Compute first five Adomian polynomials for the nonlinearity, $Ny = 1/y$ (Adomian33.m).

Solution

Following are the output list of the MATLAB script (Adomian33.m);

```
A0= 1/y0,
A1= -y1/y0^2,
A2= y1^2/y0^3 - y2/y0^2,
A3= (2*y1*y2)/y0^3 - y3/y0^2 - y1^3/y0^4,
A4= x1^4/(5*y0^5)+ (2*(y2^2/2+ x3*y1))/y0^3- y4/y0^2- (3*y1^2*y2)/y0^4
```

```
%Adomian33.m
clc;syms y0 y1 y2 y3 y4
f0=1/y0;
A(1)=(f0);
A(2)=y1*diff(f0,1,'y0');
A(3)=y2*diff(f0,1,'y0')+(y1.^2/2)*diff(f0,2,'y0');
A(4)=y3*diff(f0,1,'y0')+y1*y2*diff(f0,2,'y0')+(y1.^3/6)*diff(f0,3,'y0');
A(5)=y4*diff(f0,1,'y0')+((1/2)*y2.^2+y1*x3)*diff(f0,2,'y0')...
     +(1/2)*y1.^2*y2*diff(f0,3,'y0')+(1/120)*x1.^4*diff(f0,4,'y0');
```

Problem 7.1.5 Compute first five Adomian polynomials for the nonlinearity, $Ny = 1/(y+1)$ (Adomian333.m).

Solution
Following are the output list of the MATLAB script (Adomian333.m);

```
A0= 1/(y0 + 1),
A1= -y1/(y0 + 1)^2,
A2= y1^2/(y0 + 1)^3 - y2/(y0 + 1)^2,
A3= (2*y1*y2)/(y0 + 1)^3 - y1^3/(y0 + 1)^4 - y3/(y0 + 1)^2,
A4= y1^4/(5*(y0 + 1)^5) - y4/(y0 + 1)^2 + (2*(y2^2/2
    + y1*y3))/(y0 + 1)^3 - (3*y1^2*y2)/(y0 + 1)^4
```

```
%Adomian333.m
clc;syms y0 y1 y2 y3 y4
f0=1/(1+y0);
A(1)=(f0);
A(2)=y1*diff(f0,1,'y0');
A(3)=y2*diff(f0,1,'y0')+(y1.^2/2)*diff(f0,2,'y0');
A(4)=y3*diff(f0,1,'y0')+y1*y2*diff(f0,2,'y0')+(y1.^3/6)*diff(f0,3,'y0');
A(5)=y4*diff(f0,1,'y0')+((1/2)*y2.^2+y1*y3)*diff(f0,2,'y0')...
     +(1/2)*y1.^2*y2*diff(f0,3,'y0')+(1/120)*y1.^4*diff(f0,4,'y0');A
```

Problem 7.1.6 Compute first seven Adomian polynomials for the nonlinearity, $Ny = y^5$ (Adomian22.m).

Solution
Following is the edited output list of the MATLAB script (Adomian22.m);

```
A0 = y0^5,
A1 = 5*y0^4*y1,
A2 = 5*y2*y0^4 + 10*y0^3*y1^2,
A3 = 5*y3*y0^4 + 20*y2*y0^3*y1 + 10*y0^2*y1^3,
A4 = 20*y0^3*(y2^2/2 + y1*y3) + y0*y1^4 + 5*y0^4*y4 + 30*y0^2*y1^2*y2,
A5 = 20*y0^3*(y1*y4 + y2*y3) + 5*y0^4*y5 + y1^5 + 60*y0^2*((y3*y1^2)/2
    + (y1*y2^2)/2) + 20*y0*y1^3*y2,
```

```
A6 = 60*y0^2*((y4*y1^2)/2 + y3*y1*y2 + y2^3/6) + 20*y0^3*(y3^2/
2 + y1*y5 + y2*y4) + 5*y1^4*y2 + 5*y0^4*y6 + 120*y0*((y3*y1^3)/6
    + (y1^2*y2^2)/4), 5*y0^4*y6 + 120*y0*((y3*y1^3)/6 +
(y1^2*y2^2)/4)
```

MATLAB script (Adomian22.m);

```
%Adomian22.m
clc;syms x0 x1 x2 x3 x4 x5 x6  y0 y1 y2 y3 y4 y5 y6
f0=x0^(5);
A(1)=(f0);
A(2)=x1*diff(f0,1,'x0');
A(3)=x2*diff(f0,1,'x0')+(x1.^2/2)*diff(f0,2,'x0');
A(4)=x3*diff(f0,1,'x0')+x1*x2*diff(f0,2,'x0')+(x1.^3/6)*diff(f0,3,'x0');

A(5)=x4*diff(f0,1,'x0')+((1/2)*x2.^2+x1*x3)*diff(f0,2,'x0')...
    +(1/2)*x1.^2*x2*diff(f0,3,'x0')+(1/120)*x1.^4*diff(f0,4,'x0');

A(6)=x5*diff(f0,1,'x0')+ (x1*x4 + x2*x3)*diff(f0,2,'x0')...
    +((x3*x1^2)/2 + (x1*x2^2)/2)*diff(f0,3,'x0')...
    +((x1^3*x2)/6)*diff(f0,4,'x0')+(x1^5/120)*diff(f0,5,'x0');

A(7)=x6*diff(f0,1,'x0')+ (x3^2/2 + x1*x5 + x2*x4)*diff(f0,2,'x0')...
    +((x4*x1^2)/2 + x3*x1*x2 + x2^3/6)*diff(f0,3,'x0')...
    +((x3*x1^3)/6 + (x1^2*x2^2)/4)*diff(f0,4,'x0')...
    +((x1^4*x2)/24)*diff(f0,5,'x0')+(x1^6/720)*diff(f0,6,'x0');

A1=subs(A,[x0,x1,x2,x3,x4,x5,x6],[y0,y1,y2,y3,y4,y5,y6])
```

Problem 7.1.7 Another algorithm for calculating Adomian polynomials for non-linear operators is established by parametrization, which can also be used to generate Adomian polynomials for non-linear terms of several variables [23]. The following is the algorithm for calculating $A_0, A_0, \ldots, A_n$ where n is the order of Adomian polynomials:

1. Input nonlinear term $Nu = F(u), n$
2. Set $u = u_0 + \lambda u_1 + \lambda^2 u_2 + \lambda^3 u_3 + \cdots + \lambda^n u_n$
3. Let

$$\sum_{k=0}^{n} \lambda^k A_k = F\left(u_0 + \lambda u_1 + \lambda^2 u_2 + \lambda^3 u_3 + \cdots + \lambda^n u_n\right) \qquad (7.6)$$

4. For $i = 0, 1, 2, \cdots, n$, take the ith order derivative of both sides of (7.6) with respect to λ
5. Let $\lambda = 0$ and determine A_i by solving the equation with respect to A_i.

Apply this method to compute first four Adomian polynomials for $F(u) = uu'$.

Solution
We first set

$$F(u(\lambda)) = \sum_{k=0}^{\infty} \lambda^k A_k = \left(\sum_{k=0}^{\infty} \lambda^k u_k\right)\left(\sum_{k=0}^{\infty} \lambda^k u_k\right)' \qquad (7.7)$$

Letting $\lambda = 0$, we find $A_0 = u_0 u_0'$.

Take first order derivative of both sides of (7.7) with respect to λ and let $\lambda = 0$,

$$\frac{\partial(A_0 + \lambda A_1)}{\partial \lambda}\bigg|_{\lambda=0} = \partial\left(\frac{(u_0 + \lambda u_1)(u_0 + \lambda u_1)'}{\partial \lambda}\right)\bigg|_{\lambda=0} \qquad (7.8)$$

Solving the above equation with respect to A_1 we obtain,

$$A_1 = u_1 u_0' + u_0 u_1'$$

Take second order derivative of both sides of (7.7) with respect to λ and let $\lambda = 0$, then solving the equation obtained with respect to A_2 we get,

$$A_2 = u_2 u_0' + u_1 u_1' + u_0 u_2'$$

In similar way, Take third order derivative of both sides of (7.7) with respect to λ and let $\lambda = 0$, then solving the equation obtained with respect to A_3 we get,

$$A_3 = u_3 u_0' + u_2 u_1' + u_1 u_2' + u_0 u_3'$$

Problem 7.1.8 Use ADM to find the solution of $f(x) = e^{-x} + 5 - x = 0$, with four digit accuracy (Adomian3.m).

Solution
Isolating the unknown variable results in

$$x = 5 + e^{-x} = 5 + \sum_{i=0}^{\infty} A_i(e^{-x})$$

we compute the Adomian polynomials $A_i(e^{-x})$ using MATLAB script (Adomian3. m) and find the solution as $x = 5.0049$.

Following list is an edited output of this code.

```
A1=exp(-x0),
A2=-x1*exp(-x0),
A3=(exp(-x0)*x1^2)/2 - x2*exp(-x0),
A4=x1*x2*exp(-x0) - (x1^3*exp(-x0))/6 - x3*exp(-x0),
A5=(x1^4*exp(-x0))/120 - x4*exp(-x0) + exp(-x0)*(x2^2/
2 + x1*x3) - (x1^2*x2*exp(-x0))/2]
```

zero = 5.0049

MATLAB code given below is written to show all steps of computations. First part computes the Adomian polynomials for exponential function, while the second part computes the numerical values of Adomian polynomials for given constant term and then finds the combined solution with the constant term.

Note that this result is obtained by the sum of first six terms of Adomian polynomials. The exact solution of the given equation (with four digit accuracy) is $x = 5.0066$.

```
%Adomian3.m                      f(x)=exp(-x)+k-x=0      k>0
clc;syms x0 x1 x2 x3 x4
f0=exp(-x0);
A(1)=(f0);   A(2)=x1*diff(f0,1,'x0');
A(3)=x2*diff(f0,1,'x0')+(x1.^2/2)*diff(f0,2,'x0');
A(4)=x3*diff(f0,1,'x0')+x1*x2*diff(f0,2,'x0')+(x1.^3/6)*diff(f0,3,'x0');
A(5)=x4*diff(f0,1,'x0')+((1/2)*x2.^2+x1*x3)*diff(f0,2,'x0')...
     +(1/2)*x1.^2*x2*diff(f0,3,'x0')+(1/120)*x1.^4*diff(f0,4,'x0');
A1=A
k=5;
X(1)=vpa(subs(A(1),x0,k));   X(2)=subs(A(2),[x0,x1],[1,X(1)]);
X(3)=subs(A(3),[x0,x1,x2],[1,X(1),X(2)]);
X(4)=subs(A(4),[x0,x1,x2,x3],[1,X(1),X(2),X(3)]);
X(5)=subs(A(5),[x0,x1,x2,x3,x4],[1,X(1),X(2),X(3),X(4)]);
z1=sum(X);z1=vpa(z1);  zero=z1+k
```

7.2 Solution of Initial Value Problems Using ADM

Problem 7.2.1 (*Solution of IVPs*) Let the form of an ODE be

$$Ly + Ry + N(y) = f(x) \tag{7.9}$$

where L is the highest order invertible derivative operator, R is a linear differential operator, N is an analytic nonlinear operator, $f(x)$ is a forcing function term, and y is the solution of the equation.

The basic idea of ADM is to apply the inverse operator, L^{-1} to the expression

$$Ly(x) = f(x) - Ry(x) - N(y(x)) \tag{7.10}$$

which gives the solution

$$y(x) = \Psi_0(x) + L^{-1}f(x) - L^{-1}Ry(x) - L^{-1}N(y(x)) \tag{7.11}$$

and $\Psi_0(x)$ includes the terms arising from using the given initial conditions,

$$\Psi_0(x) = \begin{cases} y(0) & \textit{for } L = d/dx \\ y(0) + xy'(0) & \textit{for } L^2 = d^2/dx^2 \\ y(0) + xy'(0) + \frac{1}{2}x^2y''(0) & \textit{for } L^3 = d^3/dx^3 \\ y(0) + xy'(0) + \frac{1}{2}x^2y''(0) + \frac{1}{3!}x^3y'''(0) & \textit{for } L^4 = d^4/dx^4 \end{cases}$$

or

$$\sum_{n=0}^{\infty} y_n(x) = \Psi_0(x) + L^{-1}f(x) - L^{-1}R\sum_{n=0}^{\infty} y_n(x) - L^{-1}\sum_{n=0}^{\infty} A_n(x)$$

where $A_n(x)$ are Adomian polynomials.

If $y_0(x)$ is identified as $\Psi_0(x) + L^{-1}f(x)$, the remaining components are determined by the following recursion equations,

$$\begin{aligned} y_0(x) &= \Psi_0(x) + L^{-1}f(x) \\ y_k(x) &= -L^{-1}Ry_{k-1}(x) - L^{-1}A_{k-1}(x), \quad k = 1, 2, \ldots \end{aligned} \tag{7.12}$$

The ADM decomposes the solution $y(x)$ into an infinite series [24, 25],

$$\varphi_n(x) = \sum_{k=0}^{n-1} y_k(x), \quad y(x) = \lim_{n\to\infty} \varphi_n(x)$$

(a) Using the information given above, solve the first order IVP, $y' = y^2$, $y(0) = 1$ employing the ADM.

(b) Solve the IVP, $y' = y^p$, $(p \geq 1)$ $y(0) = 1$, using ADM.

(c) Compute the approximate the solution of the IVP, $y' = -y + y^2$, $y(0) = 4$, using a third order polynomial equation via ADM.

Solution

$$Ly = y', \; Ry = 0, \; N(y) = -y^2, \; f = 0,$$
$$y = \Psi_0 + L^{-1}f - L^{-1}Ry - L^{-1}N(y) = y(0) + 0 - 0 - L^{-1}N(y)$$

(a)

$$= 1 - L^{-1}N(y) = 1 + \int_0^x y(\tau)^2 d\tau$$

First four Adomian polynomials for $N(y) = y^2$ are,

$$A_0 = y_0^2, \quad A_1 = 2y_0y_1, \quad A_2 = 2y_0y_2 + y_1^2, \quad A_3 = 2y_0y_3 + 2y_1y_2$$
$$y_0 = y(0) + L^{-1}f = y(0) + 0 = 1 + 0 = 1$$

$$y_1 = \int_0^x A_0(\tau)d\tau = \int_0^x y_0^2 d\tau = \int_0^x d\tau = x$$

$$y_2 = \int_0^x A_1(\tau)d\tau = \int_0^x 2y_0y_1 d\tau = 2\int_0^x (1)(\tau)d\tau = x^2$$

$$y_3 = \int_0^x A_2(\tau)d\tau = \int_0^x (2y_0y_2 + y_1^2)d\tau = \int_0^x \left[(2)(1)(\tau^2) + (\tau)^2\right]d\tau = \int_0^x 3\tau^2 d\tau = x^3$$

$$y_4 = \int_0^x A_3(\tau)d\tau = \int_0^x (2y_0y_3 + 2y_1y_2)d\tau = \int_0^x \left[(2)(1)(\tau^3) + 2(\tau)(\tau^2)\right]d\tau = x^4$$

$$y = y_0 + y_1 + y_2 + y_3 + y_4 + \cdots = 1 + x + x^2 + x^3 + x^4 + \cdots = \frac{1}{1-x}$$

(b) Adomian method gives

$$y(x) = 1 + \int_0^x y^p(\tau)d\tau$$

Adomian polynomials for y^p,

$$A_0 - y_0^p$$
$$A_1 = py_0^{p-1}y_1$$
$$A_2 = \frac{p(p-1)}{2}y_0^{p-2}y_1^2 + py_0^{p-1}y_2$$
$$A_3 = \frac{p(p-1)(p-2)}{6}y_0^{p-3}y_1^3 + p(p-1)y_0^{p-2}y_1y_2 + py_0^{p-1}y_3$$

This yields,

$$y_0(x) = 1,$$

$$y_1(x) = x,$$

$$y_2(x) = \frac{p}{2}x^2,$$

$$y_3(x) = \frac{p(2p-1)}{3!}x^3,$$

$$y_4(x) = \frac{p(6p^2 - 7p + 2)}{4!}x^4,$$

$$y(x) = 1 + x + \frac{p}{2}x^2 + \frac{p(2p-1)}{3!}x^3 + \frac{p(6p^2 - 7p + 2)}{4!}x^4 + \cdots, \quad p \geq 1$$

(c) Using operator equations,

$$Ly = -y + \sum_{n=0}^{\infty} A_n$$

$$y = y(0) - L^{-1}\sum_{n=0}^{\infty} y_n + L^{-1}\sum_{n=0}^{\infty} A_n$$

Applying Adomian polynomials for the nonlinearity $N(y^2)$, we obtain

$$y_0 = y(0)$$

$$y_1 = -L^{-1}y_0 + L^{-1}A_0 = -L^{-1}y_0 + L^{-1}y_0^2 = -\int_0^x y_0 d\tau + \int_0^x y_0^2 d\tau = -4x + 16x = 12x$$

$$y_2 = -L^{-1}y_1 + L^{-1}A_1 = -L^{-1}y_1 + L^{-1}(2y_0y_1) = -\int_0^x 12\tau d\tau + \int_0^x (2)(4)(12\tau) d\tau$$

$$y_2 = 42x^2$$

$$y_3 = -L^{-1}y_2 + L^{-1}A_2 = -L^{-1}y_2 + L^{-1}(2y_0y_2 + y_1^2)$$

$$y_3 = -\int_0^x 42\tau^2 d\tau + \int_0^x [(2)(4)(42\tau^2) + (12\tau^2)] d\tau = \frac{-42x^3 + 336x^3 + 144x^3}{3}$$

$$y_3 = 146x^3$$

We stop here, since a third order term is reached. Hence,

$$y = y_0 + y_1 + y_2 + y_3 + \cdots = 4 + 12x + 42x^2 + 146x^3 + \cdots$$

Exact solution is

$$y = -\frac{1}{\exp\left[x + \ln\left(\frac{3}{4}\right)\right] - 1}, \quad x < \ln\left(\frac{4}{3}\right)$$

$$\cong 4 + 12x + 42x^2 + 146x^3 + \frac{1015}{2}x^4 + \frac{17641}{10}x^5 + \cdots$$

Problem 7.2.2 Use ADM to solve the IVP [26], $yy' - x = 0$, $y(0) = 1$ (Adomian8. m), (Adomian33.m).

Solution

$$y' - \frac{x}{y} = 0,$$

First three Adomian polynomials for $Ny = 1/y$ are computed as (Adomian33. m),

$$A_0 = \frac{1}{y_0}, \quad A_1 = -\frac{y_1}{y_0^2}, \quad A_2 = \frac{y_1^2}{y_0^3} - \frac{y_2}{y_0^2},$$

$$y = \psi_0 - L^{-1}Ny, \quad \psi_0 = y(0)$$

$$y_0 = \psi_0 = y(0) = 1$$

$$y_1 = -L^{-1}N(-xA_0) = L^{-1}N\left(\frac{x}{y_0}\right) = \int_0^x x\,dz = \frac{x^2}{2}$$

$$y_2 = -L^{-1}N(-xA_1) = L^{-1}N\left(-x\frac{y_1}{y_0^2}\right) = -\int_0^x x\frac{y_1}{y_0^2}\,dz = -\int_0^x x\frac{x^2}{2}\,dz = -\frac{x^4}{8}$$

$$y_3 = -L^{-1}N(-xA_2) = L^{-1}N\left(x\frac{y_1^2}{y_0^3} + x\frac{y_2}{y_0^2}\right) = \int_0^x \left(x\frac{x^4}{4} + x\frac{x^4}{8}\right)dz = \frac{x^6}{16}$$

$$\psi_3 = y_0 + y_1 + y_2 + y_3 = 1 + \frac{x^2}{2} - \frac{x^4}{8} + \frac{x^6}{16} - \cdots$$

$$y = \sqrt{x^2 + 1} = 1 + \frac{x^2}{2} - \frac{x^4}{8} + \frac{x^6}{16} - \frac{5x^8}{128} + \cdots$$

Therefore, $y = \sqrt{x^2 + 1}$.

```
%Adomian8.m        y'-x/y=0 , y(0)=1
clc;clear;syms x y0 y1 y2
y_0=1; %initial condition
psi0=(y_0); L_1f=int(f,x,0,x); y0=psi0+L_1f
A0=1/y0 ; L_1Ny0=int(-x*A0,x,0,x); y1=-L_1Ny0;
A1=-y1/y0^2; L_1Ny1=int(-x*A1,x,0,x);  y2=-L_1Ny1;
A2=y1^2/y0^3-y2/y0^2; L_1Ny2=int(-x*A2,x,0,x); y3=-L_1Ny2;
y=y0+y1+y2+y3; y=expand(y)
```

```
%Adomian33.m
clc;syms y0 y1 y2
f0=1/y0;
A(1)=(f0) %A0
A(2)=y1*diff(f0,1,'y0') %A1
A(3)=y2*diff(f0,1,'y0')+(y1.^2/2)*diff(f0,2,'y0') %A2
```

Problem 7.2.3 Use ADM to solve the IVP, $y' + yy' - x = 0$, $y(0) = 1$ (Adomian333.m).

Solution

$$y'(1+y) - x = 0, \; \rightarrow \; y' - \frac{x}{(1+y)} = 0$$

First three Adomian polynomials for

$$Ny = N\left(\frac{1}{(1+y)}\right)$$

are computed as

$$A_0 = \frac{1}{1+y_0}, \quad A_1 = -\frac{y_1}{(1+y_0)^2}, \quad A_2 = \frac{y_1^2}{(1+y_0)^3} - \frac{y_2}{(1+y_0)^2}$$

$$y = \psi_0 - L^{-1}Ny, \quad \psi_0 = y(0)$$
$$y_0 = \psi_0 = y(0) = 1$$

$$y_1 = -L^{-1}N(-xA_0) = \int_0^x zA_0 dz = \int_0^x \frac{z}{1+y_0} dz = \int_0^x \frac{z}{2} dz = \frac{x^2}{4},$$

$$y_2 = -L^{-1}N(-xA_1) = \int_0^x zA_1 dz = \int_0^x -z\frac{y_1}{(1+y_0)^2} dz = -\int_0^x z\frac{z^2}{4}\frac{1}{4} dz = -\int_0^x \frac{z^3}{16} dz,$$

$$y_2 = -\frac{x^4}{64}$$

$$\varphi_2 = y_0 + y_1 + y_2 = (1) + \left(\frac{x^2}{4}\right) + \left(-\frac{x^4}{64}\right)$$

Exact solution:

$$y = -1 + \sqrt{x^2 + 4} = 1 + \frac{x^2}{4} - \frac{x^4}{64} + \frac{x^6}{512} - \frac{5x^8}{16384} + \cdots$$

Following are the output list of the MATLAB script (Adomian333.m);

```
A0= 1/(y0 + 1),
A1= -y1/(y0 + 1)^2,
A2= y1^2/(y0 + 1)^3 - y2/(y0 + 1)^2,
```

```
%Adomian333.m
clc;syms y0 y1 y2 y3 y4
f0=1/(1+y0);
A(1)=(f0)  %A0
A(2)=y1*diff(f0,1,'y0')  %A1
A(3)=y2*diff(f0,1,'y0')+(y1.^2/2)*diff(f0,2,'y0')  %A2
```

Problem 7.2.4 Use ADM to solve the IVP, $y'' = y + \cos(x)$, $y(0) = 1$, $y'(0) = 0$. Write down the solution as a fourth order polynomial.

Solution
We bring the equation in standard operator form,

$$y'' - y = \cos(x) \quad \rightarrow \quad Ly + Ry + Ny = f$$
$$Ly = y'', \quad Ry = -y, \quad f = \cos(x), \quad Ny = 0$$
$$y_0 = 1 + L^{-1}f = 1 + \int_0^x \int_0^t \cos(\tau)d\tau dt = 1 + \int_0^x \sin(t)dt = 2 - \cos(x),$$
$$y_1 = -L^{-1}R(-y_0) = L^{-1}R(y_0) = \int_0^x \int_0^t (2 - \cos(\tau))d\tau dt$$
$$y_1 = \int_0^x (2t - \sin t)dt = x^2 + \cos(x) - 1,$$
$$y_2 = -L^{-1}R(-y_1) = L^{-1}R(y_1) = \int_0^x \int_0^t [\tau^2 + \cos(\tau) - 1]d\tau dt = \frac{x^4}{12} - \frac{x^2}{2} - \cos x + 1$$
$$y = y_0 + y_1 + y_2 = 2 - \cos(x) + \frac{x^2}{2} + \frac{x^4}{12}$$

Problem 7.2.5 Solve the following IVP using ADM,

$$y'' + x^2 y - 1 = 0, \quad y(0) = 1, \quad y'(0) = 0$$

Solution

$$Ly + Ry + N(y) = f$$
$$Ly = y'', \quad Ry = x^2 y, \quad N(y) = 0, \quad f = 1$$
$$y = \psi_0 + L^{-1}f - L^{-1}Ry$$

$$y = \sum_{n=0}^{\infty} y_n = [y(0) + xy'(0)] + \frac{x^2}{2} - \iint \left(\tau^2 \sum_{n=0}^{\infty} y_n \right) d\tau dt$$

Recursion equations are

$$y_0 = \psi_0 + L^{-1}f = [1 + x(0)] + \frac{x^2}{2} = 1 + \frac{x^2}{2}$$
$$y_k = -L^{-1}Ry_{k-1}, \quad k \geq 1$$

$$y_1 = -L^{-1}Ry_0 = -\int_0^x \int_0^\tau \tau^2 y_0 d\tau dt = -\int_0^x \int_0^t \tau^2 \left(1 + \frac{\tau^2}{2} \right) d\tau dt = -\frac{x^4}{12} - \frac{x^6}{60}$$

$$y_2 = -L^{-1}Ry_1 = -\int_0^x \int_0^t \tau^2 \left(-\frac{\tau^4}{12} - \frac{\tau^6}{60} \right) d\tau dt = \frac{x^8}{672} + \frac{x^{10}}{5400}$$

$$y \cong y_1 + y_2 + y_3$$
$$y = 1 + \frac{x^2}{2} - \frac{x^4}{12} - \frac{x^6}{48} + \frac{x^8}{672} + \frac{x^{10}}{5400}$$

Problem 7.2.6 Solve the second order IVP, $y'' + y' = 0$, $y(0) = 1, y'(0) = -1$, using the ADM

Solution

$$Ly + Ry + N(y) = f(x)$$

where L is the second order invertible derivative operator, R is a linear differential operator of order 2, N is an analytic nonlinear operator, $f(x)$ is a forcing function term. Here last two of these terms are zero. Applying the inverse operator, L^{-1} to the expression,

$$L^{-1}Ly(x) = -L^{-1}Ry'(x)$$

gives

$$y(x) = \Psi_0(x) - L^{-1}Ry'(x)$$

$\Psi_0(x)$ includes the terms arising from using the given initial conditions,

$$\Psi_0(x) = y(0) + xy'(0) = 1 - x$$

Recursion equations are

$$y_0(x) = \Psi_0(x) = 1 - x$$
$$y_k(x) = -L^{-1}Ry'_{k-1}(x), \quad k = 1, 2, \ldots$$

The ADM decomposes the solution $y(x)$ into an infinite series,

$$\varphi_n(x) = \sum_{k=0}^{n-1} y_k(x), \quad y(x) = \lim_{n \to \infty} \varphi_n(x)$$

We perform recursion operations as follows;

$$y_0(x) = \Psi_0(x) = 1 - x$$
$$y'_0(x) = (1 - x)' = -1,$$

$$y_1(x) = -L^{-1}Ry'_0(x) = -\int_0^x \int_0^t (-1)d\tau dt = \frac{x^2}{2}$$

$$y_2(x) = -L^{-1}Ry'_1(x) = -\int_0^x \int_0^t (\tau)d\tau dt = -\frac{x^3}{6}$$

$$y_3(x) = -L^{-1}Ry'_2(x) = -\int_0^x \int_0^t \left(-\frac{\tau^2}{2}\right)d\tau dt = \frac{x^4}{24}$$

$$y \cong \varphi_3(x) = y_0(x) + y_1(x) + y_2(x) + y_3(x) = 1 - x + \frac{x^2}{2} - \frac{x^3}{6} + \frac{x^4}{24} - \cdots = e^{-x}$$

Problem 7.2.7 Consider the IVP, $y'' + \alpha y' + \gamma y^m = 0$, $y(0) = c_0$, $y'(0) = c_1$.
  This equation can be written in operator form as $Ly + Ry + Ny = 0$ where, $L = d^2/dt^2$, $R = \alpha(d/dt) + \beta$, $Ny = \gamma y^m$. The solution is obtained as

$$y = \sum_{n=0}^{\infty} y_n, \quad Ny = \sum_{n=0}^{\infty} A_n$$

where A_n terms donate Adomian polynomials for y'', and $y_0 = y(0) + ty'(0) = c_0 + tc_1$. Solve the IVP, $y'' + 4y = 0$, $y(0) = 1$, $y'(0) = 0$.

Solution

This ODE neither includes any nonlinear terms, nor a first derivative term. We write it in operator form as

$$Ly = -Ry = -4y$$
$$y(0) = c_0 = 1, \quad y'(0) = c_1 = 0$$
$$y_0 = y(0) + ty'(0) = c_0 = 1$$

$$y_1 = \int_0^t \int_0^t (-Ry_0) d\tau d\tau = -4 \int_0^t \int_0^t d\tau d\tau = -2t^2$$

$$y_2 = \int_0^t \int_0^t (-Ry_1) d\tau d\tau = -4 \int_0^t \int_0^t (-2\tau^2) d\tau d\tau = \frac{2t^4}{3}$$

$$y = y_0 + y_1 + y_2 + \cdots = 1 - 2t^2 + \frac{2t^4}{3} - \cdots$$

This is equivalent to the series representation of

$$\cos(x) = 1 - \frac{1}{2}x^2 + \frac{1}{24}x^4 - \cdots$$

with $x = 2t$,

$$\cos(2t) = 1 - \frac{1}{2}(2t)^2 + \frac{1}{24}(2t)^4 - \cdots = 1 - 2x^2 + \frac{2}{3}x^4 - \cdots$$

Hence, the solution is $y(t) = \cos(2t)$.

Problem 7.2.8 Mathieu's equation is a linear second-order ODE,

$$y'' + [a - b\cos(2t)]y = 0 \tag{7.13}$$

where a, b are constants. Mathieu's equation has no closed-form analytic solution but can be solved numerically. It is well known in applied mathematics, theoretical physics (Josephson junctions), astronomy and in theory of vibrations [27–30]. This equation is associated with ODEs derived for systems with periodic forcing, and in stability studies of periodic motions in nonlinear autonomous systems.

Use ADM to approximate the solution of (7.13) for $a = 1$, $y(0) = 1$, $y'(0) = 0$, while $b = 1, 2, \ldots, 6$. Plot the solution curves on the same figure (Adomian5.m).

Solution

Operator form of given equation is

$$Ly + Ry + N(y) = f, \quad N(y) = f = 0, \quad Ly + Ry = 0$$
$$y = \psi_0 - Ry,$$
$$y_0 = \psi_0 + L^{-1}f = \psi_0 = y(0) + ty'(0) = 1$$
$$y_{n+1} = -L^{-1}(Ry_n), \quad n \geq 0$$

Approximate solutions are plotted in Fig. 7.1 using MATLAB script (Adomian5.m).

```
%Adomian5.m      y"+[a-b*cos(2*t)]*y=0  Mathieu's eq.
Mathieu's eq.
clc;clear;syms t
for b=1:6
a=1; y_0=1; dy_0=0;
psi0=y_0+t*dy_0; y0=psi0; u1=(a-b*cos(2*t))*y0;
y1=-int(int(u1,t,0,t),t,0,t); y2=-int(int(y1,t,0,t),t,0,t);
y3=-int(int(y2,t,0,t),t,0,t); y=y0+y1+y2+y3;
Y=ezplot(y,[0 2]);hold on;set(Y,'linewidth',2);
end
ylabel('y(t)');grid ; legend('b=1','b=2','b=3','b=4','b=5','b=6');
```

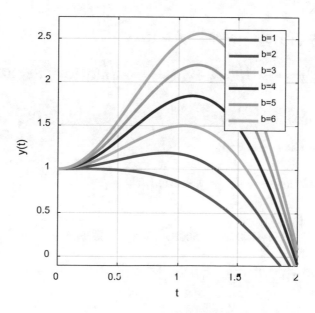

Fig. 7.1 Approximate solutions for $y'' + [1 - b\cos(2t)]y = 0$, $b = 1, 2, \ldots, 6$

Problem 7.2.9 Current flow due to a discharging capacitor through a series connected resistor and inductor circuit is described as a linear IVP [31],

$$y'' + c_1 y' + c_2 y = 0, \quad y(0) = 0, \quad y'(0) = 1000 \tag{7.14}$$

where $c_1 = 10^3$, $c_2 = 10^5$.

(a) Calculate first three terms of the series representation of the current.
(b) Plot the current waveform as a function of time, using ADM (after 50 iterations) in the same figure with the actual solution curve.

Solution
(a)

$$Ly + Ry + N(y) = f(t) = 0$$
$$Ly = f - Ry - N(y), \quad N(y) = f(t) = 0 \tag{7.15}$$

$$Ly = -Ry \tag{7.16}$$

Recursion equations:

$$y_0 = \Psi_0 = y(0) + ty'(0) \tag{7.17}$$

$$y_k = -L^{-1} R y_{k-1} \quad k = 1, 2, 3, \dots \tag{7.18}$$

Therefore,

$$y_0 = 0 + (t)(1000) = 1000t, \quad y_0' = 1000,$$

$$y_1 = L^{-1} R y_0 = -\iint_0^t (c_1 y_0' + c_2 y_0) d\tau = -\iint_0^t [c_1(1000) + c_2(1000\tau)] d\tau$$

$$y_1 = -500 c_1 t^2 - \frac{500}{3} c_2 t^3,$$

$$y_1' = -1000 c_1 t - 500 c_2 t^2,$$

$$y_2 = -L^{-1} R y_1 = -\iint_0^t \left(c_1 y_1' + c_2 y_1 \right) d\tau$$

$$y_2 = -\iint_0^t \left[c_1 (-1000 c_1 t - 500 c_2 t^2) + c_2 \left(-500 c_1 t^2 - \frac{500}{3} c_2 t^3 \right) \right] d\tau$$

$$y_2 = \frac{500}{3} c_1^2 t^3 + \frac{250}{3} c_1 c_2 t^4 + \frac{25}{3} c_2^2 t^5$$

(b) The current waveform as a function of time, using ADM (after 50 iterations) are displayed in Fig. 7.2 alongwith the actual solution curve.

Fig. 7.2 The natural response (current waveform) in an electric RLC circuit due to a discharging capacitor as obtained by using ADM (after 50 iterations) alongwith the actual solution curve

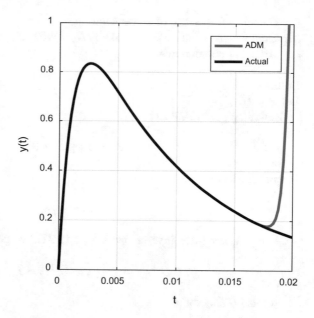

```
%Adomian_RLC.m       y"+c1*y'+c2*y=0   y(0)=0,  y'(0)=1000 c1=1e3,  c2=1e5
clc;clear;close; syms t %c1 c2
tic;
tf=0.02;         %final time of plot
c1=1e3; c2=1e5;% constants
y_0=0;           % initial function value
dy_0=1000;       % initial derivative value
psi0=y_0+t*dy_0;
y0=psi0;         % start recursion
y(1)=psi0; sum=y(1);
for j=1:50
dy(j)=diff(y(j)); Ry(j)=(c1*dy(j)+c2*y(j));
L_1Ry(j)=int(int(Ry(j),t,0,t),t,0,t); y(j+1)=-L_1Ry(j);
sum=sum+y(j+1);
end
Y=expand(sum); t=linspace(0,tf); i=eval(vectorize(Y));
plot(t,i,'linewidth',2);ylim([0 1]);hold on;
syms y(t) % Actual solution
eq='D2y+1000*Dy+100000*y=0'; ic='y(0)=0,Dy(0)=1000';
z=dsolve(eq,ic,t); t=linspace(0,tf); i=eval(vectorize(z));
plot(t,i,'linewidth',2); legend('ADM','Actual');
grid on; xlabel('t');ylabel('y(t)');toc
```

Problem 7.2.10 (*Logistic growth equation*) Solve the following IVP,

$$\frac{dy}{dt} = ry - \left(\frac{r}{N_c}\right)y^2, \quad y(0) = y_0 \tag{7.19}$$

where $y(t)$ is the number of cells at time t, r is the net rate of change of cell population, N_c is the carrying capacity of the medium.

(a) Calculate few terms of series solution,
(b) Compute the solution for $r = 0.23$, $N_0 = 400$, $N_c = 1000$ employing first five Adomian polynomials.

Solution

Let $g = r/N_c$,

Rewriting given IVP as

$$y' - ry + gy^2 = 0 \tag{7.20}$$

in operator form,

$$Ly = ry - gy^3 \tag{7.21}$$

Applying nonlinear inverse operator to (7.21), we get

$$L^{-1}Ly = y - y_0 = L^{-1}ry - L^{-1}gy^2 \tag{7.22}$$

Solving (7.22) for y,

$$y = y_0 + rL^{-1}y - gL^{-1}y^2 \tag{7.23}$$

Recursion equations,

$$y_0 = \psi_0 = y(0)$$
$$y_1 = rL^{-1}y_0 - gL^{-1}(A_0)$$
$$y_2 = rL^{-1}y_1 - gL^{-1}(A_1)$$
$$y_3 = rL^{-1}y_2 - gL^{-1}(A_2)$$
$$y_4 = rL^{-1}y_3 - gL^{-1}(A_3)$$

where Adomian polynomials are

$$A_0 = y_0^2, \quad A_1 = 2y_0y_1, \quad A_2 = y_1^2 + 2y_0y_2, \quad A_3 = 2y_1y_2 + y_0y_3$$

$$y_1 = r\int_0^t y_0\,dx - g\int_0^t y_0^2\,dx = ry_0t - gy_0^2t$$

$$y_2 = r\int_0^t \left(ry_0t - gy_0^2t\right)dx - g\int_0^t 2y_0\left(ry_0t - gy_0^2t\right)dx$$

$$y_2 = y_0\frac{r^2t^2}{2} - \frac{gy_0^2t}{2} - gy_0^2r\frac{t^2}{2} + 2g^2y_0^3\frac{t^2}{2} = \frac{r^2y_0t^2}{2} - \frac{3gry_0^2t^2}{2} + g^2y_0^3t^2$$

(b) Given, $r = 0.23$, $N_0 = 400$, $N_c = 1000$,

Exact solution is known to be [32]

$$y(k) = \frac{N_c y_0}{y_0 + (N_c - y_0)e^{-rt}}$$

We expect cell population growth rather than decay, since $N_c > y_0, r > 0$.

Following is the edited output for the MATLAB script (adomian_logistic.m). Note that the solution procedure employs first five Adomian polynomials, while Fig. 7.3 shows the computed approximation to Verhulst (Logistic) equation using Adomian Decomposition Method.

```
y = 0.00093866*t^5 - 0.010725*t^4 - 0.21723*t^3 + 1.2818*t^2
    + 55.464*t + 400.0
yV = 400000.0/(600.0*exp(-0.2311*t) + 400.0)
```

Fig. 7.3 Approximation to Verhulst (logistic) equation using Adomian decomposition method

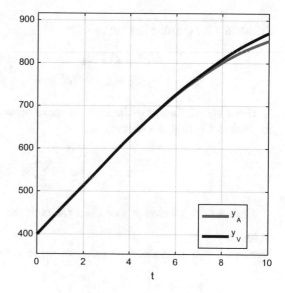

```
%adomian_logistic.m                   f(x)=exp(-x)+k-x=0    k>0
clc;clear;syms y0 y1 y2 y3 y4 t r g
L_1y0=int(y0,t,0,t); A0=y0^2;  L_1A0=int(A0,t,0,t);
y1=r*L_1y0-g*L_1A0;

L_1y1=int(y1,t,0,t);A1=2*y0*y1; L_1A1=int(A1,t,0,t);
y2=r*L_1y1-g*L_1A1;y2=expand(y2);

L_1y2=int(y2,t,0,t); A2=y1^2 + 2*y0*y2;  L_1A2=int(A2,t,0,t);
y3=r*L_1y2-g*L_1A2;y3=expand(y3);

L_1y3=int(y3,t,0,t); A3=2*y0*y3 + 2*y1*y2; L_1A3=int(A3,t,0,t);
y4=r*L_1y3-g*L_1A3;  y4=expand(y4);

L_1y4=int(y4,t,0,t); A4=y2^2 + 2*y0*y4 + 2*y1*y3; L_1A4=int(A4,t,0,t);
y5=r*L_1y4-g*L_1A4;  y5=expand(y5);

y=y0+y1+y2+y3+y4+y5;
R=0.2311;Y0=400;Nc=1000;G=R/Nc;y=subs(y,[r,y0,g],[R,Y0,G]);y=vpa(y,5)
ezplot(y,[0 10]);hold on; %Adomian solution
yV=Nc*Y0/(Y0+(Nc-Y0)*exp(-R*t)); yV=vpa(yV,5)
ezplot(yV,[0 10]); %actual solution
legend('y_A','y_V');grid on;
```

Problem 7.2.11 Solve the IVP,

$$y''' + y' = -1,$$
$$y(0) = 1, \quad y'(0) = -1, \quad y''(0) = 0$$

Using ADM. Compute the error of approximation in the sense of infinite norm for different values of elements in

$$\varphi_k = \sum_{n=0}^{N} y_k$$

Note Actual solution is $y = 1 - x$ (Adomian6.m).

Solution
Since nonlinear term does not exist, we write the equation in operator form as

$$Ly + Ry = f$$
$$Ly = f - Ry \;\rightarrow\; L^{-1}Ly = L^{-1}f - L^{-1}Ry, \quad L^{-1} = d^3/dx^3,$$
$$y = \Psi_0 + L^{-1}(f) - L^{-1}(Ry')$$

where

$$\Psi_0 = y(0) + xy'(0) + \frac{1}{2}x^2 y''(0) = 1 + x \cdot (-1) - \frac{1}{2}x^2 \cdot (0) = 1 - x$$

Recursion:

$$y_0 = \Psi_0 + L^{-1}f = (1-x) + \int\!\!\!\int\!\!\!\int_0^x \frac{d}{dx}(-1)dxdxdx = 1 - x - \frac{x^3}{6}$$

$$y_k = -L^{-2}Ry'_{k-1}$$

$$y_1 = -L^{-1}Ry'_0 = \int\!\!\!\int\!\!\!\int_0^x \left(1 - x - \frac{x^3}{6}\right)' dxdxdx$$

$$\varphi_1 = y_0 + y_1 = 1 - x + \frac{x^5}{120}$$

$$\varphi_2 = y_0 + y_1 + y_2 = 1 - x - \frac{x^7}{5040}$$

$$\varphi_3 = y_0 + y_1 + y_2 + y_3 = 1 - x + \frac{x^9}{362880}$$

$$\varphi_4 = y_0 + y_1 + y_2 + y_3 + y_4 = 1 - x - \frac{x^{11}}{39916800}$$

Infinite norm of the error of approximation for φ_4 is 2.5×10^{-8}.
Following is an output list of MATLAB script (Adomian6.m).

```
N = 2,  Y = x^5/120 - x + 1,        norminf = 0.0083333333
N = 3,  Y = - x^7/5040 - x + 1,      norminf = 0.0001984127
N = 4,  Y = x^9/362880 - x + 1,     norminf = 0.0000027557319

N = 5,  Y = - x^11/39916800 - x + 1, norminf = 0.000000025052108
```

```
%Adomian6.m        y'''+y'=-1 , y(0)=1, y'(0)=-1,y''(0)=0
clc;clear;syms x
N=3;%  number of y terms
f=-1;
y_0=1;  dy_0=-1;  d2y_0=0;
psi0=(y_0)+x*(dy_0)+x^2*(d2y_0)/2
L_1f=int(int(int(f,x,0,x),x,0,x),x,0,x)
y(1)=psi0+L_1f
Y=y(1);
for k=1:N-1
Dy(k)=diff(y(k))
L_1RDy=int(int(int(Dy(k),x,0,x),x,0,x),x,0,x)
y(k+1)=-L_1RDy
Y=Y+y(k+1)
end
Y=simplify(Y)
r=0:.1:1; Yact=1-r; Y=subs(Y,x,r); X=vpa(abs(Y-Yact),4)
norminf=vpa(norm(X,inf),8)
```

7.3   Solution of BVPs Using ADM

Problem 7.3.1 In the original (classic) ADM "recursion with undetermined coefficients", we obtain n-term approximations with the undetermined coefficients first, and then match the other boundary condition. Finally, we solve the resulting sequence of nonlinear algebraic equations.

Use this method to determine an approximate solution for the nonlinear BVP,

$$y'' = e^y, \quad y(0) = y(1) = 0 \tag{7.24}$$

including only the terms $y(x) = y_0 + y_1$ and $y(x) = y_0 + y_1 + y_2$.

Plot these two solutions in the same figure along with the exact solution.

The exact solution is

$$y(x) = 2\ln\left[k\sec\frac{k(2x-1)}{4}\right] - \ln 2, \quad k = 1.33605569 \tag{7.25}$$

(adomian_bvp1.m).

Solution

Applying the inverse linear operator to both sides of (7.24). we obtain,

$$L^{-1}(y'') = \int_0^x \int_0^x (e^\tau) d\tau d\tau \tag{7.26}$$

$$y = Bx + L^{-1}e^y, \quad B = y'(0) \tag{7.27}$$

$$y(x) = \sum_{n=0}^{\infty} y_n(x), \quad e^y = \sum_{n=0}^{\infty} A_n$$

A_n terms within the summation are the Adomian Polynomials for nonlinearity $N(y) = e^y$.

First few of these polynomials are given below.

$$A_0 = e^{y_0}$$
$$A_1 = e^{y_0} y_1$$
$$A_2 = e^{y_0} \left(\frac{y_1^2}{2} + y_2 \right)$$
$$A_3 = e^{y_0} \left(\frac{y_1^3}{6} + y_1 y_2 + y_3 \right)$$
$$A_4 = e^{y_0} \left(\frac{y_1^4}{24} + \frac{1}{2} y_1^2 y_2 + \frac{1}{2} y_2^2 + y_1 y_3 + y_4 \right)$$

$$y_o = Bx$$

$$y_1 = L^{-1} A_0 = \int_0^x \int_0^x e^{\tau_0} d\tau d\tau = \int_0^x \left(\frac{e^{B\tau}}{B} - \frac{1}{B} \right) d\tau = \frac{e^{Bx}}{B^2} - \frac{x}{B} - \frac{1}{B^2} = \frac{e^{Bx} - Bx - 1}{B^2}$$

$$\varphi_1 = Bx$$

$$\varphi_2 = Bx + \frac{e^{Bx} - Bx - 1}{B^2}$$

$$\varphi_3 = Bx + \frac{e^{Bx} - Bx - 1}{B^2} + Bx + \frac{e^{Bx}(4 - 4Bx) - 2Bx + e^{2Bx} - 5}{4B^4}$$

Matching at the other boundary condition, $x = 1$,

$$\varphi_2(1) = 0 = B + \frac{e^B - B - 1}{B^2} = 0 \;\rightarrow\; B = -0.434775$$

$$\varphi_2 = B + \frac{1}{B^2} e^{Bx} - \frac{1}{B} x - \frac{1}{B^2} = \left(B - \frac{1}{B} \right) x + \frac{1}{B^2} e^{Bx} - \frac{1}{B^2}$$

$$\varphi_2 = 1.86526x + 5.29017 e^{-0.434755x} - 5.29017$$

In similar way, we proceed to determine B value in φ_3 from the boundary condition at $x = 1$, and obtain $B = -0.46044864$.

Following is the edited output list for the MATLAB script (adomian_bvp1.m). Note that given problem requires the implementation of the first three Adomian polynomials only, but ADM still provides a fast convergence as shown in Fig. 7.4, which displays the resulting solution curves for the given BVP.

Fig. 7.4 Approximate
solutions and the exact
solution for the nonlinear
BVP, $y'' = e^y$,
$y(0) = y(1) = 0$, using ADM
(recursion with undetermined
coefficients)

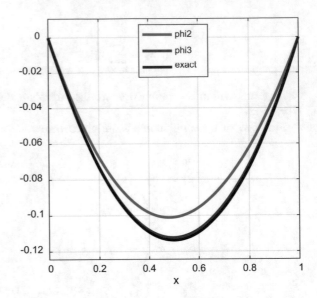

```
phi1 = b*x
phi2 = b*x - (b*x - exp(b*x) + 1)/b^2
A = [exp(b*x), -(exp(b*x)*(b*x - exp(b*x) + 1))/b^2]
phi3 = b*x + (exp(b*x) + exp(2*b*x)/4 - b*(x/2 + x*exp(b*x)) - 5/4)/b^4
        - (b*x - exp(b*x) + 1)/b^2
y2 = b*x - x/b + exp(b*x)/b^2 - 1/b^2
y3 = b*x - x/b - x/(2*b^3) + exp(b*x)/b^2 + exp(b*x)/b^4
      + exp(2*b*x)/(4*b^4) - 1/b^2 - 5/(4*b^4) - (x*exp(b*x))/b^3
phi2 = b - (b - exp(b) + 1)/b^2
B = -0.43477548
y2 = 1.8653*x + 5.2902*exp(-0.43478*x) - 5.2902
phi3 = b - (b - exp(b) + 1)/b^2 + (exp(2*b)/4 + exp(b) - b*(exp(b) + 1/2)
        - 5/4)/b^4
B = -0.46044864
y3 = 6.8332*x + 26.964*exp(-0.46045*x) + 5.5618*exp(-0.9209*x)
      + 10.244*x*exp(-0.46045*x) - 32.526
k = 1.3361
yact = 2.0*log(1.3360557/cos(0.66802785*x - 0.33401392)) - 0.69314718
```

```
%adomian_bvp1.m
clc; clear; syms y0 b x Y1
f0=exp(y0);
phi1=vpa(b*x)
A(1)=(f0); A(2)=Y1*diff(f0,1,'y0'); A=subs(A,y0,phi1);
phi20=int(int(A(1),x,0,x),x,0,x); phi2=phi1+phi20
A(2)=subs(A(2),Y1,phi20)
phi30=int(int(A(2),x,0,x),x,0,x); phi3=phi2+phi30
y2=expand(phi2),y3=expand(phi3)
%y=y0+y1
phi2=subs(phi2,x,1)
B=solve(phi2); B=vpa(B,8)
y2=vpa(subs(y2,b,B),5)
app1=ezplot(y2,[0 1]); set(app1,'linewidth',2); hold on;
%y=y0+y1+y2
phi3=subs(phi3,x,1)
B=solve(phi3); B=vpa(B,8)
y3=vpa(subs(y3,b,B),5)
app2=ezplot(y3,[0 1]);set(app2,'linewidth',2);hold on;
%exact solution
k=1.33605569
yact=2*log(k*sec(k*(2*x-1)/4))-log(2); yact=vpa(yact,8)
ya=ezplot(yact,[0 1]); set(ya,'linewidth',2); grid;
legend('phi2','phi3','exact');
```

Problem 7.3.2 The multivariable Adomian polynomials can be used for the solution of simultaneous nonlinear equations [33]. Compute first four Adomian polynomials of a function of two variables, $f(x,y) = xy^2$ (AdomianXY1.m).

Solution

A computational form for Adomian polynomials associated to a function of two variables is given below [34],

$$A_0 = f_{0,0},$$

$$A_1 = x_1 f_{1,0} + y_1 f_{0,1},$$

$$A_2 = x_2 f_{1,0} + y_2 f_{0,1} + \frac{1}{2!} x_1^2 f_{2,0} + \frac{1}{2!} y_1^2 f_{0,2} + x_1 y_1 f_{1,1},$$

$$A_3 = x_3 f_{1,0} + y_3 f_{0,1} + x_1 y_2 f_{2,0} + (x_1 y_2 + x_2 y_1) f_{1,1} + y_1 y_2 f_{0,2}$$
$$+ \frac{1}{3!} x_1^3 f_{3,0} + \frac{1}{3!} y_1^3 f_{0,3} + \frac{1}{2!} x_1^2 y_1 f_{2,0} + \frac{1}{2!} x_1 y_1^2 f_{0,2}$$

$$A_4 = x_4 f_{1,0}(x_0) + y_4 f_{0,1} + \left(\frac{1}{2!} x_2^2 + x_1 x_3\right) f_{2,0} + \frac{1}{2!} x_1^2 y_2 f_{3,0} + \left(\frac{1}{2!} y_2^2 + y_1 y_3\right) f_{0,2}$$

$$+ (x_1 y_3 + x_2 y_2 + x_3 y_1) f_{1,1} + \left(\frac{1}{2!} x_1^2 y_2 + x_1 x_2 y_1\right) f_{2,1} + \left(x_1 y_1 y_2 + \frac{1}{2!} x_2 y_1^2\right) f_{1,2}$$

$$+ \frac{1}{2!} y_1^2 y_2 f_{0,3} + \frac{1}{4!} x_1^4 f_{4,0} + \frac{1}{3!} x_1^3 y_1 f_{3,1} + \frac{1}{2!} \frac{1}{2!} x_1^2 y_1^2 f_{2,2} + \frac{1}{3!} x_1 y_1^3 f_{1,3} + \frac{1}{4!} f_{0,4}$$

where $f_{i,j}$ denotes partial derivatives with respect to x and y.

Following is the edited output list of MATLAB script (AdomianXY1.m) for the first four polynomials of the nonlinearity function, $f(x, y) = xy^2$ derived from the computational form given above.

```
A0 = x0*y0^2

A1 = x1*y0^2 + 2*x0*y1*y0

A2 = x2*y0^2 + 2*x1*y0*y1 + 2*x0*y2*y0 + x0*y1^2

A3 = x1*y1^2 + x3*y0^2 + 2*y0*(x1*y2 + x2*y1) + 2*x0*y0*y3 + 2*x0*y1*y2

A4 = x0*y2^2 + x2*y1^2 + x4*y0^2 + 2*x0*y0*y4 + 2*x0*y1*y3 + 2*x1*y0*y3
     + 2*x1*y1*y2 + 2*x2*y0*y2 + 2*x3*y0*y1
```

MATLAB script (AdomianXY1.m);

```
%AdomianXY1.m
clc;clear;
syms x0 y0 x1 x2 x3 x4 y1 y2 y3 y4
f=x0*y0^2
f10=diff(f,x0);f01=diff(f,y0);f11=diff(f10,y0);f20=diff(f,2,x0);
f21=diff(f20,y0); f22=diff(f20,2,y0); f12=diff(f10,2,y0);
f02=diff(f,2,y0); f30=diff(f,3,x0); f31=diff(f30,y0);f13=diff(f10,3,y0);
f03=diff(f,3,y0); f40=diff(f,4,x0); f04=diff(f,4,y0);
A0=f
A1=x1*f10+y1*f01
A2=x2*f10+ y2*f01+ (1/2)*x1^2*f20 +(1/2)*y1^2*f02+ x1*y1*f11
A3=x3*f10+ y3*f01+ x1*x2*f20+ (x1*y2+x2*y1)*f11+ y1*y2*f02...
  +(1/6)*x1^3*f30+ (1/6)*y1^3*f03+ (1/2)*x1^2*y1*f21+ (1/2)*x1*y1^2*f12
A4=x4*f10+ y4*f01+ (x2^2/2+ x1*x3)*f20+ (1/2)*x1^2*x2*f30...
+(y2^2/2+y1*y3)*f02+(x1*y3+x2*y2+x3*y1)*f11...
+(x1^2*y2/2+ x1*x2*y1)*f21+ (x1*y1*y2+ x2*y1^2/2)*f12+ (1/2)*y1^2*y2*f03...
+(1/24)*x1^4*f40+ x1^3*y1*f31/6+(1/4)*x1^2*y1^2*f22...
    +(1/6)*x1*y1^3*f13+ (1/24)*y1^4*f04; A4=expand(A4)
```

Notes The same group of authors also give new formulae for the calculation of Adomian polynomials for a multidimensional operator, and then use them for solving the Navier-Stokes equations in [35]. Another approach for the calculation of multivariable Adomian polynomials is presented by Duan [36].

Problem 7.3.3 The chemical reaction mechanism so called the Brusselator, or Belousov–Zhabotinsky Reaction [37, 38] is an autocatalytic chemical reaction in which a component acts to increase the rate of its producing reaction. In this kind of systems complex dynamics such as multiple steady-states and periodic orbits can be seen.

(a) Set up recursion equations for the following system of ODEs using ADM,

$$x' = a + x^2 y - (b+1)x$$
$$y' = bx - x^2 y \tag{7.28}$$
$$x(0) = c, \quad y(0) = d$$

where x, y are concentration values of chemical constituents at time t.

(b) Solve the following system of ODEs and plot the solution curves on the same figure using first five Adomian Polynomials;

$$x' = x^2 y + \frac{1}{2}x$$
$$y' = -\frac{3}{2}x - x^2 y,$$
$$x(0) = y(0) = c = d = 1$$

(adomian_sys1.m), (AdomianXY2.m), (ode_BZ2.m).

Solution

(a) Using operators,

$$L = d/dt, \quad L^{-1} = \int_0^t (\cdot) dt$$

We write given system (7.28) in operator form,

$$Lx = L[a + x^2 y - (b+1)x]$$
$$Ly = L[bx + x^2 y] \tag{7.29}$$

Applying inverse operator to (7.29), we get

$$x = x(0) + L^{-1}(a) - L^{-1}[(b+1)x + L^{-1}(x^2 y)]$$
$$y = y(0) + L^{-1}(bx) - L^{-1}(x^2 y) \tag{7.30}$$

$$x = x(0) + at - \int_0^t (b+1)x \, dt + \int_0^t (x^2 y) \, dt$$

$$y = y(0) + \int_0^t bx \, dt - \int_0^t (x^2 y) \, dt \tag{7.31}$$

$$x = \sum_{n=0}^{\infty} x_n, \quad y = \sum_{n=0}^{\infty} y_n$$

First five Adomian polynomials for the nonlinearity function $(x^2 y)$ are,

$$A_0 = x_0^2 y_0$$
$$A_1 = 2x_0 y_0 x_1 + x_0^2 y_1$$
$$A_2 = x_1^2 y_0 + 2x_0 y_1 x_1 + 2x_0 y_0 x_2 + x_0^2 y_2$$
$$A_3 = 2x_1 y_0 x_2 + x_1^2 y_1 + 2x_0 y_2 x_1 + 2x_0 y_1 x_2 + 2x_0 y_0 x_3 + x_0^2 y_3$$
$$A_4 = x_2^2 y_0 + 2x_1 y_1 x_2 + 2x_1 y_0 x_3 + x_1^2 y_2 + 2x_0 y_3 x_1$$
$$\quad + 2x_0 y_2 x_2 + 2x_0 y_1 x_3 + 2x_0 y_0 x_4 + x_0^2 y_4$$

Recursion equations,

$$x_0 = x(0) + at = c + at$$
$$y_0 = y(0) = d$$

$$x_{n+1} = \int_0^t A_n dt - (b+1) \int_0^t x_n dt \quad n = 0, 1, 2, \ldots$$

$$y_{n+1} = b \int_0^t x_n dt - \int_0^t A_n dt \quad n = 0, 1, 2, \ldots$$

(b) Here $a = 0$, $b = -1.5$, $c = d = 1$

We compute the Adomian polynomials $A_i(x^2 y)$ using MATLAB script (AdomianXY2.m).

Following is the output of the code (adomian_sys1.m), these functions are plotted in Fig. 7.5, while the MATLAB code (BZ2.m) is used to present an alternative solution.

```
x = - 1.205*t^5 - 3.435*t^4 - 1.688*t^3 + 0.625*t^2 + 1.5*t + 1.0
```

```
y = 1.892*t^5 + 3.857*t^4 + 1.479*t^3 - 1.375*t^2 - 2.5*t + 1.0
```

Fig. 7.5 Belousov–
Zhabotinsky reaction
(concentration) curves

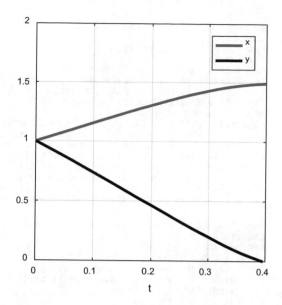

```
%adomian_sys1.m      2x2 ODE system solution via ADM
clc;clear;syms  x1 x2 x3 x4 y1 y2 y3 y4 c  d a b t
a=0,b=-3/2,c=1,d=1
x0=c+a*t;
y0=d;
A0 =x0^2*y0;
L_1A0=int(A0,t,0,t); L_1R0=int(x0,0,t);
x1=L_1A0-(b+1)*L_1R0; y1=b*L_1R0-L_1A0; x1=expand(x1);y1=expand(y1);

A1 =y1*x0^2 + 2*x1*y0*x0;
L_1A1=int(A1,t,0,t); L_1R1=int(x1,0,t);
x2=L_1A1-(b+1)*L_1R1; y2=b*L_1R1-L_1A1; x2=expand(x2);y2=expand(y2);

A2 =y2*x0^2 + 2*y1*x0*x1 + 2*x2*y0*x0 + y0*x1^2;
L_1A2=int(A2,0,t); L_1R2=int(x2,0,t);
x3=L_1A2-(b+1)*L_1R2; y3=b*L_1R2-L_1A2;x3=expand(x3);y3=expand(y3);

A3 =x1^2*y1 + x0^2*y3 + 2*x0*(x1*y2 + x2*y1) + 2*x0*x3*y0 + 2*x1*x2*y0;
L_1A3=int(A3,0,t); L_1R3=int(x3,0,t);
x4=L_1A3-(b+1)*L_1R3; y4=b*L_1R3-L_1A3; x4=expand(x4);y4=expand(y4);

A4 =x2^2*y0 + x1^2*y2 + x0^2*y4 + 2*x0*x1*y3 + 2*x0*x2*y2...
    + 2*x0*x3*y1 + 2*x0*x4*y0 + 2*x1*x2*y1 + 2*x1*x3*y0;
L_1A4=int(A4,0,t); L_1R4=int(x4,0,t);
x5=L_1A4-(b+1)*L_1R4; y5=b*L_1R4-L_1A4; x5=expand(x5);y5=expand(y5);

x=x0+x1+x2+x3+x4+x5; x=vpa(x,4)
X=ezplot(x,[0 0.4 0 2]);set(X,'linewidth',2); hold on;
y=y0+y1+y2+y3+y4+y5; y=vpa(y,4)
Y=ezplot(y,[0 0.4 0 2]);set(Y,'linewidth',2);grid; legend('x','y');
```

```
%AdomianXY2.m
clc;clear;
syms x0 y0 x1 x2 x3 x4 y1 y2 y3 y4
f=y0*x0^2
f10=diff(f,x0);f01=diff(f,y0);f11=diff(f10,y0);f20=diff(f,2,x0);
f21=diff(f20,y0);f22=diff(f20,2,y0);f12=diff(f10,2,y0);
f02=diff(f,2,y0);f30=diff(f,3,x0);f31=diff(f30,y0);f13=diff(f10,3,y0)
f03=diff(f,3,y0);f40=diff(f,4,x0);f04=diff(f,4,y0);
A0=f; A1=x1*f10+y1*f01
A2=x2*f10+ y2*f01+ (1/2)*x1^2*f20 +(1/2)*y1^2*f02+ x1*y1*f11
A3=x3*f10+ y3*f01+ x1*x2*f20+ (x1*y2+x2*y1)*f11+ y1*y2*f02...
   +(1/6)*x1^3*f30+ (1/6)*y1^3*f03+ (1/2)*x1^2*y1*f21+ (1/2)*x1*y1^2*f12
A4=x4*f10+ y4*f01+ (x2^2/2+ x1*x3)*f20+ (1/2)*x1^2*x2*f30...
+(y2^2/2+y1*y3)*f02+(x1*y3+x2*y2+x3*y1)*f11...
+(x1^2*y2/2+ x1*x2*y1)*f21+ (x1*y1*y2+ x2*y1^2/2)*f12+ (1/2)*y1^2*y2*f03...
+(1/24)*x1^4*f40+ x1^3*y1*f31/6+(1/4)*x1^2*y1^2*f22...
+(1/6)*x1*y1^3*f13+ (1/24)*y1^4*f04; A4=expand(A4)

%ode_BZ2.m      Belousov-Zhabotinsky reaction, 2x2 system
clc;clear;close;
a=0,b=-3/2;
tf=0.4; tspan=[0 tf];
ic=[1 1]; %initial conditions [x(0) y(0)]
f = @(t,x) [a+x(1)^2*x(2)-x(1)*(b+1); b*x(1)-x(1)^2*x(2)]
[t, Y]=ode45(f,tspan, ic);
x = Y(:,1); y = Y(:,2);
plot(t,Y,'linewidth',2);xlabel('t');legend('x','y');grid on;
```

Problem 7.3.4 Solve the BVP, $y'' = e^y, y(0) = y(1) = 0$ using ADM, with the following recursion scheme [39],

$$y_0 = 0$$
$$y_1 = Bx + L^{-1}A_0$$
$$y_n = L^{-1}A_{n-1}, \quad n = 2, 3, \ldots$$

and calculate the sequence of first three approximate solutions starting from $\varphi_1(x) = 0$. Determine the error of approximation for the second and third approximate series solutions in the sense of infinite norm (adomian_bvp2.m).

Solution
First three Adomian polynomials for the nonlinearity $N_y = e^y$,

$$A_0 = e^{y_0}, \quad A_1 = e^{y_0} \cdot y_1, \quad A_2 = e^{y_0}\left(\frac{y_1^2}{2} + y_2\right)$$

We calculate the solution components using the given recursion scheme as

$$y_0 = 0,$$

$$y_1 = Bx + L^{-1}A_0 = Bx + \int_0^x \int_0^x e^{\tau_0} d\tau d\tau = Bx + \frac{1}{2}x^2,$$

$$y_2 = L^{-1}A_1 = L^{-1}\left(e^{y^0}y_1\right) = \int_0^x \int_0^x (e^{\tau_0}y_1 d\tau)d\tau = \int_0^x \left(\frac{B\tau^2}{2} + \frac{\tau^3}{6}\right)d\tau = B\frac{x^3}{6} + \frac{x^4}{24}$$

$$y_3 = L^{-1}A_2 = \int_0^x \int_0^x \left[\frac{1}{2}(B^2\tau^2 + \frac{\tau^4}{4} + \frac{2B\tau^3}{2}) + \left(\frac{B\tau^3}{6} + \frac{\tau^4}{24}\right)d\tau\right] \cdot d\tau$$

$$\varphi_1(x) = y_0 = 0$$

$$\varphi_2(x) = y_0 + y_1 = Bx + \frac{x^2}{2}$$

$$\varphi_3(x) = y_0 + y_1 + y_2 = Bx + \frac{x^2}{2} + \frac{Bx^3}{6} + \frac{x^4}{24}$$

We match these approximate solutions at $x = 1$, and then find the sequence of the approximate solutions without involving any undetermined constants,

$$\varphi_2(1) = 0 = B + \frac{1}{2} \rightarrow B = -\frac{1}{2}$$

$$\varphi_2(x) = -\frac{x}{2} + \frac{x^2}{2}$$

$$\varphi_3(1) = 0 = B + \frac{1}{2} + \frac{B}{6} + \frac{1}{24} = \frac{7B}{6} + \frac{13}{24} \rightarrow B = -\frac{13}{28}$$

$$\varphi_3(x) = -\frac{13}{28}x + \frac{1}{2}x^2 - \frac{13}{168}x^3 + \frac{1}{24}x^4$$

The error of approximation for $\varphi_2(x)$ and $\varphi_3(x)$ within given boundaries are computed in the sense of infinite norm using a MATLAB script (adomian_bvp2.m) and found to be 0.01130 and 0.00073764, respectively, which shows that ADM provides rapid convergence.

```
%adomian_bvp2.m
clc;clear;
k=1.33605569;
x=0:0.1:1;
phi2=-x/2+x.^2/2;  phi3=-13*x/28 + x.^2/2 - x.^3*13/168 + x.^4/24;
yact= 2*log(k*sec(k*(2*x-1)/4))-log(2);
X2=abs(phi2-yact); X3=abs(phi3-yact);
norm2inf=norm(X2,Inf), norm3inf=norm(X3,Inf)
```

Problem 7.3.5 (*Adomian-Rach modified recursion for double decomposition method* [40, 41]) The BVPs of the form

$$Ly = Ny + f, \quad y(a) = \alpha, \quad y(b) = \beta$$

can be solved using the inverse operator,

$$L^{-1}(\cdot) = c_0 + c_1 x + I_x^2(\cdot)$$

$$I_x^2(\cdot) = \int_0^x \int_0^x (\cdot) d\tau d\tau$$

$$y = c_0 + c_1 x + I_x^2 Ny + I_x^2 f$$

$$y = \sum_{n=0}^{\infty} y_n, \quad Ny = \sum_{n=0}^{\infty} A_n, \quad c_0 = \sum_{n=0}^{\infty} c_{0,n}, \quad c_1 = \sum_{n=0}^{\infty} c_{1,n}$$

Recursion equations are

$$y_0 = c_{00} + c_{10} + I_x^2 f$$
$$y_n = c_{0n} + c_{1n} x + I_x^2 A_{n-1}, \quad n \geq 1$$

Use Adomian-Rach modified recursion for double decomposition method (including the fourth order term in the solution) to solve the BVP, $y'' = e^y, y(0) = y(1) = 0$.
Compute the error of approximation for the solution within given boundaries in the sense of infinite norm, using the exact solution,

$$y(x) = 2 \ln \left[k \sec \frac{k(2x - 1)}{4} \right] - \ln 2, \quad k = 1.33605569$$

Solution
Explicit solution for this BVP could not be found using dsolve of MATLAB.
    Since $g = 0, \to I_x^2 f = 0$,

$$y_0 = c_{00} + c_{10}x$$

$$y_n = c_{0n} + c_{1n}x + I_x^2 A_{n-1}, \quad n \geq 1$$

$$\varphi_1(0) = 0 = c_{00} + c_{10}(0) \rightarrow c_{00} = 0$$

$$\varphi_1(1) = 0 = c_{00} + c_{10}(1) \rightarrow c_{01} = 0$$

$$\varphi_1(x) = y_0 = 0$$

$$\varphi_2(x) = y_0 + y_1 = c_{01} + c_{11}x + I_x^2 A_0 = c_{01} + c_{11}x + \int_0^x \int_0^x e^{\tau_0} d\tau d\tau = c_{01} + c_{11}x + \frac{x^2}{2}$$

$$\varphi_2(0) = c_{01} = 0$$

$$\varphi_2(1) = c_{11} + \frac{1}{2} \rightarrow c_{11} = -\frac{1}{2}$$

$$y_1 = -\frac{x}{2} + \frac{x^2}{2}$$

$$y_2 = c_{02} + c_{12}x + I_x^2 A_1 = c_{02} + c_{12}x + I_x^2(e^{y_0} y_1) = c_{02} + c_{12}x + \int_0^x \int_0^x y_1 d\tau d\tau$$

$$y_2 = c_{02} + c_{12}x + \int_0^x \int_0^x \left(-\frac{\tau}{2} + \frac{\tau^2}{2} \right) d\tau d\tau = c_{02} + c_{12}x - \frac{x^3}{12} + \frac{x^4}{24}$$

$$\varphi_3(0) = 0 = c_{02} \rightarrow c_{02} = 0$$

$$\varphi_3(1) = c_{12} - \frac{1}{12} + \frac{1}{24} = 0 \rightarrow c_{12} = \frac{1}{24}$$

$$y_2 = \frac{x}{24} - \frac{x^3}{12} + \frac{x^4}{24}$$

$$y(x) = \sum_{n=0}^{\infty} y_n = 0 + \left(-\frac{x}{2} + \frac{x^2}{2} \right) + \left(\frac{x}{24} - \frac{x^3}{12} + \frac{x^4}{24} \right) + \cdots$$

$$y(x) = -\frac{11}{24}x + \frac{1}{2}x^2 - \frac{1}{12}x^3 + \frac{1}{24}x^4 - \cdots$$

The error of approximation is computed (in the sense of infinite norm) using a MATLAB script (adomian_rach1.m) as 0.0017, which shows that ADM provides rapid convergence.

```
%adomian_rach1.m
clc;clear;
k-1.33605569;
x=0:0.1:1;
phi3=-11*x/24+  x.^2/2  -  x.^3/12   + x.^4/24 ;
yact=2*log(k*sec(k*(2*x-1)/4))-log(2);
X3=abs(phi3-yact); norm3=norm(X3,Inf)
```

Problem 7.3.6 (*Duan-Rach modified decomposition method*) Consider the following second order BVP with Dirichlet boundary conditions (in operator form),

$$Ly + Ry + Ny = f, \quad y(a) = \alpha, \; y(b) = \beta \tag{7.32}$$

From the first boundary condition, the solution can be written as

$$y(x) = \alpha + y'(a)(x - a) + L^{-1}f - L^{-1}(Ry) - L^{-1}(Ny) \tag{7.33}$$

We solve (7.32) for $y'(a)$ by applying the second boundary condition,

$$y'(a) = \frac{1}{b-a}\left[\beta - \alpha - \left(L^{-1}f\right)\big|_{x=b} + \left(L^{-1}(Ry)\right)\big|_{x=b} + \left(L^{-1}(Ny)\right)\big|_{x=b}\right] \tag{7.34}$$

Substituting (7.34) into (7.33), we get

$$\begin{aligned}
y(x) = {} & \alpha + \frac{\beta - \alpha}{b - a}(x - a) - d\left(L^{-1}f\right)\big|_{x=b} + d\left(L^{-1}(Ry)\right)\big|_{x=b} \\
& + d\left(L^{-1}(Ny)\right)\big|_{x=b} + L^{-1}f - L^{-1}(Ry) - L^{-1}(Ny)
\end{aligned} \tag{7.35}$$

where $d \to d(x) = (x - a)/(b - a)$

Duan-Rach modified recursion scheme [42] is

$$\begin{aligned}
y_0 = {} & \alpha + \frac{\beta - \alpha}{b - a}(x - a) - d\left(L^{-1}f\right)\big|_{x=b} + L^{-1}f \\
y_{n+1} = {} & d\left(L^{-1}(Ry)\right)\big|_{x=b} + d\left(L^{-1}A_n\right)_{x=b} - L^{-1}(Ry) - L^{-1}(A_n), \quad n \geq 0
\end{aligned} \tag{7.36}$$

Note that evaluation of inverse operator at the boundaries allows one to get the solution without computing the values of undetermined coefficients of integration. Duan-Rach method does not require solving a system of nonlinear algebraic equations originating from the n-term approximation (which can be imaginary or multiple roots).

Solve the BVP,

$$y'' = -(y')^2, \quad y(0) = 0, \; y(1) = 1$$

Using ADM applying Duan Rach approach.

Hint: The exact solution is $y = \ln[(e - 1)x + 1]$ [43] (aduan_rach1.m).

Solution

First four Adomian Polynomials for the nonlinearity $(y')^2$ are

$$A = \left[(y_0')^2, \quad 2y_0'y_1', \quad 2y_0'y_2' + (y_1')^2, \quad 2y_0'y_3' + 2y_1'y_2'\right]$$

Duan-Rach modified recursion scheme given in Eq. (7.36) is applied to (7.32). The output list of MATLAB script (aduan_rach1.m) is given below. Solution curves of the Actual and ADM (Duan-Rach approach) for the given BVP are plotted in Fig. 7.6. It is seen that using Duan-Rach ADM provides a good approximation to the actual solution for the given BVP.

```
d = x

y0 = x

A0 = 1

y1 = - x^2/2 + x/2

A1 = 1 - 2*x

y2 = x/6 + (x^2*(2*x - 3))/6

A2 = (2*x*(2*x - 3))/3 + (x - 1/2)^2 + (2*x^2)/3 + 1/3
```

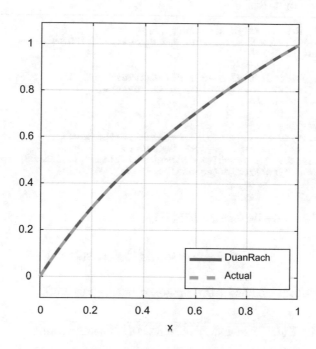

Fig. 7.6 Solution curves of the actual and ADM (Duan-Rach approach) for the BVP, $y'' = -(y')^2$, $y(0) = 0$, $y(1) = 1$

y3 = x/24 - (x^2*(6*x^2 - 12*x + 7))/24

A3 = 1/12 - (x*(6*x^2 - 12*x + 7))/6 - (x^2*(12*x - 12))/12
     - (2*x - 1)*((x*(2*x - 3))/3 + x^2/3 + 1/6)

y4 = x/120 + (x^2*(24*x^3 - 60*x^2 + 50*x - 15))/120

yDR = x^5/5 - (3*x^4)/4 + (5*x^3)/4 - (17*x^2)/12 + (103*x)/60
yact = log(1.7182818*x + 1.0)

```
%aduan_rach1.m        Ly+Ry+Ny=f,        y(a)=?,    y(b)=?
% y"+(y')^2=0,   y(0)=0, y(1)=1
clc; clear; syms  y0 y1 y2 y3 x
a=0; b=1; alpha=0; beta=1;
d=(x-a)/(b-a)
y0=alpha + d*(beta-alpha)

A0=diff(y0,1,x)^2
L_1A0=int(int(A0,x,0,x),x,0,x),  L_1A0_b=subs(L_1A0,x,b)
y1=d*(L_1A0_b)-L_1A0

A1=2*diff(y1,x)*diff(y0,x)
L_1A1=int(int(A1,x,0,x),x,0,x),  L_1A1_b=subs(L_1A1,x,b)
y2=d*(L_1A1_b)-L_1A1

A2=2*diff(y0,x)*diff(y2,x)+diff(y1,x)^2
L_1A2=int(int(A2,x,0,x),x,0,x),  L_1A2_b=subs(L_1A2,x,b)
y3=d*(L_1A2_b)-L_1A2

A3=2*diff(y0,x)*diff(y3,x)+2*diff(y1,x)*diff(y2,x)
L_1A3=int(int(A3,x,0,x),x,0,x),  L_1A3_b=subs(L_1A3,x,b)
y4=d*(L_1A3_b)-L_1A3

y=y0+y1+y2+y3+y4; %yDR=expand(y)
yDR=ezplot(y,[0 1]);set(yDR,'linewidth',2);hold on;
yact=log((exp(1)-1)*x+1); yact=vpa(yact,8)
ya=ezplot(yact,[0 1]);set(ya,'linewidth',2);set(ya,'lineStyle','--');
set(ya,'Color','g');grid;legend('DuanRach','Actual');
```

Problem 7.3.7 Solve the linear BVP [44],

$$y'' - \frac{1}{x}y' + \frac{1}{x^2}y = 0, \quad y(1) = 1, \; y(2) = 1 \tag{7.37}$$

Use Duan-Rach modified ADM approach (aduan_rach2.m).

Solution
The BVP with Dirichlet boundary conditions (in operator form),

$$Ly + Ry + Ny = f, \quad y(a) = \alpha, \; y(b) = \beta \tag{7.38}$$

In this case, there is no need to compute any Adomian Polynomials, since the ODE is linear,

$$Ly + Ry = 0, \rightarrow Ly = -Ry,$$

General Duan-Rach modified ADM recursion scheme with

$$d = (\beta - \alpha)/(b - a), \quad e(x) = (x - a)/(b - a), \quad e \rightarrow e(x),$$

$$y_0 = \alpha + d(x - a) - e\left(L^{-1}f\right)\big|_{x=b} + L^{-1}f \tag{7.39a}$$

$$y_{n+1} = e\left(L^{-1}(Ry)\right)\big|_{x=b} + e\left(L^{-1}A_n\right)_{x=b} - L^{-1}(Ry) - L^{-1}(A_n), \quad n \ge 0 \tag{7.39b}$$

from given boundary conditions, $y(a) = 1, y(b) = 1, d = 0$, becomes

$$y_0 = \alpha, \; y_{n+1} = e\left(L^{-1}(Ry)\right)\big|_{x=b} - L^{-1}(Ry), \quad n \ge 0 \tag{7.40}$$

The exact analytical solution is

$$y = x - \frac{x\ln(x)}{2\ln 2}$$

Solution curves of the exact and ADM (Duan-Rach approach) for the given BVP are plotted in Fig. 7.7. The approximate solutions lie on top of the exact solution. Thus, for comparison between the approximate solutions, their maximum absolute error functions (in the sense of infinite norm) are computed to highlight the rapid rate of convergence.

Maximum absolute error functions (infinite norms) are given below.

$$err(3terms) = 5.6058 \times 10^{-4}$$
$$err(4terms) = 5.2461 \times 10^{-5}$$
$$err(5terms) = 6.2533 \times 10^{-6}$$
$$err(6terms) = 6.9164 \times 10^{-7}$$
$$err(7terms) = 6.8429 \times 10^{-8}$$

The MATLAB script (aduan_rach2.m) is the following;

Fig. 7.7 Solution curves of the actual and modified Duan-Rach ADM for the BVP, $y'' - \frac{1}{x}y' + \frac{1}{x^2}y = 0$, $y(1) = 1$, $y(2) = 1$, truncated with four terms, $y \cong y_0 + y_1 + y_2 + y_3$, absolute max. error $= 5.2461 \times 10^{-5}$

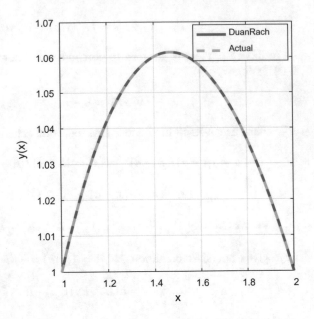

```
%aduan_rach2.m        Ly+Ry+Ny=f,         y(a)=alpha,    y(b)=beta
% y"-(y'/x)+y/x^2=0,    y(1)=1, y(2)=1
clc;clear;close;syms   y0 y1 y2 y3 x ;tic;
a=1;b=2;alpha=1;beta=1;
% note that, d=(beta-alpha)/(b-a)=0
e=(x-a)/(b-a);
Ry0=-(1/x)*diff(y0,x)+(1/x^2)*y0;
L_1Ry0=int(int(Ry0,x,a,x),x,a,x); L_1Ryb0=subs(L_1Ry0,x,b);
y1= e*L_1Ryb0 - L_1Ry0; y1=subs(y1,y0,alpha); y(2)=y1;
for j=2:6
Ry(j)=-(1/x)*diff(y(j),x)+(1/x^2)*y(j);
L_1Ry(j)=int(int(Ry(j),x,a,x),x,a,x); L_1Ryb(j)=subs(L_1Ry(j),x,b);
y(j+1)= e*L_1Ryb(j) - L_1Ry(j);
end
y(1)=alpha; Y=sum(y); yDR=expand(Y)  % approximate  solution
yact=x-(x*log(x)/(2*log(2))); yact=vpa(yact,8);%Exact solution
abserr=abs(yact-yDR); x=linspace(a,b,10); X=eval(vectorize(abserr));
maxerr=norm(X,Inf) %returns the largest element of error
Rng=[1 2 1 1.07]; % plotting ranges
yDR=ezplot(Y,Rng);set(yDR,'linewidth',2);hold on;
ya=ezplot(yact,Rng); set(ya,'linewidth',2);set(ya,'Color','g');grid;
set(ya,'lineStyle','--');legend('DuanRach','Actual');ylabel('y(x)');
toc
```

Solution using MATLAB `dsolve` function:

```
syms y(x)
eq='D2y-(Dy/x)+y/x^2=0',bc='y(1)=1,y(2)=1';
S=dsolve(eq,bc,x),simplify(S),ezplot(S,[1 2]);
```

Problem 7.3.8 Singular nonlinear BVPs arise in the modeling of many problems in engineering, and there is considerable amount of literature on the numerical treatment of such problems. Solving nonlinear two-point BVPs using classical ADM can be a computationally involved task, because it requires computation of unknown constants in a sequence of nonlinear or complicated transcendental equations, and in some cases these may not be uniquely determined. Various methods have been introduced to overcome the difficulties that occur in the classical ADM for solving nonlinear Singular BVPs. In one of these methods, Singh et al. [45] propose a recursive scheme based on the ADM for solving nonlinear singular BVP of the form

$$y'' + \frac{k}{x}y' = f(x,y), \quad y'(0) = 0, \quad ay(1) + by'(1) = c \qquad (7.41)$$

where $y \to y(x)$, a, b, c are constants. For a unique solution, the nonlinear function $f(x,y)$ and df/dy must be continuous and $df/dy \geq 0$ within given interval. This is a cylindrical problem when $k = 1$, and it becomes a spherical case when $k = 2$ [46].

In this method, we rewrite (7.41) in operator form as

$$Ly = x^k f(x,y), \quad y'(0) = 0, \quad ay(1) + by'(1) = c \qquad (7.42)$$

where $L = d/dx(x^k d/dx)$ is a linear second order differential operator. Inverse operator is proposed as a twofold integral operator,

$$L^{-1}y = \int_x^1 s^{-k} \int_0^s (\cdot) dx ds \qquad (7.43)$$

Applying (7.43) on both sides of (7.42), using boundary conditions and after performing some operations involving ADM, following recursion equations are obtained;

$$y_0 = \frac{c}{a}, \qquad (7.44a)$$

$$y_n = -\frac{b}{a} \int_0^1 x^k A_{n-1} dx - \int_x^1 s^{-k} \int_0^s x^k A_{n-1} dx ds, \quad n \geq 1 \qquad (7.44b)$$

The truncated -term series solution can then be written as

$$\varphi_n = \sum_{j=0}^{n} y_j$$

Use this method to solve the nonlinear singular BVP [47, 48] in the study of isothermal gas sphere,

$$y'' + \frac{2}{x}y' = -y^5, \quad y'(0) = 0, \quad y(1) = \sqrt{3}/2 \qquad (7.45)$$

which is also known as the Emden-Fowler equation of the first kind[2]. Exact solution of this singular nonlinear BVP is $y(x) = \sqrt{3/(x^2+1)}$ (adomian22.m), (adomian_sbvp1.m).

Solution
Explicit solution for this BVP could not be found using dsolve function.
   Following recursion equations are to be used;

$$y_0 = \sqrt{3}/2, \qquad (7.46a)$$

$$y_n = -\int_x^1 s^{-2} \int_0^s x^2 A_{n-1} dx ds, \quad n \geq 1 \qquad (7.46b)$$

Adomian polynomials for the nonlinearity y^5 are computed by MATLAB script (adomian22.m) and substituted into the code (adomian_sbvp1.m).The computed truncated 6-term series solution is listed below. Approximate and exact solutions are also displayed in Fig. 7.8.

```
y = 0.00000846*x^12 - 0.00014504*x^10 + 0.00125283*x^8 - 0.00740931*
x^6 + 0.03538366*x^4 - 0.15986519*x^2  + 0.9968
```

Maximum error of approximation always occurs at $y(0)$. Its value for truncated 6-term series solution is found to be $\varepsilon \cong 0.0032$.
   Following is the MATLAB script (adomian_sbvp1.m) used for the solution of the problem.

---

[2]Note that in the related paper by Singh et al., there is (most possibly) a typing error; The minus sign in right hand side of Eq. (7.45) (in Eq. 59 of their paper) is missing.

Fig. 7.8 Approximate and exact solutions of the BVP, $y'' + (2/x)y' = -y^5$, $y'(0) = 0$, $y(1) = \sqrt{3}/2$, using the truncated 6-term series solution

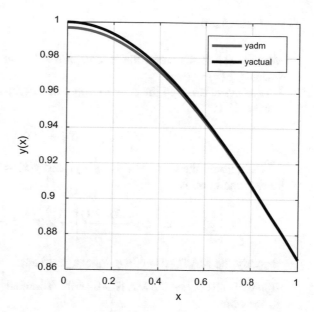

```
%adomian_sbvp1.m       y"+(k/x)*y'=f,     y'(0)=0,   a*y(1)+b*y'(1)=c
clc;clear;close;syms x s y0 y1 y2 y3 y4 y5 y6
% eq='D2y+(2/x)*(Dy)=y^5', bc='Dy(0)=0, y(1)=sqrt(3)/2';
% S=dsolve(eq,bc,x),simplify(S) % No Explicit solution
k=2;a=1;b=0;c=sqrt(3)/2;
% Adomian Polynomials for the nonlinearity N(-y^5)
A=-[ y0^5, 5*y0^4*y1, 5*y2*y0^4 + 10*y0^3*y1^2,...
    5*y3*y0^4 + 20*y2*y0^3*y1 + 10*y0^2*y1^3,...
    20*y0^3*(y2^2/2 + y1*y3) + y0*y1^4 + 5*y0^4*y4 + 30*y0^2*y1^2*y2, ...
    20*y0^3*(y1*y4 + y2*y3) + 5*y0^4*y5 + y1^5 + 60*y0^2*((y3*y1^2)/2 ...
    + (y1*y2^2)/2) + 20*y0*y1^3*y2, 60*y0^2*((y4*y1^2)/2 + y3*y1*y2...
    + y2^3/6) + 20*y0^3*(y3^2/2 + y1*y5 + y2*y4) + 5*y1^4*y2 +...
    5*y0^4*y6 + 120*y0*((y3*y1^3)/6 + (y1^2*y2^2)/4)];
Y0=c/a;
Y1=-int(s^(-k)*(int(x^k*A(1),x,0,s)),s,x,1); A(2)=subs(A(2),y1,Y1);
Y2=-int(s^(-k)*(int(x^k*A(2),x,0,s)),s,x,1);
A(3)=subs(A(3),[y1,y2],[Y1,Y2]);
Y3=-int(s^(-k)*(int(x^k*A(3),x,0,s)),s,x,1);
A(4)=subs(A(4),[y1,y2,y3],[Y1,Y2,Y3]);
Y4=-int(s^(-k)*(int(x^k*A(4),x,0,s)),s,x,1);
A(5)=subs(A(5),[y1,y2,y3,y4],[Y1,Y2,Y3,Y4]);
Y5=-int(s^(-k)*(int(x^k*A(5),x,0,s)),s,x,1);
A(6)=subs(A(6),[y1,y2,y3,y4,y5],[Y1,Y2,Y3,Y4,Y5]);
Y6=-int(s^(-k)*(int(x^k*A(6),x,0,s)),s,x,1); Y=[Y0,Y1,Y2,Y3,Y4,Y5,Y6];
for j=1:7,Y(j)=subs(Y(j),y0,Y0); y(j)=vpa(expand(Y(j)),8), end; y=sum(y)
rng=[0 1 .86 1];fi=ezplot(y,rng);set(fi,'linewidth',2); hold on;
yact=(3/(x^2+3))^(1/2);ya=ezplot(yact,rng);set(ya,'linewidth',2)
legend('yadm','yactual');ylabel('y(x)');grid;
```

7.4   Exercises

1. Write a MATLAB script file to compute Adomian polynomials for the non-linearity $F(u) = uu'$.
2. Use ADM to describe the error of approximation in the sense of infinite norm for the BVP,

$$y'' = e^y, \quad y(0) = y(1) = 0$$

including only the terms $y(x) = y_0 + y_1$ and $y(x) = y_0 + y_1 + y_2$.
The exact solution is

$$y(x) = 2\ln\left[k\sec\frac{k(2x-1)}{4}\right] - \ln 2, \quad k = 1.33605569$$

Modify the MATLAB code (adomian_bvp1.m).

3. Improve and shorten the MATLAB script (adomian_sbvp1.m) using "for loops" in the code.
4. In the solution of Problem (7.3.8), include an Adomian Polynomial computation scheme to extend the number of Adomian polynomials for the nonlinearity term of the problem. Then, compute the truncated 10-terms and 15-terms series solutions of the BVP and determine error of approximation, in each case.

   Use ADM to solve the following BVPs.

5. $y'' + (1/x)y' + y = 4 - 9x + x^2 - x^3, \quad x \in [0,1], \ y(0) = y(1) = 0$ [49, 50].
6. $y'' - 1.5y^2 = 0, \quad 0 \le x \le 1, \ y(0) = 4, \ y(1) = 1$ [51]
7. $y'' = y^3 - yy', \quad y(1) = 1/2, \ y(2) = 1/3$
   Exact solution: $y = 1/(x+1)$ [52]
8. $y'' = y^3/2, \quad y(1) = -2/3, \ y(2) = -1$
   Exact solution: $y = 2/(x-4)$ [52]
9. $(1+x^2)y'' + 4xy' + 2y = 0, \quad y(0) = 1, \ y(2) = 0.2$
   Exact solution: $y = (1+x^2)^{-1}$ (Runge function) [53]
10. $y'' + (2/x)y' + y^5 = 0, \quad y(0) = 1, \ y(1) = \sqrt{3}/2$
    Exact solution: $y = 1/(1+x^2/3)$ [54]
11. Solve the BVP, $y'' = y^3 - (1 + \sin^2 t)\sin t, \quad t \in [0,\pi], \ y(0) = y(\pi) = 0$
    Exact solution: $y(t) = \sin(t)$ [55]
12. Use ADM to solve the initial value problem [56],

$$y'' - 2e^y = 0, \ y(0) = y'(0) = 0, \quad 0 < x < 1$$

without using Taylor series for the nonlinear term. Obtain first four nonzero terms of the solution.

Answer: $x^2 + x^4/6 + 2x^6/45 + 17x^8/1260 + \cdots$

13. Use ADM to solve the following BVP [57] and plot the solution along with the analytical solution in the same figure. Determine infinite norm of error. Let $k = 0.1$.

$$y'' - \frac{yy'}{k} = 0, \quad y(0) = 1, \ y(1) = 0.$$

Exact solution: $2/\{1 + \exp[(x-1)/k]\} - 1$

14. Consider the fourth order Sturm-Liouville BVP given by Attili and Lesnic [58],

$$y^{(4)} + \lambda y = 0, \quad x \in (0,1), \quad y(0) = y'(0) = y(1) = y''(1) = 0$$

Use ADM to compute the first eigenvalue of this equation.

15. Use ADM to compute the solution of the fourth order BVP,

$$y^{(4)} + xy = -(8 + 7x + x^3)e^x, \quad y(0) = y(1) = 0, \ y'(0) = 1, \ y'(1) = -e$$

Compare their run times, under the same initial mesh values. Determine the error in computing the value of $y(0.5)$ for each solver.
Exact solution: $y(x) = x(1-x)e^x$ [59].

References

1. Abbaoui K, Cherruault Y (1994) Convergence of Adomian's method applied to nonlinear equations. Math Comput Model 20(9):69–73
2. Adomian G (1994) Solving frontier problems of physics: the decomposition method. Kluwer Academic Publishers, Boston
3. Fatoorehchi H, Abolghasemi H (2011) On calculation of Adomian polynomials by MATLAB. J Appl Comput Sci Math 11(5):85–88
4. Choi H-W, Shin J-G (2003) Symbolic implementation of the algorithm for calculating Adomian polynomials. Appl Math Comput 146:257–271
5. Duan JS (2011) Convenient analytic recurrence algorithms for the Adomian polynomials. Appl Math Comput 217(13):6337–6348
6. Babolian E, Javadi S (2004) New method for calculating Adomian polynomials. Appl Math Comput 153:253–259
7. Biazar J, Babolian E, Kember G, Nouri A, Islam R (2003) An alternate algorithm for computing Adomian polynomials in special cases. Appl Math Comput 138:523–529
8. Wazwaz A (2000) A new algorithm for calculating Adomian polynomials for nonlinear operators. Appl Math Comput 111:53–69
9. Guellal S, Cherruault Y (1994) Practical formulae for calculation of Adomians polynomials and application to the convergence of the decomposition method. Int J Biomed Comput 36:223–228

10. Chen W, Lu Z (2004) Symbolic implementation of the algorithm for calculating Adomian polynomials. Appl Math Comput 159:221–235
11. Li J-L (2009) Adomian's decomposition method and homotopy perturbation method in solving nonlinear equations. J Comput Appl Math 228:168–173
12. Kaliyappan M, Hariharan S (2015) Symbolic computation of Adomian polynomials based on Rach's rule. Br J Math Comput Sci 5(5):562–570
13. Pamuk S (2005) An application for linear and nonlinear heat equations by Adomian's decomposition method. Appl Math Comput 163:89–96
14. Adjedj B (1999) Application of the decomposition method to the understanding of HIV immune dynamics. Kybernetes 28(3):271–283
15. Bozyigit B, Yesilce Y, Catal S (2018) Free vibrations of axial-loaded beams resting on viscoelastic foundation using Adomian decomposition method and differential transformation. Eng Sci Technol Int J Jestech 21(6):1181–1193
16. Adair D, Jaeger M (2018) Vibration analysis of a uniform pre-twisted rotating Euler-Bernoulli beam using the modified Adomian decomposition method. Math Mech Solids 23(9):1345–1363
17. Moradweysi P, Ansari R, Hosseini K et al (2018) Application of modified Adomian decomposition method to pull-in instability of nano-switches using nonlocal Timoshenko beam theory. Appl Math Model 54:594–604
18. Daoud Y, Khidir AA (2018) Modified Adomian decomposition method for solving the problem of boundary layer convective heat transfer. Propul Power Res 7(3):231–237
19. Turkyilmazoglu M (2018) A reliable convergent Adomian decomposition method for heat transfer through extended surfaces. Int J Numer Methods Heat Fluid Flow 28(11):2551–2566
20. Lisboa TV, Marczak RJ (2018) Adomian decomposition method applied to anisotropic thick plates in bending. Eur J Mech A Solids 70:95–114
21. Lin Y (2018) Numerical prediction of the energy efficiency of the three-dimensional fish school using the discretized Adomian decomposition method. Results Phys 9:1677–1684
22. Alizadeh A, Effati S (2018) Modified Adomian decomposition method for solving fractional optimal control problems. Trans Inst Meas Control 40(6):2054–2061
23. Zhu Y, Chang Q, Wu S (2005) A new algorithm for calculating Adomian polynomials. Appl Math Comput 169:402–416
24. Adomian G (1988) A review of the decomposition method in applied mathematics. J Math Anal Appl 135:501–544
25. Hermann M, Saravi M (2016) Nonlinear ordinary differential equations: analytical approximation and numerical methods. Springer, New York, pp 44–60
26. Keskin AU (2019) Ordinary differential equations for engineers, problems with MATLAB Solutions. Springer, New York, p 19
27. Mathieu E (1868) Mémoire sur Le Mouvement Vibratoire d'une Membrane de forme Elliptique. J Math Pures Appl 13:137–203
28. McLachlan NW (1947) Theory and applications of Mathieu functions. Clarendon Press, Oxford, UK
29. Ruby L (1996) Applications of the Mathieu equation. Am J Phys 64(1):39–44
30. Rand RH, Ramani DV, Keith WL, Cipolla KM (2000) The quadratically damped Mathieu equation and its application to submarine dynamics. In: Control of noise and vibration: new millennium, AD-vol 61. ASME, New York, pp 39–50
31. Keskin AU (2017) Electrical circuits in biomedical engineering, problems with solutions. Springer, p 307 (problem 5.1.3)
32. Keskin AU (2019) Ordinary differential equations for engineers, problems with MATLAB solutions. Springer, pp 81–83
33. Babolian E, Biazar J, Vahidi AR (2004) Solution of a system of nonlinear equations by Adomian decomposition method. Appl Math Comput 150:847–854
34. Abboui K, Cherruault Y, Seng V (1995) Practical formulae for the calculus of multivariable Adomian polynomials. Math Comput Model 22(1):89–93
35. Seng V, Abbaoui K, Cherruault Y (1996) Adomian's polynomials for nonlinear operators. Math Comput Model 24(1):59–65

36. Duan J-S (2010) An efficient algorithm for the multivariable Adomian polynomials. Appl Math Comput 217:2456–2467
37. Keskin AU (2019) Ordinary differential equations for engineers, problems with MATLAB solutions. Springer, pp 710–713
38. Biazar J, Ayati Z (2007) An approximation to the solution of the Brusselator system by Adomian decomposition method and comparing the results with Runge-Kutta method. Int J Contemp Math Sci 2(20):983–989
39. Duan J-S, Rach R, Baleanu D, Wazwaz AM (2012) A review of the Adomian decomposition method and its applications to fractional differential equations. Commun Frac Calc 3(2):73–99
40. Adomian G, Rach R (1993) A new algorithm for matching boundary conditions in decomposition solutions. Appl Math Comput 58:61–68
41. Adomian G, Rach R (1994) Modifed decomposition solution of linear and nonlinear boundary-value problems. Nonlinear Anal 23:615–619
42. Duan JS, Rach R (2011) A new modification of the Adomian decomposition method for solving boundary value problems for higher order nonlinear differential equations. Appl Math Comput 218:4090–4118
43. Epperson JF (2007) An introduction to numerical methods and analysis. Wiley, p 405
44. Benabidallah M, Cherruault Y (2004) Application of the Adomain method for solving a class of boundary problems. Kybernetes 33(1):118–132
45. Singh R, Kumar J, Nelakanti G (2014) Approximate series solution of nonlinear singular boundary value problems arising in physiology. Sci World J (Article ID 945872). http://dx.doi.org/10.1155/2014/945872
46. Xie L-J, Zhou C-L, Xu S (2016) An effective numerical method to solve a class of nonlinear singular boundary value problems using improved differential transform method, vol 5. SpringerPlus, p 1066. https://doi.org/10.1186/s40064-016-2753-9
47. Chawla MM, Subramanian R, Sathi HL (1988) A fourth order method for a singular two-point boundary value problem". BIT Numer Math 28(1):88–97
48. Ravi Kanth ASV, Aruna K (2010) He's variational iteration method for treating nonlinear singular boundary problems. Comput Math Appl 60(3):821–829
49. Chun C, Ebaid A, Lee MY, Aly E (2012). An approach for solving singular two-point boundary value problems: analytical and numerical treatment. ANZIAM J 53(E):E21–E43
50. Cui M, Geng F (2007) Solving singular two-point boundary value problem in reproducing kernel space. J Comput Appl Math 205:6–15
51. Lesnic DA (2007) Nonlinear reaction-diffusion process using the Adomian decomposition method. Int Commn Heat Mass Trans 34:129–135
52. Matinfar M, Ghasemi M (2013) Solving BVPs with shooting method and VIMHP. J Egypt Math Soc 21:354–360
53. Fyfe DJ (1969) The use of cubic splines in the solution of two point boundary value problems. Comput J 12:188–192
54. De Hoog FR, Weiss R (1978) Collocation methods for singular BVPs, SIAM. J Numer Anal 15:198–217
55. Rández L (1992) Improving the efficiency of the multiple shooting technique. Comput Math Appl 24(7):127–132
56. Wazwaz AM (2005) Adomian decomposition method for a reliable treatment of the Bratu-type equations. Appl Math Comput 166:652–663
57. Liu Z, Yang Y, Cai Q (2019) Neural network as a function approximator and its application in solving differential equations. Appl Math Mech Engl Ed 40(2):237–248. https://doi.org/10.1007/s10483-019-2429-8
58. Attili BS, Lesnic D (2006) An efficient method for computing eigen elements of Sturm-Liouville fourth-order boundary value problems. Appl Math Comput 182:1247–1254
59. Islam S, Tirmizi IA, Ashraf S (2006) A class of methods based on non-polynomial spline functions for the solution of a special fourth-order boundary-value problems with engineering applications. Appl Math Comput 174:1169–1180

Chapter 8
Collocation, Galerkin, and Rayleigh–Ritz Methods

Weighted residual methods (WRM) (also called Petrov-Galerkin methods) provide simple and highly accurate solutions of BVPs. Collocation, Galerkin, and Rayleigh–Ritz methods are examples of the WRMs.[1] They can be used in solving the nonlinear problems of differential equations [1, 2], and involve a finite dimensional trial solutions and test (or weight) functions which can be piecewise continuous and differentiable basis polynomials. Any polynomial can be used provided it satisfies the boundary conditions. Besides spline functions [3] and Bernstein polynomials, Bernoulli polynomials and Legendre polynomials can also be used as basis functions in the approximation [4]. The basis functions are linearly independent, but they may not be orthonormal.

In the method of weighted residuals, weighted integrals of the residual are forced to be zero. A weighted residual is defined as the integral over the domain of the residual multiplied by a weight function. By choosing N weight functions, and setting these N weighted residuals to zero, we obtain N equations which we solve to determine the N unknown values of coefficients of trial polynomial.

The primary goal of this chapter is to serve as an introduction to one dimensional collocation, Galerkin and Rayleigh-Ritz methods for the solution of BVPs. Considering its wider applications in the field of Finite Elements (FEM), this chapter is focussed on Galerkin Weighted Residual Method more than the others.

8.1 Collocation Method

Problem 8.1.1 Another approach for solving BVPs is based on approximating the solution by a linear combination of specific trial functions (which are usually polynominals). These functions are linearly independent and they satisfy the

---

[1]Here, we do not deal with the other types of WRMs, such as Subdomain Method, Least-squares Method and Method of Moments.

© Springer Nature Switzerland AG 2019
A. Ü. Keskin, *Boundary Value Problems for Engineers*,
https://doi.org/10.1007/978-3-030-21080-9_8

boundary conditions. Then, the problem becomes to determine unknown coefficients of these trial functions.

In residual methods, such as the collocation method,[2] an approximate form of the solution $y(x)$ is assumed, and the residual $R(x)$ is defined by substituting the approximate solution into the given ODE. The approximate solution is chosen as the sum of a number of linearly independent trial functions. These coefficients are then chosen to minimize the residual. The residual is set equal to zero at selected locations. Then number of locations is the same as the number of unknown coefficients in the approximate solution $y(x)$.

Consider the solution of one-dimensional BVP specified by

$$y'' + qy = f, \quad y(0) = 0, \quad y(1) = Y \tag{8.1}$$

where $q = q(x)$, $f = f(x)$, [5].
The collocation method steps (applied to this BVP) are as follows:

1. Assume that the functional form of the approximate solution $y(x)$ is given by

$$y(x) = \sum_{i=1}^{n} c_i y_i(x) \tag{8.2}$$

   Choose the functional form of the trial functions $y_i(x)$. Make sure that they are linearly independent and satisfy the boundary conditions.
2. Substitute the approximate solution $y(x)$ into the ODE. Then, define the residual as

$$R(x) = y'' + qy - f = R(x, c_i) \tag{8.3}$$

3. Set $R(x, c_i) = 0$, and solve the system of residual equations for the coefficients c_i. Apply this method for the solution of (8.1) with $q = -25$, $f = 0$, $y(1) = Y = 50$.
   Plot the solution curve (collocation1.m).

Solution
In step 1, we assume that the functional form of the approximate solution is

$$y(x) = Yx + c_2 x(x - 1) + c_3 x^2(x - 1) \tag{8.4}$$

---

[2]The collocation method is used as an introduction of the concept of a residual, which leads to the Galerkin weighted residual method. This method is usually less accurate than the Galerkin or Rayleigh-Ritz Methods. If the governing ODE is known, then we apply the Galerkin (weighted residual) approach, as in fluid mechanics and heat transfer problems. The Galerkin's approach is usually more straightforward than the Rayleigh-Ritz approach.

These functions are linearly independent and satisfy the boundary conditions in (8.1).

In step 2, we substitute $y'' = 2c_2 + c_3(6x - 2)$ and Eq. (8.4) into (8.3), and get

$$R(x) = 2c_2 + c_3(6x - 2) + q[Yx + c_2x(x - 1) + c_3x^2(x - 1)] - f \qquad (8.5)$$

In the third step, $R(x)$ is set equal to zero at two arbitrary locations of $x(x = 0.25, x = 0.5)$.

Following is the edited output list of the MATLAB script (collocation1.m). Figure 8.1 displays the resulting solution curve for the given BVP.

```
y = 50*x + c2*x*(x - 1) + c3*x^2*(x - 1)

R = 2*c2-1250*x+4*c3*x+2*c3*(x - 1)-25*c2*x*(x - 1)-25*c3*x^2*(x - 1)

R1 =(107*c2)/16 + (43*c3)/64 - 625/2

R2 =(33*c2)/4 + (33*c3)/8 - 625

c2 =39.4294,   c3 =72.6564

y = 72.6564*x^3 - 33.227*x^2 + 10.5706*x

%collocation1.m    y"-q*y=f, y(0)=0, y(1)=Y=50
clc;clear;syms y(x) c2 c3
f=0; Y=50; q=-25;
y=Y*x+c2*x*(x-1)+c3*x*x*(x-1)% functional form of approx.solution
R=diff(y,2,x)+q*y-f            % Residual
R1=subs(R,x,1/4)              % Residual at x1=1/4
R2=subs(R,x,2/4)              % Residual at x1=1/2
[Sc2,Sc3] = (solve(R1==0,  R2 == 0));
Sc2=double(Sc2)
Sc3=double(Sc3)
y=subs(y,[c2,c3],[Sc2,Sc3]); y=expand(y); y=vpa(y,6)
p=ezplot(y, [0 1]);set(p,'linewidth',2); grid;
```

Problem 8.1.2 Use MATLAB "dsolve" function to obtain the analytical solution of the BVP,

$$y'' + qy = 0, \quad y(0) = 0, \quad y(1) = Y = 50, \quad q = -25,$$

then, plot this solution and the solution of the same BVP obtained by using the Collocation Method, in the same figure (dsolve_collocation1.m).

Solution

The anaytical solution is found to be

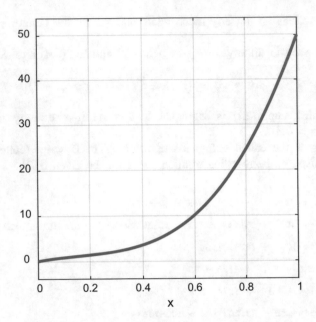

Fig. 8.1 Solution curve of the BVP, $y'' - 25y = 0, y(0) = 0, y(1) = 50$, using the collocation method

$$y(x) = \frac{50e^{-5}}{e^{10} - 1}\left(e^{5x} - e^{-5x}\right)$$

This solution and the solution of the same BVP obtained by using the Collocation Method, are shown in Fig. 8.2.

```
%dsolve_collocation1.m  dsolve & collocation, D2y-25*y=0, y(0)=0, y(1)=50
clc;clear;close;syms  y(x) c2 c3
f=0; Y=50; q=-25;
eq='D2y-25*y=0';
bc='y(0)=0, y(1)=50';
y=dsolve(eq,bc,x) ; pretty(y)
p=ezplot(y);set(p,'linewidth',2);hold on;
%collocation
y=Y*x+c2*x*(x-1)+c3*x*x*(x-1);% functional form of approx.solution
R=diff(y,2,x)+q*y-f;            % Residual
R1=subs(R,x,1/4);               % Residual at x1=1/4
R2=subs(R,x,2/4);               % Residual at x1=1/2
[Sc2,Sc3] = (solve(R1==0, R2 == 0));
Sc2=double(Sc2); Sc3=double(Sc3);
y=subs(y,[c2,c3],[Sc2,Sc3]);y=expand(y);y=vpa(y,6);
p=ezplot(y);set(p,'linewidth',2); grid;
xlim([0 1]);ylim([0 50]);legend('exact','collocation');
```

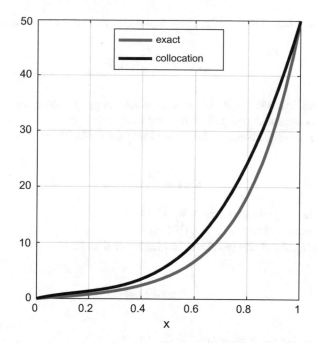

Fig. 8.2 Analytical and the Collocation solutions for the BVP, $y'' + qy = 0, y(0) = 0,$ $y(1) = Y = 50, q = -25$

8.2 Galerkin's Weighted Residual Approach

Problem 8.2.1 (Galerkin[3] Weighted Residual Approach) The Galerkin's Method is a residual method. However, unlike the Collocation Method, it is based on the integral of the residual over the domain of interest. The residual $R(x)$ is weighted over the domain of interest by multiplying $R(x)$ by weighting functions. Any functions can be used as the weighting functions, $W_j(x)$ (the use of Dirac delta function as the weighting function yields the collocation method). Then, integrating them over the range of integration, and setting the integrals of the weighted residuals equal to zero to give equations for the evaluation of the coefficients c_i of the trial functions $y_i(x)$.

Consider the solution of one-dimensional BVP specified by

$$y'' + qy = f, \quad y(0) = 0, \quad y(1) = Y \tag{8.6}$$

where $q = q(x), f = f(x)$.

The steps in the Galerkin weighted residual method are the following:

---

[3]Boris Grigorievich Galerkin (1871–1945) Russian Mathematician. Galerkin's method was published in Russian in 1915 [6]. The detailed history of this method is found in [7, 8].

1. Assume that the functional form of the approximate solution $y(x)$ is given by

$$y(x) = \sum_{i=1}^{n} c_i y_i(x) \qquad (8.7)$$

Choose the functional form of the trial functions $y_i(x)$. Make sure that they are linearly independent and satisfy the boundary conditions.

2. Substitute the approximate solution $y(x)$ into the ODE. Then, define the residual as

$$R(x) = y'' + qy - f \qquad (8.8)$$

3. Choose the weighting functions $W_j(x), j = 1, 2, \ldots$
4. Find the weighted residuals $W_j(x)R(x)$,
5. Set the integrals of the weighted residuals equal to zero;

$$\int_{x_1}^{x_2} W_j(x)R(x)dx = 0 \qquad (8.9)$$

which is called the "weak form of the BVP".

Integrate (8.9), and solve the system of weighted residual integral equations for the coefficients c_i, $i = 1, 2, \ldots n$.

Apply this method for the solution of (8.6) with $q = -25, f = 0, y(1) = Y = 50$.

Plot the solution curve (galerkin1.m). How can the accuracy of approximation be improved in this method?

Solution

In step 1, we assume that the functional form of the approximate solution is

$$y(x) = c_1 y_1(x) + c_2 y_2(x) + c_3 y_3(x) = c_1 x + c_2 x(x-1) + c_3 x^2(x-1) \qquad (8.10)$$

These functions are linearly independent and satisfy the boundary conditions in (8.6). Using given Boundary Conditions in Eq. (8.6) we obtain $c_1 = Y$, then,

$$y(x) = c_1 x + c_2 x(x-1) + c_3 x^2(x-1) \qquad (8.11)$$

In step 2, we substitute $y'' = 2c_2 + c_3(6x - 2)$ and Eq. (8.11) into (8.8), and get

$$R(x) = 2c_2 + c_3(6x - 2) + q[Yx + c_2 x(x-1) + c_3 x^2(x-1)] - f \qquad (8.12)$$

In step 3, using the form of the approximate solution $y(x)$, we choose two weighting functions, $W_2(x) = y_2(x)$, $W_3(x) = y_3(x)$ (that is the weighting functions are chosen as the trial functions),

$$W_2(x) = y_2(x) = x(x-1), \quad W_3(x) = y_3(x) = x^2(x-1)$$

In step 4, we find the weighted residuals $W_j(x)R(x)$,

$$W_2(x)R(x) = x(x-1)\left[2c_2 + c_3(6x-2) + q\left[Yx + c_2x(x-1) + c_3x^2(x-1)\right] - f\right]$$
$$W_3(x)R(x) = x^2(x-1)\left[2c_2 + c_3(6x-2) + q\left[Yx + c_2x(x-1) + c_3x^2(x-1)\right] - f\right]$$

In step 5, we set the integrals of the weighted residuals equal to zero,

$$\int_0^1 x(x-1)\left[2c_2 + c_3(6x-2) + q\left[Yx + c_2x(x-1) + c_3x^2(x-1)\right] - f\right] = 0$$

$$\int_0^1 x^2(x-1)\left[2c_2 + c_3(6x-2) + q\left[Yx + c_2x(x-1) + c_3x^2(x-1)\right] - f\right] = 0$$

Taking these two integrals, evaluating the results and collecting terms yield the solution. Following is the edited output list for the MATLAB script (galerkin1.m). Figure 8.3 displays the resulting solution curve for the given BVP.

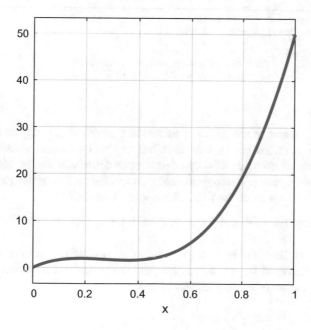

Fig. 8.3 The solution curve for the given BVP, $y'' - 25y = 0$, $y(0) = 0$, $y(1) = 50$, using Galerkin Weighted Residual Method

```
y =50*x + c2*x*(x - 1) + c3*x^2*(x - 1)

R = 2*c2 - 1250*x + 4*c3*x + 2*c3*(x-1) - 25*c2*x*(x-1) - 25*c3*x^2*(x-1)

W2R = x*(x-1)*(2*c2 - 1250*x + 4*c3*x + 2*c3*(x-1) - 25*c2*x*(x-1) -
25*c3*x^2*(x-1))

W3R =x^2*(x-1)*(2*c2-1250*x + 4*c3*x + 2*c3*(x-1) - 25*c2*x*(x-1) -
25*c3*x^2*(x-1))

I2 =625/6 - (7*c3)/12 - (7*c2)/6
I3 =125/2 - (13*c3)/35 - (7*c2)/12

c2 = 23.9872, c3 = 130.5970
```

y = 130.597*x^3 - 106.61*x^2 + 26.0128*x

```
%galerkin1.m    y"-q*y=f, y(0)=0=yi, y(1)=Y=50
clc;clear;syms y(x) c2 c3
f=0; Y=50; q=-25;
xi=0; xf=1;              % boundaries
y=Y*x+c2*x*(x-1)+c3*x*x*(x-1) % functional form of approx.solution
R=diff(y,2,x)+q*y-f   %Residual
W2=x*(x-1);            % Weight function1
W3=x*x*(x-1);          % Weight function2
W2R=W2*R              % Weighted Residual1
W3R=W3*R              % Weighted Residual2
I2 = int(W2R,xi,xf)  % Integral of Weighted Residual1
I3 = int(W3R,xi,xf)  % Integral of Weighted Residual2
[Sc2,Sc3] = (solve(I2==0,  I3 == 0));
Sc2=double(Sc2)      % c2 coefficient
Sc3=double(Sc3)      % c3 coefficient
y=subs(y,[c2,c3],[Sc2,Sc3]); y=expand(y); y=vpa(y,6)
p=ezplot(y, [0 1]); set(p,'linewidth',2); grid;
```

In order to improve the accuracy of residual methods, the degree of the polynominal trial functions can be increased, but this will also lead to increased complexity. Improved accuracy of polynominal approximations can be obtained more
easily by applying lower degree polynominals to subdomains of the global domain,
which is the basic idea of the finite element method (FEM).

Problem Solve the IVP, $y' + y = x, y(0) = 0$, using Galerkin's weighted residual
method.

Let the test function be $y = ax^2 + bx^4 = x^2(a + bx^2)$. Evaluate the solution
within the interval of $0 \leq x \leq p = 1$. Determine the error of approximation in the
sense of infinite norm (Galerkinfirstorder.m).

Solution

Error function,

$$\varepsilon_\Omega(x) = y' + y - x = ax^2 + bx^4 + 2ax + 4bx^3 - x$$

Weighted residual function:

$$\omega = \int_0^p \varepsilon_\Omega(x)\psi_{1,2}dx = 0$$

Substituting $\psi_{1,2}$ = basis function = $\{x^2, x^4\}$ in this integral, we get

$$\omega_1 = \int_0^p \left(bx^4 + ax^2 + 2ax + 4bx^3 - x\right)x^2 dx = 0$$

$$\omega_2 = \int_0^p \left(bx^4 + ax^2 + 2ax + 4bx^3 - x\right)x^4 dx = 0$$

Solution of these two integral equations for a, b yields,

$$a = 0.422252, \quad b = -0.056300$$

Hence, approximate solution of the given IVP is

$$y(x) = 0.422252x^2 - 0.056300x^4$$

Exact analytical solution,

$$y(x) = x - 1 + e^{-x}$$

Infinite norm of the error of approximation within specified interval is $n_\infty = 0.0045$.

```
%Galerkinfirstorder.m
clc;clear; syms y(x)
eq='Dy+y=x'; ic='y(0)=0';
Y=0.422252*x.^2-0.0563*x.^4;% Galerkin WR solution
S=dsolve(eq,ic,x);%Analytical solution
x=linspace(0,1);
z1 = eval(vectorize(S));
z2 = eval(vectorize(Y));
e=abs(z1-z2); % error of approximation
infnorm = norm(e,inf)% infinite norm of vector e.
```

Problem 8.2.2 Use MATLAB "dsolve" function to obtain the anaytical solution of the BVP,

$$y'' + qy = 0, \quad y(0) = 0, \quad y(1) = Y = 50, \quad q = -25,$$

then, plot this solution and the solution of the same BVP obtained by using the Galerkin Weighted Residual Method, in the same figure (dsolve_galerkin1.m).

Solution
The anaytical solution is

$$y(x) = \frac{50e^{-5}}{e^{10} - 1} \left(e^{5x} - e^{-5x} \right)$$

This solution and the solution of the same BVP obtained by using the Galerkin Weighted Residual Method are shown in Fig. 8.4.

```
% dsolve_galerkin1.m  dsolve & Galerkin, D2y-25*y=0, y(0)=0, y(1)=50
clc;clear;close; syms  y(x) c2 c3
eq='D2y-25*y=0'; bc='y(0)=0, y(1)=50';
y=dsolve(eq,bc,x) ; pretty(y)
p=ezplot(y);set(p,'linewidth',2);hold on;
%Galerkin
f=0; Y=50; q=-25; xi=0; xf=1;
y=Y*x+c2*x*(x-1)+c3*x*x*(x-1); R=diff(y,2,x)+q*y-f; %Residual
W2=x*(x-1); W3=x*x*(x-1);   % Weight functions
W2R=W2*R ; W3R=W3*R ;        % Weighted Residuals
I2=int(W2R,xi,xf); I3=int(W3R,xi,xf); [Sc2,Sc3]=(solve(I2==0,I3==0));
Sc2=double(Sc2);Sc3=double(Sc3);
y=subs(y,[c2,c3],[Sc2,Sc3]); y=expand(y); y=vpa(y,6);
p=ezplot(y);set(p,'linewidth',2); grid;
xlim([0 1]);ylim([0 50]);legend('Exact','Galerkin');
```

Problem 8.2.3 Solve the BVP, $y'' + y = -x, \quad y(0) = y(1) = 0$ using Galerkin Weighted Residual Method and plot the solution (galerkin2.m).

Solution
Assume that an approximate solution is of the form

$$y = c_0 + c_1 x + c_2 x^2 + c_3 x^3 \tag{8.13}$$

Using the boundary conditions, we find

$$y(0) = 0 = c_0 + 0 + 0 + 0 \rightarrow c_0 = 0 \tag{8.14}$$

$$y(1) = 0 = 0 + c_1 + c_2 + c_3 \rightarrow c_1 = -(c_2 + c_3) \tag{8.15}$$

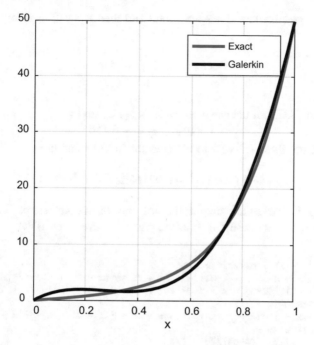

Fig. 8.4 Analytical and the Galerkin weighted residual solutions for the BVP, $y'' + qy = 0$, $y(0) = 0$, $y(1) = Y = 50$, $q = -25$

Substituting these values of c_0, c_1 into (8.13),

$$y = -(c_2 + c_3)x + c_2 x^2 + c_3 x^3 = c_2 x(x-1) + c_3 x(x^2 - 1) \tag{8.16}$$

$$y' = (2x - 1)c_2 + (3x^2 - 1)c_3$$

$$y'' = 2c_2 + 6x c_3 \tag{8.17}$$

Substituting y'', y into the given ODE, we get the residual,

$$R = 2c_2 + 6x c_3 - (c_2 + c_3)x + c_2 x^2 + c_3 x^3 + x$$
$$R = (2 + x^2 - x)c_2 + (6x - x + x^3)c_3 + x$$

Weight functions are found from Eq. (8.16) as

$$W_1 = x^2 - x, \quad W_2 = x^3 - x$$

Weighted residuals: $W_1 \cdot R, W_2 \cdot R$

Integral of weighted residuals are found and then equated to zero:

$$\int_0^1 W_1 R \, dx = 0, \quad \int_0^1 W_2 R \, dx = 0 \tag{8.18}$$

Solution of (8.18) for unknown coefficients c_2, c_3 gives

$$c_2 = -0.1924, \quad c_3 = -0.1707$$

We substitute these values into (4) and obtain the solution as

$$y = -0.170732x^3 - 0.0216802x^2 + 0.192412x$$

Following is the edited output list for the MATLAB script (galerkin2.m). Figure 8.5 displays the resulting solution curve for the given BVP.

```
y = c2*x*(x - 1) + c3*x^2*(x - 1)
R = 2*c2 + x + 4*c3*x + 2*c3*(x - 1) + c2*x*(x - 1) + c3*x^2*(x - 1)
W2R = x*(x - 1)*(2*c2 + x + 4*c3*x + 2*c3*(x - 1) + c2*x*(x - 1)
      + c3*x^2*(x - 1))
W3R = x^2*(x - 1)*(2*c2 + x + 4*c3*x + 2*c3*(x - 1) + c2*x*(x - 1)
      + c3*x^2*(x - 1))
I2 = - (3*c2)/10 - (3*c3)/20 - 1/12
I3 = - (3*c2)/20 - (13*c3)/105 - 1/20
Sc2 =   -0.1924
Sc3 =   -0.1707
y = - 0.170732*x^3 - 0.0216802*x^2 + 0.192412*x
```

```
%galerkin2.m  solve bvp, y"+q*y=-x, y(0)=0, y(1)=Y=0
clc;clear;syms y(x) c2 c3
f=-x; Y=0; q=1; xi=0; xf=1;    % boundaries
y=Y*x+c2*x*(x-1)+c3*x*x*(x-1) % functional form of approx.solution
R=diff(y,2,x)+q*y-f            % Residual
W2=x*(x-1); W3=x*x*(x-1);      % Weight functions
W2R=W2*R ,   W3R=W3*R          % Weighted Residuals
I2 = int(W2R,xi,xf),I3 = int(W3R,xi,xf)% Integrals of Weighted Residuals
[Sc2,Sc3] = (solve(I2==0,   I3 == 0));
Sc2=double(Sc2)     % c2 coefficient
Sc3=double(Sc3)     % c3 coefficient
y=subs(y,[c2,c3],[Sc2,Sc3]); y=expand(y); y=vpa(y,6)
p=ezplot(y, [0 1]); set(p,'linewidth',2); grid;
```

Problem 8.2.4 Solve the BVP, $y'' - 4x^2 = 2$, $y(0) = y(1) = 0$ using Galerkin Weighted Residual Method, with one parameter, and two parameters. Plot these solutions along with the exact solution in the same figure (dsolve_galerkin4.m).

Solution

(i) "One parameter" means that the solution is approximated by a second order polynomial,

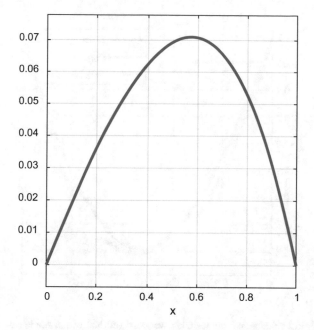

Fig. 8.5 The solution curve for the given BVP, $y'' + y = -x, y(0) = 0 = y(1) = 0$, using Galerkin Weighted Residual Method

$$y = c_0 + c_1 x + c_2 x^2 \qquad (8.19)$$

Using boundary conditions,

$$y(0) = 0 = c_0 + 0 + 0 \rightarrow c_0 = 0, \quad y(1) = 0 = 0 + c_1 + c_2 \rightarrow c_1 = -c_2,$$

Substituting these into (8.19),

$$y = c_2(x^2 - x) \qquad (8.20)$$

Residual: $R = y'' - 4x^2 - 2$
Weight function: $W = x^2 - x$

$$I = \int_0^1 W \cdot R \, dx - 0 \qquad (8.21)$$

Solution of (8.21) using MATLAB script (given below) for c_2 yields, $c_2 = 1.6$.
From (8.20), the solution of the given BVP is

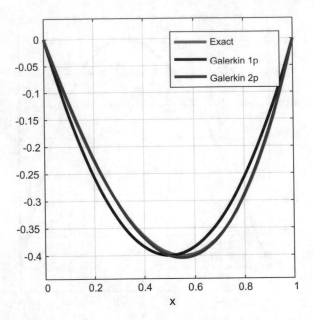

Fig. 8.6 Analytical solution and the Galerkin weighted residual with a single parameter and two parameter approximations for the BVP, $y'' - 4x^2 - 2 = 0, y(0) = 0 = y(1) = 0$

$$y(x) = 1.6(x^2 - x) \tag{8.22}$$

(ii) Two parameters approximation means that the solution is approximated by a second order polynomial,

$$y = c_0 + c_1 x + c_2 x^2 + c_3 x^3$$
$$y(0) = 0 \rightarrow c_0 = 0$$
$$y(1) = 0 = 0 + c_1 + c_2 + c_3 \rightarrow c_1 = -c_2 - c_3$$
$$y = (-c_2 - c_3)x + c_2 x^2 + c_3 x^3 = c_2 x(x - 1) + c_3 x(x^2 - 1)$$

Exact solution: $y(x) = (1/3)x(x^3 + 3x - 4)$.

Following is the edited output list for the MATLAB script (dsolve_galerkin4.m). Exact solution curve and the solution curves of the same BVP obtained by using the Galerkin Weighted Residual Method with a single parameter and two parameters are displayed in Fig. 8.6. It is apparent that the two parameters Galarkin Weighted Residual solution is a much better approximation to exact result than the one using a single parameter based Galarkin Weighted Residual Method.

```
One Parameter:
y = c2*x*(x - 1)
R = - 4*x^2 + 2*c2 - 2
W2R = -x*(x - 1)*(4*x^2 - 2*c2 + 2)
I2 = 8/15 - c2/3
Sc2 = 1.6000
y = 1.6*x^2 - 1.6*x

Two parameters:
y = c2*x*(x - 1) + c3*x*(x^2 - 1)
R = - 4*x^2 + 6*c3*x + 2*c2 - 2
W2R = x*(x - 1)*(- 4*x^2 + 6*c3*x + 2*c2 - 2)
W3R = x^2*(x - 1)*(- 4*x^2 + 6*c3*x + 2*c2 - 2)
I2 = 8/15 - c3/2 - c2/3
I3 = 3/10 - (3*c3)/10 - c2/6
Sc2 =  0.6000
Sc3 =  0.6667
y = 0.666667*x^3 + 0.6*x^2 - 1.26667*x
```

```
% dsolve_galerkin4.m  dsolve & Galerkin, y"=4*x^2+2=f, y(0)=y(1)=0
clc;clear;close;
syms  y(x) c2 c3
eq='D2y-4*x^2-2=0';
bc='y(0)=0, y(1)=0';
y=dsolve(eq,bc,x) ; y=simplify(y); pretty(y)
p=ezplot(y);set(p,'linewidth',2);hold on;
%Galerkin Weighted Residual Method 1 parameter solution
f=4*x^2+2;
xi=0; xf=1;              % boundaries
y=c2*x*(x-1)             % approx.solution
R=diff(y,2,x)-f          % Residual
W2=x*(x-1);              % Weight function1
W2R=W2*R                 % Weighted Residual1
I2 = int(W2R,xi,xf)      % Integral of Weighted Residual1
Sc2= solve(I2==0); Sc2=double(Sc2)      % c2 coefficient
y=subs(y,c2,Sc2); y=expand(y); y=vpa(y,6)
p=ezplot(y, [0 1]); set(p,'linewidth',2);hold on;
%Galerkin Weighted Residual Method  2 parameters solution
y=c2*x*(x-1)+c3*x*(x*x-1) % functional form of approx.solution
R=diff(y,2,x)-f          % Residual
W2=x*(x-1);              % Weight function1
W3=x*x*(x-1);            % Weight function2
W2R=W2*R                 % Weighted Residual1
W3R=W3*R                 % Weighted Residual2
I2 = int(W2R,xi,xf)      % Integral of Weighted Residual1
I3 = int(W3R,xi,xf)      % Integral of Weighted Residual2
[Sc2,Sc3] = (solve(I2==0,  I3 == 0));
Sc2=double(Sc2)         % c2 coefficient
Sc3=double(Sc3)         % c3 coefficient
y=subs(y,[c2,c3],[Sc2,Sc3]); y=expand(y); y=vpa(y,6)
p=ezplot(y, [0 1]); set(p,'linewidth',2);
grid;legend('Exact','Galerkin 1p','Galerkin 2p');
```

Problem 8.2.5 Deflection of a simply supported beam of length L is given by the BVP,

$$y'' = \frac{M}{EI} = f, \quad y(0) = y'\left(\frac{L}{2}\right) = 0$$

M, E, I are bending moment, elastic modulus and moment of inertia, respectively.

(a) Find the analytical solution, manually.

(b) Assume that the solution is of the form

$$y = k\sin(\pi x L) \tag{8.23}$$

Solve the given BVP to find the value of constant term, k in Eq. (1) using Galerkin Weighted Residual Method.

(c) If the approximate solution is

$$y = c_1 x + c_2 x^2 + c_3 x^3$$

and the weight functions are

$$W_1 = x, \quad W_2 = x^2, \quad W_3 = x^3$$

what are the three equations to be solved for the same Residual determined in previous part of the problem?

Solution

(a) Integrate given ODE once,

$$y' = \frac{M}{EI}x + c_1 \tag{8.24}$$

Apply boundary conditions,

$$c_1 = -\frac{ML}{2EI}$$

substitute c_1 into (2)

$$y' = \frac{M}{EI}x - \frac{ML}{2EI}$$

Integrate this result with respect to x,

$$y = \frac{M}{EI}\frac{x^2}{2} - \frac{ML}{2EI}x + c_2$$

Applying boundary condition $y(0) = 0$ will give $c_2 = 0$, then we get the solution as

$$y(x) = \frac{M}{2EI}\left(x^2 - Lx\right) \tag{8.25}$$

(b) We arrange the governing ODE for residual,

$$y'' - f = R \tag{8.26}$$

determine y'' from (1),

$$y'' = -\frac{k\pi^2}{L^2}\sin\left(\frac{\pi x}{L}\right)$$

substitute this result into (8.26),

$$R = -\frac{k\pi^2}{L^2}\sin\left(\frac{\pi x}{L}\right) - \frac{M}{EI} \tag{8.27}$$

Let the weight function be

$$w = \sin\left(\frac{\pi x}{L}\right)$$

Multiply Residual with the weight function then integrate and equate to zero,

$$I - \int\limits_0^L WRdx = \int\limits_0^L \sin\left(\frac{\pi x}{L}\right)\left[-\frac{k\pi^2}{L^2}\sin\left(\frac{\pi x}{L}\right) - \frac{M}{EI}\right]dx = 0$$

Solve for unknown parameter k,

$$k = -\frac{4ML^2}{\pi^3 EI}$$

Substituting this parameter value into (8.23) we get the solution,

$$y = -\frac{4ML^2}{\pi^3 EI} \sin\left(\frac{\pi x}{L}\right) \tag{8.28}$$

Keeping M, E, I fixed and substituting different values of x (in terms of L) into the Eqs. (8.25) and (8.28) may help to analyze the accuracy of Galerkin Weighted Residual method.

(c) Three equations to be solved for c_1, c_2, c_3 are

$$\int_0^L W_1 R dx = \int_0^L W_2 R dx = \int_0^L W_3 R dx$$

Problem 8.2.6 Write a MATLAB script to compute and plot the solutions of second order, linear BVPs with Dirichlet boundary conditions using Galerkin Weighted Residual Method. The code must accept input of variable ODE coefficients, boundary conditions and the number of basis functions (N). Then, test the script for the following BVPs.

(a) $xy'' + 2y' + 2xy = 5\cos(x), y(-1) = 5, y(4) = 3, N = 5$
(b) $xy'' + y = \sin(x)\cos(x), y(0) = 1, y(2) = 3, N = 4$

(galerkinL2.m)

Solution
(a) $y = 35.47^*x^2{}^*(x+1.0)^*(x-4.0) - 16.051^*(x+1.0)^*(x-4.0)$
      $-12.241^*x^*(x+1.0)^*(x-4.0) - 0.4^*x - 16.436^*x^3{}^*(x+1.0)^*$
      $(x-4.0) + 2.195^*x^4{}^*(x+1.0)^*(x-4.0) + 4.6$
(b) $y = x + 1.7203^*x^2{}^*(x-2.0) - 0.80329^*x^3{}^*(x-2.0)$
      $+ 0.16562^*x^4{}^*(x-2.0) - 3.1105*x^*(x-2.0) + 1.0$
Both of these solutions are plotted in Fig. 8.7.

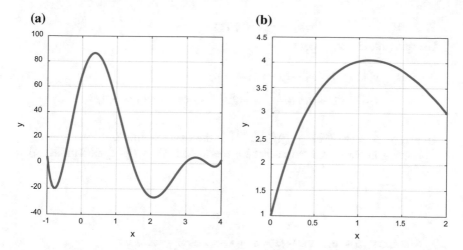

Fig. 8.7 Solution curves for the BVP. **a** $xy'' + 2y' + 2xy = 5\cos(x), y(-1) = 5, y(4) = 3, N = 5$,
b $xy'' + y = \sin(x)\cos(x), y(0) = 1, y(2) = 3, N = 4$

```
% galerkinL2.m
% Solves 2nd order,Linear BVPs with Dirichlet boundary cond.,
% employs Galerkin Weighted Residual Method
clc;clear;syms x;
% p=[x   2   2*x   5*cos(x)]  % ODE coefficients
% bc=[-1   5   4   3]          % boundary conditions,
% N=5                          % number of base functions.
p=[x   0   1   sin(x)*cos(x)]  % ODE coefficients
bc=[0   1   2   3]             % boundary conditions,
N=4                            % number of base functions.
a2=p(1);a1=p(2);a0=p(3);f=p(4);
x1=bc(1);      y1=bc(2);      x2=bc(3);      y2=bc(4);
phi=(y2-y1)/(x2-x1)*(x-x1)+y1;
dphi1=diff(phi,1); dphi2=diff(phi,2);
for i=1:N; W(1,i)=x^(i-1); end
W-(x-x1)*(x x2)*W;
dW1=diff(W,1); dW2=diff(W,2);
E1=a2*dW2+a1*dW1+a0*W;
E2=a2*dphi2+a1*dphi1+a0*phi-f;
c= -(int(W'*E1,x,x1,x2))\int(W'*E2,x,x1,x2);
y=phi+W*c; y=vpa(y,5)
del=(x2-x1)/100; x=x1:del:x2; ya=subs(y);
plot(x,ya,'linewidth',2);xlabel('x');ylabel('y');grid on
```

Problem 8.2.7 Solve the BVP, $2x^2y'' + 3xy' - y = 0, x > 0, y(1) = 0, y(4) = 1$
using Galerkin Weighted Residual method for one and two parameters polynomial
approximations. Plot these solutions and the exact solution curve (galerkinL20.m).

Solution

The ODE is a second order Euler's equation which has the general solution [9]

$$y(x) = c_1 x^{1/2} + c_2 x^{-1}, \quad x > 0$$

Using boundary conditions,

$$y(1) = 0 = c_1 + c_2 \rightarrow c_1 = -c_2,$$
$$y(4) = 1 = 2c_1 + \frac{c_2}{4} \rightarrow -2c_2 + \frac{c_2}{4} = 1 \rightarrow c_2 = -4/7,$$

Hence, $y(x) = (4/7)x^{1/2} - (4/7)(x^{-1})$ is the exact solution.
Following is the edited output list for the MATLAB script (galerkinL20.m).

```
p = [ 2*x^2, 3*x, -1, 0]

bc = 1      0     4     1

N = 1

y = 0.33333*x - 0.063898*(x - 1.0)*(x - 4.0) - 0.33333

y = - 0.063898*x^2 + 0.65282*x - 0.58892

N = 2

y = 0.33333*x - 0.13747*(x - 1.0)*(x - 4.0)

    + 0.02338*x*(x - 1.0)*(x - 4.0) - 0.33333

y = 0.02338*x^3 - 0.25437*x^2 + 1.1142*x - 0.88322
```

Solution curves of the given BVP for different number of parameters, and the exact solution curve are displayed in Fig. 8.8.

```
% galerkinL20.m
clc;clear;syms x;
p=[2*x*x  3*x  -1  0]          % ODE coefficients
bc=[1  0  4  1]                % boundary conditions,
N=2      % number of base functions.
a2=p(1);a1=p(2);a0=p(3);f=p(4);
x1=bc(1);     y1=bc(2);     x2=bc(3);     y2=bc(4);
phi=(y2-y1)/(x2-x1)*(x-x1)+y1;
dphi1=diff(phi,1); dphi2=diff(phi,2);
for i=1:N; W(1,i)=x^(i-1); end
W=(x-x1)*(x-x2)*W;
dW1=diff(W,1); dW2=diff(W,2);
E1=a2*dW2+a1*dW1+a0*W; E2=a2*dphi2+a1*dphi1+a0*phi-f;
c= -(int(W'*E1,x,x1,x2))\int(W'*E2,x,x1,x2);
y=vpa(phi+W*c,5), y=expand(y); y=vpa(y,5)
del=(x2-x1)/100; x=x1:del:x2;
ya=subs(y); plot(x,ya,'linewidth',2);hold on;
%yex=(4/7)*x.^(1/2)-(4/7)*x.^(-1);plot(x,yex,'linewidth',2);
xlabel('x');ylabel('y');grid on;legend('N=1','N=2','exact');
```

Problem 8.2.8 What are the limitations of the Traditional Galerkin Method?

Solution
With the application of Galerkin-WRM an acceptable accuracy can be achieved with less than five terms in the trial solution. However, there can be difficulties that

Fig. 8.8 The solution curves of the BVP, $2x^2y'' + 3xy' - y = 0, x > 0, y(1) = 0, y(4) = 1$, for different number of parameters, and the exact solution curve

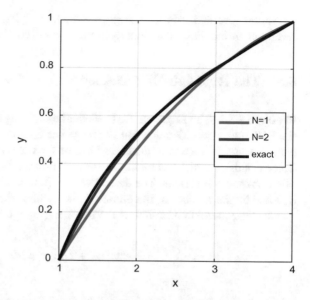

may arise if a traditional Galerkin method were applied to complex problems (the complexity might be associated with an irregular boundary, difficult boundary conditions, or complicated physical processes). Often a problem will be characterized by large gradients in a small part of the solution domain and small gradients elsewhere [10].

On the other hand, the need to achieve solutions of acceptable accuracy manifests itself as a need to retain a large number of coefficients in the trial solution.

Consider for example, the ODE, $y' - y = 0, y(0) = 1$. Application of a traditional Galerkin method with trial functions in the form of a general polynomial produces a system of linear algebraic equations that can be written as $= D$, an element of D is $d_k = 1/k, k = 1, 2, \ldots, N$; An element of M is

$$m_{kj} = \frac{j}{j+k-1} - \frac{1}{j+k}$$

for large k (=large N),

$$m_{kj} \approx \frac{j-1}{j+k}$$

and the difference in the magnitude between corresponding terms in the $(k-1)$th and kth equations will be very small, hence matrix M can be ill conditioned for large N which implies that the corresponding solution A becomes very sensitive to small changes in D or M In other words, for large N, successive weight functions produce almost linearly dependent algebraic equations. (A condition number μ quantifies the condition of the matrix M).

Another problem arises for large N if the underlying equation is nonlinear. This results in an increased computational time for solving a BVP.

8.3   The Rayleigh-Ritz Method

Problem 8.3.1 The Rayleigh–Ritz[4] Method of solving BVPs relies on the fact that the BVP under consideration can be restated as a variational problem involving the extremization of a certain quadratic functional over a function space [11].
The calculus of variations minimizes or maximizes a functional, which depends on other unknown functions. The derivative of a functional is called a variation and is denoted by the symbol δ, in contrast to the symbol d.
   Consider the extremization of the following integral:

$$I[y(x)] = \int_a^b G(x,y,y')dx \tag{8.29}$$

where $G(x,y,y')$ is the fundamental function. The square bracket notation, $[y(x)]$ emphasizes that I is a function of the function $y(x)$ and it does not depend on x. The objective of the calculus of variations is to determine the particular function $y(x)$ which extremizes the functional I [12].
   Computing the first variation δI and then setting $\delta I = 0$ gives

$$\delta I = \int_a^b \left[\frac{\partial G}{\partial y} - \frac{d}{dx}\left(\frac{\partial G}{\partial y'}\right) \right] \delta y \delta x = 0 \tag{8.30}$$

This must be satisfied for arbitrary distributions of δy, which requires

$$\frac{\partial G}{\partial y} = \frac{d}{dx}\left(\frac{\partial G}{\partial y'}\right) \tag{8.31}$$

known as the Euler equation of the calculus of variations.
   Consider the solution of one-dimensional BVP specified by

$$y'' + qy = f, \quad y(x_1) = y_1, \quad y(x_2) = y_2 \tag{8.32}$$

where $q = q(x), f = f(x)$. The problem here is to determine a functional $I[y(x)]$ whose extremum is Eq. (8.32). The fundamental function is

$$G(x,y,y') = (y')^2 - qy^2 + 2fy \tag{8.33}$$

---

[4]Walther Ritz (1878–1909) was a Swiss physicist who introduced this method [13] for problems involving elastic plates.

The functional whose extremization yields Eq. (8.32) is

$$I[y(x)] = \int_a^b G(x, y, y')dx = \int_a^b \left[(y')^2 - qy^2 + 2fy \right] dx \qquad (8.34)$$

Hence, the function which extremizes the functional I also satisfies the BVP. The steps in the Rayleigh-Ritz method[5]:

1. Find the functional $I[y(x)]$ that yields the BVP when the Euler equation is applied.
2. Assume that the functional form of the approximate solution $y(x)$ is given by

$$y(x) = \sum_{i=1}^n c_i y_i(x) \qquad (8.35)$$

Choose the functional form of the trial functions $y_i(x)$. Make sure that they are linearly independent and satisfy the boundary conditions.
3. Substitute the Eq. (8.35), into the functional $I[y(x)]$, to obtain $I[c_i]$
4. Form the partial derivatives and set them equal to zero:

$$\frac{\partial I}{\partial c_i} = 0, \quad i = 1, 2, \ldots, n \qquad (8.36)$$

5. Solve Eq. (8.36) for c_i, $i = 1, 2, \ldots, n$

Apply this method for the solution of (8.32) with $q = -25, f = 0, y(0) = 0, y(1) = Y = 50$. Plot the solution curve (ritz1.m).

Solution
The functional is

$$I[y(x)] = \int_a^b G(x, y, y')dx = \int_0^1 \left[(y')^2 - qy^2 + 2fy \right] dx$$

We assume that the functional form of the approximate solution is

$$y(x) = c_1 y_1(x) + c_2 y_2(x) + c_3 y_3(x) = c_1 x + c_2 x(x - 1) + c_3 x^2(x - 1) \qquad (8.37)$$

---

[5]Note that the use of Rayleigh–Ritz Method is restricted to symmetric BVPs. The Galerkin Method is more general and does not require symmetry of the BVP.

These functions are linearly independent and satisfy the boundary conditions. Using given Boundary Conditions, $y(0) = 0, y(1) = Y$ we obtain $c_1 = Y$, then,

$$y(x) = Yx + c_2x(x - 1) + c_3x^2(x - 1) \tag{8.38}$$

Substituting the approximate solution, Eq. (8.38), into functional gives

$$I[y(x)] = \int_0^1 G(x, y, y')dx = I[c_2, c_3] \tag{8.39}$$

We then form the partial derivatives of Eq. (8.36) with respect to c_2, c_3 and set them equal to zero.

Solution of these two simultaneous equations yields the numerical values for c_2, c_3. Finally, by substituting these two values of coefficients into Eq. (8.38) and simplifying it we obtain the approximate solution $y(x)$.

Following is the edited output list for the MATLAB script (ritz1.m). Figure 8.9 displays the resulting solution curve for the given BVP.

```
y = 50*x + c2*x*(x - 1) + c3*x^2*(x - 1)

G = 25*(50*x + c2*x*(x - 1) + c3*x^2*(x - 1))^2 + (c2*x + c2*(x - 1)
    + c3*x^2 + 2*c3*x*(x - 1) + 50)^2

Iyx = (7*c2^2)/6 + (7*c2*c3)/6 - (625*c2)/3 + (13*c3^2)/35 - 125*c3
    + 70000/3

dIdc2 = (7*c2)/3 + (7*c3)/6 - 625/3

dIdc3 = (7*c2)/6 + (26*c3)/35 - 125

Sc2 =    23.9872
Sc3 =   130.5970
```

y = 130.597*x^3 - 106.61*x^2 + 26.0128*x

```
%ritz1.m    y"-q*y=f, y(0)=0=yi, y(1)=Y=50
clc;clear;syms y(x) c2 c3
f=0; Y=50; q=-25;
xi=0; xf=1;              % boundaries
y=Y*x+c2*x*(x-1)+c3*x*x*(x-1) % functional form of approx.solution
G= diff(y,1,x).^2-q*y.^2+2*f*y
Iyx=int(G,xi,xf)
dIdc2=diff(Iyx,1,c2)
dIdc3=diff(Iyx,1,c3)
[Sc2,Sc3] = (solve(dIdc2==0,  dIdc3 == 0));
Sc2=double(Sc2)        % c2 coefficient
Sc3=double(Sc3)        % c3 coefficient
y=subs(y,[c2,c3],[Sc2,Sc3]); y=expand(y); y=vpa(y,6)
p=ezplot(y, [0 1]); set(p,'linewidth',2); grid;
```

Fig. 8.9 The solution curve
for the given BVP, $y'' -$
$25y = 0, y(0) = 0, y(1) = 50$,
using Rayleigh-Ritz Method

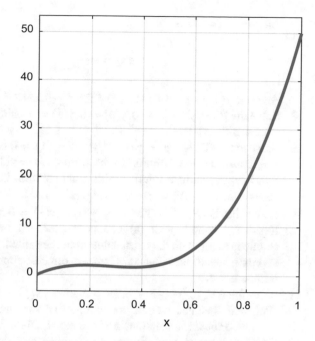

8.4   Exercises

1. What is the difference between collocation and Galerkin's Weighted Residual Method?
2. What is the general principle of "weighted residuals method"?
3. What is the principle of Rayleigh-Ritz method?
4. What are the main differences between Rayleigh-Ritz and collocation methods?
5. Describe the relation between the Galerkin method and Raylcigh-Ritz method.
6. How can the accuracy of approximation be improved in Galerkin's Weighted Residual Method?
7. Use Galerkin WRM to solve the IVP [14], $y' - y = 0, y(0) = 1, 0 \le x \le 1$, including cubic basis function. The exact solution is $y(x) = e^x$.
   Assume that a trial solution is introduced by the general polynomial,

$$y = a_0 + a_1 x + a_2 x^2 + \cdots + a_N x^N - \sum_{j=0}^{N} a_j x^j$$

where a_0 is chosen to satisfy the boundary condition ($a_0 = 1$, in this case).
Hint: Solution of a linear matrix equation $\mathbf{MA} = \mathbf{D}$ yields the coefficients of trial function, a_j.
An element of $\mathbf{D}$ is $d_k = 1/k, \quad k = 1, 2, \ldots, N$

An element of M is

$$m_{kj} = \frac{j}{j+k-1} - \frac{1}{j+k}$$

Answer: $y(x) = 1 + 1.0141x + 0.4225x^2 + 0.2817x^3$

8. Solve the BVP, $y'' - 9y = 0, y(0) = 0, y(1) = 1$ using, (a) Collocation Method, (b) Galerkin Method.

9. Solve the BVP [15], $y'' - y = -x, y(0) = y(1) = 0$ using Galerkin WRM using a second order trial solution of the form $\varphi(x) = x(1-x)$.
   Plot the approximate solution with exact solution in the same figure.
   Exact solution: $y(x) = x - (e^x - e^x)/(e^1 - e^{-1})$

10. Solve the BVP, $y'' - 25x^2 = 2$, $y(0) = y(1) = 0$ using Galerkin Weighted Residual Method, with (a) one parameter, (b) two parameters. Plot these solutions along with the exact solution in the same figure. Which method better approximates to the solution? Compute error of approximations in the sense of infinite norm.

11. Solve the BVP, $x^2 y'' + 3xy' - 2y = 0, x > 0, y(1) = 0, y(2) = 1$ using Galerkin Weighted Residual method for one and two parameters polynomial approximations. Plot these solutions and the exact solution curve in the same figure. Compute error of approximations in the sense of infinite norm.

12. Consider the linear fourth order BVP,

$$y^{(4)} + xy = -(8 + 7x + x^3)e^x, \quad 0 < x < 1$$

Subject to boundary conditions, $y(0) = y(1) = 1$, $y'(0) = 1$, $y'(1) = -e$.
Use any one of the methods described in [16–18] to solve this problem. Compute error of approximations in the sense of two's (Euler) norm.
The exact solution is $y(x) = x(1-x)e^x$.

13. Use Galerkin's WRM to solve the following BVPs. Determine maximum absolute error in each case. Indicate basis functions. [19]

   (a) $y'' = -2\sinh(x) + (1-x)\cosh(x), x \in [0, 1], y(0) = 1, y(1) = 0$.
       Exact solution: $y(x) = (1-x)\cosh(x)$
   (b) $y'' = (4x^2 - 2)y, x \in [0, 1], y'(0) = 0, y'(1) = -2e^{-1}$.
       Exact solution: $y(x) = \cosh(x^2) - \sinh(x^2)$
   (c) $y'' + y = -x, x \in [0, 2], y(0) = y'(2) = 0$.
       Exact solution: $y(x) = \sin(x)/\cos(x) - x$.

14. (Collocation-WRM) A third order BVP results in the analysis of magneto-hydrodynamic (MHD) Hamel-Jeffery flow [20–22] of an incompressible conducting viscous fluid through divergent channels,

$$y''' + Ayy' + By' = 0, \quad y(0) = 1, \quad y'(0) = y(1) = 0 \qquad (8.40)$$

$$A = 2\alpha R_e, \quad B = (4 - H_a)\alpha^2$$

where R_e and H_a are Reynolds number and Hartmann number, α is the angle unit in radians, y is dimensionless parameter. Solve the BVP assuming a trial function consisting of a cubic polynomial, $y(x) = a_0 + a_1 x + a_2 x^2 + a_3 x^3$, $\alpha = 5^o, Re = 10, Ha = 1000$.

Hint: Force trial function to satisfy the boundary conditions and substitute the trial function into (8.40), then solve the resulting equations for unknown parameters.

15. Consider the fourth order Sturm-Louville BVP given by

$$y^{(4)} + \lambda y = 0, \quad x \in (0,1), \quad y(0) = y'(0) = y(1) = y''(1) = 0$$

Use Galerkin WRM to compute the smallest eigenvalue of this BVP [23].

16. Consider an iron spherical particle which is combusted in the gaseous oxidizing medium. It is assumed that particle's temperature is only a function of time and not a function of radial coordinate and there are no interactions with other particles, while the particle burns in a quiescent, infinite ambient medium and thermo-physical properties for the particle and the ambient gaseous oxidizer are constant, the convection heat coefficient is constant.

Consequently, the nonlinear IVP can be expressed in the dimensionless form as [24]

$$Ayy' + (1 - Ak)y' + B(y^4 - p) + y - \varphi - k = 0$$

$$y(0) = 1$$

where A, k, p, φ have constant values.

Use trial function, $Y = 1 + c_1 t + c_2 t^2 + c_3 t^3 + c_4 t^4 + c_5 t^5$ to solve this problem via Collocation and Galerkin methods.

Let $A = 0.051595, B = 0.00263, \varphi = 0.98579, k = 1.2, p = 0.5$.

17. Squeezing fluid flow between two parallel plates occurs in industrial applications like moving pistons in engines and hydraulic brakes. Consequently, we solve the following nonlinear ODE in axisymmetric case [25],

$$f^{(4)} + S(-xf''' - 3f'' + ff''') = 0$$

Squeeze number S is the non-dimensional parameter under which the flow is characterized.

The boundary conditions are such that on the plates the lateral velocities are zero and the normal velocity f is equal to the velocity of the plate, that is,

$$f(0) = f''(0) = f'(1) = 0, \quad f(1) = 1$$

Solve this BVP via Galerkin WRM. Use a general polynomial of order 13 as the trial function φ and assume $S = -5$.

$$\varphi(x) = \sum_{i=0}^{13} a_i x^i$$

Answer:

$$F(x) = 1.535702961x + 0.5712593134x^3 + 0.03521724945x^5$$
$$+ 0.0002645873264x^6 + 0.001843382027x^7 + 0.003338395464x^8$$
$$+ 0.005168645285x^9 + 0.005454162841x^{10} + 0.003387547401x^{11}$$
$$+ 0.001183239062x^{12} + 0.0001799583891x^{13}$$

References

1. Finlayson BA (1972) The method of weighted residuals and variational principles with application in fluid mechanics, heat and mass transfer. Elsevier, Amsterdam
2. Hatami M (2017) Weighted residual methods, principles, modifications and applications. Academic Press, London
3. Lamnii A, Mraoui H (2013) Spline collocation method for solving boundary value problems. Int J Math Model Comput 3(1):11–23
4. Islam MS, Hossain MB (2013) On the use of piecewise standard polynomials in the numerical solutions of fourth order boundary value problems. GANIT J Bangladesh Math Soc 33:53–64
5. Hoffman JD (1992) Numerical methods for engineers and scientists, 2nd edn. Marcel Dekker, Inc., p 720
6. Galerkin BG (1915) Series occurring in various questions concerning the elastic equilibrium of rods and plates (Russian). Eng Bull (Vestn Inzh Tech) 19:897–908
7. Gander M, Wanner G (2012) From Euler, Ritz, and Galerkin to modern computing. SIAM Rev 54:627–666
8. Fletcher C (1984) Computational Galerkin methods. Springer, Berlin
9. Boyce WE, DiPrima RC (2012) Elementary differential equations and boundary value problems. Wiley, New York, p 273
10. Fletcher C (1984) Computational Galerkin methods. Springer, Berlin, p 72
11. Suli E, Mayers DF (2003) An introduction to numerical analysis (Chap. 14.2). Cambridge University Press, Cambridge
12. Hoffman JD (1992) Numerical methods for engineers and scientists (Sect. 12.2.1), 2nd edn. Marcel Dekker, Inc.
13. Ritz W (1908) Über eine neue Methode zur Lösung gewisser Variationsprobleme der mathematischen Physik. J Reine Angew Math 135:1–61
14. Fletcher C (1984) Computational Galerkin Methods. Springer, p. 5

15. Cicelia JE (2014) Solution of weighted residual problems by using Galerkin's method. Indian J Sci Technol 7(3S):52–54
16. Viswanadham KNSK, Krishna PM, Koneru RS (2010) Numerical solutions of fourth order boundary value problems by Galerkin method with quintic B-splines. Int J Nonlinear Sci 10 (2):222–230
17. Smith RC, Bogar GA, Bowers KL, Lund J (1991) The Sinc-Galerkin method for fourth-order differential equations, Siam. J. Numer. Anal. 28:760–788
18. Hossain MB, Islam MS (2014) Numerical solutions of general fourth order two point boundary value problems by Galerkin method with Legendre polynomials. Dhaka Univ J Sci 62(2):103–108
19. Khayyari OE, Lamnii A (2014) Numerical solutions of second order boundary value problem by using hyperbolic uniform B-splines of order 4. Int J Math Model Comput 4(1):25–36
20. Khan U, Ahmed N, Zaidi ZA, Mohyuddin ST (2013) On Jeffery-Hamel flows. International journal of modern Mathematical Sciences. 7:236–247
21. Abbasbandy S, Shivinian E (2012) Exact analytical solution of MHD Jeffery-Hamel flow problem. Meccanica 47:1379–1389
22. Alao S, Akinola EI, Salaudeen KA, Oderinu RA, Akinpelu FO (2017) On the solution of MHD Jeffery–Hamel flow by weighted residual method. Int J Chem Math Phys (IJCMP) 1(1) 80–85
23. Farzana H, Islam S, Bhowmik SK (2015) Computation of eigenvalues of the fourth order Sturm-Liouville BVP by Galerkin weighted residual method. Br J Math Comput Sci 9(1):73–85, Article no. BJMCS.2015.188
24. Hatami M, Ganji DD, Jafaryar M, Farkhadnia F (2014) Transient combustion analysis for iron micro-particles in a gaseous media by weighted residual methods (WRMs). Case Stud Therm Eng 4:24–31
25. Alao S, Salaudeen KA, Akinola EI, Akinboro FS, Akinpelu FO (2017) Weighted residual method for the squeezing flow between parallel walls or plates. Am Int J Res Sci Techn Eng Math 17(309):42–46

Chapter 9
Approximating Solutions of BVPs Using Cubic B-Splines

Cubic B-spline function is a piecewise third degree polynomial function, constructed from a linear combination of some recursive functions, called cubic B-spline basis. The derivation of B-spline basis and the construction of B-spline function are discussed elsewhere [1].

The use of the elementary cubic splines in solving BVPs was first reported by Bickley in 1968 [2] and followed by the works of Albasiny and Hoskins [3] and others [4, 5].

On the other hand, application of cubic B-splines for the solution of BVP was proposed by Caglar et al. in 2006 [6] which was tested on the second order linear two-point BVPs with Dirichlet boundary conditions. In [7, 8], the same procedure was implemented by an extended version of cubic B-spline.

Among a number of numerical methods used to solve two-point singular BVPs, spline methods (including cubic splines, non-polynomial splines, parametric splines and B-splines) provide efficient tools [9]. Various numerical treatments based on splines for Singular BVPs can be found in literature, such as the papers of Kanth and Bhattacharya [10], Kadalbajoo and Aggarwal [11], and also references [12–17].

Recently, there has been a considerable amount of interest on certain types of Singular BVPs that are based on the Michaelis-Menten kinetics for the steady state oxygen diffusion in cells or in the study of tumor growth and the distribution of heat sources in the human head. In this book, we will also consider problems with more general nonlinear singular nature where the removal of the singularity is achieved by using L'Hopital's rule when a Neumann boundary condition exist at the origin in addition to Robin boundary condition at the other boundary, then a cubic B-spline collocation approach is used to approximate the solution.

B-spline methods can be used for the numerical solution of linear system of BVPs[1] [18].

---

[1] In these problems, the method is more involved with intense algebraic computations of relatively large sized matrices. Therefore, they are not included in this chapter.

© Springer Nature Switzerland AG 2019
A. Ü. Keskin, *Boundary Value Problems for Engineers*,
https://doi.org/10.1007/978-3-030-21080-9_9

An advantage of B-spline methods for the solution of BVPs is that the solution can be estimated within the complete boundary interval in continuous manner, while other methods (such as the finite difference method) can be used to approximate the solution solely at the nodal (grid) points.

9.1   Solutions of Linear BVPs Using Cubic B-Splines

Problem 9.1.1 (B-Splines). Restricting our attention to equally spaced knots with $h = x_{i+1} - x$ for any $0 \leq i \leq N$, we define the uniform cubic B-spline $B_i(x)$ as

$$B_i(x) = \frac{1}{6h^3} \begin{cases} (x - x_{i-2})^3 & x_{i-2} \leq x < x_{i-1} \\ -3(x - x_{i-1})^3 + 3h(x - x_{i-1})^2 + 3h^2(x - x_{i-1}) + h^3, & x_{i-1} \leq x < x_i \\ -3(x_{i+1} - x)^3 + 3h(x_{i+1} - x)^2 + 3h^2(x_{i+1} - x) + h^3, & x_i \leq x < x_{i+1} \\ (x_{i+2} - x)^3 & x_{i+1} \leq x < x_{i+2} \end{cases}$$

What will be $B_0(x)$, if $h = 1$, $x \in [-2, 2]$?
  Evaluate $B_0(-2), B_0(-1), B_0(0), B_0(1), B_0(2)$.

Solution
$B_0(x), i = 0, x[-2, -1, 0, 1, 2]$, (Centered at $x_i = x_0 = 0$)
  $x_{0-2} \leq x \leq x_{0-1} = (-2 \leq x \leq -1)$:

$$B_0(x) = \frac{(x - x_{0-2})^3}{6(1)^3} = \frac{[x - (-2)]^3}{6} = \frac{(x+2)^3}{6}$$

$-1 \leq x \leq 0$:

$$B_0(x) = \frac{1}{6}\left\{ -3(x+1)^3 + 3(x+1)^2 + 3(x+1) + 1 = 3(x+1)\left[-(x+1)^2 + x + 2 \right] + 1 \right\}$$
$$= \frac{3(-x^3 - x^2 + x - x^2 - x + 1) + 1}{6} = \frac{-3x^3 - 6x^2 + 4}{6}$$

$0 \leq x \leq 1$:

$$B_0(x) = \frac{1}{6}\left[-3(1-x)^3 + 3(1-x)^2 + 3(1-x) + 1 \right] = \frac{1}{6}(3x^3 - 6x^2 + 4)$$

$1 \leq x \leq 2$:

$$B_0(x) = \frac{1}{6}(2 - x)^3$$

For $x \geq 3$ or $x \leq 3$, $B_0(x) = 0$
In summary; for $h = 0$, $x \in [-2, -1, 0, 1, 2]$,

$$B_0(x) = \frac{1}{6} \begin{cases} (x+2)^3 & -2 \le x \le -1 \\ -3x^3 - 6x^2 + 4 & -1 \le x \le 0 \\ 3x^3 - 6x^2 + 4 & 0 \le x \le 1 \\ (2-x)^3 & 1 \le x \le 2 \\ 0 & \text{otherwise} \end{cases}$$

Substituting x values,

$$B_0(-2) = 0, \quad B_0(-1) = \frac{1}{6}, \quad B_0(0) = \frac{4}{6}, \quad B_0(1) = \frac{1}{6}, \quad B_0(2) = 0$$

Problem 9.1.2 Consider the second order linear BVP of the form

$$a_1 y'' + a_2 y' + a_3 y = f, \quad y(a) = \alpha, y(b) = \beta \tag{9.1}$$

where $a_1 \ne 0, a_2, a_3$ and f are continuous real-valued functions.

Let Y be a cubic spline with knots Δ. The solution Y can be written as

$$Y = \sum_{i=-1}^{N+1} c_i B_i \tag{9.2}$$

where c_i are the coefficients to be computed and the B_i are cubic B-splines defined as

$$B_i = \frac{1}{6h^3} \begin{cases} (x - x_{i-2})^3 & x_{i-2} \le x < x_{i-1} \\ -3(x - x_{i-1})^3 + 3h(x - x_{i-1})^2 + 3h^2(x - x_{i-1}) + h^3, & x_{i-1} \le x < x_i \\ -3(x_{i+1} - x)^3 + 3h(x_{i+1} - x)^2 + 3h^2(x_{i+1} - x) + h^3, & x_i \le x < x_{i+1} \\ (x_{i+2} - x)^3 & x_{i+1} \le x < x_{i+2} \\ 0 & \text{otherwise} \end{cases} \tag{9.3}$$

The computation requires the solution of $(N + 1)$ linear equations with $N + 1$ unknown c_i parameters,

$$\begin{bmatrix} s_1 & s_2 & 0 & 0 & \cdots & 0 & 0 & 0 & 0 \\ p_1 & q_1 & r_1 & 0 & \cdots & 0 & 0 & 0 & 0 \\ 0 & p_2 & q_2 & r_2 & \cdots & 0 & 0 & 0 & 0 \\ \vdots & \vdots & \vdots & \vdots & \ddots & \vdots & \vdots & \vdots & \vdots \\ 0 & 0 & 0 & 0 & \cdots & p_{N-2} & q_{N-2} & r_{N-2} & 0 \\ 0 & 0 & 0 & 0 & \cdots & 0 & p_{N-1} & q_{N-1} & r_{N-1} \\ 0 & 0 & 0 & 0 & \cdots & 0 & 0 & s_3 & s_4 \end{bmatrix} \begin{bmatrix} c_0 \\ c_1 \\ c_2 \\ \vdots \\ c_{N-2} \\ c_{N-1} \\ c_N \end{bmatrix} = 6 \begin{bmatrix} z_0 \\ h^2 f(x_1) \\ h^2 f(x_2) \\ \vdots \\ h^2 f(x_{N-2}) \\ h^2 f(x_{N-1}) \\ z_N \end{bmatrix} \tag{9.4}$$

where $p_i, q_i, r_i, z_N, s_1, s_2, s_3, s_4$ are

$$p_i = 6a_1 - 3a_2h + a_3h^2 \tag{9.5a}$$

$$q_i = -12a_1 + 4a_3h^2 \tag{9.5b}$$

$$r_i = 6a_1 + 3a_2h + a_3h^2 \tag{9.5c}$$

$$s_1 = q_0 - 4p_0 \tag{9.5d}$$

$$s_2 = r_0 - p_0 \tag{9.5e}$$

$$s_3 = p_N - r_N \tag{9.5f}$$

$$s_4 = q_N - 4r_N \tag{9.5g}$$

$$z_0 = h^2 f(x_0) - \alpha p_0 \tag{9.5h}$$

$$z_N = h^2 f(x_N) - \beta r_N \tag{9.5i}$$

Since the limits of summation in (9.2) are from $-1 \le i \le N + 1$,

$$c_{-1} = 6\alpha - 4c_0 + c_1 \tag{9.6a}$$

$$c_{N+1} = 6\beta - c_{N-1} - 4c_N \tag{9.6b}$$

Corresponding values of $B_i(x)$ at $x = x_0$, $x = x_N$ are

$$B_{-1}(x_0) = B_{N+1}(x_N) = \frac{1}{6} \tag{9.7}$$

Use this information [19] and approximate the solution of the following linear constant coefficient nonhomogeneous BVP,

$$y'' + 2y + 10y = 10x, \quad y(0) = 0, \quad y(1) = 1,$$

for a step size of $h = 0.05$.
    Exact solution is

$$y_{exact}(x) = x + \frac{\cos(3x)e^{-x}}{5} - \frac{\sin(3x)e^{-x}(\cos(3) - e^1)}{5\sin(3)} - \frac{1}{5}$$

When $h = 0.05$, determine the value of $y(0.58)$ using spline interpolation function, interp1 in MATLAB (bspline_bvp1.m).

Solution

A step size of $h = 0.05$ corresponds to $N = 20$ Substituting the values of given parameters in the code (bspline_bvp1.m) provides the following edited elements of coefficients vector c_i in Eq. (9.2), (read raw-wise),

```
C =  -0.848171649274928    0.013680187891531    0.793450897708803
      1.481620016925809    2.071742578977744    2.560310212321723
      2.946555772645204    3.232211498646018    3.421230835651197
      3.519483970672061    3.534436793184888    3.474822467478676
      3.350314106247076    3.171206203640282    2.948111551787326
      2.691679359751553    2.412339248547199    2.120074738835420
      1.824228805408503    1.533343067909628    1.255031240657850
      0.995886592892615    0.761422387771690
```

Cubic B-spline approximation to the solution of the BVP as well as exact solution values at each sampling points are plotted in Fig. 9.1 and also given below:

x	CubicBspline	Exact	AbsError
0	0	0	0
0.05	0.778183965942092	0.785177337241055	0.00699337129896316
0.1	1.46527892406496	1.47819113302894	0.0129122089639715
0.15	2.05481675752642	2.07254603305502	0.0177292755286005
0.2	2.54325653348497	2.5647070714838	0.0214505379988288
0.25	2.92979080025809	2.9539018768298	0.0241110765717081
0.3	3.21610543381341	3.24187661331904	0.0257706795056287
0.35	3.40610313532048	3.43261241406588	0.0265092787454049
0.4	3.50560058525405	3.53202295312263	0.0264223678685727
0.45	3.52200893514838	3.54762546368212	0.0256165285337424
0.5	3.46400679489111	3.48821196943455	0.0242051745434408
0.55	3.34121418268454	3.36351878720575	0.0223046045212119
0.6	3.16387507876592	3.18390551440936	0.0200304356434353
0.65	2.94255529509019	2.96004976749556	0.0174944724053692
0.7	2.68786137322346	2.70266341974569	0.0148020465222385
0.75	2.4101851821293	2.42223502931495	0.012049847185656
0.8	2.11947783848829	2.12880208024133	0.00932424535843568
0.85	1.82505550472984	1.83175560659804	0.00670010186819803
0.9	1.53543871961748	1.53967875656938	0.00424003695190711
0.95	1.25822577057227	1.26021989931093	0.00199412873865779
1	1	1	0

```
infNorm = 0.026509278745405
```

When $h = 0.05$, the value of $y(0.58)$ is computed using spline interpolation function, interp1 of MATLAB as $y(0.58) = 3.24068344$.

```
0.55    3.34121418268454
0.58    3.24068344568957
0.60    3.16387507876592
```

Following is the MATLAB script (bspline_bvp1.m) for the computation of approximate solution of this problem, using B-splines.

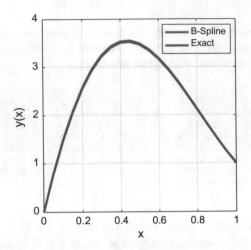

Fig. 9.1 Cubic B-spline approximation to the solution of the BVP, $y'' + 2y' + 10y = 10x$, $y(0) = 0$, $y(1) = 1$, as well as exact solution ($N = 20$). Infinite norm is computed as 0.02651

```
%bspline_bvp1.m   linear nonhom. const. coeff. BVP,   a1*y"+a2*y'+a3*y=f
clc;clear;format long
a=0; b=1; alpha=0; beta=1; a1=1; a2=2; a3=10; N=20; f=@(x) 10*x ;
h=(b-a)/N, x=linspace(a, b, N+1);
p0=6*a1-3*a2*h+a3*h^2; q0=-12*a1+4*a3*h^2; r0=6*a1+3*a2*h+a3*h^2;
pN=p0;qN=q0;rN=r0;  % Constant coefficient type of ODE!
s1 = q0 - 4*p0;   s2 = r0 - p0;   s3 = pN - rN;  s4=qN-4*rN;
z0=h^2* f(x(1))-alpha*p0;   zN=h^2* f(x(N+1))-beta*rN;
Z=zeros(size(x)); Z(1)=z0; Z(N+1)=zN;% Z vector setting
for j=2:N; Z(j)=h^2*f(x(j)); end;
Z=6*Z';
A=zeros(N+1); A(1,1)=s1; A(1,2)=s2; A(N+1,N)=s3; A(N+1,N+1)=s4;
for j=2:N
A(j,j-1)=p0; A(j,j)=q0; A(j,j+1)=r0; end;
A;
c=A\Z ;%sub-solution vector
c_1=6*alpha-4*c(1)-c(2) ; %c_1
cN2=6*beta-4*c(N+1)-c(N); %c_N+2
C=[c_1 c' cN2]'
B=[1/6  4/6  1/6  0];
for j=1:N;    D=[C(j) C(j+1) C(j+2) C(j+3)]; Y(j)= dot(D,B); end;
Y=[Y beta]';
U = x + (cos(3*x).*exp(-x))/5 - ...
(sin(3*x).*exp(-x)*(cos(3) - exp(1)))/(5*sin(3)) - 1/5;% Exact solution
plot(x,Y,x,U,'linewidth',2);grid;xlabel('x');ylabel('y(x)');
legend('B-Spline','Exact');
err=abs(Y-U'); infNorm=norm(err,Inf)
x=x'; CubicBspline=Y; Exact=U'; AbsError=err;
T = table(x, CubicBspline, Exact, AbsError)
vY = interp1(x,Y,0.58,'spline')% Spline interpolation
```

Problem 9.1.3 Use cubic B-splines to find an approximate the solution of the following linear, constant coefficient, nonhomoheneous BVP,

$$y'' + y + y = 10 \sin(\pi x), \quad y(0) = 4, \quad y(1.5) = 0,$$

for equal number of samples, $N = 20$ within the interval $0 \le x \le 1.5$. Compute analytical solution using dsolve function, then plot exact and approximate solutions in the same figure. Determine the infinite norm of this approximation (bspline_bvp2.m).

Solution
Substituting the values of given parameters in the code (bspline_bvp2.m) provides the elements of coefficients vector c_i in the following equation,

$$Y = \sum_{i=-1}^{N+1} c_i B_i$$

where the B_i are cubic B-splines.
Read raw-wise, from c_{-1} to c_{N+1},

```
C =   4.627277943374607    3.996008805568603    3.388686834350982
      2.819517458850477    2.300774693585880    1.842213983054474
      1.450616190916100    1.129488063095737    0.878935880760865
      0.695719478738553    0.573483864799656    0.503155884734571
      0.473484278399241    0.471693565882196    0.484215927637106
      0.497460944737295    0.498580982957990    0.476190252387856
      0.420998139145800    0.326322145935482    0.188452431807237
      0.006848141075522   -0.215844996109323
```

Cubic B-spline approximation to the solution of the BVP as well as the exact solution are plotted in Fig. 9.2. Infinite norm of the approximation is computed as

```
infNorm = 0.005560566825077
```

Following is the MATLAB script (bspline_bvp2.m) for the computation of approximate solution of this problem, using B-splines.

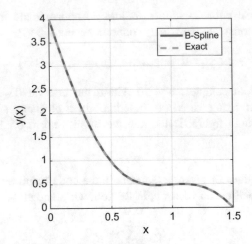

Fig. 9.2 Cubic B-spline approximation to the solution of the BVP, $y'' + y' + y = 10\sin(\pi x)$, $y(0) = 4$, $y(1.5) = 0$, as well as the exact solution (N = 20). Infinite norm is computed as 0.00556

```
%bspline_bvp2.m  linear nonhom. const. coeff. BVP,  a1*y"+a2*y'+a3*y=f
clc;clear;format long
a=0; b=1.5; alpha=4; beta=0; a1=1; a2=1; a3=1; N=20; f=@(x) 10*sin(pi*x);
h=(b-a)/N ; x=linspace(a, b, N+1);
p0=6*a1-3*a2*h+a3*h^2; q0=-12*a1+4*a3*h^2; r0=6*a1+3*a2*h+a3*h^2;
pN=p0;qN=q0;rN=r0;  % Constant coefficient type of ODE!
s1 = q0 - 4*p0; s2 = r0 - p0; s3 = pN - rN; s4=qN-4*rN;
z0=h^2* f(x(1))-alpha*p0;  zN=h^2* f(x(N+1))-beta*rN;
Z=zeros(size(x)); Z(1)=z0; Z(N+1)=zN;% Z vector setting
for j=2:N; Z(j)=h^2*f(x(j)); end;
Z=6*Z';
A=zeros(N+1); A(1,1)=s1; A(1,2)=s2; A(N+1,N)=s3; A(N+1,N+1)=s4;
for j=2:N; A(j,j-1)=p0; A(j,j)=q0; A(j,j+1)=r0; end;
A; c=A\Z ;%sub-solution vector
c_1=6*alpha-4*c(1)-c(2) ; %c_1
cN2=6*beta-4*c(N+1)-c(N); %c_N+2
C=[c_1 c' cN2]'
B=[1/6  4/6  1/6  0];
for j=1:N ;    D=[C(j) C(j+1) C(j+2) C(j+3)]; Y(j)= dot(D,B); end ;
Y=[Y beta]';
syms  U(x) % Exact solution
eq='D2y+Dy+y=10*sin(pi*x)'; bc='y(0)=4, y(1.5)=0';
U=dsolve(eq,bc,x) ; x = linspace(a, b, N+1); U=eval(vectorize(U));
plot(x,Y,x,U,'g--','linewidth',2);grid;xlabel('x');ylabel('y(x)');
legend('B-Spline','Exact');
err=abs(Y-U'); infNorm=norm(err,Inf)
```

Problem 9.1.4 Use a B-splines method to solve the following linear, nonhomogeneous, variable coefficient BVP,

$$y'' = -(4/x)y' - (2/x^2)y + (2/x^2)\ln x, \quad y(1) = 1/2, \quad y(2) = \ln 2,$$

for equal number of samples, $N = 20$, within the interval $1 \le x \le 2$. Compute analytical solution using dsolve function, then plot exact and approximate solutions in the same figure. Determine the infinite norm of this approximation (bspline_bvp3.m).

Solution

Substituting the values of given parameters in the code (bspline_bvp3.m) provides the elements of coefficients vector c_i in the following equation,

$$Y = \sum_{i=-1}^{N+1} c_i B_i$$

where the B_i are cubic B-splines.

Read raw-wise, from c_{-1} to c_{N+1} (using short format for space saving reasons),

```
C = 0.4455     0.5021     0.5461     0.5804     0.6072     0.6280     0.6442
    0.6568     0.6665     0.6739     0.6795     0.6838     0.6869     0.6891
    0.6907     0.6918     0.6925     0.6929     0.6931     0.6932     0.6932
    0.6931     0.6931

Y = 0.5000     0.5445     0.5792     0.6062     0.6272     0.6436     0.6563
    0.6661     0.6736     0.6793     0.6836     0.6867     0.6890     0.6906
    0.6917     0.6924     0.6928     0.6931     0.6932     0.6932     0.6931

infNorm =    4.8370e-04
```

The exact solution is $y = (4/x) - (2/x^2) + \ln x - (3/2)$ [20].

Cubic B-spline approximation to the solution of the BVP as well as the exact solution are plotted in Fig. 9.3. Infinite norm of the approximation is computed as 0.0004837.

Following is the MATLAB script (bspline_bvp3.m) for the computation of approximate solution of this nonhomogeneous BVP with variable coefficients, using B-splines.

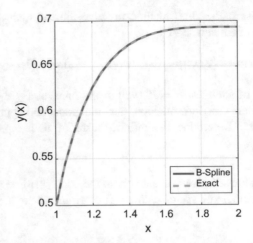

Fig. 9.3 Cubic B-spline approximation to the solution of the BVP, $y'' + (4/x)y' + (2/x^2)y = (2/x^2)\ln x, y(1) = 1/2, y(2) = \ln 2$, as well as the exact solution ($N = 20$). Infinite norm is computed as 0.0004837

```
%bspline_bvp3.m  linear nonhom. Var.coeff. BVP,  a1*y"+a2*y'+a3*y=f
clc;clear;close;
a=1; b=2; alpha=1/2; beta=log(2);
N=20; f=@(x) (2/x^2)*log(x); a1=@(x) 1 ; a2=@(x) 4/x ; a3=@(x) 2/x^2 ;
h=(b-a)/N;
x=linspace(a, b, N+1);
for j=1:N+1
p(j)=6*a1(x(j)) - 3*a2(x(j))*h + a3(x(j))*h^2;
q(j)=-12*a1(x(j)) + 4*a3(x(j))*h^2;
r(j)=6*a1(x(j)) + 3*a2(x(j))*h + a3(x(j))*h^2;
end
s1 = q(1) - 4*p(1); s2 = r(1) - p(1);
s3 = p(N+1) - r(N+1); s4=q(N+1)-4*r(N+1);
z0=h^2* f(x(1))-alpha*p(1); zN=h^2* f(x(N+1))-beta*r(N+1);
Z=zeros(size(x)); Z(1)=z0; Z(N+1)=zN; % Z vector setting
for j=2:N; Z(j)=h^2*f(x(j)); end; Z=6*Z';
A=zeros(N+1); A(1,1)=s1; A(1,2)=s2; A(N+1,N)=s3; A(N+1,N+1)=s4;
for j=2:N; A(j,j-1)=p(j); A(j,j)=q(j); A(j,j+1)=r(j); end;
A;                      % System matrix
c=A\Z;                  % sub-solution for c coefficients
c_1=6*alpha-4*c(1)-c(2) ; % c_1
cN2=6*beta-4*c(N+1)-c(N); % c_N+2
C=[c_1 c' cN2]'         % c coefficients
B=[1/6  4/6  1/6  0];   % B-spline basis
for j=1:N; D=[C(j) C(j+1) C(j+2) C(j+3)]; Y(j)= dot(D,B); end;
Y=[Y beta]'             % Solution vector
syms  U(x)              % Start Exact solution
eq='D2y+(4/x)*Dy+(2/x^2)*y=(2/x^2)*log(x)'; bc='y(1)=1/2, y(2)=log(2)';
U=dsolve(eq,bc,x) ; %ezplot(U,[a,b])
x = linspace(a, b, N+1); U=eval(vectorize(U));
plot(x,Y,x,U,'g--','linewidth',2);grid;xlabel('x');ylabel('y(x)');
legend('B-Spline','Exact');
err=abs(Y-U'); infNorm=norm(err,Inf)
```

Problem 9.1.5 Use cubic B-spline approach to solve the linear, homogeneous, variable coefficient BVP [21],

$$y'' - \frac{1}{x}y' + \frac{1}{x^2}y = 0, \quad y(1) = 1, \, y(2) = 1$$

for equal number of samples, $N = 20$, within the interval $1 \leq x \leq 2$. Compute analytical solution using dsolve function, then plot exact and approximate solutions in the same figure. Determine the infinite norm of this approximation (bspline_bvp4.m).

Solution

The exact analytical solution is

$$y = x - \frac{x\ln(x)}{2\ln2}$$

Substituting the values of given parameters in the code (bspline_bvp3.m) provides the elements of coefficients vector c_i in the following equation,

$$Y = \sum_{i=-1}^{N+1} c_i B_i$$

where the B_i are cubic B-splines.

Read raw-wise, from c_{-1} to c_{N+1} (using short format without loss of generality),

```
C = 0.9855    1.0003    1.0133    1.0247    1.0343    1.0424    1.0490
    1.0542    1.0580    1.0604    1.0616    1.0615    1.0602    1.0577
    1.0542    1.0495    1.0437    1.0370    1.0292    1.0205    1.0108
    1.0002    0.9886


Y = 1.0000    1.0130    1.0244    1.0341    1.0422    1.0488    1.0540
    1.0578    1.0602    1.0614    1.0613    1.0600    1.0576    1.0540
    1.0493    1.0436    1.0368    1.0290    1.0203    1.0106    1.0000

infNorm = 1.3336e-05
```

Solution curves of the exact and cubic B-spline approach for the given BVP are plotted in Fig. 9.4. The approximate solution lie on top of the exact solution. Thus, for comparison between the approximate and exact solutions, their maximum absolute error (in the sense of infinite norm) is computed to highlight the rate of convergence, which is found to be 0.000013336.

Following is the MATLAB script (bspline_bvp4.m) for the computation of approximate solution of this homogeneous BVP with variable coefficients, using cubic B-splines.

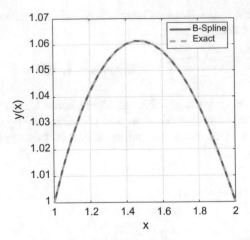

Fig. 9.4 Solution curves of the analytical and cubic B-spline approach for the given BVP, $y'' - (1/x)y' + (1/x^2)y = 0$, $y(1) = 1$, $y(2) = 1$, for equal number of grids, $N = 20$ Infinite norm of the approximation is computed as 0.000013336

Note that this code is more general than earlier scripts (bspline_bvp1.m) and (bspline_bvp2.m), therefore, it can be used for nonhomogeneous BVPs with constant coefficients, as well.

```
%bspline_bvp4.m  linear hom. Var.coeff. BVP,  a1*y"+a2*y'+a3*y=f
clc;clear;close;
a=1; b=2; alpha=1; beta=1; N=20;
f=@(x) 0; a1=@(x) 1 ; a2=@(x) -1/x ; a3=@(x) 1/x^2 ;
h=(b-a)/N; x=linspace(a, b, N+1);
for j=1:N+1
p(j)=6*a1(x(j)) - 3*a2(x(j))*h + a3(x(j))*h^2;
q(j)=-12*a1(x(j)) + 4*a3(x(j))*h^2;
r(j)=6*a1(x(j)) + 3*a2(x(j))*h + a3(x(j))*h^2;
end
s1 = q(1) - 4*p(1); s2 = r(1) - p(1);
s3 = p(N+1) - r(N+1); s4=q(N+1)-4*r(N+1);
z0=h^2* f(x(1))-alpha*p(1); zN=h^2* f(x(N+1))-beta*r(N+1);
Z=zeros(size(x)); Z(1)=z0; Z(N+1)=zN; % Z vector setting
for j=2:N; Z(j)=h^2*f(x(j)); end; Z=6*Z';
A=zeros(N+1); A(1,1)=s1; A(1,2)=s2; A(N+1,N)=s3; A(N+1,N+1)=s4;
for j=2:N; A(j,j-1)=p(j); A(j,j)=q(j); A(j,j+1)=r(j); end;
A;                        % System matrix
c=A\Z;                    % sub-solution for c coefficients
c_1=6*alpha-4*c(1)-c(2) ; % c_1
cN2=6*beta-4*c(N+1)-c(N); % c_N+2
C=[c_1 c' cN2]'           % c coefficients
B=[1/6  4/6  1/6  0];     % B-spline basis
for j=1:N; D=[C(j) C(j+1) C(j+2) C(j+3)]; Y(j)= dot(D,B); end;
Y=[Y beta]'               % Solution vector
syms  U(x)                % Start Exact solution
eq='D2y-(1/x)*Dy+(1/x^2)*y=0'; bc='y(1)=1, y(2)=1';
U=dsolve(eq,bc,x) ; %ezplot(U,[a,b])
x = linspace(a, b, N+1); U=eval(vectorize(U));
plot(x,Y,x,U,'g--','linewidth',2);grid;xlabel('x');ylabel('y(x)');
legend('B-Spline','Exact'); err=abs(Y-U'); infNorm=norm(err,Inf)
```

9.2 Approximating Solutions for Singular BVPs

Problem 9.2.1 Consider the second-order singularly perturbed BVP of the form [22, 23],

$$\varepsilon y'' - y = \cos^2(\pi x) + 2\varepsilon\pi^2\cos(2\pi x), \; y(0) = 0, \; y(1) = 0$$

Use cubic B-spline method to compute infinite norm measures for $\varepsilon = 0.1$, $\varepsilon = 0.01$, $\varepsilon = 0.001$. Plot the solution curve and its approximation using cubic B-splines in the same figure when $\varepsilon = 0.01$ and the number of nodes $N = 50$ (bspline_bvp22.m).

Solution
A second-order singularly perturbed BVP is of the form

$$\varepsilon y'' = p(x)y' + q(x)y + f(x), \; y(a) = \alpha, \; y(b) = \beta \tag{9.8}$$

where $p(x), q(x), f(x)$ are smooth, bounded, real functions and ε is a parameter such that, $0 < \varepsilon \ll 1$. It is known that (9.8) exhibits boundary layers at one or both ends of the interval depending on the choice of the function $p(x)$.

In this case, since $p(x) = 0$ and $q(x) = 1 > 0$, the boundary layer exists at both ends.

For $N = 50$, perturbation parameters versus corresponding computed infinite norm measures are as given below.

$$\varepsilon = 0.1, \quad \|y - Y\|_\infty = 0.000827$$
$$\varepsilon = 0.01, \quad \|y - Y\|_\infty = 0.000535$$
$$\varepsilon = 0.001, \quad \|y - Y\|_\infty = 0.006145$$

When $\varepsilon = 0.01$, the solution curve and its approximation using cubic B-splines are displayed in Fig. 9.5.

Following is the MATLAB script (bspline_bvp22.m) for the computation of approximate solution of this problem, using B-splines.

```
%bspline_bvp22.m  linear nonhom. const. coeff. BVP,  a1*y"+a2*y'+a3*y=f
clc;clear;format long
eps=1e-2;
a=0; b=1; alpha=0; beta=0; a1=eps; a2=0; a3=-1; N=50;
f=@(x) (cos(pi*x))^2 + 2*eps*pi^2*cos(2*pi*x);
h=(b-a)/N
x=linspace(a, b, N+1);
p0=6*a1-3*a2*h+a3*h^2;
q0=-12*a1+4*a3*h^2;
r0=6*a1+3*a2*h+a3*h^2;
pN=p0;qN=q0;rN=r0;  % Constant coefficient type of ODE!
s1 = q0 - 4*p0; s2 = r0 - p0; s3 = pN - rN; s4=qN-4*rN;
z0=h^2* f(x(1))-alpha*p0;  zN=h^2* f(x(N+1))-beta*rN;
Z=zeros(size(x)); Z(1)=z0; Z(N+1)=zN;% Z vector setting
for j=2:N; Z(j)=h^2*f(x(j)); end;
Z=6*Z';
A=zeros(N+1); A(1,1)=s1; A(1,2)=s2; A(N+1,N)=s3; A(N+1,N+1)=s4;
for j=2:N
A(j,j-1)=p0; A(j,j)=q0; A(j,j+1)=r0; end;
A;
c=A\Z ;%sub-solution vector
c_1=6*alpha-4*c(1)-c(2) ; %c_1
cN2=6*beta-4*c(N+1)-c(N); %c_N+2
C=[c_1 c' cN2]';
B=[1/6 4/6 1/6 0];
for j=1:N
    D=[C(j) C(j+1) C(j+2) C(j+3)];
Y(j)= dot(D,B);
end
Y=[Y beta]';
syms  U(x) % Exact solution
eq='0.01*D2y-y=(cos(pi*x))^2 + 2*0.01*pi^2*cos(2*pi*x)';
bc='y(0)=0, y(1)=0';
U=dsolve(eq,bc,x) ;  % ezplot(U,[0 1]);
x = linspace(a, b, N+1); U=eval(vectorize(U));
plot(x,Y,x,U,'g--','linewidth',2);grid;xlabel('x');ylabel('y(x)');
legend('B-Spline','Exact');
err=abs(Y-U'); infNorm=norm(err,Inf)
```

Problem 9.2.2 (Singular BVP). Consider a nonlinear singular BVP of the form

$$y'' + Qy' = f(x,y), \quad 0<x\le 1, \quad y'(0) = 0, \quad \alpha y(1) + \beta y'(1) = \gamma \qquad (9.9)$$

where $Q = p'/p$ and $p = x^b g$, $\alpha > 0$, $\beta \ge 0$, $\gamma \in \mathbb{R}$,

The existence and uniqueness of (9.9) is established for given boundary conditions with $b \ge 0$, provided that xp'/p is analytic within given interval of definition [24].

For the simplification of the following study, we restrict the variable coefficient function as $p = x^b$, in other words we let $g \to g(x)$ be a constant, which implies a zero derivative value.[2]

---

[2]An analogous study is described in [25].

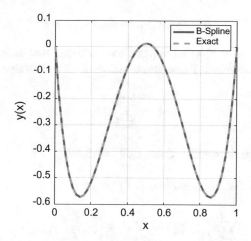

Fig. 9.5 Cubic B-spline approximation to the solution of the BVP, $\varepsilon y'' - y = \cos^2(\pi x) + 2\varepsilon\pi^2\cos(2\pi x)$, $y(0) = 0$, $y(1) = 0$, and the exact solution curve. Here, $\varepsilon = 0.01$, $\|y - Y\|_\infty = 0.000535$

Substituting the value of p into (9.9), we get[3] with $f \to f(x, y)$,

$$y'' + \left(\frac{b}{x}\right)y' = f \qquad (9.10)$$

Applying L'Hopital's rule to overcome the singularity at $x = 0$, as x approaches zero to the term $(b/x)y'$ in (9.10), one obtains

$$y'' + \left(\frac{\mu}{x}\right)y' = F \qquad (9.11)$$

where

$$F \to F(x, y) = \begin{cases} f & x \neq 0 \\ \frac{1}{b+1}f(0, y) & x = 0 \end{cases} \qquad (9.12)$$

and

$$\mu = b \text{ When } x \neq 0, \qquad (9.13a)$$

$$\mu = 0 \text{ When } x = 0. \qquad (9.13b)$$

Consider the equidistant nodal points x_i on the interval $[a, b]$, where

---

[3]Readers may refer to [24] for more detailed derivation of the method.

$$a = x_0 < x_1 < .. < x_{N-1} < x_N = b$$

and $h = (b - a)/N$. Let Y be an approximate solution that satisfies the boundary conditions and written as a linear combination of $(N + 3)$ shape functions,

$$Y = \sum_{i=-3}^{N-1} c_i B_i \qquad (9.14)$$

where B_i are the cubic B-splines, and c_i, $i = -3, -2, \ldots, N - 1$ are unknown real coefficients.

Substituting the last equation into (9.8) yields

$$\sum_{i=-3}^{N-1} c_i \left[B_i''(x_j) + \frac{\mu}{x_j} \right] = F\left(x_j, \sum_{i=-3}^{N-1} c_i B_i(j) \right) \qquad (9.15)$$

$j = 0, 1, 2, 3, \ldots, N$ and $\mu/x_0 = 0$. This system is with $(N + 1)$ equations and $(N + 3)$ unknowns. On the other hand, the boundary conditions can be expressed as

$$\sum_{i=-3}^{N-1} c_i B_i'(x_0) = 0, \qquad \sum_{i=-3}^{N-1} c_i \left(\alpha B_i(x_N) + \beta B_i'(x_N) \right) = \gamma \qquad (9.16)$$

Equations (9.15) and (9.16) can be put in matrix form as

$$Ac = Z \qquad (9.17a)$$

$$A = \begin{bmatrix} s_1 & 0 & s_2 & 0 & 0 & \cdots & 0 \\ p_0 & q_0 & r_0 & 0 & 0 & \cdots & 0 \\ 0 & p_1 & q_1 & r_1 & 0 & \cdots & 0 \\ \vdots & \vdots & \vdots & \vdots & \vdots & \vdots & \vdots \\ 0 & 0 & 0 & \cdots & p_N & q_N & r_N \\ 0 & 0 & 0 & \cdots & s_3 & s_4 & s_5 \end{bmatrix} \qquad (9.17b)$$

where,

$$s_1 = -\frac{3}{h}, \quad s_2 = \frac{3}{h}, \quad s_3 = \alpha - \frac{3\beta}{h}, \quad s_4 = \alpha + \frac{3\beta}{h}, \quad s_5 = \frac{6 - 3hG_i}{h^2},$$

$$q_i = -\frac{12}{h^2}, \quad r_i = \frac{6 + 3hG_i}{h^2}, \quad G_i = G(x_i) = \frac{\mu}{x_i}, \quad i = 0, 1, 2, \ldots, N$$

$$Z = \begin{bmatrix} 0 \\ F(x_0, c_{-3} + 4c_{-2} + c_{-1}) \\ F(x_1, c_{-2} + 4c_{-1} + c_0) \\ F(x_1, c_{-2} + 4c_{-1} + c_0) \\ \vdots \\ F(x_{N-1}, c_{N-4} + 4c_{N-3} + c_{N-2}) \\ F(x_N, c_{N-3} + 4c_{N-2} + c_{N-1}) \\ \gamma \end{bmatrix} \tag{9.17c}$$

$$c^T = [c_{-3}, c_{-2}, c_{-1}, c_0, \ldots, c_{N-3}, c_{N-2}, c_{N-1}] \tag{9.17d}$$

(a) Use this information to solve the singular BVP of the form [26],

$$y'' + \frac{2}{x}y' = 4y - 2, \quad 0 < x \le 1, \quad y'(0) = 0, \quad \alpha y(1) = \gamma = 1 \tag{9.18}$$

(bspline_sbvp1.m)
The exact solution is

$$y = \frac{1}{2} + 5\frac{\sinh(2x)}{x.\sinh(2)} \tag{9.19}$$

(b) Determine the order of convergence of this method (order of convergence1.m).

Solution

(a) Substituting the values of given parameters in (bspline_sbvp1.m) provides elements of the coefficients vector c in Eq. (9.14).

The $(N + 3)$ coefficients of the c vector and $N + 1$ solution vector elements (read raw-wise, from c_{-3} to c_{N-1}, displaying in short format for space saving reasons) are,

```
C = 0.5433    0.5410    0.5433    0.5456    0.5495    0.5549    0.5619
    0.5706    0.5810    0.5931    0.6071    0.6230    0.6409    0.6610
    0.6833    0.7079    0.7352    0.7651    0.7979    0.8339    0.8732
    0.9160    0.9627

Y = 3.2508    3.2600    3.2753    3.2985    3.3310    3.3732    3.4254
    3.4877    3.5606    3.6445    3.7400    3.8476    3.9680    4.1019
    4.2502    4.4137    4.5936    4.7908    5.0067    5.2426    5.5000
```

Solution curves of the exact and cubic B-spline approach for the given BVP are plotted in Fig. 9.6. Both solutions lie on top of each other. Thus, for comparison between these solutions, their maximum absolute error (in the sense of infinite norm) is computed to highlight the rate of convergence, which is found to be 0.00181025 (N = 20).

Following is the MATLAB script (bspline_sbvp1.m) for the computation of approximate solution of this singular BVP, using cubic B-splines.

```
%bspline_sbvp1.m  singular BVP,  p*y"+(p'/p)*y'=p*f(x,y),    g'=0
% y"+(2/x)y'= 4y-2 ,  y'(0)=0, y(1)=5.5
clc;clear;format compact
N=20;          % number of grid points
b=2;              %p=x^b
alpha=1; beta=0; gamma=5.5;    % alpha*y(1)+beta*y'(1)=gamma
h=1/N; x=linspace(0, 1, N+1);
G(1)=0;   % G0
p(1)=6/h^2; q(1)=-12/h^2; r(1)=6/h^2;
for i=2:N+1
mu=b; G(i)=mu/x(i); p(i)=(6-3*h*G(i))/h^2;
q(i)=-12/h^2;  r(i)=(6+3*h*G(i))/h^2;
end
s1 = -3/h ;s2 = 3/h;s3 = alpha-3*beta/h;s4=4*alpha;s5= alpha+3*beta/h;
A=zeros(N+1); A(1,1)=s1; A(1,3)=s2; A(N+3,N+1)=s3;
A(N+3,N+2)=s4; A(N+3,N+3)=s5;
for j=2:N+2
A(j,j-1)=p(j-1); A(j,j)=q(j-1); A(j,j+1)=r(j-1); end;
A;
z0=0;   zN=gamma; Z = sym('Z',[N+3,1]); c = sym('c',[N+3,1]);
Z(1)=z0; Z(N+3)=zN;
for j=1:N+1;
Z(j+1)=4*(c(j)+4*c(j+1)+c(j+2))-2; % f(x,y)
end;
Z; f = A*c-Z; S = solve(f == 0);  G = struct2cell(S); csol = 0;
for i=1:size(G,1)
    csol(end+1) = G{i};
end
csol(1)= []; C=csol'

B=[1  4  1  0]; %B-spline basis
for j=1:N
    D=[C(j) C(j+1) C(j+2) C(j+3)];   Y(j)= dot(D,B);
end
Y=[Y gamma];Y=Y'                    % Approx. Solution
U=5*sinh(2*x)./(x*sinh(2))+0.5;U=U' % Exact Solution
plot(x,Y,x,U,'g--','linewidth',2);grid;xlabel('x');ylabel('y(x)');
legend('Cubic B-spline', 'Exact');
Y(1)=[]; U(1)=[]; err=abs(Y-U); infNorm=norm(err,Inf) % error norm
```

(b) To examine the error of approximation, let $E(h) = \max|y(x_i) - y_i|$, $i = 1 \leq i \leq N$. If M is the order of convergence of the method, then $e(h) \approx K|h|^M$, where K is the error constant.

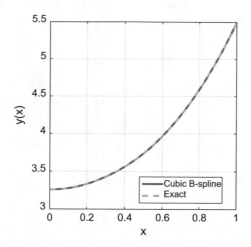

Fig. 9.6 Solution curves of the exact and cubic B-spline approach for the singular BVP, $y'' - (2/x)y' = 4y - 2$, $y'(0) = 0$, $y(1) = 1$, for equal number of samples, $N = 20$ Infinite norm of the approximation is computed as 0.00181025

We successively run the script file (bspline_sbvp.m) for different values of N and record the value of maximum absolute error (infinite norm), in each case. This yields the Table 9.1:

Table 9.1 Values of maximum absolute error for different values of N

N	Max. abs. err.
5	2.59×10^{-2}
10	7.06×10^{-3}
20	1.81×10^{-3}
40	4.56×10^{-4}
80	1.14×10^{-4}
160	2.86×10^{-5}
320	7.14×10^{-6}

We can determine the values of M and K from the equation of line [27],

$$y = Mx + \log(K) + M\log(h)$$

which fits best to the equation

Fig. 9.7 The best fit line for
the order of convergence

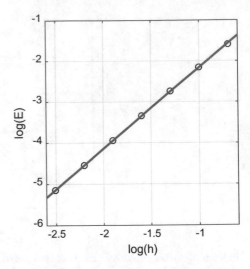

$$\log E(h) = \log(K) + M \log(h)$$

(Logarithms are base 10).
Applying this procedure in our problem, the best fit line equation is

$$y = 1.977x - 0.1832$$

where $x = \log(h)$ and $y = \log[E(h)]$. See, Fig. 9.7.
This means that the order of convergence of the method is M $\approx$ 2.

```
Linear model Poly1:   f(x) = p1*x + p2
Coefficients:   p1 =  1.977 ,    p2 = -0.1832
Goodness of fit:
SSE: 0.0009308,   R-square: 0.9999,   RMSE: 0.01364
```

```
%orderofconvergence1.m
clc;clear;
N=[5,10,20,40,80,160,320]; h=1./N; x=log10(h); x=flip(x)
E=[2.59e-2, 7.06e-3, 1.81e-3, 4.56e-4, 1.14e-4, 2.86e-5, 7.14e-6];
y=log10(E); y=flip(y)
plot(x,y,'ko');grid;hold on;
x=-2.6:0.1:-0.6;
Y=1.977*x-0.1832 % computed via MATLAB curve fitting toolbox
plot(x,Y,'linewidth',2);xlabel('log(h)');ylabel('log(E)');
```

Problem 9.2.3

(a) Use cubic B-spline approach to solve the Bessel's equation of order zero,
   containing a derivative boundary condition at the origin,

$$y'' + \frac{1}{x}y' + y = 0, \quad y'(0) = 0, \quad y(1) = 1$$

for equal number of samples, $N = 20$, within the interval $0 \le x \le 1$. Compute analytical solution, then plot exact and approximate solutions in the same figure. Determine the infinite norm of this approximation (bspline_sbvp2.m).

(b) In a second order BVP, how can the order of approximation be increased for a given set of node points (for a specified boundary condition) while using a B-spline approach?

Solution
The exact analytical solution is

$$y = \frac{J_0(x)}{J_0(1)}$$

where $J_0(x)$ is the Bessel function of the first kind of order zero.

Here, we use the method outlined in the solution of previous problem.

Substituting the values of given parameters in the code (bspline_sbvp2.m) provides the elements of coefficients vector, read raw-wise, from c_{-1} to c_{N+1}, and the solution vector,

```
C = 0.2179    0.2182    0.2179    0.2175    0.2168    0.2158    0.2146
    0.2131    0.2113    0.2093    0.2070    0.2045    0.2017    0.1987
    0.1955    0.1920    0.1883    0.1844    0.1803    0.1759    0.1714
    0.1667    0.1618


Y = 1.3087    1.3074    1.3046    1.3004    1.2946    1.2872    1.2782
    1.2676    1.2555    1.2419    1.2268    1.2102    1.1921    1.1726
    1.1518    1.1296    1.1061    1.0813    1.0554    1.0282    1.0000
```

Exact and approximate solutions are shown in Fig. 9.8. The approximate solution lie on top of the exact solution. Their maximum absolute error (in the sense of infinite norm) is computed to highlight the rate of convergence, which is found to have a value of 0.001394522.

Following is the MATLAB script (bspline_sbvp2.m);

Fig. 9.8 Exact and approximate solutions for the Bessel's equation of order zero, with boundary conditions $y'(0) = 0$, $y(1) = 1$ for equal number of grids, $N = 20$ Infinite norm of the approximation is computed as 0.00181025

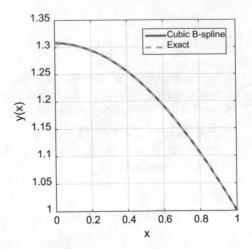

```
% %bspline_sbvp2.m  singular BVP,  p*y"+(p'/p)*y'=p*f(x,y)
% y"+(1/x)y'= -y ,  y'(0)=0, y(1)=1
clc; clear;close; format compact
N=20;                            % number of grid points
b=1;                             % p=x^b
alpha=1; beta=0; gamma=1;   % alpha*y(1)+beta*y'(1)=gamma
%--------------------------------------------------------------
h=1/N; x=linspace(0, 1, N+1);
p(1)=6/h^2; q(1)=-12/h^2; r(1)=6/h^2;
for i=2:N+1
mu=b;
G(i)=mu/x(i); p(i)=(6-3*h*G(i))/h^2;
q(i)=-12/h^2; r(i)=(6+3*h*G(i))/h^2;
end
s1 = -3/h ;s2 = 3/h;s3 = alpha-3*beta/h;s4=4*alpha;s5= alpha+3*beta/h;
A=zeros(N+1); A(1,1)=s1; A(1,3)=s2; A(N+3,N+1)=s3;
A(N+3,N+2)=s4; A(N+3,N+3)=s5;
for j=2:N+2
A(j,j-1)=p(j-1); A(j,j)=q(j-1); A(j,j+1)=r(j-1); end;
A;
z0=0;  zN=gamma; Z = sym('Z',[N+3,1]);  c = sym('c',[N+3,1]);
Z(1)=z0; Z(N+3)=zN;
for j=1:N+1;Z(j+1)=-(c(j)+4*c(j+1)+c(j+2));  % f(x,y)
end;
f = A*c-Z; S = solve(f == 0);G = struct2cell(S); csol = 0;
for i=1:size(G,1); csol(end+1) = G{i};   end
csol(1)= []; C=csol'
B=[1  4  1  0]; %B-spline basis
for j=1:N; D=[C(j) C(j+1) C(j+2) C(j+3)]; Y(j)= dot(D,B); end
Y=[Y gamma];Y=Y'  % Approx. Solution
J = besselj(0,x); % Bessel functions of the 1st kind of order 0
J01=besselj(0,1);
U=J/J01; U=U'  % Exact Solution
plot(x,Y,x,U,'linewidth',2);grid;xlabel('x');ylabel('y(x)');
legend('Cubic B-spline', 'Exact');
Y(1)=[];U(1)=[]; err=abs(Y-U); infNorm=norm(err,Inf) %error norm
```

(c) In a second order BVP, the order of convergence can be increased by applying a higher degree B-spline method (such as quartic B-splines) rather than using cubic B-splines. See, for example, [28]. Note, however, that when the sixth-degree B-spline method is considered for the numerical solution of a fifth-order BVP, it is seen to exhibit a first-order convergence [29].

9.3 Exercises

1. Solve the following BVP, [30] using cubic B-splines method.

$$y'' = y^3 - (1 + \sin^2 t)\sin t, \quad t \in [0, \pi]$$

$$y(0) = y(\pi) = 0$$

Exact solution: $y(t) = \sin(t)$.

2. Solve the following BVP, [31] using cubic B-splines method.

$$y'' = \mu\sinh(\mu y), \ y(0) = 0, \ y(1) = 1, \ \mu = 10, 20, 30, 40.$$

3. Use cubic B-splines method to solve the following singular BVP, [25]

$$y'' + \frac{2}{x}y' = e^y, \quad 0 < x \le 1, \quad y'(0) = 0, \quad y(1) = 0$$

Compute the observed infinite norm values (maximum absolute errors) for various values h.

The exact solution is $y(x) = 2\ln[(B+1)/(Bx^2+1)], \quad B = 3 \pm 2\sqrt{2}$.

4. Use cubic B-splines method to solve the following strongly nonlinear singular BVP, [32]

$$y'' + \frac{2}{x}y' = -e^{-10y}, \quad 0 < x \le 1, \quad y'(0) = 0, \quad 4y(1) + y'(1) = 0$$

Determine the order of convergence of this method.

5. Use cubic B-splines method to solve the following singular BVP [33],

$$y'' + \frac{1}{x}y' = \left(\frac{8}{8 - x^2}\right)^2, \quad 0 < x \le 1, \quad y'(0) = y(1) = 0$$

Exact solution: $y(x) = 2\ln[7/(8 - x^2)]$.

6. Solve the following singular BVP [34] using cubic B-splines method,

$$y'' + \frac{2}{x}y' - 4y = -2, \quad 0 < x \leq 1, \quad y'(0) = 0, \quad y(1) = 5.5$$

Exact solution: $y(x) = 0.5 + 5\sinh 2x/(x\sinh 2)$.

References

1. Salomon D (2006) Curves and surfaces for computer graphics. Springer Science + Business Media Inc, New York
2. Bickley WG (1968) Piecewise cubic interpolation and two-point boundary problems. Comput J 11(2):206–208
3. Albasiny EL, Hoskins WD (1969) Cubic spline solutions to two-point boundary value problems. Comput J 12(2):151–153
4. Al-Said EA (1998) Cubic spline method for solving two-point boundary-value problems. J Appl Math Comput 5(3):669–680
5. Khan A (2004) Parametric cubic spline solution of two point boundary value problems. Appl Math Comput 154(1):175–182
6. Caglar H, Caglar N, Elfaituri K (2006) B-spline interpolation compared with finite difference, finite element and finite volume methods which applied to two-point boundary value problems. Appl Math Comput 175(1):72–79
7. Xu G, Wang G-Z (2008) Extended cubic uniform B-spline and [alpha]-B-spline. Acta Autom Sinica 34(8):980–984
8. Hamid NNA, Majid AA, Ismail AIM (2011) Extended cubic B-spline method for linear two-point boundary value problems. Sains Malaysiana 40(11):1285–1290
9. Kumar M, Gupta Y (2010) Methods for solving singular boundary value problems using splines: a review. J Appl Math Comput 32:265–278
10. Kanth ASVR, Bhattacharya V (2006) Cubic spline for a class of nonlinear singularboundary value problems arising in physiology. Appl Math Comput 174:768–774
11. Kadalbajoo MK, Aggarwal VK (2005) Numerical solution of singular boundary value problems via Chebyshev polynomial and B-spline. Appl Math Comput 160:851–863
12. Chawla MM, Subramanian R, Sathi HL (1988) A fourth-order spline method for singular two-point boundary-value problems. J Comput Appl Math 21:189–202
13. Fyfe DJ (1989) The use of cubic splines in the solution of two-point boundary value problems. Comput J 12:188–192
14. Guoqiang H (1993) Spline finite difference methods and their extrapolation for singular two-point boundary value problems. J Comput Math 11:289–296
15. Iyengar SRK, Jain P (1987) Spline finite difference methods for singular two point boundary value problems. Numer Math 50:363–376
16. Kadalbajoo MK, Kumar V (2007) B-spline method for a class of singular two-point boundary value problems using optimal grid. Appl Math Comput 188:1856–1869
17. Kumar M (2007) Higher order method for singular boundary value problems by using spline function. Appl Math Comput 192(1):175–179
18. Caglar N, Caglar H (2009) B-spline method for solving linear system of second order boundary value problems. Comput Math Appl 27:757–762
19. Munguia M, Bhatta D (2015) Use of cubic B-Spline in approximating solutions of boundary value problems. Int J Appl Appl Math 10(2):750–771
20. Faires JD, Burden RL (2003) Numerical Methods, Thomson/Brooks/Cole, p 531
21. Benabidallah M, Cherruault Y (2004) Application of the adomian method for solving a class of boundary problems. Kybernetes 33(1):118–132

22. Doolan EP, Miller JJH, Schilders WHA (1980) Uniform numerical methods for problems with initial and boundary layers. Boole Press, Dublin, Ireland
23. Aziz T, Arshad Khan A (2002) A spline method for second-order singularly perturbed boundary-value problems. J Comput Appl Math 147:445–452
24. Abukhaled M, Khuri SAA, Sayfy A (2011) A numerical approach for solving a class of singular boundary value problems arising in physiology. Int J Num Anal Model 8(2):353–363
25. Caglar SH, Caglar HN, Ozer M (2009) B-spline solution of non-linear singular boundary value problems arising in physiology. Chaos, Solitons Fractals 39(3):1232–1237
26. Caglar HN, Caglar SH (2006) B-spline solution of singular boundary value problems. Appl Math Comput 182:1509–1513
27. Sayfy A, Khuri S (2008) A generalized algorithm for the order verification of numerical methods. Far East J Appl Math 33(2):295–306
28. Thula K, Roul P (2018) A high-order B-Spline collocation method for solving nonlinear singular boundary value problems arising in engineering and applied science. Mediterr J Math 15:176
29. Caglar HN, Caglar SH, Twizell EH (1999) The numerical solution of fifth-order boundary-value problems with sixth-degree B-spline functions. Appl Math Lett 12:25–30
30. Rández L (1992) Improving the efficiency of the multiple shooting technique. Comput Math Appl 24(7):127–132
31. Cash JR, Wright MH (1991) A deferred correction method for nonlinear two point boundary value problems: implementation and numerical evaluation. SIAM J Numer Anal 12:971–989
32. Raul P (2019) A fast and accurate computational technique for efficient numerical solution of nonlinear singular boundary value problems. Int J Comput Math 96(1):51–72
33. Ascher UM, Mattheij RMM, Russell RD, (1995) Numerical solution of boundary value problems for ordinary differential equations, SIAM, p 192
34. Goh J, Majid AA, Ismail AIM (2011) Extended cubic uniform B-spline for a class of singular boundary value problems. Science Asia 37:79–82

Chapter 10
Solution of BVPs Using bvp4c and bvp5c of MATLAB

MATLAB provides a platform to solve BVPs which consist of two residual control based, adaptive mesh solvers named as bvp4c and bvp5c. Kierzenka and Shampine [1] developed these codes for solving BVPs for ordinary differential equations, which can be used to solve a large class of two-point boundary value problems of the form

$$y'(x) = f(x, y(x), p)$$
$$g(x_L, x_R, y(x_L), y(x_R), p) = 0$$

where f is continuous and Lipschitz function in y and p is a vector of unknown parameters. The bvp5c function can also be used exactly like bvp4c, with the exception of the meaning of error tolerances between the two solvers.

An adaptive mesh solver adjusts the mesh points at each stage in the iterative procedure, distributing them to points where they are most needed. This can be advantageous in terms of computational and storage costs as well as allowing control over the grid resolution. The concept of a residual is the cornerstone of the bvp4c framework; being responsible for both errors control and mesh selection [2, 3].

The code bvp4c can solve explicit nonlinear systems of order one with nonlinear boundary conditions and unknown parameters. The basic solution method is based on polynomial collocation with four Lobatto points for bvp4c, and five Lobatto points for bvp5c. The order of the method is fixed to four, or five. The quantity to be estimated and controlled is the residual for bvp4c, and residual and error in case of bvp5c.

Not only for the computation of the solution of interest, but also whether any solution is achieved or not depends strongly on the initial guess. The quality of the initial guess can also be critical to the solver performance, which reduces or augments the run time. However, to make a good guess usually is the most challenging part of solving a BVP. MATLAB bvp4c and bvp5c solvers take an unusual approach to the control of error in case of having poor guesses for the mesh and

© Springer Nature Switzerland AG 2019
A. Ü. Keskin, *Boundary Value Problems for Engineers*,
https://doi.org/10.1007/978-3-030-21080-9_10

solution, especially for the nonlinear BVP. If any guess values works for the range of length, the rest of the length may be extended using continuation which exploits the fact that the solution obtained for one input will serve as the initial guess for the next value tried. In case of any difficulty in finding a guess for the interval of interest, generally it will be easier to solve the problem on a shorter interval. [4]. How the guess value good is, the less computation time it takes with the continuation method. This is due the fact that, the remaining length depends of the convergence length (based on the guess value) reduces the computation time.

The bvp4c (and bvp5c) framework uses a number of functions required for the user to enter the ODE function, initial data and (if any) some parameters for a given problem. By way of many examples presented in this chapter we will demonstrate how a BVP is supplied and solved by bvp4c.

10.1   Basics of Using bvp4c and bvp5c for the Solution of BVPs

Problem 10.1.1 Describe briefly the functions used for solving boundary value problems in MATLAB, other than dsolve function. Describe output terms in the basic syntax for calling this solver.

Solution
As long as there exists an analytical solution, dsolve is a helpful function for the solution of many BVPs. However, when an explicit solution could not be found, a function used to solve boundary value problems for ordinary differential equations in MATLAB is bvp4c which integrates a system of ODEs of the form $y' = f(x, y)$ on the interval $[a, b]$ subject to two-point boundary value conditions $bc(y(a), y(b)) = 0$. It produces a solution that is continuous on $[a, b]$ and has a continuous first derivative there. This function can also solve multipoint boundary value problems.

Before using bvp4c to solve the problem, there is a need to rewrite the given second or higher order ODE as a system of first-order ODEs.

The bvp5c function can also be used exactly like bvp4c, with the exception of the meaning of error tolerances between the two solvers. If $S(x)$ approximates the solution $y(x)$, bvp4c controls the residual $|S'(x) - f(x, S(x))|$. This controls indirectly the true error $|y(x) - S(x)|$, while bvp5c controls the true error directly.

This function is a finite difference code that implements the three-stage Lobatto IIIa collocation formula and the collocation polynomial provides a solution that is fourth-order accurate uniformly in $[a, b]$. Mesh selection and error control are based on the residual of the continuous solution.

Basic syntax;

$$\mathtt{sol = bvp4c(odefun, bcfun, solinit)}$$

The solver bvp4c produces a solution that is continuous on [a, b] and has a continuous first derivative there. Use the function deval and the output sol of bvp4c to evaluate the solution at specific points xint in the interval [a, b],

$$\mathtt{sxint = deval(sol, xint)}$$

The structure sol returned by bvp4c has the following fields:

sol.x: Mesh selected by bvp4c
sol.y: Approximation to $y(x)$ at the mesh points of sol.x
sol.yp: Approximation to $y'(x)$ at the mesh points of sol.x
sol.parameters: Values returned by bvp4c for the unknown parameters (if any).

Problem 10.1.2 Write the following boundary conditions for bvp4c in MATLAB, $0 \le x \le 1$.

(a) $y(0) = y(1) = 0$
(b) $y'(0) = y'(1) = 0, y(0) = 1$
(c) $y'(0) = 0, y(1) = 1$
(d) $y_1(0) = y_2(0) = y_3(0) = 1, y_4(0) = -5, y_3(1) = y_5(1)$
(e) $y(0) = y'(0) = 0, y(1) = 1, y'(1) = 0$
(f) $y_1(0) = 0, y_4(1) = 1, y_1(1) = y_3(0), y_2(1) = y_4(0)$.

Solution

First derivative terms $\rightarrow y_x(2)$, function terms $\rightarrow y_x(1)$.

(a) [ya(1); yb(1)]
(b) [ya(2); yb(2); ya(1)-1]
(c) [ya(2); yb(1)-1]
(d) [ya(1)-1; ya(2)-1; ya(3)-1; ya(4)+5; yb(3)-yb(5)]
(e) [ya(1); ya(2); yb(1)-1; yb(2)]
(f) [ya(1); yb(4)-1; yb(1)-ya(3); yb(2)-ya(4)]

Note: A BVP in standard form consists of second order linear ODE with $a \le x \le b$,

$$y'' + P(x)y' + Q(x)y = f(x)$$

and boundary conditions,

$$\alpha_1 y(a) + \beta_1 y'(a) = \gamma_1$$
$$\alpha_2 y(a) + \beta_2 y'(a) = \gamma_2$$

where P, Q, f are continuous in $[a, b]$ and $(\alpha_1, \alpha_2, \beta_1, \beta_2, \gamma_1, \gamma_2) \in \mathbb{R}$.

It is also required that α_1, β_1 are note both zero and α_2, β_2 are not both zero.

Problem 10.1.3 Write a stand-alone MATLAB m-file to solve the BVP, $y'' + y' = U; y(a) = e, y(b) = h; (U, a, b, e, h) \in \mathbb{R}$, which does not refer to any other external MATLAB functions. Then, list the solution of the BVP, x vs $y(x)$, for $y'' + y' = 1$, $y(0) = 1, y(2.5) = 0$ in 13 equally spaced points within the interval of definition. Plot the solution x vs $y(x)$. (bvp_1.m)

Solution

Graph of the solution for $y'' + y' = 1$, $y(0) = 1, y(2.5) = 0$ is displayed in Fig. 10.1.

Following is the output list obtained by running the m-file bvp_1.m.

x	y
0	1.0000
0.2083	0.6544
0.4167	0.3238
0.6250	0.0224
0.8333	-0.2367
1.0417	-0.4423
1.2500	-0.5856
1.4583	-0.6603
1.6667	-0.6632
1.8750	-0.5942
2.0833	-0.4562
2.2917	-0.2552
2.5000	0

Fig. 10.1 Graph of the solution for $y'' + y' = 1$, $y(0) = 1$, $y(2.5) = 0$

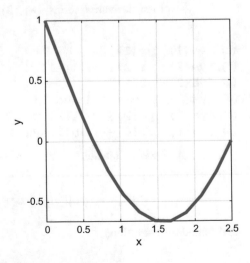

MATLAB m-file to solve the BVP, $y'' + y' = U; y(a) = e, y(b) = h;$ $(U, a, b, e, h) \in \mathbb{R}$, (which does not refer to any other external MATLAB functions, but implementing in-line autonomous equations) is listed below.

```
%bvp_1.m      2 point nonhomogeneous bvp  y"+y=U; y(a)=e, y(b)=h
clc;clear;
U=1; e=1; h=0;
a=0; b=2.5; N=13; %interval end points and number of data samples
f = @(x,y)    [ y(2) ; U-(y(1)) ]; % y1=y , y2=y'=z,  z'=1-y=1-y1,  [z  z']
bc=@(ya,yb)  [ ya(1)-e ; yb(1)-h ];%A1.y(a)+B1.y'(a)=e, A2.y(b)+B2.y'(b)=h
%Form a guess structure by an initial mesh of 5  points in [a,b]
% and a guess of the constant values [1 ; 1]
Si = bvpinit(linspace(a,b,5),[1 ; 1]);
S = bvp4c(f,bc,Si); %Solver  bvp4c
x = linspace(a,b,N);% n equally spaced points
y = deval(S,x);% Evaluation of y at x
%[x' y']
disp('      x            y ')
[x'  y(1,:)']  % tabulate x-y
plot(x,y(1,:),'linewidth',2.5);axis tight;xlabel('x');ylabel('y');grid;
```

Problem 10.1.4 Use bvp4c to approximate the solution and its first derivative for the linear BVP, $y'' + y' + 2y = 10x^3, 0 \le x \le 1, y(0) = y(1) = 0$. For 10 equal steps within the given interval, tabulate x, y, y' and plot them in a single figure. Do not use any MATLAB external "function" functions, but only anonymous functions. (bvp4c01.m).

Solution
For 10 equal steps within the given interval, the solution of given BVP and its derivative are listed below. Graphs of these results are displayed in Fig. 10.2.

X	Y	dYdX
0	0	-0.92016
0.1	-0.087271	-0.82377
0.2	-0.16444	-0.71752
0.3	-0.23027	-0.59577
0.4	-0.28273	-0.448
0.5	-0.31851	-0.25951
0.6	-0.33266	-0.01208
0.7	-0.31827	0.3154
0.8	-0.26616	0.74619
0.9	-0.16476	1.3051
1	0	2.0177

Fig. 10.2 The solution curve
and its first derivative for the
linear BVP,
$y'' + y' + 2y = 10x^3$,
$0 \le x \le 1$, $y(0) = y(1) = 0$

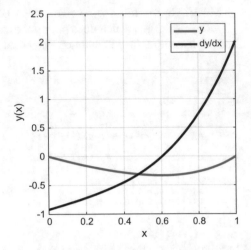

```
%bvp4c1.m    y"+y'+2*y=10*x^3,    y(0) = y(1) = 0
clc;clear;close;
ode = @(x,y)  [ y(2) ; -y(2)-2*y(1)+10*x^3];
bc = @(ya,yb)[ ya(1); yb(1)];
solinit = bvpinit(linspace(0,1,5),[1 0]);
sol = bvp4c(ode,bc,solinit);
x = linspace(0,1,11); y = deval(sol,x);
X=x'; Y=y(1,:)'; dYdX=y(2,:)'; A = table(X,Y,dYdX)
plot(x,y,'linewidth',2); xlabel('x'); ylabel('y(x)'); grid;
legend('y','dy/dx');
```

Problem 10.1.5 Use bvp4c to approximate the solution and its first derivative
for the nonlinear BVP, $y'' + y - 2(y')^2/y = 0$, $-1 \le x \le 1$, $y(-1) = y(1) =$
0.324027137 [5].

For 10 equal steps within the given interval, tabulate x, y, y' in long format, and
plot them in a single figure. Do not use any MATLAB external "function" functions, but only anonymous functions. (bvp4c02.m).

Solution
For 10 equal steps within the given interval, the solution of given BVP and its
derivative are listed below, in long MATLAB format. Graphs of these results are
displayed in Fig. 10.3.

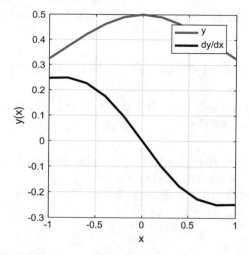

Fig. 10.3 The solution curve and its first derivative for the nonlinear BVP, $y'' + y - 2(y')^2/y = 0$, $-1 \leq x \leq 1$, $y(-1) = y(1) = 0.324027137$

X	Y	dYdX
-1	0.324027137	0.246780576371077
-0.8	0.373850869063608	0.24825594184809
-0.6	0.421776199364775	0.226519893301862
-0.4	0.462502898682855	0.175730515779181
-0.2	0.490160545911545	0.096745733135697
0	0.499996411800013	8.95731208997056e-15
0.2	0.490160545911545	-0.0967457331357116
0.4	0.462502898682852	-0.175730515779198
0.6	0.421776199364768	-0.226519893301872
0.8	0.373850869063603	-0.248255941848059
1	0.324027137	-0.246780576371109

```
%bvp4c02.m    y"+y-2*(y')^2/y=0,       y(-1) = y(1) = 0.324027137
%ex.sol: y=1/(exp(x)+exp(-x))
clc;clear;close;format long
ode = @(x,y) [ y(2) ; -y(1)+2*y(2)^2/y(1)];
bc = @(ya,yb)[ ya(1)-0.324027137; yb(1)-0.324027137];
solinit = bvpinit(linspace(-1,1,11),[1 0]);
sol = bvp4c(ode,bc,solinit);
x = linspace(-1,1,11); y = deval(sol,x);
X=x'; Y=y(1,:)'; dYdX=y(2,:)'; A = table(X,Y,dYdX)
plot(x,y,'linewidth',2);xlabel('x');ylabel('y(x)');grid;
legend('y','dy/dx');
```

Problem 10.1.6 Axial dispersion in a chemical tubular reactor is modeled using the following BVP,

$$\varepsilon y'' - y' - ky^m = 0, \quad y'(0) = \frac{y(0) - 1}{\varepsilon}, \quad y'(1) = 0$$

Solve this BVP if $k = 3, m = 1.3, \quad \varepsilon \in [0.3 \quad 0.1 \quad 0.03]$ and x is the tubular axis. (bvp4c_tubulareactor1.m).

Solution
We can use MATLAB script file (bvp4c_tubulareactor1.m) to solve this problem. Three solution curves are displayed in Fig. 10.4.

MATLAB script (bvp4c_tubulareactor1.m) is listed below.

```
%bvp4c_tubulareactor1.m
clc;clear;close;tic;
k=3; m=1.3;
epsilon=[0.3 0.1 0.03];
for i=1:3 %epsilon
e=epsilon(i);
ode=@(x,y) [y(2); (k*y(1)^m + y(2))/e];
bc=@(ya,yb) [ya(2)+(1-ya(1))/e; yb(2)];
x=linspace(0,1,2); guessy=@(x) [x; 0];%constant guess value
solinit=bvpinit(x,guessy); sol=bvp4c(ode,bc,solinit)
X=linspace(0,1); Y=deval(sol,X,1);
plot(X,Y,'linewidth',2);grid;hold on;
end
xlabel('x');ylabel('y(x)'); legend('e=0.3','e=0.1','e=0.03');toc;
```

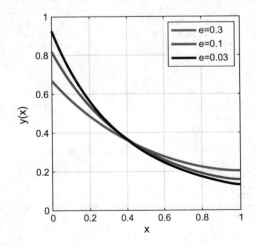

Fig. 10.4 Solution curves for the BVP, $\varepsilon y'' - y' - ky^m = 0, \quad y'(0) = (y(0) - 1)/\varepsilon, \quad y'(1) = 0, \, m = 1.3, \, k = 3$

Problem 10.1.7 The second-order nonlinear homogeneous ODE, $y'' + |y| = 0$ has two solutions that satisfy Dirichlet boundary conditions, $y(0) = 0, y(4) = -2$. Solve this problem [6] using the bvp4c solver in MATLAB (bvp_abs.m).

Solution

For the nonlinear problem, $y'' + |y| = 0, y(0) = 0, y(b) = B$, if $y(x) \le 0$, the solution is found by solving $z'' - z = 0$, otherwise[1] the solution is found by solving $z'' + z = 0$.

If $b < \pi$, the BVP may be solved by using $z'' - z = 0$ or $z'' + z = 0$,

$$y = \begin{cases} c \cdot \sin(x), & B > 0 \\ -c \cdot \sinh(x), & B < 0 \\ 0, & B = 0 \end{cases}$$

Here c is chosen such that $y(0) = 0$ condition is satisfied, by taking

$$c = \begin{cases} B/\sin(b), & B > 0 \\ -B/\sin h(b), & B < 0 \end{cases}$$

Proving a unique solution of the BVP when $b < \pi$.

$$y(x) = \left(\frac{B}{\sin h(\pi)}\right) \cdot \sin h(x), \ B < 0.$$

If $b = \pi$, and $B > 0$, $y(x) = c \sin(x), (c \ge 0)$ has infinite number of solutions but when $b = \pi$ and $B > 0$ there are no solutions of the BVP.

If $b > \pi$ and $B > 0$, there is no solution, while for $B = 0, y(x) = 0$. On the other hand, for $B < 0$, there are exactly two solutions:

$$y_1(x) = \left(\frac{B}{\sin h(b)}\right), \sin h(x)$$

$$y_2(x) = \begin{cases} c_1 \sin(x) & 0 \le x < \pi \\ c_2[\sin h(x) - \tan h(\pi) \cdot \cos h(x)] & \pi \le x < b \end{cases}$$

$$c_2 = B[\sin h(b) - \tan h(\pi) \cdot \cos h(b)]^{-1}, \quad c_1 = -B[\sin h(b - \pi)]^{-1}$$

Rearranging,

$$y_2(x) = \begin{cases} -\{B/[\sin h(b - \pi)]\} \cdot \sin(x) & 0 \le x < \pi \\ \{B/[\sin h(b - \pi)]\} \cdot \sin h(x - \pi) & \pi \le x < b \end{cases}$$

In brief, the solution of the BVP $y'' + |y| = 0, y(0) = 0, y(b) = B$ depends upon the value of b and the sign of B:

---

[1]MATLAB demo program twoode solves given BVP twice, using the guess $y(x) = 1, y'(x) = 0$ and then $y(x) = -1, y'(x) = 0$.

Fig. 10.5 Solution curves for the BVP, $y'' + |y| = 0; y(0) = 0$, $y(4) = -2$ using bvp4c

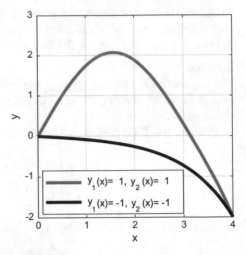

There is one solution if $b < \pi; y_1(x)$ *or* $y_2(x)$

There is one solution if $b > \pi, B = 0; y(x) = 0$

There is no solution if $b \geq \pi, B > 0$

There are infinitely many solutions if $b = \pi, B = 0$

There are two solutions if $b \geq \pi, B < 0; y_1(x)$ *and* $y_2(x)$.

This problem demonstrates that (while a linear BVP has infinitely many solutions if it has more than one possible solution) a nonlinear BVP can have a finite number of solutions. Therefore, when solving nonlinear BVPs one should indicate which solution is the focus of interest.

Numerical solution via MATLAB: We use bvp4c solver along with an anonymous function which defines the given ODE in reduced order form. In this case, first we rewrite the ODE as a system of two first-order ODEs, then form a guess structure in [0, 4] and a guess of the constant values. Following the first solution, we use the second set of initial guess to obtain the other solution of this problem. The two solution curves are displayed in Fig. 10.5.

Note If we use dsolve function (such as in the code dsolve_abs.m), it provides only one solution curve and corresponding solution using bvp_abs.m is not identical with this solution (For example, first maximum is at $x = 2.6427$ for dsolve, but other solution yields maximum point at $x = 2.0666$). Because dsolve solves problems analytically, but this particular BVP does not have smooth coefficients (the $|y|$ term is not differentiable as y passes through the origin), we should not expect a meaningful result from the application of dsolve function.

MATLAB m-file is given below.

```
%bvp_abs.m       (2 point bvp) Solution for  y"+|y|=0; y(0)=y(4)=-2
clc;clear;
f = @(x,y)   [ y(2); -abs(y(1)) ];   % y1=y , y2=y'
bc=@(ya,yb)  [ ya(1); yb(1) + 2 ];% =[ 0; -2 + 2 ]
%Form a guess structure by an initial mesh of five  points in [0,4]
% and a guess of the constant values [1 1]
Si = bvpinit(linspace(0,4,5),[1 1]);
S = bvp4c(f,bc,Si); %Solver  bvp4c
x = linspace(0,4,100);% 100 equally spaced points
y = deval(S,x);% Evaluation of y at x
%[x' y']
plot(x,y(1,:),'linewidth',2.5); hold on;
%To obtain the other solution of this problem, use guess values [-1 -1]
Si = bvpinit(linspace(0,4,5),[-1 -1]);
S = bvp4c(f,bc,Si);
x = linspace(0,4,100);
y = deval(S,x); %[x' y']
plot(x,y(1,:),'k','linewidth',2.5);grid;xlabel('x');ylabel('y');
legend('y_1(x)=  1, y_2(x)=  1','y_1(x)= -1, y_2(x)= -1');hold off
```

Problem 10.1.8 Use MATLAB solver bvp4c to solve the BVP,

$$y'' - xy = -1, \quad 0 < x < 1, \quad y(0) = y(1) = 1.$$

First compute the solution at 5 equally space points within the given interval, indicate these values by black circles in the figure. Then, plot the solution curve by computing the solution at 60 equally spaced points (bvp_2.m).

Solution

Following are the results of computed solution of the BVP at 5 equally spaced points within the given interval. Figure 10.6 displays the computed solution at 5 equally spaced points within the given interval and the solution curve at 60 points.

Fig. 10.6 The results of computed solution at 5 equally spaced points within the given interval and the solution curve at 60 points

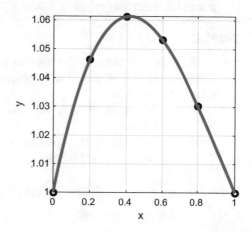

```
x =     0   0.2000   0.4000   0.6000   0.8000   1.0000
y = 1.0000  1.0464   1.0613   1.0531   1.0301   1.0000
```

MATLAB m-file (bvp_2.m) is listed below.

```
%bvp_2.m   (2-point bvp) Solution for  y"-xy=-1; y(0)=y(1)=1
clc;clear;close;
f = @(x,y)    [ y(2); -1+x*y(1) ];   % y(1)=y , y(2)=y'
bc=@(ya,yb)  [ ya(1)-1; yb(1)-1 ];% =[ 0; -2 + 2 ]
%Form a guess structure by an initial mesh of five  points in [0,1]
% and a guess of the constant values [1 1]
Si = bvpinit(linspace(0,1,5),[1 1]);
S = bvp4c(f,bc,Si);   %Solver  bvp4c
x = linspace(0,1,6); % 6 equally spaced points
y = deval(S,x);      % Evaluation of y at x
x,y(1,:)
plot(x,y(1,:),'ko','markersize',5,'linewidth',2); hold on;
x = linspace(0,1,60);% 60 equally spaced points
y = deval(S,x);       % Evaluation of y at x
plot(x,y(1,:),'linewidth',2.5);axis tight;
grid on;xlabel('x');ylabel('y');
```

Problem 10.1.9 Use a MATLAB built in solver to solve the following BVP;

$$y'' - y = f(x) \quad 0 < x < 1, \quad y(0) = 1, \quad y'(1) = -1$$
$$f(x) = \begin{cases} 0 & 0 < x < 1/2 \\ -5 & x = 1/2 \\ -10 & (1/2) < x < 1 \end{cases}$$

(a) For five equally spaced points in $0 < x < 1$. List the output and plot the solution using circles at the solution points.
(b) Plot the solution on the same figure as in part (a), for 60 equally spaced points within the same interval (bvp_3.m).

Solution

(a) MATLAB m-file (bvp_3.m) computes the solution of the BVP, $y'' - y = f(x)$ for any n and $f(x)$ as shifted step (Heaviside) function as well as for any given boundary conditions.

   Figure 10.7 displays the solution curve of $y'' - y = -10u(x - 1/2)$, $y(0) = -y'(1) = 1$.

   The circles indicate the solution values at five equally spaced points in $0 < x < 1$. These are listed below;

```
x =       0.0000   0.2000   0.4000   0.6000   0.8000   1.0000

y =       1.0000   1.4162   1.8891   2.3879   2.6312   2.5788
```

Fig. 10.7 The solution of
$y'' - y = -10u(x - 1/2)$,
$y(0) = -y'(1) = 1$, within
$0 < x < 1$

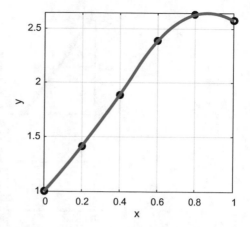

(b) The solid line in Fig. 10.7 displays the solution of $y'' - y = -10u(x - 1/2)$,
$y(0) = -y'(1) = 1$, evaluated at 60 equally spaced points within $0 < x < 1$.

MATLAB m-file (bvp_3.m) is given below.

```
%bvp_3.m       (2 point bvp) Solution for  y"-y=f;  y(0) = - y'(1)=1
% Forcing function is piecewise continuous
clc;clear;close;tic;
syms f(x)
f(x)=-10*heaviside(x-1/2);        % shifted step forcing function
D = @(x,y)   [ y(2); f(x)+y(1) ] ;   % y(1)-y , y(2)=y'
bc=@(ya,yb)  [ ya(1)-1; (yb(2))+1 ]; % y(a)-1=0, y'(b)+1=0
% Form a guess structure by an initial mesh of five  points in [0,1]
% and a guess of the constant values [1 1]
Si = bvpinit(linspace(0,1,5),[1 1]);
S = bvp4c(D,bc,Si);    % Solver  bvp4c
x = linspace(0,1,6);   % 6 equally spaced points between 0 and 1
y = deval(S,x);  % Evaluation of y at x
x,y(1,:)          % Output list
plot(x,y(1,:),'ko','markersize',5,'linewidth',2);hold on;
x = linspace(0,1,60);% 60 equally spaced points
y = deval(S,x);% Evaluation of y at x
plot(x,y(1,:),'linewidth',2.5);axis tight;grid on;xlabel('x');ylabel('y');
toc;
```

Problem 10.1.10 Voltage across an inductor changes as $v(t) = Lq'' = t$, $\quad 0 < t < 1$,
where q, t are the charge flow and time, respectively.

   The boundary conditions are defined as $q(0) - q'(0) = 1$, $\quad q(1) = 0$. Let
$L = 1H$.

   Use a MATLAB solver to compute $q(t)$ and plot the solution for $0 < t < 1$
(bvp_4.m), (dsolve_bvp2.m).

Solution

MATLAB m-file (bvp_4.m) computes the solution of the BVP, $q'' = v(t) = t$
for $n = 50$, while Fig. 10.8 displays the solution curve of the BVP, $v(t) = q'' = t$, $\quad 0 < t < 1$, $\quad q(0) - q'(0) = 1$, $\quad q(1) = 0$.

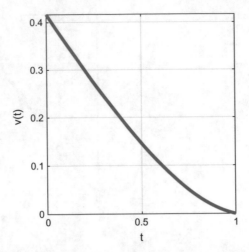

Fig. 10.8 The solution curve of the BVP, $v(t) = q'' = t$, $0 < t < 1$, $q(0) - q'(0) = 1$, $q(1) = 0$.

MATLAB m-file (bvp_4.m) is listed below.

```
%bvp_4.m      (2 point bvp) Solution for  q"=t;  q(0)-q'(0)=1, q(1)=0
clc;clear;close;tic;
%v(t)=t;             % forcing function (voltage)
D = @(t,q)    [ q(2); t ] ; % q(1)=q , q(2)=q'
bc=@(ya,yb)   [ ya(1)- ya(2)-  1; yb(1) ]; % y(a)-y'(a)-1=0, y(b)=0
Si = bvpinit(linspace(0,1,5),[1 1]);% Guess [1 1] in initial mesh
S = bvp4c(D,bc,Si);    % Solver  bvp4c
t = linspace(0,1,50);   % 50 equally spaced points between 0 and 1
q = deval(S,t);   % Evaluation of y at x
% [t',q(1,:)']            % Output list
plot(t,q(1,:),'linewidth',2.5);axis tight;grid on;
xlabel('t');ylabel('v(t)');toc;
```

Alternative MATLAB solver which can be used for the solution of this BVP is dsolve.

An m-file (dsolve_bvp2.m) using this solver is also given below. Analytical solution provided by running this code is $q(t) = (t^3 - 3.5t + 2.5)/6$.

```
% dsolve_bvp2.m, Solve BVP using dsolve and plot the solution.
clc;clear; syms  q(t)
eq='D2q=t'; %  ODE
bc1='q(0)-Dq(0)=1';   bc2='q(1)=0';    % boundary conditions
S=dsolve(eq,bc1,bc2,'t'); pretty(S)
ezplot(S,[0 1]);grid;
```

Problem 10.1.11

(a) Use bvp4c to solve the BVP for $y(x)$,

$$y'' - 50sn(\alpha, \beta)y = 0, \quad y(0) = y(1) = 1, \quad 0 \le x \le 1$$

where $\beta = 0.7$, $\alpha = 1$ and sn is the Jacobi elliptic function. Determine the values of derivatives of the function at two boundaries. (bvp4c_ellipj1.m).

(b) For the integer values of the parameter $3 \le \alpha \le 5$, determine the values of α for which $y(0.5)$ will have a minimum and a maximum, while all else remaining the same in the previous part of the problem. (bvp4c_ellipj2.m).

(c) Determine the value of α (with two digit accuracy) for which $y(0.5)$ has a positive minimum value, with a tolerance of $\varepsilon < 0.06$ (bvp4c_ellipj3.m).

Solution

(a) The Jacobi elliptic functions are integral equations of the form [7]

$$\alpha = \int_0^\phi \frac{d\theta}{\sqrt{1 - \beta \sin^2 \alpha}}, \quad sn(\alpha) = \sin \phi$$

MATLAB script file (bvp4c_ellipj1.m) is used to solve the problem using constant guess values of zero at equally spaced five initial mesh points within the interval of definition. Edited results are given below. The solution curve is plotted in Fig. 10.9a. The values of derivatives of the function at two boundaries are $y'(0) = -y'(1) = 6.2484$.

```
sn = 0.7868
sol = solver: 'bvp4c', x: [1x17 double],y: [2x17 double], yp: [2x17 double]
Y = 1.0000    -6.2484
    1.0000     6.2484
Elapsed time is 0.200276 seconds.
```

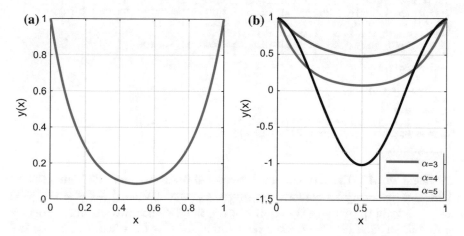

Fig. 10.9 **a** The solution curve for the BVP, $y'' - 50sn(\alpha, \beta)y = 0$, $y(0) = y(1) = 1$. **b** Solution curves for different values of parameter α

(b) From the curves plotted for different values of parameter α, we observe the minimum value of $y(0.5)$ for $\alpha = 5$, and maximum value of $y(0.5)$ for $\alpha = 4$, as shown in Fig. 10.9b.

(c) We use the script file (bvp4c_ellipj3.m) and compute the minimum value of $y(0.5) = 0.0589$ for $\alpha = 1.88$.

MATLAB script files used in the solution of this problem are listed below.

```
%bvp4c_ellipj1.m
clc; clear; tic;
beta=0.7; alpha=1;
sn = ellipj(alpha, beta);          % Jacobi elliptic function
ode=@(x,y) [y(2); y(1)*sn*50]; bc=@(ya,yb) [ya(1)-1;  yb(1)-1];
x=linspace(0,1,5);
guessy = @(x) [x; 0];              % constant guess value
solinit = bvpinit(x,guessy); sol=bvp4c(ode,bc,solinit)
X = linspace(0,1); Y = deval(sol,X); Y = Y'
plot(X,Y(:,1),'linewidth',2);
xlabel('x'); ylabel('y(x)'); grid; toc;


%bvp4c_ellipj2.m
clc;clear;tic;
beta=0.7;
for alpha=3:5
sn=ellipj(alpha,beta);                  % Jacobi elliptic function
ode=@(x,y) [y(2); y(1)*sn*50];bc=@(ya,yb) [ya(1)-1;  yb(1)-1];
x=linspace(0,1,5); guessy=@(x) [x; 0];  % constant guess value
solinit=bvpinit(x,guessy);sol=bvp4c(ode,bc,solinit)
X=linspace(0,1);Y=deval(sol,X);Y=Y';
plot(X,Y(:,1),'linewidth',2);hold on;
end
xlabel('x');ylabel('y(x)');grid;
legend('\alpha=3','\alpha=4','\alpha=5');
toc;
%bvp4c_ellipj3.m
clc;clear;tic;
beta=0.7;alpha=0; ymin=1; epsilon=0.059
while (ymin>epsilon)
sn=ellipj(alpha,beta);          % Jacobi elliptic function
ode=@(x,y) [y(2); y(1)*sn*50];bc=@(ya,yb) [ya(1)-1;  yb(1)-1];
x=linspace(0,1,5); guessy=@(x) [x; 0]; % constant guess value
solinit=bvpinit(x,guessy);sol=bvp4c(ode,bc,solinit);
X=linspace(0,1); Y=deval(sol,X,1);Y=Y';
if min(Y)<0, disp('negative minimum observed'); return
else
 ymin=min(Y);
end
alpha=alpha+0.01;
end
alpha,ymin
plot(X,Y(:,1),'linewidth',2);xlabel('x');ylabel('y(x)');grid;
legend(sprintf('alpha = %.3f', alpha));toc;
```

Problem 10.1.12 The classical one dimensional Bratu problem with homogeneous Dirichlet boundary conditions is given by the BVP, $y'' + Ce^y = 0, \quad C \in \mathbb{R}^+, \quad x \in [0, 1], y(0) = y(1) = 0$. If $C < C_C$, the BVP has two solutions. There is only one solution for $C = C_C$ and no solution for $C > C_C$ where C_C is a critical bifurcation point which is known as Frank-Kamenetskii parameter in chemistry.

The 1D Bratu problem has the exact solution of

$$y = 2 \, ln\left\{\frac{\cos h(a)}{\cos h[a(1-2x)]}\right\}, \quad x \in [0, \quad 1]$$

In this case, the critical point is $C_C \cong 3.51383$.

The Bratu problem appears in a variety of application areas such as the fuel ignition model of thermal combustion, chemical reactor theory, radiative heat transfer, thermal reaction, and the Chandrasekhar model of the expansion of the universe [8–10].

Write a MATLAB m-file to solve Bratu problem and observe the effect of critical point on the solution of this BVP (bvp_bratu1.m).

Solution
The effect of critical point on the solution of Bratu BVP is illustrated in Fig. 10.10.

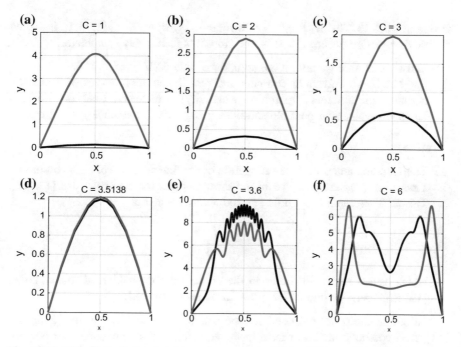

Fig. 10.10 The effect of critical point on the solution of Bratu BVP. **a** : C = 1, **b** : C = 2, **c** : C = 3, **d** : C = Cc **e** : C = 3.6, **f** : C = 6

```
C = 7.0276
Error using bvp4c
Unable to solve the collocation equations -- a singular Jacobian
encountered.

%bvp_bratu1.m   Bratu's problem   y" + C*exp(y) = 0,  y(0) = y(1)= 0
% C<Cc  two solutions,   C=Cc one solution,     C>Cc no solution
clc;clear;close;tic;
Cc=3.5138;
C=3;
%C=Cc;
%C=2*Cc
bc =@(ya,yb)  [ya(1);   yb(1)];
f =@(x,y) [ y(2) ; -C*exp(y(1))];
solinit = bvpinit(linspace(0,1,5),[0.1 0]);
s1 = bvp4c(f,bc,solinit);
% Change initial guess and run again
solinit = bvpinit(linspace(0,1,5),[2.5 0]);
s2 = bvp4c(f,bc,solinit);
plot(s1.x,s1.y(1,:),'k',s2.x,s2.y(1,:),'linewidth',2.5)
%plot(s1.x, s1.y, s2.x, s2.y, 'linewidth',2);
title(['C = ' num2str(C)]); xlabel('x');ylabel('y'); grid; toc
```

Problem 10.1.13 The BVP, $y'' = \mu . \sin h(\mu y)$, $y(0) = 0$, $y(1) = 1$, is called Troesch's equation, where $\mu > 0$ is the Troesch's (sensitivity) parameter.

(a) Make a literature survey on the solution of this BVP.
(b) Use MATLAB solvers to compute and plot the solution curves of Troesch's equation for different Troesch's (sensitivity) parameters. Find the limiting sensitivity values for possible solutions using default tolerance parameters.

Solution

(a) This equation arises in the investigation of confinement of a plasma column by a radiation pressure and also in the theory of gas porous electrodes [11, 12]. This BVP problem has a pole approximately located at

$$x = \frac{1}{\mu} \ln \left(\frac{8}{y'(0)} \right)$$

which causes some difficulties in the numerical computation of this equation [13] Branching or bifurcation behavior is also observed.

Various methods are proposed for the solution of this BVP. In one approach [14], the hyperbolic nonlinear term in the equation is first converted into polynomial nonlinear terms by variable transformation, and a simple shooting method is then used directly to solve this transformed problem. Other solutions of this BVP obtained by using different methods (such as variational iteration method, decomposition method approximation, Laplace transform decomposition method, spline method, homotopy analysis method, homotopy perturbation method [15], the sinc–Galerkin method [16], Padé Approximant [17], using Christov rational functions [18], successive complementary expansion method [19]) have also been reported in literature.

(b) Explicit solution could not be found using `dsolve` built in solver.

MATLAB `bvp4c` solver is unable to solve the collocation equations when $\mu > 7.3922$, while `bvp5c` solver is unable to solve the collocation equations when $\mu > 8.9718$, because a singular Jacobian is encountered in both cases when a guess structure of five initial mesh points is used. Inspection of the curves shown in Fig. 10.11 shows that `bvp5c` solver does not provide a smooth pattern of the solution curve at the limiting sensitivity value. The number of equally spaced points within the interval of $0 \leq x \leq 1$ does not influence this pattern. However, increasing the guess structure from five to ten initial mesh points yields solution curves with an increased limiting sensitivities for both of the solvers as shown in Fig. 10.12.

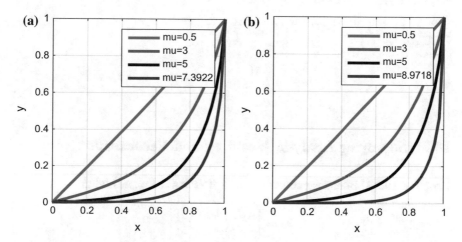

Fig. 10.11 Solution of Troesh equation via, **a** bvp4c, **b** bvp5c MATLAB solver for different sensitivity values in $0 \leq x \leq 1$. Guess structure is formed by an initial mesh of five points

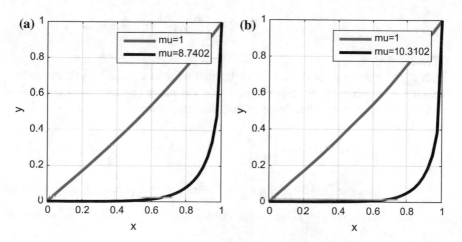

Fig. 10.12 Solution of Troesh equation via, **a** bvp4c, **b** bvp5c MATLAB solver for different sensitivity values in $0 \leq x \leq 1$. Guess structure is formed by an initial mesh of 10 points

```
%bvp_Troesh1.m  Solution for  y"=m*sinh(my); y(0)=0,y(1)=1
clc;clear;close;tic
for m=[0.5  3  5  8.9718]     % 8.9718-bvp5c
%for m=[0.5  3  5  7.3922]     % 7.3922-bvp4c
%for m=[1  10.3102]     %10.3102-bvp5c
%for m=[1  8.7402]% 8.7402-bvp4c;
f = @(x,y)   [ y(2); m*sinh(m*y(1)) ];   % y(1)=y , y(2)=y'
bc=@(ya,yb)  [ ya(1); yb(1)-1 ];% =[ 0; -2 + 2 ]
%Form a guess structure by an initial mesh of M(=5,10)  points in [0,1]
% and a guess of the constant values [1 1]
Si = bvpinit(linspace(0,1,10),[1 1]);
%S = bvp4c(f,bc,Si); %Solver  bvp4c
S = bvp5c(f,bc,Si); %Solver  bvp5c
x = linspace(0,1,40);% 40 equally spaced points
y = deval(S,x);% Evaluation of y at x
plot(x,y(1,:),'linewidth',2.5);hold on
end
axis tight;grid on;xlabel('x');ylabel('y');
%legend('mu=0.5','mu=3','mu=5','mu=7.3922');toc
legend('mu=0.5','mu=3','mu=5','mu=8.9718');toc
%legend('mu=1','mu=8.7402');toc
%legend('mu=1','mu=10.3102');toc
```

10.2   Supplying Analytic Jacobians and Vectorization

Problem 10.2.1 Boundary conditions of a BVP can be expressed as

$$bc = \begin{bmatrix} \alpha y(a) + \beta y'(a) \\ \alpha y(b) + \beta y'(b) \end{bmatrix}$$

Find analytic partial derivatives $\partial bc/\partial ya$ and $\partial bc/\partial yb$ of the following expressions.

$$i)\, bc = \begin{bmatrix} y(0)+1 \\ y(1) \end{bmatrix}, \quad ii)\, bc = \begin{bmatrix} y(0)-1 \\ y'(1) \end{bmatrix}, \quad 0 \le x \le 1$$

Solution

Each partial derivative is a 2-by-2 matrix, first column of which indicates differentiation with respect to function itself, while the second column indicates differentiation with respect to derivative of the function. Hence,

$$i)\, \partial bc/\partial ya = \begin{bmatrix} 1 & 0 \\ 0 & 0 \end{bmatrix}, \quad \partial bc/\partial yb = \begin{bmatrix} 0 & 0 \\ 1 & 0 \end{bmatrix}$$

$$ii)\, \partial bc/\partial ya = \begin{bmatrix} 1 & 0 \\ 0 & 0 \end{bmatrix}, \quad \partial bc/\partial yb = \begin{bmatrix} 0 & 0 \\ 0 & 1 \end{bmatrix}$$

Problem 10.2.2 (Perturbations and supplying Jacobian of ODEs) By default, the bvp4c solver approximates all partial derivatives with finite differences. It can be more efficient if analytical partial derivatives $\partial f / \partial y$ of the differential equations, and analytical partial derivatives, $\partial bc / \partial ya$ and $\partial bc / \partial yb$ of the boundary conditions are provided.

If the problem involves unknown parameters, one also provides partial derivatives,$\partial f / \partial p$ and $\partial bc / \partial p$, with respect to the parameters.

Although user supplied Jacobian matrices are not necessary, they can speed up numerical computations when supplied.

FJacobian is a function handle that computes the analytical partial derivatives of $f(x, y)$

BCJacobian is a function handle that computes the analytical partial derivatives of $bc(ya, yb)$. User can supply either one of these Jacobian matrices or both of them.

(a) Solve the BVP, $yy'' = -1, y(0) = y'(1/2) = 0$, using bvp4c with default tolerances and by moving the boundary condition from the singularity at the origin to a point $d \leq 0.1$.
   What is the minimum value of $d \leq 0.001$, so that BVP can be solved without encountering a singular Jacobian?
   What is the effect of reducing the value of perturbation on boundary point d on run time?
   Repeat the computations using options statement for analytical Jacobian of the given ODE. (bvp4c_perturb2.m) (bvp4c_perturb22.m).
(b) Compute run times of the code when we use analytical partial derivatives for the boundary conditions in the same problem alone, for different values of perturbations about the origin (bvp4c_bc.m).

Solution

(a) Reducing the value of boundary point d increases the run time of the code. Minimum value of this parameter is computed to be $d \leq 10^{-12}$, so that BVP can be solved without encountering a singular Jacobian.
   When analytical Jacobian matrix of the ODE is input into options statement, run time is substantially reduced. Both cases are displayed in Fig. 10.13, while the solution is shown in Fig. 10.14.
(b) When we use analytical partial derivatives for the boundary conditions in the same problem alone, we obtain the run time plot as shown in Fig. 10.15. It is noted here that the use of analytical partial derivatives for the boundary conditions alone has minor run time advantages as compared to the use of Jacobian for the ODE.

Following are the script files (bvp4c_perturb2.m), (bvp4c_perturb22.m), (bvp4c_bc.m).

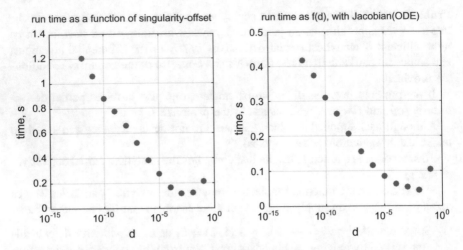

Fig. 10.13 Plots of elapsed time as a function of perturbation for the BVP, $yy'' = -1$, $y(0) = y'(1/2) = 0$, $d \leq 0.1$ Left:Using default parameters. Right:Options set for analytic Jacobian of the ODE

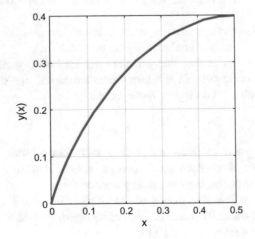

Fig. 10.14 Solution curve for the BVP, $yy'' = -1, y(0) = y'(1/2) = 0, d \leq 0.1$

```
% bvp4c_perturb2.m
clc;clear;close;
p=zeros(1,m);q=zeros(1,m);%preallocations
for n=1:m
tic;
p(n) = 10.^(-n); % - power of 10
d=p(n)
solinit = bvpinit(linspace(d,0.5,2),[1,0]);
ode =@(x,y) [y(2); -1/y(1) ];
bc= @(ya,yb) [ ya(1)- d; yb(2)];
sol = bvp4c(ode,bc,solinit);
plot( [ 0 sol.x], [ 0 sol.y(1,:)], 'linewidth',2 );% approximate solution
axis( [ 0 0.5 0 0.4 ] ); grid on; xlabel('x');ylabel('y(x)');
q(n)=toc;
end
figure;
semilogx(p,q,'.','markersize',20);grid;xlabel('d');ylabel('time, s');
title('run time as a function of singularity-offset');

% bvp4c_perturb22.m (reducing run time using analytical Jacobien of ODE)
clc;clear;close;
m=12;% maximum (absolute) power
p=zeros(1,m);q=zeros(1,m);%preallocations
for n=1:m
tic;
p(n) = 10.^(-n); % - power of 10
d=p(n);
solinit = bvpinit(linspace(d,0.5,2),[1,0]) ;
ode =@(x,y) [y(2); -1/y(1) ];odeJ=@(z,y)[0, 1; 1/y(1)^2, 0];
options=bvpset('fjacobian',odeJ);
bc= @(ya,yb) [ ya(1)- d; yb(2)];
sol = bvp4c(ode,bc,solinit,options);
plot( [ 0 sol.x], [ 0 sol.y(1,:)], 'linewidth',2 );% approximate solution
axis( [ 0 0.5 0 0.4 ] ); grid on; xlabel('x');ylabel('y(x)');
q(n)=toc;
end
figure;
semilogx(p,q,'.','markersize',20);grid;xlabel('d');ylabel('time, s');
title('run time as f(d), with Jacobian(ODE)');

function bvp4c_bc
clc;clear;close;
m=12;% maximum (absolute) power
p=zeros(1,m);q=zeros(1,m);%preallocations
for n=1:m
tic;
p(n) = 10.^(-n); % - power of 10
d=p(n);
solinit = bvpinit(linspace(d,0.5,2),[1,0]) ;
ode =@(x,y) [y(2); -1/y(1) ];
options=bvpset('bcjacobian',@bcJ); bc= @(ya,yb) [ ya(1)- d; yb(2)];
sol = bvp4c(ode,bc,solinit,options);
plot( [ 0 sol.x], [ 0 sol.y(1,:)], 'linewidth',2 );% approximate solution
axis( [ 0 0.5 0 0.4 ] ); grid on; xlabel('x');ylabel('y(x)');
q(n)=toc;
end
figure;
semilogx(p,q,'.','markersize',20);grid;xlabel('d');ylabel('time, s');
title('run time as f(d), with Jacobian(bca,bcb)');

function [dBCdya,dBCdyb] = bcJ(ya,yb)
% Evaluate the partial derivatives of the boundary conditions
dBCdya = [ 1 0; 0 0 ];dBCdyb = [ 0 0; 0 1 ],end
end
```

Fig. 10.15 Plot of elapsed time as a function of perturbation for the BVP, $yy'' = -1, y(0) = y'(1/2) = 0, d \leq 0.1$ using analytic Jacobian matrices of the boundary conditions

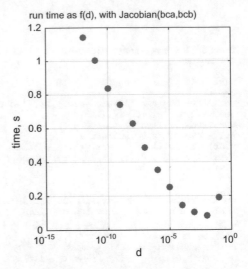

Problem 10.2.3 Many physical, chemical and biological phenomena can be described by nonlinear reaction-diffusion models. Typical examples are given by the Fisher equation which was suggested as a deterministic version of a stochastic model for the spatial spread of a favored gene in a population [20]. The Fisher equation is the basis for a variety of models for spatial spread [21].

Use bvp4c and analytic Jacobian of the ODE to solve the Fisher equation (BVP),

$$y'' + cy' + y(1 - y) = 0, \quad y(-\infty) = 1, \quad y(\infty) = 0$$

where c is the wave speed, y is a chemical concentration.

Let the initial guess of the solution vector be [0, 0] for 10 equally spaced mesh points within the boundaries, $\infty \to z = 10c, x \in [-z, z]$.

For a unique solution, re-define boundary conditions as [22],

$$\frac{y'(-z)}{y(-z) - 1} - A = 0, \quad A = \frac{-c + \sqrt{c^2 + 4}}{2}$$

$$\frac{y(z)}{e^{Bz}} - 1 = 0, \quad B = \frac{-c + \sqrt{c^2 - 4}}{2}$$

Plot the solution curves for $c \in [3, 6, 9, 12, 15]$ (bvp4c_Fisher.m).

Solution

The MATLAB script file (bvp4c_Fisher.m) is used for the solution of this problem by the application of three anonymous functions within the code. Five solution curves are plotted and displayed in Fig. 10.16.

Fig. 10.16 Solution curves
for $c \in [3, 6, 9, 12, 15]$ of
Fisher's BVP

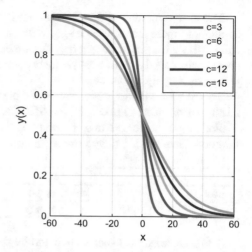

```
% bvp4c_Fisher.m   y"+cy'+y(1-y)=0 y(-inf)=1, y(inf)=0.
clc;clear;close;
for c=3:3:15
A=(-c+sqrt(c^2+4))/2; B=(-c+sqrt(c^2-4))/2;
z=10*c % assignment for infinity
solinit = bvpinit(linspace(-z,z,10),[0,0]);
ode =@(x,y) [y(2);  -(c*y(2)+y(1)*(1-y(1))) ];
odeJ=@(z,y)[0, 1; -1+2*y(1),  -c];
options=bvpset('fjacobian',odeJ);
bc= @(ya,yb) [ ya(2)/(ya(1)-1)-A;   yb(1)/exp(B*z)-1];
sol = bvp4c(ode,bc,solinit,options)
%[sol.x' sol.y(1,:)']
plot( sol.x,   sol.y(1,:), 'linewidth',2 );hold on;
end
axis( [ -60 60 0  1 ] ); grid on; xlabel('x');ylabel('y(x)');
legend('c=3','c=6','c=9','c=12','c=15');
```

Problem 10.2.4 Use an existing problem to test if bvp4c/bvp5c vectorization option shortens the elapsed time, radically.

Solution

ODE vectorization option in bvp4c and bvp5c can be coded by changing scalar quantities in an ODE like $x(1)$ into arrays as $x(1,:)$ and changing from scalar operations to array operations by replacing multiplication sign "*" with ".*", and exponent sign " ^ " with " .^ ", respectively. Note that, the solver must be informed about the presence of vectorization at the beginning using the options statement, for example,

```
options = bvpset('stats','on', 'vectorized','on');
```

The execution time of MATLAB codes was dominated by the cost of function evaluations more than a decade ago. With 'vectorization', there were fewer function calls, which translated to shorter run times. Since then, MATLAB's execution engine improved significantly, incorporating optimizations like code inline, just-in-time compilation, etc.

To see where things stand now, a simple experiment was conducted, using the latest version of MATLAB and the SHOCKBVP demo (with plotting removed).

For various solver options, a loop of 100 calls was timed to the simplified demo function. Here are the tic/toc times (in seconds)[2]:

	A	B	C	D
bvp4c:	14.395	18.885	6.390	4.300
bvp5c:	21.396	24.393	12.810	11.297

The comparison between bvp4c and bvp5c is not that critical, as the solvers use different algorithms. However, the relative timing of A, B, C, D seem fairly similar for either of the solvers.

The solver options were:

A: The defaults

```
Options   = bvpset('FJacobian',[],'BCJacobian',[],'Vectorized',
'off');
```

B: Defaults + vectorization

```
Options   = bvpset('FJacobian',[],'BCJacobian',[],'Vectorized',
'on');
```

C: Analytical Jacobians, no vectorization

```
Options = bvpset('FJacobian',@shockJac,'BCJacobian',@shockBCJac,
'Vectorized','off');
```

D: Analytical Jacobians + vectorization

```
Options        =       bvpset('FJacobian',@shockJac,'BCJacobian',
@shockBCJac,'Vectorized','on');
```

In particular, the difference between A and B is quite counter intuitive: Vectorization appears to slow things down. Yet, the number of ODE function evaluations is reduced from about 5700 to around 900. An explanation to that can be that MATLAB is able to more aggressively optimize the execution of non-vectorized version of this ODE function, which compensates for any savings

---

[2]The numbers (even their ratios) are likely to change with MATLAB version/operating system/ processor model. The numbers were obtained using MATLAB version 2016a, 64 bit, run on a computer with 2.5 GHz Intel Core[TM] i5-2430 M processor.

due to vectorization. However, that is not always the case. The difference between C and D shows that vectorization could actually help.

Therefore, in can be concluded that[3] if the execution time is dominated by ODE function evaluations, then vectorization option may shorten the elapsed time radically, hence the vectorization is likely to help. On the other hand, vectorizing simple functions that MATLAB optimizes aggressively may actually slow things down. Therefore, whether bvp4c vectorization option shortens the elapsed time or not depends on profiling a particular code available in hand.

```
>> tic, for i=1:100, shockbvpA(@bvp5c), end, toc
Elapsed time is 21.396079 seconds.
>> tic, for i=1:100, shockbvpB(@bvp5c), end, toc
Elapsed time is 24.392697 seconds.
>> tic, for i=1:100, shockbvpC(@bvp5c), end, toc
Elapsed time is 12.810138 seconds.
>> tic, for i=1:100, shockbvpD(@bvp5c), end, toc
Elapsed time is 11.297051 seconds.
>>
```

10.3 Parametric Problems and Periodic Boundary Conditions

Problem 10.3.1 Compute the value of parameter λ for which there is a solution of the nonlinear first order parametric BVP,

$$\varepsilon y' = \cos(x) - \frac{\lambda \cos^3(x)}{y}, \quad \varepsilon = 0.1, \quad y\left(-\frac{\pi}{2}\right) = 0, \quad y\left(\frac{\pi}{2}\right) = 2.6.$$

List first and last few terms of the solution. Plot the solution curve within the interval of definition. (bvp4c04.m).

Solution
The first five and last five terms of the solution are listed below, along with solver output.

The solution curve within the interval of definition is shown in Fig. 10.17.

---

[3]Private communication with Dr. Jacek Kierzenka, Mathworks Inc.

Fig. 10.17 The solution
curve for the parametric BVP

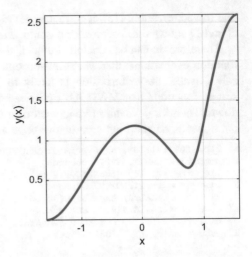

```
sol =
solver: 'bvp4c'
x: [1x45 double], y: [1x45 double], yp: [1x45 double]
parameters: 1.201033147621122,   stats: [1x1 struct]
L = 1.201033
```

X	Y
-1.57079632679490	3.72529029846191e-09
-1.53906306766773	0.00198618122951
-1.50732980854056	0.00781420430310
-1.47559654941339	0.01728597731936
-1.44386329028622	0.03019320245282
1.44386329028622	2.51984639342526
1.47559654941339	2.55481439360541
1.50732980854056	2.57988205814819
1.53906306766773	2.59496167237509
1.57079632679490	2.59999999403954

```
%bvp4c04.m  parametric BVP,   eps*y'=cos(x)-L*cos(x)^3/y
% bc:  y(-pi/2)=0  , y(pi/2)=2.6
clc;clear;close;format long
eps=1e-1;
solinit = bvpinit(linspace(-pi/2,pi/2,10),1, 2);
ode = @(x,y,L) (cos(x)-L*cos(x)^3/y)/eps;
bc = @(ya,yb,L)[ ya(1);  yb(1)-2.6];
sol = bvp4c(ode,bc,solinit)
fprintf('L = %f\n',sol.parameters)
% x=sol.x; y=sol.y;
x = linspace(-pi/2,pi/2); y = deval(sol,x);
X=x'; Y=y(1,:)'; A = table(X,Y)
plot(x,y,'linewidth',2);xlabel('x');ylabel('y(x)');grid;
axis([-pi/2, pi/2, min(y), max(y)]);
```

Problem 10.3.2 Use bvp4c to solve the BVP,

$$y'' - \lambda y^{-1/2} = 0, \quad y(0) = y(1) = 1$$

and determine the value of λ and $y'(1)$.

Hint: In order to relieve the BVP from undetermined system of equations, impose the initial (launch) angle value as 0.2 rad. (bvp4c_rediff1.m).

Solution
This parametric problem is solved using MATLAB script file (bvp4c_rediff1.m).

We take $y'(0) = 0.2$, and plot the solution curves as shown in Fig. 10.18. The value of the eigenvalue is computed to be $\lambda = -0.144881$ and $y'(1) = -0.2$.

X	Y	dY
0	0.1	0.2
1	0.1	-0.2

MATLAB script file (bvp4c_rediff1.m) is listed below.

```
%bvp4c_rediff1.m  parametric BVP,   y"=L/y^(1/2),  y(0)= y(1)=1
clc;clear;close;tic;
L=1; % lambda guess
guessy=@(x) [1; 0]; x=linspace(0,1,5); solinit=bvpinit(x,guessy,L);
ode = @(x,y,L) [y(2);   L*y(1).^(-1/2)];
bc = @(ya,yb,L)[ ya(1)-0.1; ya(2)-0.2; yb(1)-0.1];
sol = bvp4c(ode,bc,solinit);
fprintf('L = %f\n',sol.parameters)
x = linspace(0,1); y = deval(sol,x);
X=x'; Y=y(1,:)';   dY=y(2,:)';
A = table(X,Y,dY)
plot(X,Y,X,dY,'linewidth',2);xlabel('x');grid;
legend('y(x)','dy(x)/dx'); toc;
```

Problem 10.3.3 Use bvp4c to solve the nonlinear eigenvalue problem,

$$y'' + (5 - \beta)y + y^5 = 0, \quad y(-1) = y(1) = 0$$

and determine the value of β.

Hint: In order to relieve the BVP from undetermined system of equations, impose the initial (launch) angle value as $1/10$ rad. (bvp4c_eigen1.m).

Solution
This parametric (eigenvalue) problem is solved using MATLAB script file (bvp4c_eigen1.m).

We take $y'(-1) = 0.1$, and plot the solution curves as shown in Fig. 10.19. The value of β is computed to be $\beta = 2.5325$.

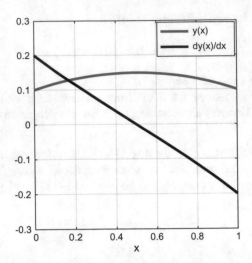

Fig. 10.18 Solution curves for the BVP, $y'' - \lambda y^{-1/2} = 0, \quad y(0) = y(1) = 1$

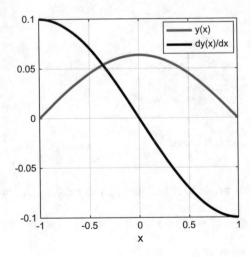

Fig. 10.19 Solution curves for the BVP, $y'' + (5 - \beta)y + y^5 = 0, \quad y(-1) = y(1) = 0$

```
sol = solver: 'bvp4c'
             x: [1x13 double]
             y: [2x13 double]
            yp: [2x13 double]
    parameters: 2.5325
Elapsed time is 0.133532 seconds.
```

```
%bvp4c_eigen1.m
clc;clear;tic;
B=10; % beta guess
guessy=@(x) [1; 0]; x=linspace(-1,1,5);
solinit=bvpinit(x,guessy,B);
ode=@(x,y,B) [y(2); (B-5)*y(1)-y(1)^(5)];
bc=@(ya,yb,B) [ya(1);ya(2)-0.1;  yb(1)];
sol=bvp4c(ode,bc,solinit)
X=linspace(-1,1);Y=deval(sol,X);
plot(X,Y,'linewidth',2);xlabel('x');grid;
legend('y(x)','dy(x)/dx'); toc;
```

Problem 10.3.4 Use bvp5c solver to solve the fourth order eigenvalue BVP,

$$y^{(4)} - \lambda y = 0, \quad 0 \le x \le 1, \quad y(0) = y'(0) = y(1) = y''(1) = 0$$

Plot the solution curves $y(x)$ and its first and second derivatives on the same figure. Determine error of approximation in computing the first eigenvalue of the BVP.

Hint: Exact solution is obtained by solving the equation,

$$\tanh\left(\lambda^{1/4}\right) - \tan\left(\lambda^{1/4}\right) = 0$$

(bvp5c_4th_eigen.m).

Solution

A MATLAB script file (bvp5c_4th_eigen.m) is written for the solution of this problem.

In order to relieve the BVP from undetermined system of equations, we impose the initial (launch) angle value as 0.1 rad.

Exact eigenvalue is computed as $\lambda = 237.7210675311166$. When bvp4c solver is used with default settings, first eigenvalue is computed as $\lambda = 237.744613$ When bvp5c solver is used with tolerance settings for relative error and absolute error both at 10^{-13}, first eigenvalue is computed as $\lambda = 237.7210675311167$. This yields the relative error of computation as 10^{-16}.

Edited output list of the MATLAB script (bvp5c_4th_eigen.m) is given below[4]. Three solution curves of the BVP are shown in Fig. 10.20.

---

[4](Elapsed time) The code was run on a laptop with 2.5 GHz Intel Core i5™ 2430 M single core computer.

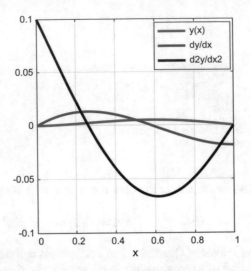

Fig. 10.20 Solution curves $y(x)$ and its first and second derivatives for the BVP (using the smallest eigenvalue), $y^{(4)} - \lambda y = 0$, $0 \le x \le 1$, $y(0) = y'(0) = y(1) = y''(1) = 0$

```
     solver: 'bvp5c'
 parameters: 2.377210675311167e+02
          x: [1x403 double]
          y: [4x403 double]
      idata: [1x1 struct]
      stats: [1x1 struct]

r = 2.377210675311166e+02
error = 8.369e-16
Elapsed time is 1.35 seconds.
```

```
%bvp5c_4th_eigen.m      y^(4)-L*y=0,   y(0)= y(1)=y'(0)=y'(1)=0
clc;clear;close;tic;format long;
L=1; % lambda guess
options=bvpset('reltol',1e-13,'abstol',1e-13);
guessy=@(x) [1; 0; 0; 0;]; x=linspace(0,1,5); solinit=bvpinit(x,guessy,L);
ode = @(x,y,L) [y(2); y(3); y(4);   L*y(1)];
bc = @(ya,yb,L) [ ya(1); ya(2); yb(1); yb(3);ya(3)-.1];
%sol = bvp4c(ode,bc,solinit,options)
sol = bvp5c(ode,bc,solinit,options)
%fprintf('L = %3.12f\n',sol.parameters)
x = linspace(0,1); y = deval(sol,x);
X=x'; Y=y(1,:)';   dY=y(2,:)';d2Y=y(3,:)';
%A = table(X,Y,dY,d2Y)
plot(X,Y,X,dY,X,d2Y,'k','linewidth',2);xlabel('x');grid;
legend('y(x)','dy/dx','d2y/dx2');
r = fzero(@(L) tanh(sqrt(sqrt(L)))-tan(sqrt(sqrt(L))), 230)%actual L
error=abs(sol.parameters-r)/r
toc
```

Problem 10.3.5 (Periodic boundary conditions) Although most BVP solvers require that the boundary conditions be separated, bvp4c can operate with non-seperated boundary conditions.

Use bvp4c solver to find the solution of the following system of ODEs,

$$y_1' = \frac{dy_1}{dt} = 2\left[y_1\left(1 - y_1^2\right) + y_2\right]$$

$$y_2' = \frac{dy_2}{dt} = -\frac{2(y_1 + y_2)}{3}$$

Subject to boundary conditions,

$$y_1(0) = y_1(T), \quad y_2(0) = y_2(T)$$

Determine the unknown period T. What is the minimum number of initial mesh points for the solution to converge? Observe the effect of varying the number of mesh points on the period.

(bvp4c_per02.m).

Solution

We need another boundary condition since the value of period is also unknown. We may use $y_1(0) = 0$ which means that we solve the given ODEs using new set of boundary conditions,

$$y_1(0) = 0, \quad y_1(T) = 0, \quad y_2(0) = y_2(T)$$

Because the interval is not numerically defined, we first transform the problem to one formulated on a fixed interval by changing the independent variable from t to $x = t/T$, which transforms the ODEs into the following form:

$$y_1' = \frac{dy_1}{dx} = 2T\left[y_1\left(1 - y_1^2\right) + y_2\right]$$

$$y_2' = \frac{dy_2}{dx} = -\frac{2T(y_1 + y_2)}{3}$$

subject to new boundary conditions on the interval $0 \leq x \leq 1$,

$$y_1(0) = 0, \quad y_1(1) = 0, \quad y_2(0) = y_2(1)$$

The MATLAB code is written using following guessed functions,

$$y_1(x) = \sin(2\pi x), \quad y_2(x) = \cos(2\pi x), \quad T = 2\pi$$

One needs to rescale the variable x to $t = Tx$, before the solution curves are plotted. This is performed by the following operation in the script file;

```
X =T*sol.x ;        % x-axis values
```

Computed period is $T = 8.054787$ units. Two solution curves are displayed in Fig. 10.21.

The minimum number of initial mesh points for the solution to converge is found to be 3.

```
sol =    solver: 'bvp4c'
              x: [1x31 double]
              y: [2x31 double]
             yp: [2x31 double]
     parameters: 8.0548
          stats: [1x1 struct]
T=8.054787.
Elapsed time is 0.306083 seconds.
```

Increasing the number of initial mesh points slightly changes the period, for example for 30 mesh points $T = 8.054755$, while for 100 equal mesh points, $T = 8.055057$ units.

MATLAB script file (bvp4c_per02.m) is listed below.

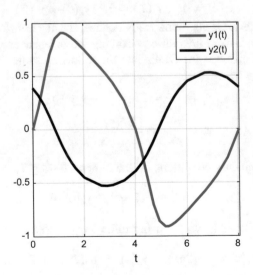

Fig. 10.21 Periodic solution curves of the system of ODEs, $y_1' = \frac{dy_1}{dt} = 2[y_1(1 - y_1^2) + y_2]$, $y_2' = \frac{dy_2}{dt} = -\frac{2(y_1 + y_2)}{3}$ Subject to boundary conditions, $y_1(0) = y_1(T)$, $y_2(0) = y_2(T)$

```
%bvp4c_per02.m    nonlinear periodic BVP
clc;clear;close;tic
g=@(x) [sin(2*pi*x);cos(2*pi*x)]; % guessed solution
solinit = bvpinit(linspace(0,1,3), g, 2*pi);
ode = @(x,y,T) [ 2*T*(y(1)- y(1)^3 +y(2) )
               -(T/1.5)*(y(1) + y(2)) ];
bc = @(ya,yb,T) [ ya(1); yb(1); ya(2)- yb(2)];
sol = bvp4c(ode,bc,solinit)
T=sol.parameters; % computed value of the period
fprintf('T=%f.\n',T);
X =T*sol.x ;       % x-axis values
plot(X,sol.y(1,:),X,sol.y(2,:),'k','linewidth',2);
xlabel('x');legend('y1(x)','y2(x)');grid; xlim([0 T]);toc;
```

Problem 10.3.6 (Periodic boundary conditions) In an epidemiology model for the spread of an infectious disease, it is assumed that a given population of constant size N can be divided into four categories: Susceptibles, whose number at time t is $S(t)$, latents, $L(t)$, infectives $I(t)$, and immunes $M(t)$. Under certain assumptions, its dynamics is expressed as

$$y_1' = \mu - \beta(t)y_1 y_3$$
$$y_2' = \beta(t)y_1 y_3 - y_2/\lambda$$
$$y_3' = y_2/\lambda - y_3/\eta$$

where,

$$y_1 = S/N, \quad y_2 = L/N, \quad y_3 = I/N, \quad \mu = 0.01, \quad \lambda = 0.02, \quad \eta = 0.03,$$
$$\beta(t) = \beta_0(1 + \cos 2\pi f t), f = 0.25, \quad \beta_0 = 2500.$$

The solution sought is periodic; that is, the boundary conditions are [23],

$$y_1(0) = y_1(1), \quad y_2(0) = y_2(1), y_3(0) = y_3(1)$$

Use bvp4c and guess vector $= [1, 1, 1]$ for y(0) to solve this BVP (bvp4c_per01.m).

Solution

The BVP has nonseparated boundary conditions,

$$y_4(0) - y_1(1), \quad y_5(0) = y_2(1), \quad y_6(0) = y_3(1),$$

which means that

$$y_4(0) = y_1(0), \quad y_5(0) - y_2(0), \quad y_6(0) = y_3(0),$$

as well as

$$y_4(1) = y_1(1), \quad y_5(1) = y_2(1), \quad y_6(1) = y_3(1),$$

creating a new guess vector,

$$\text{new guess vector} = [1, 1, 1, 1, 1, 1]$$

The solution curves are obtained via the MATLAB script file (bvp4c_per01.m) and displayed in Fig. 10.22.

```
%bvp4c_per01.m
clc;clear;close;
mu = 0.01; lambda = 0.02; eta = 0.03; beta0 =2500; % constants
options = bvpset( 'Vectorized', 'on' );
solinit = bvpinit(linspace(0,1,50),ones(1,6)); % guess structure
bc=@(ya,yb,mu,lambda,eta,beta0)   [ ya(1)-ya(4); ya(2)-ya(5);ya(3)-ya(6);
    yb(1)-yb(4);yb(2)-yb(5); yb(3)-yb(6) ];      % boundary conditions
ode =@(x,y,mu,lambda,eta,beta0)...               % x --> t
    [ mu - beta0*(1+cos(0.5*pi*x)).* y(1,:).* y(3,:);
    beta0*(1+cos(0.5*pi*x)) .* y(1,:) .* y(3,:) - y(2,:)/lambda;
    y(2,:)/lambda - y(3,:)/eta;
    zeros(size(x)); zeros(size(x)); zeros(size(x))];
sol = bvp4c(ode,bc,solinit,options,mu,lambda,eta,beta0);
k=.1;  % display scale factor for S
plot(sol.x,k*sol.y(1,:), sol.x, sol.y(2,:),'k', sol.x, sol.y(3,:),...
'r','linewidth',2); legend('S','L','I');grid; xlabel('t');
```

10.4 Solution of BVPs with Free (Unkown) Boundaries Using bvp4c

Problem 10.4.1 Use MATLAB `bvp4c` solver to solve the following BVP,

$$y'' + y = 1, \quad 0 \leq x \leq b, \quad y(0) = 0, \quad y(b) = 1, \quad y'(b) = 2,$$

where the value of right-end boundary b is unknown (bvp4c_unknownbndry1.m).

Fig. 10.22 Solution curves of an epidemiology model for the spread of an infectious disease

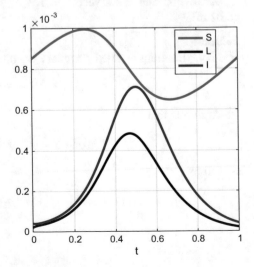

Solution

The MATLAB code to numerically solve the given BVP is listed below. This script file (as it is the case for most of the codes presented in the book) does not employ external function files for defining boundary conditions and the ODE, but use simple anonymous functions.

Note that we must also provide an initial guess for the unknown right-most boundary value, even though it is a free variable.

We introduce the right-most boundary as an unknown parameter and remove any explicit dependencies on it. We accomplish this goal by using scaling, so that

$$X = x/x_f, \quad b = x_f, \quad dX = dx/x_f, \quad d/dX = x_f(d/dx), \quad 0 \leq X \leq 1.$$

This parameterization allows the solver to solve the problem with assumed initial and final values $X_0 = 0$, $X_f = 1$. Actual boundary x_f is treated as an unknown parameter to be found by the solver.

Solutions are influenced by the guessed values for unknown boundary value. If the guessed value $0.09 \leq b_g \leq 0.767$, the unknown boundary value $b = 0.5236$ is determined by the code which also gives $y'(0) \approx \sqrt{3}$, the value of the derivative function at point $x = 0$.

However, if the guessed value $0.79 \leq b_g \leq 2.767$, the unknown boundary value is computed as $b = 2.6180$ and $y'(0) \approx -\sqrt{3}$.

Edited output lists of the MATLAB script (bvp4c_unknownbndry1.m) are shown below, for both cases.

x	y	dy/dx	x	y	dy/dx
0.0000	0.0000	1.7321	0.0000	0.0000	-1.7321
0.0524	0.0920	1.7820	0.2618	-0.4142	-1.4142
0.1047	0.1865	1.8271	0.5236	-0.7321	-1.0000
0.1571	0.2833	1.8672	0.7854	-0.9319	-0.5176
0.2094	0.3820	1.9021	1.0472	-1.0000	-0.0000
0.2618	0.4824	1.9319	1.3090	-0.9319	0.5176
0.3142	0.5842	1.9563	1.5708	-0.7321	1.0000
0.3665	0.6871	1.9754	1.8326	-0.4142	1.4142
0.4189	0.7909	1.9890	2.0944	-0.0000	1.7321
0.4712	0.8953	1.9973	2.3562	0.4824	1.9319
0.5236	1.0000	2.0000	2.6180	1.0000	2.0000

BVP solver bvp4c solves the system using the default relative error tolerance of 10^{-3} for each component. One can increase the accuracy of the solution by using "options" and "long format" in the script. Solution curves are shown in Fig. 10.23.

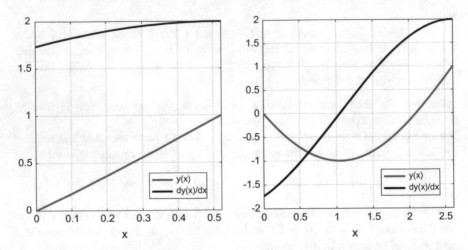

Fig. 10.23 Left: Solution curves for the BVP, $y'' + y = 1$, $y(0) = 0$, $y(b) = 1, y'(b) = 2$, for guessed value of $0.09 \leq b_g \leq 0.767$, where the unknown boundary value is determined by the code using the bvp4c solver as $b = 0.5236$. This point also gives $y'(0) \approx \sqrt{3}$, the value of derivative function at point $x = 0$. Right: Solution curves for the BVP when guessed value of $0.79 \leq b_g \leq 2.76$, where the unknown boundary value is computed as $b = 2.6180$ and $y'(0) \approx -\sqrt{3}$

```
%bvp4c_unknownbndry1.m     y"+y=1,      y(0)=0,   y(b)=1,   y'(b)=2
clc; clear; close;tic
x0 = 0;  % initial (left-most boundary) value
yinit = [x0  0 ]; % guess  initial conditions, insert zero if unknown
xf_guess = 0.8; % sec, initial guess for final time
Nx = 11;
X = linspace(0,1,Nx)'; % nondimensional time vector
solinit = bvpinit(X,yinit,xf_guess);
bc = @(Y0,Yf,xf) [Y0(1) - x0; Yf(1) - 1;  Yf(2) - 2 ];
ode= @ (x,y,xf ) xf*[y(2); 1-y(1)];
sol = bvp4c(ode, bc, solinit);
xf = sol.parameters; % Extract final x (right boundary) value
% Evaluate the solution at all x in X and store variables in U
U = deval(sol,X);
x = x0 + X.*(xf-x0); % Rescale for plotting
y = U(1,:)'  ;        % Extract the solution from  matrix U
dy = U(2,:)' ;        % Extract the derivative of solution from U
A=[x  y  dy], b=xf
plot(x,y,x,dy,'k','linewidth',2);  xlim([0 xf]);grid
xlabel('x'); legend('y(x)','dy(x)/dx')
toc
```

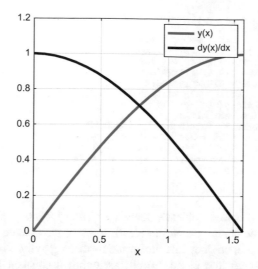

Fig. 10.24 Solution curves for the BVP, $y'' + y = 0$, $y(0) = 0$, $y(b) = 1$, $y'(b) = 0$, for guessed value of $b_g = 1$, where the unknown boundary value is determined by the code using the bvp4c solver as $b = 1.5708$. This point also gives $y'(0) = 1$, the value of derivative function at point $x = 0$. Observe the entries in guessed initial conditions vector

Problem 10.4.2 Use MATLAB bvp4c solver to solve the BVP,

$$y'' + y = 0, \quad y(0) = 0, \ y(b) = 1, \ y'(b) = 0.$$

where b is an unknown boundary point. (bvp4c_unknownbndry2.m).

Solution
Solution curves for the given BVP are displayed in Fig. 10.24 for guessed value of $b_g = 1$, where the unknown boundary value is determined by the code using the bvp4c solver as $b = 1.5708$. This point also gives $y'(0) = 1$. The "guessed initial conditions vector" must be a nonzero vector to avoid singular Jacobian matrix formation. Computed initial and final data are listed below.

x	y	dy
0.0000	0.0000	1.0000
1.5708	1.0000	0.0000

b = 1.5708

Elapsed time is 0.226205 seconds.

```
%bvp4c_unknownbndry2.m     y"+y=0,      y(0)=0,  y(b)=1,   y'(b)=0
clc; clear; close;tic
x0 = 0;   % initial left boundary value
yinit = [x0  0.3 ]; % guessed initial conditions
xf_guess = 1; % initial guess for final boundary
Nx = 21; X = linspace(0,1,Nx)'; % scaled x vector
solinit = bvpinit(X,yinit,xf_guess);
bc = @(Y0,Yf,xf) [Y0(1) - x0; Yf(1) - 1;   Yf(2) - 0 ];
ode= @ (x,y,xf ) xf*[y(2); -y(1)];
sol = bvp4c(ode, bc, solinit);
xf = sol.parameters; % right boundary value
U = deval(sol,X); x = x0 + X.*(xf-x0); % rescale
y = U(1,:)'  ;  dy = U(2,:)'  ;   A=[x  y  dy],     b=xf
plot(x,y,x,dy,'k','linewidth',2);  xlim([0 xf]); grid
xlabel('x'); legend('y(x)','dy(x)/dx'); toc
```

Problem 10.4.3 A space vehicle's launch trajectory from a planet without an atmosphere is to be computed. In the idealized case, we assume constant acceleration (from thrusters), no drag, and uniform flat-planet gravity. We try to maximize the payload delivered into circular orbit, assuming a constant burn rate for the propellant. Minimizing the time to reach orbit is an equivalent goal. So, the problem can be expressed by using a state vector:

$$X = \begin{bmatrix} x & y & v_x & v_y & \lambda_2 & \lambda_4 \end{bmatrix}^T \tag{10.1}$$

where x, y denote position components of the vehicle, while $v_x = x'$, $v_y = y'$ are the velocity components, T indicates transpose operation. The λ_2, λ_4 are the optimization constants. Acceleration is expressed as

$$a = (F/m)g \tag{10.2}$$

The system of ODEs can be described by using first order differential equations,[5]

$$X' = \begin{bmatrix} v_x & v_y & v'_x & v'_y & 0 & \lambda'_2 \end{bmatrix}^T = \begin{bmatrix} X_3 & X_4 & v'_x & v'_y & 0 & -X_5 \end{bmatrix}^T \tag{10.3}$$

where the acceleration components are defined by

$$v'_x = a\left(\frac{1}{\sqrt{1+\lambda_4}}\right) \tag{10.4}$$

$$v'_y = a\left(\frac{\lambda_4}{\sqrt{1+\lambda_4}}\right) - g \tag{10.5}$$

---

[5]For more detailed analysis and derivations of these equations, readers may refer to the book by Longuski et al. [24].

Non-dimensional time is represented by $\tau = t/t_f$ where t_f is the final time, and

$$dX/d\tau = t_f X' \tag{10.6}$$

Since the final time is the unknown boundary, parameterizing the problem by the free final time, t_f creates a time vector, τ, with N linearly spaced elements ($\tau = t/t_f$). The independent variable is the non-dimensional time, τ. One can implement the free final time τ in the code as an unknown parameter and the code solves for the actual final time.

Compute the position and velocity components of the space vehicle for the following conditions:

Final orbiting altitude: $h = 200\,\text{km}$
Circular speed at this altitude: $v_c = 1000\,\text{m/s}$
Gravitational acceleration in the planet: $g = 2\,\text{ms}^{-2}$
Thrust to Weight ratio for space vehicle (in planet G's): $TWR = 5$
Initial time: $t_0 = 0$,
Initial and final component values:

$$x_0 = 0, \quad y_0 = 0, \quad v_{x0} = 0, \quad v_{y0} = 0, \quad y_f = h, \quad v_{xf} = v_c, \quad v_{yf} = 0$$

(rocket1.m).

Solution
Applying given data into Eq. (10.3) and running the MATLAB script (rocket1.m) yields the position and velocity profiles as displayed in Fig. 10.25. We use guessed

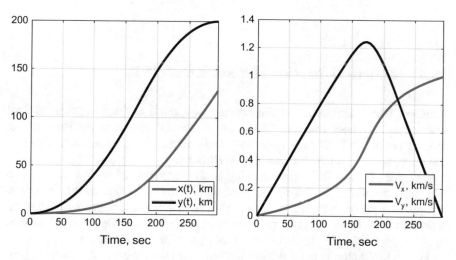

Fig. 10.25 Left: Space vehicle's altitude and horizontal distance from take-off point as a function of time. Right: Horizontal and vertical speed as a function of time. It takes 293.96 s for the space vehicle to reach its orbital elevation

value for the final time as $t_{fg} = 100$ s and 60 grid points. We include two anonymous functions in the code for computing boundary conditions and the ODEs (as it is usual throughout the book).

```
% rocket1.m
 clc;close; clear;tic;
 h = 200e3; %  final altitude, m
 Vc = 1000; %  Circular speed at elevation  h, m/s
 g = 2; % m/s/s, gravitational acceleration of planet
 TWR = 5; % Thrust to Weight ratio for space vehicle, (in planet G's)
 % Initial and final conditions
 x0 = 0; y0 = 0; Vx0 = 0; Vy0 = 0;yf = h; Vxf = Vc; Vyf = 0;
 t0 = 0;% initial time
 yinit = [x0 y0 Vx0 Vy0 0 0];%guess initial conditions, use zero if unknown
 tfg = 100; % sec, initial guess for final time
 N = 60;
 tau = linspace(0,1,N)'; % nondimensional time vector
 solinit = bvpinit(tau,yinit,tfg);
 a = TWR*g;% Acceleration (=Force/mass) , m/s/s
 ode=@(tau,X,tf)  tf*[X(3); X(4); a*(1/sqrt(1+X(6)^2));
     a*(X(6)/sqrt(1+X(6)^2)) - g;   0;   -X(5)];
 bc = @(Y0,Yf,tf) [Y0(1) - x0; Y0(2) - y0;
     Y0(3) - Vx0; Y0(4) - Vy0 ; Yf(2) - yf; Yf(3) - Vxf ; Yf(4) - Vyf ];
 sol = bvp4c(ode, bc, solinit);
 tf = sol.parameters(1);% Extract the final time from the solution:
 S = deval(sol,tau);% store the state variables in the matrix S.
 time = t0 + tau.*(tf-t0); % Rescale
 % Extract the solution for each state variable from the matrix S
 x = S(1,:); y = S(2,:); vx = S(3,:); vy = S(4,:);
 plot(time,x/1000,'linewidth',2); hold on;
 plot(time,y/1000,'k','linewidth',2);
 grid on; xlim([0 tf]);legend('x(t), km','y(t), km');figure;
 plot(time,vx/1000,'linewidth',2);  hold on;
 plot(time,vy/1000,'k','linewidth',2);xlabel('Time, sec');
 grid on; xlim([0 tf]);legend('V_x, km/s','V_y, km/s' );toc;
```

10.5   Numerical Solution of a System of BVPs

Problem 10.5.1 Brillouin scattering occurs as a result of the interaction of light with an optical fiber material. The result of this interaction is a change in the momentum of the light wave (frequency and energy) along preferential angles. It can occur spontaneously, but can also be stimulated.

The physics behind Brillouin scattering is the tendency of optical materials to become compressed in the presence of an electric field (electrostriction). The compression of the medium changes the index of refraction which leads to some reflection or scattering (From a quantum point of view, the process is considered as the interaction of light photons with acoustic or vibrational phonons).

For an oscillating electric field at the laser pump frequency, this process generates an acoustic wave at some other frequency. Spontaneous Brillouin scattering can be viewed as scattering of the pump wave from this acoustic wave, resulting in creation of a new wave at a different pump frequency [25]. The difference of these two frequencies is called the Stokes shift.

Steady-state Brilloin intensity equations are

$$I'_P = -g_B I_P I_S - \alpha I_P$$
$$I'_S = -g_B I_P I_S - \alpha I_S$$

subject to boundary conditions,

$$I_P(0) = I_{P0}, \quad I_S(L) = I_{SL}$$

where $0 \leq z \leq L$, I_P and I_S are the laser pump and stokes fields, α and g_B are fiber loss coefficient and Brillouin gain coefficient, respectively, while L is the total length of the optical fiber.

(a) At which value of L a singular Jacobian is encountered while computing the signal intensities along the fiber length for different values of L, $1 \leq L \leq 20$? For simplicity of computations, let all parameters (α, I_{P0}, I_{SL}) be scaled to unity (brilloin1.m).

(b) MATLAB bvp4c and bvp5c solvers aim to use smallest number of mesh points and the step size is incremented automatically depending on the previous mesh points. In bvpxtend, after obtaining convergence for the mesh, the codes adapt the mesh to get a solution with a modest number of mesh points. Hence, using bvpxtend function also reduces the computation time.

Improve the previous solution using MATLAB command bvpxtend for $1 \leq L \leq 20$.

Solution

We write a MATLAB code using bvp4c with two associated anonymous files (brilloin1.m). The two solution curves (normalized Pumped and Stokes Brilloin intensity curves) along the length of an optical fiber are depicted in Fig. 10.26.

(a) Keeping all other parameters the same and changing the value of L each time for $1 \leq L \leq 20$ in unit increments, it is observed that when $L = [9, 17, 18]$ the system is unable to solve the collocation equations due to encountering a singular Jacobian.

On the other hand, when $L = [8, 11 - 15]$ following warning message is prompted;

```
Warning: Unable to meet the tolerance without using more than 5000
mesh points. The last mesh of 4564 points and the solution are
available in the output argument. The maximum residual is 45.6388,
while requested accuracy is 0.001.
```

Fig. 10.26 Normalized
Pumped and Stokes Brilloin
intensity curves along the
length of an optical fiber
when $L = 6.0$

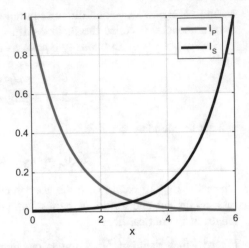

```
% brilloin1.m
clc; clear; close;
gb=1;alpha=1;Ip0=1;IsL=1;Pp0=1;PsL=1;Aeff=1;
L=6
ode =@(x,y) [-gb*y(1)*y(2)-alpha*y(1);-gb*y(1)*y(2)+alpha*y(2)];
bc=@(ya,yb) [ya(1)- Ip0; yb(2)- IsL ];
solinit = bvpinit(linspace(0,L,2), [Ip0,IsL]);
sol = bvp4c(ode, bc, solinit);
x=linspace(0,L,41); Y = deval(sol,x);
Y=Y'; plot(x,Y,'linewidth',2); grid; legend('I_P','I_S');xlabel('x');
```

(b) We use the MATLAB command bvpxtend as shown in the script file
(brilloin2.m). This eliminates previously observed problems for $L = [9, 17, 18]$
(due to encountering a singular Jacobian) and $L = [8, 11 - 15]$. Solution of the
BVP is possible even for much larger values of L (we test $L = 2000$ without
any warning messages, higher values are also possible) while these (higher)
values could not be used in part (a) of this problem (limited to $L = 20$).

```
% brilloin2.m
clc;clear;close;tic
gb=1;alpha=1;Ip0=1;IsL=1;Pp0=1;PsL=1;Aeff=1;
L=7
ode =@(x,y) [-gb*y(1)*y(2)-alpha*y(1);-gb*y(1)*y(2)+alpha*y(2)];
bc=@(ya,yb) [ya(1)- Ip0; yb(2)- IsL ];
solinit = bvpinit(linspace(0,L,2), [Ip0,IsL]);
sol = bvp4c(ode,bc,solinit);
for U=L:20
U=U
solinit=bvpxtend(sol,U);sol = bvp4c(ode,bc,solinit)
x=linspace(0,U,40); Y = deval(sol,x); Y=Y';figure;
plot(x,Y,'linewidth',2);grid;legend('I_P','I_S');xlabel('x');
end
```

Problem 10.5.2 In 1691, James Bernoulli provided a mathematical formulation for a bending lamina fixed at one end point and subject to a load at the other end point. Assuming a lamina AB of uniform thickness and width and negligible weight of its own, supported on its lower perimeter at A, and with a weight hung from its top at B, the force F from the weight at B to ground (let it be BC) is sufficient to bend the lamina perpendicular.

The rectangle formed by the tangent between the axis and its own tangent has a constant area. This poses one specific instance of the general elastica problem, now generally known as the rectangular elastica, because the force applied to one end of the curve bends it to a right angle with the other end held fixed. The fundamental idea is that at every point along the curve, the product of the radius of curvature and the distance from the line BC is a constant, i.e. the two quantities are inversely proportional.[6]

Consider an elastica in the (x, y) plane, with natural parameterisation $0 \leq s \leq 1/2$. We track an angle φ (in place of derivatives for the x and y components) with the curvature $\kappa = \varphi' = d\varphi/ds$. The solution can be found by the five dimensional first order system of ODEs,

$$x' = \cos(\varphi)$$
$$z_2' = \sin(\varphi)$$
$$\varphi' = \kappa$$
$$\kappa' = F\cos(\varphi) = Fx'$$
$$F' = 0$$

subject to following boundary conditions,

$$x(0) = y(0) = \kappa(0) = y(0.5) = 0, \quad \varphi(0.5) = -\pi/2.$$

Use bvp4c (or bvp5c) to solve this BVP, and compute $\varphi(0)$, $x(0.5)$, $\kappa(0.5)$, F (bvp4c_elastica1.m).

Solution
Making the assignments of $z_1 = x$, $z_2 = y$, $z_3 = \varphi$, $z_4 = \kappa$, $z_5 = F$, we arrive at the following system of ODEs,

---

[6]In 1744, Euler wrote his treatise on variational techniques in which he devoted an entire chapter, "De Curvis Elasticis", working out a complete characterization of those curves which are solutions to the elastica problem, and these are nowadays known as "elasticae". Remarkably, the elastica appears as yet another shape of the solution of a fundamental physics problem, the capillary. Laplace investigated the equation for the shape of the capillary in 1807. Since then, the subject has attracted many researchers and it is still an active field of investigation.

In a broad sense elasticae are curves which are stationary points of the elastic energy functional. The elastic energy of a smooth curve is the integral of its squared curvature.

The closed-form solutions of the elastica rely heavily on Jacobi elliptic functions [26–28.

$$z'_1 = \cos(z_3)$$
$$z'_2 = \sin(z_3)$$
$$z'_3 = z_4$$
$$z'_4 = z_5 \cos(z_3) = z_5 z'_1$$
$$z'_5 = 0$$

This system is solved using the MATLAB script file (bvp4c_elastical.m). Following are the output list of this code. It is found that,

$$\varphi(0) = 0.7105\, rad, \quad x(0.5) = 0.3916, \quad \kappa(0.5) = -8.4385, \quad F = -21.5491.$$

```
guess =  1      1      1      1      1

solinit = solver: 'bvpinit',
   x: [0 0.5000], y: [5x2 double], yinit: [1 1 1 1 1]

sol = solver: 'bvp4c'
   x: [1x51 double], y: [5x51 double], yp: [5x51 double]
```

x	y	phi	kappa	F
0.0000	0.0000	0.7105	0.0000	-21.5491
0.3916	0.0000	-1.5708	-8.4385	-21.5491

Solution curves are displayed in Fig. 10.27.

```
%bvp4c_elastical.m
clc;clear;close;
guess = ones(1,5)
solinit = bvpinit(linspace(0,0.5,2),guess)
ode=@(x,z)  [ cos(z(3)); sin(z(3)); z(4); z(5)*cos(z(3)); 0];
options = bvpset('RelTol',1e-6,'AbsTol',1e-6);
bc=@(za,zb) [za(1); za(2); zb(2); zb(3)+ pi/2 ;za(4) ];
sol = bvp4c(ode,bc,solinit,options)
%[sol.x' , sol.y']
x=linspace(0,0.5); Z = deval(sol,x);
% Extract the solution for each state variable from [Z]:
x_sol = Z(1,:); y_sol = Z(2,:);
phi = Z(3,:);    kappa = Z(4,:);
subplot(2,2,1);plot(x,x_sol,'linewidth',2);xlabel('x');xlim([0 0.5]);
grid on;title(('x-component'));
subplot(2,2,2);plot(x,y_sol,'linewidth',2);xlabel('x');xlim([0 0.5]);
grid on;title(('y-component'));ylim([0 0.15]);
subplot(2,2,3);plot(x,phi,'linewidth',2);xlabel('x');xlim([0 0.5]);
grid on;title(('change of \Phi'));
subplot(2,2,4);plot(x,kappa,'linewidth',2);xlabel('x');xlim([0 0.5]);
grid on;title(('change of \kappa'));
Z=Z'
```

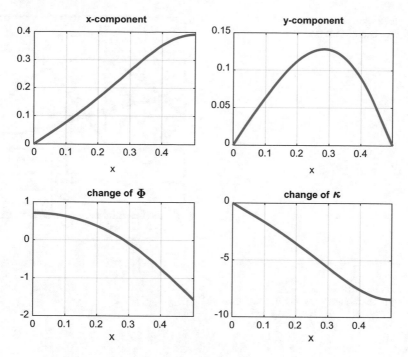

Fig. 10.27 Solution curves for the elastica problem

Problem 10.5.3 A study models the behavior of a system consisting of four first-order coupled ODEs in one space variable,

$$y_1' = 3y_2$$
$$y_2' = e^1 - 2y_1$$
$$y_3' = 10y_4$$
$$y_4' = y_1 + 0.5y_3$$

These equations are to be solved subject to boundary conditions,

$$y_1(0) = y_1(1) = y_1(0) = y_3(0) = y_3(1) = 0$$

Use MATLAB to solve this BVP and plot the solutions in a single figure (bvp_sys1.m).

Solution

We want to solve the BVP on [0, 1], then we specify $x(1) = 0$ and $x(end) = 1$. Here, x is a vector that specifies an initial mesh. We choose five equally spaced points on this mesh. However, we provide equally spaced 10 points of output solution, as listed below. Figure 10.28 displays the solution curves.

Fig. 10.28 Solution curves
for the system of ODEs

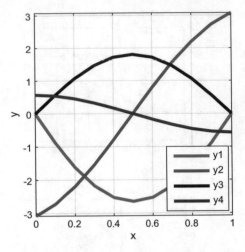

```
          init =  solver: 'bvpinit'
     x: [0 0.2000 0.4000 0.6000 0.8000 1]
              y: [4x6 double]
              yinit: [0 0 0 0]
```

x	y1	y2	y3	y4
0	0	-3.0777	0	0.5719
0.1000	-0.8735	-2.7168	0.5616	0.5415
0.2000	-1.6138	-2.1936	1.0648	0.4568
0.3000	-2.1766	-1.5394	1.4609	0.3293
0.4000	-2.5282	-0.7934	1.7135	0.1722
0.5000	-2.6479	-0.0000	1.8003	0.0000
0.6000	-2.5282	0.7934	1.7135	-0.1722
0.7000	-2.1766	1.5394	1.4609	-0.3293
0.8000	-1.6138	2.1936	1.0648	-0.4568
0.9000	-0.8735	2.7168	0.5616	-0.5415
1.0000	0	3.0777	0	-0.5719

MATLAB code (bvp_sys1.m) is the following:

```
%bvp_sys1.m        4 coupled  equations  in a 2-points BVP
clc;clear;close;
a=0;b=1;                   % initial-final (boundary) points
x=a:0.2:b;                 % x = vector that specifies an initial mesh.
yinit=[0 0 0 0];           % Guess  vector for the solution
init = bvpinit(x,yinit)    % forms  the initial guess for the bvp solver.
f = @(x,y)    [ 3*y(2)
                exp(1)-2*y(1)
                10*y(4)
                y(1)+.5*y(3)] ;
bc=@(ya,yb) [ ya(1); yb(1) ; ya(3); yb(3) ];% Boundary conditions
% A1.y(a)+B1.y'(a)=e, A2.y(b)+B2.y'(b)=h
S=bvp4c(f,bc,init);
x=S.x;%
y=S.y;% Evaluation of y at x (min 11 points)
%y=deval(S,x);% Evaluation of y at x (specified number of points by x)
disp('      x         y1        y2        y3        y4 ')
[x' y']
plot(x,y,'linewidth',2.5);legend('y1','y2','y3','y4');
axis tight; xlabel('x'); ylabel('y'); grid;
```

Problem 10.5.4 Coolants like water, oil and ethylene glycol have low thermal conductivities. Researchers recently try to enhance their properties by adding nano-sized metallic particles [29]. Nano-fluid is the combination of base-fluid (water and kerosene oil) and nanoparticles having diameter 10–100 nm, made of metal nitrides (SiN), oxide ceramics (Al2O3, CuO), carbide ceramics, metals (Cu, Ag, Au) and graphite, carbon nanotubes and fullerene.

These studies require theoretical prediction of the electro-magneto squeezing flow of nano-fluid between two stretchable Riga plates [30]. A Riga plate is an electromagnetic actuator which is the combination of permanent magnets and a span-wise aligned array of alternating plus-minus electrodes mounted on a plane surface so that magnetic and electric fields generate Lorentz force parallel to the wall which controls the flow of fluid. For a more detailed study and possible applications of this subject, please refer to articles referenced at the end of this chapter, [31–37].

Numerical solutions of the dimensionless problems are constructed (for example), by using shooting method in [38]. Cartesian coordinate system is selected such that x-axis is in the direction of the stretched Riga plate while y-axis is perpendicular to the x-axis.

In [39], the convective heat transfer of the two-dimensional unsteady squeezing flow past a Riga plate is investigated using Cattaneo-Christov heat flux model. With the help of suitable similarity transformation, the governing partial differential equations (PDEs) are converted into the ODEs that are further reduced to first order system of ODEs consisting of eight equations given below;

$$f_1' = f_2,$$
$$f_2' = f_3,$$
$$f_3' = f_4,$$
$$f_4' = f_2 f_3 - f_1 f_4 + \frac{S}{2}(3f_3 + \eta f_4) + ZBe^{-B\eta},$$
$$f_5' = f_6,$$
$$f_6' = (P_r A - \beta_e B)/(1 + R_d - P_r \beta_e S^2 \eta^2 / 4 - P_r f_1^2 \beta_e + P_r S \beta_e \eta f_1),$$
$$f_7' = f_8,$$
$$f_8' = -S_c \left(f_1 f_8 - \frac{S}{2} \eta f_8 + K_r f_7 \right),$$
$$A = \frac{S}{2} \eta f_6 - f_1 f_6, \quad B = \frac{3}{4} S^2 \eta f_6 + f_1 f_2 f_6 + \frac{S}{2} \eta f_2 f_6 + \frac{3}{2} f_1 f_6$$

Use MATLAB bvp4c to solve this system of ODEs for the following boundary conditions,[7] and coefficient values, $0 \leq \eta \leq 1$. Then, plot velocity profile $f_2(\eta)$ as functions of squeezing parameter $0 \leq S \leq 0.6$, $0.5 \leq \eta \leq 1$.

---

[7]These boundary conditions and used coefficient values differ from those reported in [39].

$$f_2(0) = 1, \quad f_4(0) = 0, \quad f_6(0) = -B_{i1}(1 - f_5(0)), \quad f_8(0) = -B_{i2}(1 - f_7(0))$$
$$f_1(1) = \tfrac{S}{2}, \quad f_2(1) = 0, \quad f_5(1) = 0, \quad f_7(1) = 0$$
$$Z = 1.5, \quad B = 10, \quad K_r = 0.2, \quad P_r = 1, \quad B_e = 0.1, \quad R_d = 0.2, \quad S_c = 0.5,$$
$$B_{i1} = B_{i2} = 0.2$$

(bvp4c_sys5.m).

Solution

We use MATLAB script (bvp4c_sys5.m) to solve this system of ODEs for given boundary conditions and coefficient values. For convenience, $f(\eta)$ is replaced by $y(x)$. Curves showing velocity profile $f_2(\eta)$ as functions of squeezing parameter $0 \le S \le 0.6$ are displayed in Fig. 10.29.

```
%bvp4c_sys5.m        8 coupled 1st order ODEs  in a 2-point BVP
clc;clear;close;
Z=0; B=10;  Kr=0.2; Pr=1; Be=0.1; Rd=0.2; Sc=0.5; Bi1=0.2; Bi2=0.2;
x=linspace(0,1,5);      % initial mesh.
yinit=ones(1,8);        % Guess  vector for the solution
solinit = bvpinit(x,yinit) % initial guess
for S=0:0.2:0.6
ode = @(x,y)   [ y(2); y(3); y(4); % diff. equations
y(2)*y(3)-y(1)*y(4)+(S/2)*(3*y(3)+x*y(4))+Z*B*exp(-B*x);y(6);
Pr*(S*x*y(6)/2-y(1)*y(6))-...
Be*(3*S^2*x*y(6)/4+y(1)*y(2)*y(6)+S*x*y(2)*y(6)/2 + 3*S*y(1)*y(6)/2)/...
(1+Rd-Pr*Be*S^2*x^2/4-Pr*y(1)^2*Be+Pr*S*Be*x*y(1));
y(8); -Sc*y(1)*y(8) ];
bc=@(ya,yb) [ ya(2)-1; ya(4); ya(6)+Bi1*(1-ya(5)); ya(8)+Bi2*(1-ya(7));
     yb(1)-S/2; yb(2); yb(5); yb(7) ]; % Boundary conditions
sol=bvp4c(ode, bc, solinit); x=linspace(0,1,41);
y = deval(sol,x);% Evaluation of y at x
[x' y'];
plot(x, y(2, :),'linewidth',2); hold on
end
grid; legend('S=0','S=0.2','S=0.4','S=0.6');
xlabel('\eta'); ylabel('f_2(\eta)');
```

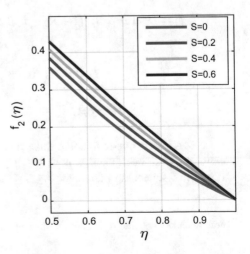

Fig. 10.29 Velocity profile change as function of squeezing parameter

10.6 Continuation Methods for Solving BVPs

Problem 10.6.1 (Nonlinear perturbation problem) Use bvp4c to solve the following BVP for $y(t)$,

$$\varepsilon y'' + yy' - y = 0, \quad y(0) = 1, \quad y(1) = -1/3, \quad \varepsilon \in [0.1 \quad 0.005]$$

(bvp4c_perturb1.m).

Solution

We first solve the problem for a larger value of ε, then apply continuation by using the result of the first solution as a guessed value for the second solution while tighten the relative tolerance to resolve the boundary layer better and reduce the perturbation parameter in second solution.

Solution curves of the BVP for $\varepsilon = 0.1$ and $\varepsilon = 0.005$ are shown in Fig. 10.30. The solution has boundary layers at $t = 0$, and $t = 1$ [40].

```
%bvp4c_perturb1.m ,   e*y'' +yy'-y = 0 , BCs y(0)=1,y(1)=-1/3
clc;clear;close;
%A nonlinear problem with a boundary layer at the origin and x=1.
e = 0.1 ; % perturbation
options =[]; % default
% bvpinit is used to specify an initial guess for the mesh of 2
% equally spaced points.  A constant guess based on a straight line
% between the boundary values for y is 0 for y(t) and 1 for y'(t).
solinit = bvpinit(linspace(0,1,2),[0  1]);
ode=@(x,y) [ y(2); -y(1)*y(2)/e + y(1)/e];% ODE, y(1) = y, y(2) = y'
bc=@(ya,yb)[ ya(1)-1; yb(1) + 1/3 ];
sol = bvp4c(ode,bc,solinit, options);
t = sol.x; y = sol.y;
clf reset
plot(t,y(1,:),'linewidth',2);axis([0  1  -0.4  1]);
title('RelTol = 1e-3, \epsilon = 0.1');xlabel('t');ylabel('y');grid;
% A smaller RelTol is used to resolve  the boundary layer better
% The previous solution provides a guess.
options = bvpset(options,'RelTol',1e-4);
e = 0.005; % new perturbation
ode=@(x,y) [ y(2); -y(1)*y(2)/e + y(1)/e];
sol = bvp4c(ode,bc,sol,options);
t = sol.x;y = sol.y;
figure
plot(t,y(1,:),'linewidth',2); axis([0  1 -0.4  1]);
title('RelTol = 1e-4, \epsilon = 0.005'); xlabel('t');ylabel('y');grid;
```

Problem 10.6.2 (Continuation in a perturbed equation) Allen-Cahn equation [41] is a reaction–diffusion equation which describes the process of phase separation in multi-component alloy systems. This ODE also admits periodic solutions, which could be expressed as a suitable scaling of the Jacobi elliptic functions. It is well accepted that the Allen-Cahn equations can be used to describe population dynamics, as well [42, 43].

In one dimensional case, the Allen-Cahn equation reduces to a second order ODE which can be written as

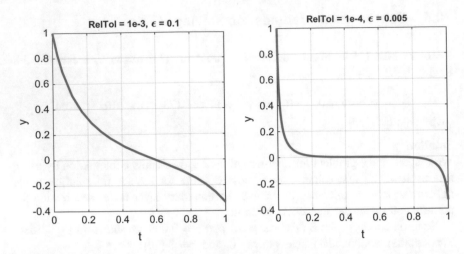

Fig. 10.30 Solution curves for $\varepsilon = 0.1$(left), and $\varepsilon = 0.005$ (right) of the BVP, $\varepsilon y'' + yy' - y = 0$, $y(0) = 1$, $y(1) = -1/3$

$$\epsilon y'' + y(1 - y^2) = 0,\ 0 < x < 1$$

subject to boundary conditions,

$$y(0) = -1, \quad y(1) = 1$$

Both $y \equiv -1$ and $y \equiv 1$ are solutions of the problem, and it models the change between these extreme solutions. When ϵ is small, the curve becomes more steep. The unique odd heteroclinic solution is $y = \tan h(x/\sqrt{2})$, $\epsilon = 1$.

(a) Use bvp4c to solve the BVP for $\epsilon = 0.1$, 0.01, $0.001, 0.0001$ and only 2 initial mesh points for the guessed solution having a constant value. Comment on the results (bvp4c11.m), (bvp4c12.m).

(b) Plot the solution for the initial "guessed solution vector" consisting of different number of mesh points within the interval $0 \le x \le 1$, $mp = [2, 9, 10, 14, 16]$ using perturbation parameter of $\epsilon = 0.001$ (bvp4c13.m).

(c) Use the method of continuation to obtain the solution of steady-state Allen-Cahn equation for $\epsilon = 0.001$ and for $\epsilon = 0.0001$ when the number of guessed mesh points within the interval $0 \le x \le 1$ is 2. Let the guessed solutions have a constant value of zero (bvp4c14.m)

Compute the mean elapsed time of the script for four successive runs, disregarding the result of first elapsed time measurement (bvp4c15.m).

Solution

(a) We write the first MATLAB script (bvp4c11.m). We obtain following output when 2 mesh points are used as the initial guess vector;

```
solver: 'bvp4c'
     x: [1x49 double]
     y: [2x49 double]
    yp: [2x49 double]
 stats: [1x1 struct]
```

Use of different ϵ values influence the number of solution mesh points, as described in Table 10.1.

The structure solution means that the initial number of 2 mesh points have been replaced by unequally spaced 15, 28 or 49 mesh points. These three solution curves are shown in Fig. 10.31.

For $\epsilon = 0.0001$, following message is received;

```
Unable to solve the collocation equations - a singular Jacobian
encountered.
```

This happens because as ϵ gets smaller, the horizontal transition region also becomes smaller, and constant initial guess is not close enough to the correct form. One may then use $\epsilon = 10^{-3}$ as the initial guess and apply continuation (in

Table 10.1 Number of solution points obtained for different perturbations and initial guess mesh points (mp)

eps	mp
0.1	15
0.01	28
0.001	49

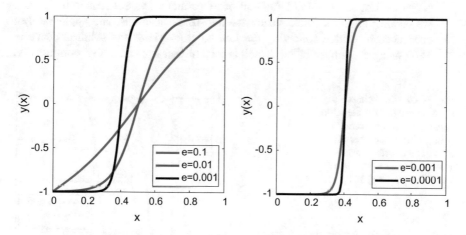

Fig. 10.31 Left: Three solution curves for different perturbations. Right: Solution curves for smaller perturbations and post continuation. For both cases, the number of initial (guessed) mesh points is two (with constant value of zero at these points). Note that, the solution curves for $\epsilon = 0.001$ and $\epsilon = 0.0001$ do not pass through the origin

which the initial guess for each iteration is the previous solution) with some minor changes in the code [44] for $\epsilon = 10^{-4}$ using the modified script (bvp4c12.m), which yields the two curves of Fig. 10.31, right. The structure solution means that the initial number of 2 mesh points have been replaced by unequally spaced 49 mesh points, first. Then, the solution is extended for $\epsilon = 10^{-4}$, yielding 82 new mesh points.

```
sol1 =
    solver: 'bvp4c'
         x: [1x82 double]
         y: [2x82 double]
        yp: [2x82 double]
     stats: [1x1 struct]
Elapsed time is 0.535877 seconds.
```

It should be noted here that, the solution curves for $\epsilon = 0.001$ and $\epsilon = 0.0001$ do not pass through the origin.

(b) The solution curves for the "guessed solution vector" consisting of $mp = [2, 9, 10, 14, 16]$ (number of mesh points) using perturbation parameter of $\epsilon = 0.001$ are displayed in Fig. 10.32. This curve indicates that there is something "wrong" in the previous computation of the solution using 2 mesh points when $\epsilon = 0.001$, and needs to be corrected. Once this is done, the solution for $\epsilon = 0.0001$ will also be improved.

(c) After some tests performed on MATLAB script files, it has been observed that computed solution curves depart from the "expected" solution for the values of perturbation parameter when $\epsilon < 0.006$. Therefore, one can choose $\epsilon = 0.007$ and then apply continuation method successively to get the solution curves plotted for $\epsilon = 0.001$ and also for $\epsilon = 0.0001$. We note that when the number of guessed mesh points is 2 (and all other parameter values remain the same), we obtain the first solution vector having 42 mesh points, and application of continuation method following the first solution yields the solution curve in 1469 unequal number of mesh points and further application of continuation

Fig. 10.32 The "ghost solution curves" for the initial "guessed solution vector" using different number of mesh points with constant values of $y_{guess} = 0$, when $\epsilon = 0.001$, $0 \leq x \leq 1$

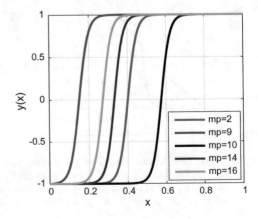

Fig. 10.33 Three solution curves for smaller different perturbations (pre and post continuation). For all cases, the number of initial mesh points is two (with constant value is assumed for the solution at these points). Note that, all solution curves pass through the origin

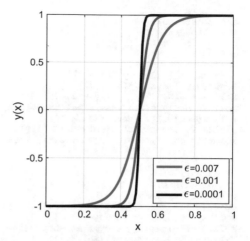

method following the second solution yields the third solution curve in 688 unequal number of mesh points as shown in Fig. 10.33. Program statistics are as follows:

$\epsilon = 0.007$:

The solution was obtained on a mesh of 42 points. The maximum residual is 9.481e-04.

There were 16,220 calls to the ODE function, 534 calls to the BC function.

$\epsilon = 0.001$:

The solution was obtained on a mesh of 1469 points. The maximum residual is 8.681e-04.

There were 228,215 calls to the ODE function, 385 calls to the BC function.

$\epsilon = 0.0001$:

The solution was obtained on a mesh of 688 points. The maximum residual is 7.353e-04.

There were 72,171 calls to the ODE function. There were 103 calls to the BC function.

The mean elapsed time for four successive runs, disregarding the result of first elapsed time measurement (five measurements, in total) is found to be 5.5080 s (MATLAB version 2016a, 64 bit, running on a computer employing an intel Core^{TM} – i5 processor, 1 GB VRAM).

All MATLAB scripts used in the solution of this problem are listed below.

```
%bvp4c11.m
clc;clear;tic;
%e=0.1; %epsilon
% e=0.01;
e=0.001;
%e=0.0001
ode=@(x,y) [y(2); (y(1)^3-y(1))/e];
bc=@(ya,yb) [ya(1)+1;  yb(1)-1];
x=linspace(0,1,2); guessy=@(x) [x; 0];%constant guess value
solinit=bvpinit(x,guessy); sol=bvp4c(ode,bc,solinit)
X=linspace(0,1);Y=deval(sol,X,1);
plot(X,Y,'linewidth',2);xlabel('x');ylabel('y(x)');hold on;
legend('e=0.1','e=0.01','e=0.001'); toc;


%bvp4c12.m
clc;clear;tic;
e=0.001; %epsilon
ode=@(x,y) [y(2); (y(1)^3-y(1))/e]; bc=@(ya,yb) [ya(1)+1;  yb(1)-1];
x=linspace(0,1,2); guessy=@(x) [x; 0];%constant guess value
solinit=bvpinit(x,guessy); sol=bvp4c(ode,bc,solinit)
X=linspace(0,1);Y=deval(sol,X,1);
plot(X,Y,'linewidth',2);xlabel('x');ylabel('y(x)');hold on;
ep=0.0001;
ode=@(x,y) [y(2); (y(1)^3-y(1))/ep];
sol1=bvp4c(ode, bc, sol) % continuation
X=linspace(0,1);Y=deval(sol1,X,1);
plot(X,Y,'linewidth',2);xlabel('x');ylabel('y(x)');
legend('e=0.001','e=0.0001');ylim([-1 1]); toc;

%bvp4c13.m
clc; clear; tic;
e=0.001;
for p=[2,9,10,14,16]
p=p
ode=@(x,y) [y(2); (y(1)^3-y(1))/e]; bc=@(ya,yb) [ya(1)+1;  yb(1)-1];
x=linspace(0,1,p); guessy=@(x) [x; 0];%constant guess value
solinit=bvpinit(x,guessy); sol=bvp4c(ode,bc,solinit)
X=linspace(0,1); Y=deval(sol,X,1);
plot(X,Y,'linewidth',2);xlabel('x');ylabel('y(x)');hold on;
end
legend('mp=2','mp=9','mp=10','mp=14','mp=16'); grid; toc;

%bvp4c14.m  continuation method for eps=0.001 and eps=0.0001
clc;clear;tic;
e=0.007 % epsilon with "working solution"
ode=@(x,y) [y(2); (y(1)^3-y(1))/e];bc=@(ya,yb) [ya(1)+1;  yb(1)-1];
x=linspace(0,1,2); guessy=@(x) [x; 0];%constant guess value
solinit=bvpinit(x,guessy);sol=bvp4c(ode,bc,solinit)
X=linspace(0,1);Y=deval(sol,X,1);plot(X,Y,'linewidth',2);hold on;
%
ec=0.001 % epsilon for the first continuation run
ode=@(x,y) [y(2); (y(1)^3-y(1))/ec];sol1=bvp4c(ode,bc,sol)
X=linspace(0,1);Y=deval(sol1,X,1);plot(X,Y,'linewidth',2);
%
ed=0.0001 % epsilon for the second continuation run
ode=@(x,y) [y(2); (y(1)^3-y(1))/ed];sol2=bvp4c(ode,bc,sol1)
```

```
X=linspace(0,1);Y=deval(sol2,X,1);plot(X,Y,'linewidth',2);
xlabel('x');ylabel('y(x)');
legend('\epsilon=0.007','\epsilon=0.001','\epsilon=0.0001');grid
toc;

% bvp4c15.m  steady-state Allen-Cahn eqn.
% Continuation and mean elapsed time measurement
clc;close; clear;tic;
M = 3 ;                     % number of times to run the code
elapsed_time = zeros(M,1); % initialize
result = zeros(M,1) ;      % initialize the result for every run
options=bvpset('stats','on');
for i = 1:M
t = tic ;
e=0.007;
ode=@(x,y) [y(2); (y(1)^3-y(1))/e];bc=@(ya,yb) [ya(1)+1;  yb(1)-1];
x=linspace(0,1,2); guessy=@(x) [x; 0];%constant guess value
solinit=bvpinit(x,guessy); sol=bvp4c(ode,bc,solinit,options);
X=linspace(0,1);Y=deval(sol,X,1);plot(X,Y,'linewidth',2);hold on;
%
ec=0.001;
ode=@(x,y) [y(2); (y(1)^3-y(1))/ec];sol1=bvp4c(ode,bc,sol,options);
X=linspace(0,1);Y=deval(sol1,X,1);plot(X,Y,'linewidth',2);
%
ed=0.0001;
ode=@(x,y) [y(2); (y(1)^3-y(1))/ed];sol2=bvp4c(ode,bc,sol1,options)
X=linspace(0,1);Y=deval(sol2,X,1);plot(X,Y,'linewidth',2);
t = toc(t);   % get time elapsed
elapsed_time(i) = t;   % store the time elapsed for the run
end
xlabel('x');ylabel('y(x)');
legend('\epsilon=0.007','\epsilon=0.001','\epsilon=0.0001');grid
mean_elapsed_time_1  = mean(elapsed_time); % avarage of M runs
mean_elapsed_time_2 = (sum(elapsed_time)-elapsed_time(1))/(M-1)%avg,M-1
```

Problem 10.6.3 Certain classes of two-point boundary value problems are particularly sensitive to the initial conditions and cause troubles, numerically. Roberts and Shipman describe a continuation method which they applied to shooting methods to solve problems normally beyond the capability of shooting methods.

Suppose we have the two-point BVP, and suppose we guess the missing initial conditions. If the problem is very sensitive to the initial conditions, or to the boundary values, our initial guesses may be sufficiently in error that the equations can be integrated only within a limited interval, beyond which the solution blows up. Other guesses can be tried, for the missing initial conditions and then we try to integrate once again the ODE. The cycle of guessing the missing initial conditions and trying to integrate the ODE may continue until these operations can be quite exhaustive.

In this problem we illustrate a case study for a problem presented by Holt and later by Roberts and Shipman [45, 46] which is a set of five first order nonlinear ODEs with some given boundary values,

$$y_1' = y_2$$
$$y_2' = y_3$$
$$y_3' = -\left(\tfrac{3-n}{2}\right)y_1y_3 - ny_2^2 + 1 - y_4^2 + sy_2$$
$$y_4' = y_5$$
$$y_5 = -\left(\tfrac{3-n}{2}\right)y_1y_5 - (n-1)y_2y_4 + s(y_4 - 1)$$

with the following initial and terminal conditions: $y_1(0) = y_2(0) = y_4(0) = 0 = y_2(t_f)$, $y_4(t_f) = 1$ and coefficient values, $n = -0.1$, $s = 0.2$, $t_0 = 0$, $t_f = 11.2$.

Such problems arise in many engineering areas (bvp_sys2.m), (bvp_sys3.m).

Solution
This problem is not solvable by convential shooting methods and can be solved using combination of quasilinearization and finite difference methods as well as by continuation in shooting method. Here, $y_3(0)$, $y_5(0)$, $y_1(t_f)$, $y_3(t_f)$, $y_5(t_f)$ are not given and they need to be estimated. The values of the first two functions are crucial to judge the performance of the solver, in particular.

Although the final values of y_3 and y_5 attain the same levels at $x = \infty$, their initial estimates are different than their computed values by this code, therefore (as a remedy) one is forced to reduce the interval length to increase the accuracy of initial estimates. However, this is still not sufficient to reach a plausible level in both parameters.

Relatively good agreement is reached for the value of y_3 when initial guess vector is taken as `yinit=[1 1 0 1 1]`, while the interval is again extended. However, this operation changes all other curves. See the results, below. Reference values are obtained from [45].

x	y1	y2	y3	y4	y5
0	0	0	-0.8581	0	0.8281

```
Reference values:y3(0)= -0.9663, y5(0)= 0.6529
Computed values:y3(0)= -0.8581, y5(0)= 0.8281  (interval=11.2)
Computed values:y3(0)= -0.9067, y5(0)= 0.7842  (interval= 2.4)
Computed values:y3(0)= -0.8222, y5(0)= 0.7171  (interval= 2.0)

Computed values:y3(0)= -0.9652, y5(0)= 0.2256  (interval=10.0)
yinit=[1 1 0 1 1];
```

So far this particular BVP is solved by `bvp4c` without application of continuation. Next, the application of continuation process using MATLAB `bvp4c` function will be demonstrated in the solution this problem. In this process we do not use MATLAB "function" functions.[8] Instead, we prefer to implement "anonymous" functions.

---

[8]See, [47] for the application of this code using "function" functions, which is somehow seems an accustomed way of writing BVP codes using `bvp4c` (Incidently, the reader may notice that their equations contain minor mistakes—subscript of y5 is written twice as y3 in the equations -, but it is in correct form within the code).

Solution with continuation:

Since we are having trouble solving the BVP due to convergence problem of the solver, we can approximate the BVP by a linear one. In linearized problems there is no iteration in bvp4c solver. Here our objective is to continue from the solution of the easy problem to the solution of the problem in hand. A way of doing this is to use a dummy parameter and solve the family of BVPs.

Rewriting the given ODEs as the sum of their linear and nonlinear terms, we obtain the following equation:

$$
y' = \begin{bmatrix} y_2 \\ y_3 \\ 1 + sy_2 \\ y_5 \\ s(y_4 - 1) \end{bmatrix} + \delta \begin{bmatrix} 0 \\ 0 \\ -(1.5 - 0.5n)y_1 y_3 - ny_2^2 - y_4^2 \\ 0 \\ -(1.5 - 0.5n)y_1 y_5 - (n - 1)y_2 y_4 \end{bmatrix}
$$

In writing this equation, we first assumed that the multiplier $\delta = 1$. Note that a linear approximation to given BVP is obtained when $\delta = 0$. This approximating linear BVP can be solved with a nominal guess along with the given boundary conditions. In the next step, we use previously computed values as the new guesses but include nonlinear part of ODE multiplied by some fraction of δ. We proceed in this manner until we reach to $\delta = 1$.

The modified MATLAB code bvp_sys3.m (listed below) works in this way. It is observed that the agreement between the estimated values for $y_3(0)$ and $y_5(0)$ of the computed solution and referenced values reported in [45] is acceptable. See, Fig. 10.34.

```
%bvp_sys2.m         5 coupled 1st order ODEs  in a 2-point BVP
%Roberts-Shipman and Holt problem
clc;clear;close;
n=-0.1;s=0.2;
a=0;b=11.2;               % initial-final (boundary) points
x=a:0.2:b;               % x = vector that specifies an initial mesh.
yinit=[0 0 0 0 0];       % Guess vector for the solution
init = bvpinit(x,yinit); % forms  the initial guess for the bvp solver.
f = @(x,y)   [ y(2)
               y(3)
               -(1.5-n/2)*y(1)*y(3)-n*y(2)*y(2)+1-y(4)*y(4)+s*y(2)
               y(5)
               -(1.5-n/2)*y(1)*y(5)-(n-1)*y(2)*y(4)+s*(y(4)-1) ] ;
bc=@(ya,yb) [ ya(1); ya(2) ; ya(4); yb(2); yb(4)-1 ];% Boundary conditions
% A1.y(a)+B1.y'(a)=e, A2.y(b)+B2.y'(b)=h
S=bvp4c(f,bc,init);
y = deval(S,x);% Evaluation of y at x   (specified number of points by x)
fprintf('Reference values:y3(0)=-0.9663,y5(0)=0.6529\n')
fprintf(' Computed values:y3(0)=%6.4f,y5(0)=%6.4f\n',y(3,1),y(5,1))
disp('      x         y1        y2        y3        y4        y5')
[x' y']
plot(x,y,'linewidth',2.5);legend('y1','y2','y3','y4','y5');
axis tight; xlabel('x'); ylabel('y'); grid;
```

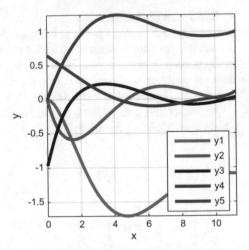

Fig. 10.34 BVP solved by bvp4c without application of continuation

Problem 10.6.4 (Singular perturbation problem) Consider the swirling flow between two rotating, coaxial disks,[9] located at $x = 0$ and at $x = 1$. The BVP consists of a system of six first order ODEs [49],

$$y_1' = y_2$$
$$y_2' = \frac{1}{\mu}(y_1 y_4 - y_2 y_3)$$
$$y_3' = y_4$$
$$y_4' = y_5$$
$$y_5' = y_6$$
$$y_6' = \frac{1}{\mu}(y_3 y_6 - y_1 y_2)$$

subject to the following boundary conditions:

$$y_3(a) = y_4(a) = y_3(b) = y_4(b) = 0, \quad y_1(a) = \omega_0, \quad y_1(b) = \omega_1$$

where ω_0, ω_1 are the angular velocities of the infinite disks, and μ is a viscosity parameter, $0 < \mu \le 1$. Taking $\omega_1 = 1$, we obtain different cases for different values of ω_0. If $\omega_0 < 0$ (with a special symmetry when $\omega_0 = -1$) then the disks are counter-rotating, if $\omega_0 = 0$ then one disk is at rest, while if $\omega_0 > 0$, then the disks are co-rotating.

(a) Let $a = 0$, $b = 1, \omega_0 = -1$, $\omega_1 = 1$. Use MATLAB bvp4c solver which yields a solution without any warning messages (in other words, find minimum possible value of μ with four decimal digits of accuracy so that no singular Jacobian is encountered during the computation of the solution or no message is received such as "unable to meet the tolerance").

---

[9]This problem is derived from [48], referred to there as Swirling Flow III (SWF-III).

Let the initial guesses for the solution components are chosen to be straight lines.
(bvp4c_swirling1.m).

(b) Apply continuation method to the solution obtained in part **(a)** of this problem, and find reduced (extended) new value of the minimum parameter μ. (bvp4c_swirling2.m).

Solution

(a) Boundary conditions are

$$y_3(0) = y_4(0) = y_3(1) = y_4(1) = 0, \quad y_1(0) = -1, \quad y_1(1) = 1$$

It is found that for the values of $\mu \geq 0.0199$ no singular Jacobian matrix is encountered during the computation of the solution. Edited output list of the MATLAB script file is given below.

```
mu = 0.0199
solinit = solver:'bvpinit'
        x:[0 1],   y:[6x2 double],   yinit: [1 1 1 1 1 1]
sol = solver: 'bvp4c'
        x: [1x480 double], y: [6x480 double], yp: [6x480 double]
```

	y1	y2	y3	y4	y5	y6
x=0:	-1.0000	2.5190	0	0	1.5779	-20.5514
x=1:	1.0000	2.5190	0	0	-1.5779	-20.5514

The six solution curves within given interval are displayed in Fig. 10.35.

Parameter $\mu \geq 0.0129$ is reached without application of the continuation procedure, if tolerances are increased, as shown below;

```
options = bvpset('RelTol',1e-1,'AbsTol',1e-3);
```

(b) Applying continuation method to the solution obtained in part **(a)** of this problem, we find reduced (extended) new value of the minimum parameter $\mu = 0.0006$. Solution curves of the BVP obtained by running the modified MATLAB script file (bvp4c_swirling2.m) are displayed in Fig. 10.36. Edited partial output list of the MATLAB script file is given below.

```
mu = 6.0000e-04
sol = solver: 'bvp4c',
x: [1x1114 double], y: [6x1114 double],  yp: [6x1114 double]
```

	y1	y2	y3	y4	y5	y6
x=0:	-1.0000	18.8565	0	0	14.3987	-779.9094
x=1:	1.0000	18.8567	0	0	-14.3984	-779.9151

```
%bvp4c_swirling1.m
clc;clear;close;
mu=0.0199
guess = [1 1 1 1 1 1];
solinit = bvpinit(linspace(0,1,2),guess)
ode=@(x,y)  [y(2);  (y(1)*y(4)-y(2)*y(3))/mu;  y(4);...
              y(5);   y(6);  (y(3)*y(6)-y(1)*y(2))/mu];
options = bvpset('RelTol',1e-3,'AbsTol',1e-6);
bc=@(ya,yb) [ya(1)+1; yb(1)-1; ya(3); ya(4) ; yb(3); yb(4)];
sol = bvp4c(ode,bc,solinit,options)
%[sol.x' , sol.y']
x=linspace(0,1); Z = deval(sol,x);
% Extract the solution for each state variable from [Z]:
Y1 = Z(1,:); Y2 = Z(2,:); Y3 = Z(3,:);
Y4 = Z(4,:); Y5 = Z(5,:); Y6 = Z(6,:);
subplot(3,2,1);plot(x,Y1,'linewidth',2);xlabel('x');xlim([0 1]);
grid on;title(('y1'));
subplot(3,2,2);plot(x,Y2,'linewidth',2);xlabel('x');xlim([0 1]);
grid on;title(('y2'));
subplot(3,2,3);plot(x,Y3,'linewidth',2);xlabel('x');xlim([0 1]);
grid on;title(('y3'));
subplot(3,2,4);plot(x,Y4,'linewidth',2);xlabel('x');xlim([0 1]);
grid on;title(('y4'));
subplot(3,2,5);plot(x,Y5,'linewidth',2);xlabel('x');xlim([0 1]);
grid on;title(('y5'));
subplot(3,2,6);plot(x,Y6,'linewidth',2);xlabel('x');xlim([0 1]);
grid on;title(('y6'));
Z=Z'


%bvp4c_swirling2.m
clc;clear;close;
mu1=0.0199    % Parameter value before applying continuation method
guess = [1 1 1 1 1 1]; solinit = bvpinit(linspace(0,1,2),guess)
ode=@(x,y)  [y(2); (y(1)*y(4)-y(2)*y(3))/mu1; y(4);...
              y(5);   y(6);  (y(3)*y(6)-y(1)*y(2))/mu1];
options =[];
bc=@(ya,yb) [ya(1)+1; yb(1)-1; ya(3); ya(4) ; yb(3); yb(4)];
sol1 = bvp4c(ode, bc, solinit, options)
mu=0.0006    % Reduced parameter
ode=@(x,y)  [y(2); (y(1)*y(4)-y(2)*y(3))/mu; y(4);...
              y(5);   y(6);  (y(3)*y(6)-y(1)*y(2))/mu];
sol = bvp4c(ode, bc, sol1, options) %continuation
x=linspace(0,1); Z = deval(sol,x);
% Extract the solution for each state variable from [Z]:
Y1 = Z(1,:); Y2 = Z(2,:); Y3 = Z(3,:);
Y4 = Z(4,:); Y5 = Z(5,:); Y6 = Z(6,:);
subplot(3,2,1);plot(x,Y1,'linewidth',2);xlabel('x');xlim([0 1]);
grid on;title(('y1'));
subplot(3,2,2);plot(x,Y2,'linewidth',2);xlabel('x');xlim([0 1]);
grid on;title(('y2'));
subplot(3,2,3);plot(x,Y3,'linewidth',2);xlabel('x');xlim([0 1]);
grid on;title(('y3'));
subplot(3,2,4);plot(x,Y4,'linewidth',2);xlabel('x');xlim([0 1]);
grid on;title(('y4'));
subplot(3,2,5);plot(x,Y5,'linewidth',2);xlabel('x');xlim([0 1]);
grid on;title(('y5'));
subplot(3,2,6);plot(x,Y6,'linewidth',2);xlabel('x');xlim([0 1]);
grid on;title(('y6'));
Z=Z'
```

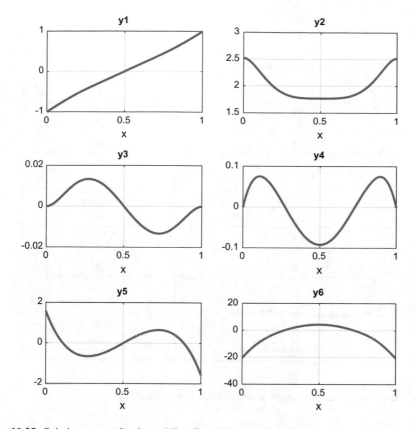

Fig. 10.35 Solution curves for the swirling flow when $\mu = 0.0199$

10.7   Numerical Solutions of Singular Boundary Value Problems

Problem 10.7.1 Consider the linear homogeneous BVP [50],

$$xy'' + \sin xy' + xy = 0, \quad -1 \le x \le 1, \quad y(-1) = A, \quad y(1) = B.$$

Is $x = 0$ an ordinary point or a singular point of this equation?

Use MATLAB function `dsolve` to solve this BVP within the interval of definition.

Solution

For a linear homogeneous ordinary differential equation of the form

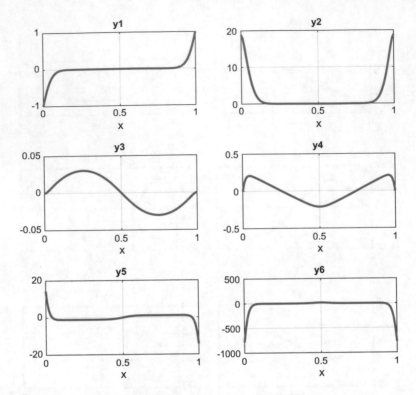

Fig. 10.36 Solution curves for the swirling flow when $\mu = 0.0006$

$$y'' + Py' + Qy = 0$$

where both P and Q are functions of x, $x = a$ is an ordinary point of the ODE if both P and Q are analytic at $x = a$. Otherwise $x = a$ is a singular point;

$$P = \frac{\sin x}{x} = \frac{1}{x}\left(x - \frac{x^3}{3!} + \frac{x^5}{5!} - \cdots\right) = 1 - \frac{x^2}{3!} + \frac{x^4}{5!}\cdots; \quad Q = 1;$$

Both P, Q are analytic, therefore $x = a = 0$ is an ordinary point of the ODE.

Application of MATLAB `dsolve` function does not give an explicit solution of this problem.

Problem 10.7.2 Consider the linear nonhomogeneous BVP,

$$y'' - \frac{1}{x}y' + \frac{1}{x^2}y = 8x - 2, \quad 0 \le x \le 1, \quad y'(0) = y(1) = c.$$

Is $x = 0$ an ordinary point, regular singular point or irregular singular point of the homogeneous part of this ODE?

Use MATLAB function `dsolve` to solve this BVP within the interval of definition for different real value s of c.

What happens if $y'(0) \neq y(1)$, in using this built-in function? (dsolvebvpN.m).

Solution

For a linear homogeneous ordinary differential equation of the form

$$y'' + Py' + Qy = 0$$

where both P and Q are functions of x, $x = a$ is an ordinary point of the ODE if both P and Q are analytic at $x = a$. [51]. In this problem, both of the functions P and Q are not analytic, therefore $x = a = 0$ is a singular point of the ODE.

The singular point is a regular singular point if both $(x - a)P$ and $(x - a)^2 Q$ are analytic at $x = a$. If not, it is an irregular singular point. In our case, both of these functions are analytic at $x = 0$, since $(x)^2 Q = -(x)P = 1$. Therefore, $x = 0$ is a regular singular point of the given ODE.

For $c = 1$, application of MATLAB `dsolve` function gives an explicit solution of this problem, $y(x) = 2x^3 - 2x^2 + x$, which is displayed in Fig. 10.37.

For any real value of c, application of MATLAB `dsolve` function gives an explicit solution of this problem as $y(x) = 2x^3 - 2x^2 + 2c$.

If $y'(0) \neq y(1)$, no explicit solution is found when using this built-in function.

```
%dsolvebvpN.m,   singular bvp
syms  y(x)
Dy = diff(y);
S0=dsolve(diff(y,2,x)-(1/x)*diff(y,1,x)+(1/x^2)*y ==8*x-2,
Dy(0)==1,y(1)==1); S=expand(S0)
p=ezplot(S,[0 1]);grid; set(p,'linewidth',2);ylabel('y(x)');
```

Fig. 10.37 Solution curve
for the BVP, $y'' - \frac{1}{x}y' +$
$\frac{1}{x^2}y = 8x - 2,$
$0 \leq x \leq 1, y'(0) = y(1) = 1$

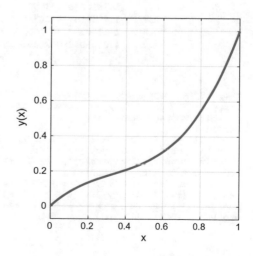

Problem 10.7.3

(a) Use MATLAB bvp4c algorithm to solve the Bessel's equation of order zero, containing a derivative boundary condition,

$$y'' + \frac{1}{x}y' + y = 0, \quad y'(0) = 0, \quad y(1) = 1$$

Compute the solution, and list for equal number of samples, $N = 10$, within the interval $0 \le x \le 1$. Then plot the solution and its derivative in the same figure (bvp4c03.m).

Solution The exact solution is

$$y = \frac{J_0(x)}{J_0(1)}$$

where $J_0(x)$ is the Bessel function of the first kind of order zero. Note that there is a regular singular point at the origin [52]. If we start very close to zero, we avoid the problem of singularity at the origin. The solution values are listed below, and Fig. 10.38 displays these curves.

X	Y	dYdX
1e-15	1.30684822433445	-1.30684821999864e-15
0.1	1.30358235376315	-0.0652607109711505
0.2	1.29381194629373	-0.130032400045544
0.3	1.2776089092504	-0.193830133359546
0.4	1.2550877140015	-0.256173993777503
0.5	1.22642340000651	-0.316601681830074
0.6	1.19184763244959	-0.374672450097297
0.7	1.15159510575274	-0.429946289347167
0.8	1.10595962897107	-0.482011621744423
0.9	1.05530526562813	-0.530508135392579
1	1	-0.57508055345353

```
%bvp4c03.m  Bessel eq. of order zero, y"+y'/x+y=0,  y'(0)=0, y(1)=1
%ex.sol: y=J0(x)/J0(1), J0:Bessel function of first kind order zero
clc;clear;close;format long
eps=1e-15;
ode = @(x,y)  [ y(2) ; -y(1)-y(2)/(x+eps)];
bc = @(ya,yb)[ ya(2);  yb(1)-1];
solinit = bvpinit(linspace(0,1,11),[1 0]);
sol = bvp4c(ode,bc,solinit);
x = linspace(eps,1,11); y = deval(sol,x);
X=x'; Y=y(1,:)'; dYdX=y(2,:)'; A = table(X,Y,dYdX)
plot(x,y,'linewidth',2);xlabel('x');ylabel('y(x)');grid;
legend('y','dy/dx');
```

Fig. 10.38 The solution
curves for the Bessel's
equation of order zero

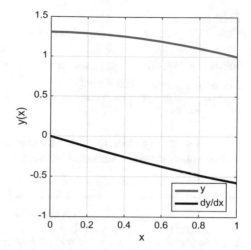

Problem 10.7.4 Consider a homogeneous second order linear differential equation
having regular singularity given by

$$y'' + fy' + gy = 0, \quad y(a) = \alpha, \quad y(b) = \beta$$

Here, α, β are constants and f, g are functions of x, also they are not analytic at
the origin.

Singular BVPs of this type have been studied by several authors for numerically
solving such problems, such as, [53] (presents a finite difference solution), [54]
(series solution in the neighborhood of the singularity), [55] (using an economized
expansion and employed deferred correction outside the range of economized
expansion), [56] (collocation method), and [57] (using Chebyshev polynomial and
B-spline).

Use MATLAB bvp4c algorithm to solve the following homogeneous second
order linear differential equation containing a Dirichlet boundary condition,

$$2x(1+x)y'' + (1+5x)y' + y = 0, \quad y(0) = y(1.5) = 1$$

List few solution values, within the interval $0 \le x \le 1.5$.
The exact analytical solution of this problem is

$$y = \frac{1 + \sqrt{3x/2}}{1 + x}$$

Then, plot the approximate solution and the exact solution in the same figure
(bvp4c06.m).

Solution

This BVP problem has earlier been studied in [55] and [57].

Note that there is a regular singular point at the origin [52]. If this is overlooked, following MATLAB error message is received;

```
Error using bvp4c (line 251)
Unable to solve the collocation equations -- a singular Jacobian
encountered.
```

If we start very close to zero, we avoid the problem of singularity at the origin. A few solution values are listed below, and Fig. 10.39 displays these curves.

X	Y	dYdX
1e-15	0.99999999999999	-1.00028503652803
0.0151515151515161	1.13360730155744	3.78331517674627
0.0303030303030313	1.17753776073665	2.27138505845516
1.46969696969697	1.00610371684134	-0.20284793456514
1.48484848484848	1.00304104716656	-0.20141799148877
1.5	1	-0.19999853527398

Using MATLAB bvp4c algorithm simplifies the numerical solution of the singular BVP.

Otherwise, analytical solution of such problems requires power series methods and Frobenius solutions, which can be quite complicated.

MATLAB script written for the solution of this singular BVP (bvp4c06.m) is given below.

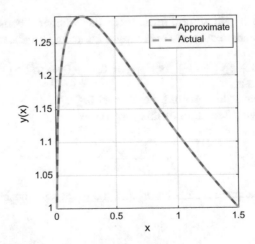

Fig. 10.39 The approximate and exact solutions for the BVP, $2x(1+x)y'' + (1+5x)y' + y = 0$, $y(0) = y(1.5) = 1$

```
%bvp4c06.m  Singular linear BVP, y"= -(1+5*x)/(2*x*(1+x))*y'-y/(2*x*(1+x))
%  y(0)= y(3/2)=1
clc;clear;close;format long
eps=1e-15;
ode = @(x,y)  [ y(2) ;
-(1+5*(x+eps))/(2*(x+eps)*(1+(x+eps)))*y(2)-y(1)/(2*(x+eps)*(1+(x+eps)))];
bc = @(ya,yb)[ ya(1)-1;   yb(1)-1];
solinit = bvpinit(linspace(0,1.5,11),[1 0]);
sol = bvp4c(ode,bc,solinit);
x = linspace(eps,1.5);
y = deval(sol,x);
X=x'; Y=y(1,:)'; dYdX=y(2,:)';
A = table(X,Y,dYdX)
plot(x,Y,'linewidth',2);xlabel('x');ylabel('y(x)');
grid; ylim([min(Y) max(Y)]);hold on;
yact=(1+sqrt(1.5*X))./(1+X); plot(X,yact,'g--','linewidth',2);
legend('Approximate','Actual');
```

Problem 10.7.5 Consider the following second order linear BVP having regular singularity at the origin,

$$\frac{d^2T}{dT^2} + \frac{k}{x}\frac{dT}{dx} = -f(T), \quad \frac{dT(0)}{dx} = 0, \quad T(1) = 1$$

which arises in the analysis of heat conduction through a solid with heat generation. The function $f(T)$ represents the heat generation within the solid, T is the temperature and the constant k is equal to 0, 1 or 2 depending on geometry of the solid.

If $f(T) = 1000$, $k = 2$, compute and plot the solution within the interval of definition. What is the value of $T(0)$? (bvp4c05.m).

Solution
The singularity at the origin is avoided by including a small perturbation at this point. The solution has the approximate value of $T(0) = 167.6667$, and it is plotted in Fig. 10.40.

Fig. 10.40 Solution curve
for the singular BVP

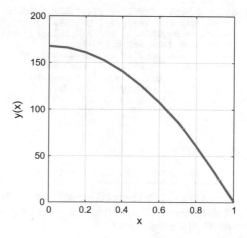

$$\frac{d^2T}{dT^2} + \frac{k}{x}\frac{dT}{dx} = -f(T), \quad \frac{dT(0)}{dx} = 0, \quad T(1) = 1$$

```
%bvp4c05.m  Singular linear BVP, y"+k*y'/x =-f(y),  y'(0)=0, y(1)=1
clc;clear;close;format long
%y"=-1-k*y'/x , y --> T
k=2; eps=1e-15;
ode = @(x,y)  [ y(2) ; -1000-k*y(2)/(x+eps)];
bc = @(ya,yb)[ ya(2); yb(1)-1];
solinit = bvpinit(linspace(0,1,11),[1 0]);
sol = bvp4c(ode,bc,solinit);
x = linspace(eps,1,11); y = deval(sol,x);
X=x'; Y=y(1,:)'; dYdX=y(2,:)';A = table(X,Y,dYdX)
plot(x,Y,'linewidth',2);xlabel('x');ylabel('y(x)');grid;
```

Problem 10.7.6 The Thomas[10]–Fermi[11] equation for the neutral atom is

$$y'' = x^{-1/2}.y^{3/2}, \quad x \in [0, \quad 1], y(0) = 1, \quad y(1) = 0$$

For large x, this equation models the charge distribution of a neutral atom. Positive ions are modeled when y becomes zero at finite x.

A compressed atom results in for the solutions where y becomes large and positive as x becomes large. In this case, $dy/dx = y/x$ [58], [59].

Use bvp4c to solve the Thomas-Fermi equation. (No analytical solution is known for this singular BVP) .Plot solution curves and their phase diagram (bvp4c_ThomasFermi.m).

Solution

We use MATLAB script file (bvp4c_ThomasFermi.m) for the solution of this BVP. The solution curves are depicted in Fig. 10.41.

```
%bvp4c_ThomasFermi.m   y"-y^(3/2)/sqrt(x)=0,  y(0)=1, y(1)=0
%exact analytical.solution is unknown
clc;clear;close;format long;tic
eps=1e-15;
ode = @(x,y)  [ y(2) ; y(1)^(3/2)/sqrt(x+eps)];
bc = @(ya,yb)[ ya(1)-1; yb(1)];
solinit = bvpinit(linspace(0,1,11),[1 0]);
sol = bvp4c(ode,bc,solinit);
x = linspace(eps,1); y = deval(sol,x);
subplot(121), plot(x,y,'linewidth',2);
xlabel('x'); grid; legend('y','dy/dx');
subplot(122), plot(y(1,:), y(2,:),'k','linewidth',2);
xlabel('y(x)'); grid; ylabel('dy(x)/dx'); toc
```

---

[10]Llewellyn Hilleth Thomas (1903–1992) was a British physicist and applied mathematician.

[11]Enrico Fermi (1901–1954) was an American physicist of Italian origin and the creator of the world's first nuclear reactor.

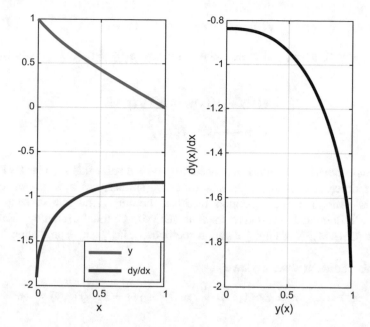

Fig. 10.41 Numerical solution curves of Thomas-Fermi equation (Left), and their phase diagram (Right)

Problem 10.7.7 Consider the following nonlinear singular BVP which arises in the theory of thermal explosion in a cylindrical vessel

$$(xy')' = xe^y, \quad y'(0) = y(1) = 0$$

The exact solution is given by

$$y = 2\log\left(\frac{s+1}{sx^2+1}\right), \quad s = 3 - 2\sqrt{2}$$

Compute the approximate solution using bvp4c, then plot the approximate and exact solutions in the same figure. Determine the maximum absolute error for 100 mesh points (bvp4c07.m).

Solution

Note that this BVP belongs to a class of problems (known as Emden-Fowler equation of second kind) of the form

$$(p(x)y')' = p(x)f(x,y), \quad 0 < x \le 1, \quad p(x) = x^\alpha g(x) \tag{10.7a}$$

Subject to boundary conditions,

$$y'(0) = 0, \quad ay(1) + by'(1) = c \tag{10.7b}$$

$\alpha \geq 0$, $a > 0$, $b \geq 0$, and c are finite constants, $p(0) \neq 0$. Using chain rule for (10.7a),

$$p(x)y'' + p'(x)y' = p(x)f(x,y)$$

$$y'' + \frac{p'(x)}{p(x)}y' = f(x,y)$$

Various numerical methods have been proposed to solve this singular BVP, such as finite difference methods, Green's function, splines, cubic B-splines, quintic B-splines, Adomian decomposition method, modified Adomian decomposition methods, Differential quadrature method (DQM), Optimal Homotopy Analysis Method (OHAM), Variational iteration methods (VIM), and reproducing kernels [60–74].

In this particular case, we have

$$p(x) = x, \quad \alpha = a = 1, \quad b = c = 0, \quad g(x) = 1, \quad f(x,y) = -e^y$$

Hence,

$$y'' + \frac{1}{x}y' = -e^y \quad \rightarrow \quad y'' = -e^y - \frac{y'}{x}$$

The approximate solution is computed by the MATLAB script (bvp4c07.m), and plotted along with the exact solution in Fig. 10.42. Note that singularity at the origin is avoided by the application of a small positive offset at this point. Infinity norm of approximation error is computed as 0.000013265 when bvpinit (linspace(0,1,**15**),[1 0]).

```
%bvp4c07.m  Singular nonlinear BVP, y"+y'/x =f(x,y),   y'(0)= y(1)=0
clc;clear;close;format long
%f(x,y)=-exp(y),    y"=-exp(y)-y'/x
eps=1e-15;
ode = @(x,y)  [ y(2) ; -exp(y(1))-y(2)/(x+eps)];
bc = @(ya,yb)[ ya(2);  yb(1)];
solinit = bvpinit(linspace(0,1,15),[1 0]);
sol = bvp4c(ode,bc,solinit);
x = linspace(eps,1);
y = deval(sol,x);
X=x'; Y=y(1,:)'; %dYdX=y(2,:)';A = table(X,Y,dYdX)
plot(x,Y,'linewidth',2);hold on;
s = 3 - 2*sqrt(2);
yact=2*log((s+1)./(s*x.^2+1)); yact=yact';
plot(x,yact,'g--','linewidth',2);
xlabel('x');ylabel('y(x)');grid;legend('Approximate','Actual');
ae=abs(yact-Y); norm(ae,Inf) % infinity norm of absolute error.
```

Fig. 10.42 Exact and approximate solutions of the singular BVP, $(xy')' = xe^y$, $y'(0) = y(1) = 0$. Infinity norm is computed as 1.3265×10^{-5}

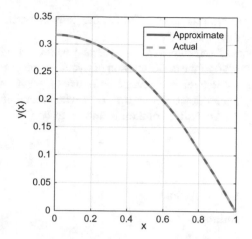

Problem 10.7.8 An important problem in chemical engineering is to predict the diffusion and reaction in a porous catalyst pellet. The goal is to predict the overall reaction rate of the catalyst pellet, which can be obtained from the solution of mass conservation equation in spherical coordinates,

$$D \frac{1}{r^2} \frac{d}{dr} \left(r^2 \frac{dc}{dr} \right) = k$$

where c is the concentration profile, k is the rate constant and D is the diffusivity [75].

The conservation of mass equation for cyclohexane within the pellet can be written as

$$\frac{d^2C}{dR^2} + \frac{2}{R} \frac{dC}{dR} = \varphi^2 \frac{\mathcal{R}(c)}{c_0}, \quad 0 < R < 1$$

Subject to boundary conditions,

$$\frac{dC(0)}{dR} = 0, \quad C(1) = 1$$

where $\mathcal{R}$ is a dimensionless radial coordinate based on the radius of the sphere, Thiele modulus φ is given by the following equation,

$$\varphi = r_p \sqrt{(k/D)}$$

while $\mathcal{R}(c)$ is the reaction rate function.

Consider a sphere (r_p = 2.5 mm in diameter) of gamma-alumina porous catalyst pellet upon which platinum is dispersed in order to catalyze the dehydrogenation of cyclohexane. At 700 K, the rate constant is $k = 4\,\mathrm{s}^{-1}$, and the diffusivity is $D = 0.05\,\mathrm{cm}^2/\mathrm{s}$.

Assuming that the spherical pellet is isothermal, and defining $C = c/c_0$ as the ratio of the concentration of cyclohexane to the concentration of cyclohexane at the surface of the sphere, use bvp4c solver to compute concentration profile of cyclohexane within the pellet when reaction rate function is $\mathcal{R}(c) = c$.

Analytical solution is known to be [76]

$$C = \frac{\sin h(\varphi R)}{R \sin h(\varphi)}$$

(bvp4c_pellet.m)

Solution

$$\frac{d^2C}{dR^2} + \frac{2}{R}\frac{dC}{dR} = \varphi^2 \frac{\mathcal{R}(c)}{c_0} = \varphi^2 C, \frac{dC(0)}{dR} = 0, \quad C(1) = 1, \quad 0 < R < 1$$

This is a linear problem with singularity at $R = 0$.

We use MATLAB script file (bvp4c_pellet.m) in the solution of this reaction-diffusion problem. Following is the edited output list obtained by running this code. Analytical and numerically computed concentration values are in good agreement. The solution curves are plotted in Fig. 10.43.

```
phi = 2.2361
```

R	C	Ca	dCdR
1e-15	0.48351	0.4835	2.4175e-15
0.1	0.48755	0.48753	0.080988
0.2	0.49978	0.49977	0.16442
0.3	0.52059	0.52058	0.25281
0.4	0.55058	0.55059	0.34885
0.5	0.59069	0.59071	0.45555
0.6	0.64218	0.64217	0.57632
0.7	0.70652	0.70657	0.71487
0.8	0.78589	0.78591	0.87597
0.9	0.88267	0.88269	1.0648
1	1	1	1.2878

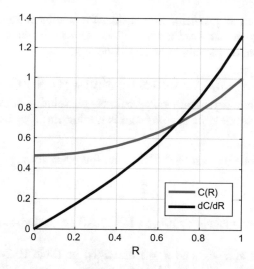

Fig. 10.43 Solution curves for the BVP, $\frac{d^2C}{dR^2} + \frac{2}{R}\frac{dC}{dR} = \varphi^2 C$, $\frac{dC(0)}{dR} = 0$, $C(1) = 1$, $0 < R < 1$

```
%bvp4c_pellet.m    y"+ 2y'/R- phi^2*y=0,    y'(0)=0, y(1)=1
%ex.sol: y=sinh(phi*R)/(R*sinh(phi))
clc;clear;close;
rp=0.25; %cm
k=4 ;%s(-1)
D=0.05 ;%cm2/s
phi=rp*sqrt(k/D)
eps=1e-15;
i=0;
for R=eps:0.1:1.1 % normalized pellet radius (offset by 0.1)
i=i+1;
Ca(i)=sinh(phi*R)/(R*sinh(phi)); % Analytical solution (concentration)
end
Ca=Ca';
ode = @(x,y)  [ y(2) ; phi^2*y(1) - 2*y(2)/(x+eps)];
bc = @(ya,yb)[ ya(2);  yb(1)-1];
solinit = bvpinit(linspace(0,1,2),[1 0]);
sol = bvp4c(ode,bc,solinit);
x =linspace(eps,1,11);
y = deval(sol,x);
R=x';
C=y(1,:)'; dCdR=y(2,:)';%computed concentration and its gradient
A = table(R,C,Ca,dCdR)% table (computed and analtical sol.)
plot(x,y,'linewidth',2);xlabel('R');grid;
legend('C(R)','dC/dR');
```

Problem 10.7.9 In this problem, we consider the nonlinear singular BVP which arises in the study of the distribution of heat sources in the human head [77], including a Neuman and Robin boundary condition;

$$\left(x^2 y'\right)' = -x^2 e^{-y}, \quad 0 < x \le 1, \quad y'(0) = y(1) + y'(1) = 0.$$

Use MATLAB bvp4c function to approximate the solution of the problem. (Note that the analytical solution of this problem is not known) (bvp4c08.m).

Solution

Using chain rule, one obtains the ODE in the following form,

$$y'' + \frac{2}{x} y' = -e^{-y}$$

A solution curve obtained by using a MATLAB script (bvp4c08.m) is shown in Fig. 10.44.

Solution values at $x = 0$ and $x = 1$ are listed in Table 10.2 along with some previously published results [73, 78].

```
%bvp4c08.m  Singular nonlinear BVP, y"+2y'/x =f(x,y)=-exp(y)
%boundary conditions: y'(0)=0,    2*y(1)+y'(1)=0
clc;clear;close;format long
eps=1e-16;
ode = @(x,y)  [ y(2) ; -exp(-y(1))-2*y(2)/(x+eps)];
bc = @(ya,yb)[ ya(2);  yb(1)+yb(2) ];
solinit = bvpinit(linspace(0,1,15),[1 0]);
sol = bvp4c(ode,bc,solinit);
x = linspace(eps,1);y = deval(sol,x); X=x'; Y=y(1,:)'; [X Y]
plot(x,Y,'linewidth',2); xlabel('x'); ylabel('y(x)'); grid;
```

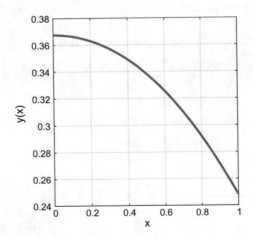

Fig. 10.44 An approximation to the solution of BVP, $y'' + \frac{2}{x} y' = -e^{-y}, 0 < x \le 1, \quad y'(0) = y(1) + y'(1) = 0$

Table 10.2 Solution values at $x = 0$ and $= 1$, along with some previously published results

x	Present method	Thula and Raul [73]	Pandey and Singh [78]
0.0	0.367518058576690	0.367516811426100	0.36752
1.0	0.247929852091527	0.247927718747438	0.24793

Problem 10.7.10 Steady-state oxygen diffusion in a spherical cell is modeled by the following singular two point BVP,

$$\left(x^2 y'\right)' = x^2 \frac{ny}{y+k}, \quad 0 \le x \le 1,$$

subject to boundary conditions, $y'(0) = 0$, $5y(1) + y'(1) = 5$, where, y represents oxygen tension and coefficient values are $n = 0.76129$, $k = 0.03119$ [79].

Use bvp4c to solve the BVP and list initial and final values of y, y', then plot the solution curve within the interval of definition (bvp4c10.m).

Solution: We run the MATLAB script (bvp4c10.m) and compute the initial and final values of y, y', as listed below;

```
x          y                   dy
0.0    0.828483380410461   0.000000000000000
1.0    0.950945887548144   0.245270562259281
```

Solution curve is shown in Fig. 10.45.

```
%bvp4c10.m  Singular nonlinear BVP, y"+2y'/x =ny/(y+k)
%boundary conditions: y'(0)=0,     5*y(1)+y'(1)=5
clc;clear;close;format long
eps=1e-16; n=0.76129; k=0.03119;
ode = @(x,y)   [ y(2) ; n*y(1)/(y(1)+k)-2*y(2)/(x+eps)];
bc = @(ya,yb)[ ya(2); 5*yb(1)+yb(2)- 5 ];%boundary conditions
solinit = bvpinit(linspace(0,1,5),[0 0]); sol = bvp4c(ode,bc,solinit);
x = linspace(eps,1); y = deval(sol,x); X=x'; Y=y(1,:)'; dY=y(2,:)';
[X Y dY]
plot(x,Y,'linewidth',2);xlabel('x');ylabel('y(x)');grid;
```

Problem 10.7.11 Use MATLAB bvp4c and singular ODE solver option assigned by the command,

```
options = bvpset('SingularTerm',A);
```

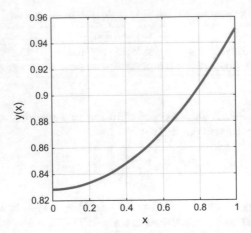

Fig. 10.45 Solution curve of the BVP, $(x^2 y')' = x^2 \frac{ny}{y+k}$, $y'(0) = 0$, $5y(1) + y'(1) = 5$

to solve the equation related to equilibrium of the isothermal gas sphere[12], (Emden-Fowler equation of the first kind)

$$(x^2 y')' = -x^2 y^5, \quad y'(0) = 0, \quad y(1) = \sqrt{3/4}, \quad 0 \le x \le 1.$$

What is the value of the derivative of the solution at $x = 1$?

Plot the exact solution and the computed solution at 11 points on the same figure.

Run the same MATLAB script using bvp5c, compare the elapsed time for each solver.

Hint: Exact solution is $y = \sqrt{3/(3+x^2)}$ (bvp4c_emden.m).

Solution

The MATLAB solver bvp4c solves singular BVPs that have the form

$$y' - A\frac{y}{x} = f(x, y)$$

where A is a constant matrix and the boundary conditions at $x = 0$ must be consistent with the necessary condition $Ay(0) = 0$.

Using the chain rule for given ODE, we find that $A = [0\ 0; 0\ 2]$.

Then, $Ay(0) = [0; -2y'] = 0$ is justified.

---

[12]This is an alternative solution of the equation which has been solved earlier in this chapter.

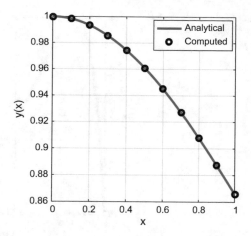

Fig. 10.46 Analytical and computed solutions for the BVP, $(x^2 y')' = -x^2 y^5$, $y'(0) = 0$, $y(1) = \sqrt{3/4}$, $0 \leq x \leq 1$

Following is the edited output list (in short format) of the code (bvp4c_emden. m). Computed solution is also plotted along with the actual solution curve in Fig. 10.46. The value of the derivative of the solution at $x = 1$ is $y'(1) = -0.2165$.

x	y	dy
0.0000	1.0000	0
0.1000	0.9983	-0.0332
0.2000	0.9934	-0.0654
0.3000	0.9853	-0.0957
0.4000	0.9744	-0.1233
0.5000	0.9608	-0.1478
0.6000	0.9449	-0.1687
0.7000	0.9271	-0.1860
0.8000	0.9078	-0.1995
0.9000	0.8874	-0.2096
1.0000	0.8660	-0.2165

The elapsed time for bvp4c solver is slightly shorter than the time it takes for bvp5c solver.

```
%bvp4c_emden.m        y" + (2/x)*y' = - y^5 , on an interval [0, 1].
clc;clear;close;
A = [0   0; 0 -2];
options = bvpset('SingularTerm',A);
guess = [sqrt(3/4);0]; %This constant guess satisfies  bndry cndtns.
solinit = bvpinit(linspace(0,1,11),guess)
bc=@(ya,yb) [ ya(2);    yb(1) - sqrt(3)/2 ];
ode=@(x,y)  [  y(2); -y(1)^5 ]; sol = bvp4c(ode,bc,solinit,options)
x = linspace(0,1);act = 1./ sqrt(1+(x.^2)/3); % Analytical solution
[(sol.x)' sol.y']
plot(x,act,sol.x,sol.y(1,:),'ko','markersize',5,'linewidth',2);
legend('Analytical','Computed');xlabel('x');ylabel('y(x)');grid;
```

Fig. 10.47 Eigenvalue
variation as a function of $y(0)$
It is found that when $y(0) =$
1.39, $\lambda_{max} \approx 2$

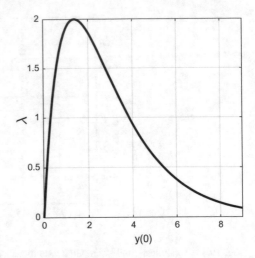

Problem 10.7.12 (Eigenvalue problem) Liouville-Bratu-Gelfand[13] equation [80],

$$y'' + (k-1)\frac{y'}{r} + \lambda e^y = 0, \quad y'(0) = y(1) = 0, \quad k \geq 1, \quad \lambda > 0,$$

appears in several contexts, such as the steady-state equation for a nonlinear heat conduction problem, combustion theory (steady-state solutions to thermal ignition problems in plane-parallel, cylindrical, and spherical vessels) [81], and chemical reactor theory, and it still continues to attract the attention of researchers.

It has strong nonlinearity, since it contains the exponential term.

Use a MATLAB built-in BVP solver to find the maximum eigenvalue $(= \lambda_{max})$ for $k = 2$ and corresponding initial value of the function. Plot $\lambda(q)$, $0 \leq q \leq 9$, $q = y(0)$ (gelfand1.m).

Solution

This equation has a singularity at $x = 0$ and a nonlinear term, e^y which requires a small perturbation of independent function in writng the code.

For this eigenvalue problem, one must include an additional boundary condition, $y(0) = q$.

Hence, following ODE and boundary conditions are at hand,

$$y'' + \frac{y'}{r} + \lambda e^y = 0, \quad y'(0) = y(1) = 0, \quad y(0) = q, \quad \lambda > 0$$

---

[13]Israel Moiseevich Gelfand (1913–2009) was a prominent Russian mathematician.

We compute λ_{max} and corresponding value of $y(0)$ by running MATLAB script file (gelfand1.m).

It is found that when $y(0) = 1.3900$, $\lambda_{max} = 1.999985410888564$.

Figure 10.47 displays the eigenvalue variation as a function of initial value of the eigenfunction.

MATLAB script (gelfand1.m) is listed below.

```
%bvp_gelfand1.m        y" + (y'/x)+ L*exp(y) = 0,    y'(0) = y(1)= 0
clc;clear;close;format long;tic;
L=1                     %lambda guess
guesy =@(x)  [ 1   0]; %initial guess
solinit = bvpinit(linspace(0,1,5), guesy ,L);
ode =@(x,y,L) [ y(2) ;   -y(2)/(x+eps) - L*exp(y(1))];
i=1;    % set index
M=9;    % y(0)max
for q=0:0.1:M
bc =@(ya,yb,L)  [ya(1)-q; ya(2);   yb(1)];
sol = bvp4c(ode,bc,solinit);
P(i)=sol.parameters;
i=i+1;
end
q=0:0.1:M;
q=q'; P=P'; A=[q P]
plot(q,P); grid; xlabel('y(0)'); ylabel('\lambda'); xlim([0 M]); toc;
```

10.8 Another MATLAB-Based BVP Solver, bvp6c

Problem 10.8.1 Make a brief literature survey on bvp6c, a MATLAB based sixth order extension to the bvp4c.

Solution

A MATLAB based two-point BVP solver which is a sixth-order extension to the bvp4c was introduced by Nick Hale and Daniel R. Moore, in 2008, and named as bvp6c [82]. The code can solve explicit nonlinear systems of order one with nonlinear boundary conditions and unknown parameters. The basic solution method is based on polynomial collocation with six Lobatto points. The order of the method is fixed to six. In this code, the quantity to be estimated and controlled is the residual.

The fundamental view of development in bvp4c is the assumption that (in general), a user is interested in only a graphical representation of a solution of a BVP. As such, a modest order solver such as the MIRK4 (mono-implicit Runge Kutta) based Simpson method is appropriate for graphical accuracy. Although MIRK formulae become more complex as order increases (which needs more function evaluations, more interior points and more arithmetic operations at each mesh point), bvp6c developers believe that a higher order method will give comparable accuracy on a sufficiently coarser mesh (or alternatively, greater accuracy on a comparable mesh) and thus prove more efficient in the majority of cases.

It is mentioned that the bvp6c solver uses the residual control framework of bvp4c (suitably modified for a more accurate finite difference approximation) to maintain a user specified accuracy, and as robust as the existing bvp4c software, requiring fewer internal mesh points and evaluations to achieve the required accuracy.

It was developed with a view that (whilst a fourth-order solver is reasonable) a sixth-order solver will supply not only greater accuracy, but also perform more efficiently. Hence, bvp6c was intended to solve the same class of problems efficiently, whilst maintaining the accuracy and robustness of the original bvp4c solver. Treatment of non-separated boundary conditions and unknown parameters without requiring further reformulations have also been maintained.

The method, like bvp4c, provides a uniform prescribed accuracy throughout the computational interval. To achieve this, it fits the MIRK6 data of [83] with the sixth-order interpolant described in [84], giving a representation of the solution throughout the computational interval. This polynomial collocates, satisfies the boundary conditions and is continuous at the ends of each subinterval (while the bvp4c can be viewed as a collocation algorithm with a piecewise cubic polynomial function used to fit the data from a three-point Lobatto IIIA approximation).

As with bvp4c, the cornerstone of bvp6c is that both error estimation and mesh selection are based on the residual.

The bvp6c is intended to be a direct extension to bvp4c, and as such implementation is almost identical, with support routines changed only where necessary to maintain sixth-order accuracy.

Note that all comparisons are made here relative to bvp4c, and not the bvp5c. The BVP solving capabilities of MATLAB have been augmented with bvp5c. Whereas the error control in bvp4c and bvp6c is based on the residual, bvp5c uses a fifth-order, four-point Lobatto IIIA formula with an error estimate combining both residual and the true error.

It is tempting to consider a further development using eighth, tenth and even twelfth-order difference methods. However, the eighth-order scheme requires a total of seven interior points per interval, and the form of the Jacobians becomes increasingly complex. It seems likely the improvement in local accuracy of an even higher order scheme may come at the expense of an unjustifiable increase in the time needed to find a global solution of the required accuracy [83].

Following are few sample articles presented in literature on the application of this solver.

In [85], authors demonstrate a numerical solution for nonlinear-coupled differential equations describing fiber amplifiers which have no explicit solution. The bvp6c solver is addressed for the solution and compared with the Newton-Raphson method.

An article in [86] simultaneously compare and discuss the characteristics of the solvers bvp4c/bvp5c/bvp6c for the numerical analysis of the high power Yb3 + -doped fiber laser. The simulation results demonstrate that the simple error control strategy can deftly adjust the initial estimate of forward signal power by a given random function. The simulation results demonstrate that these solvers can provide an easily comprehensible and rapid method for calculating the threshold pump power and minimum fiber length in Yb3 + -doped fiber lasers.

In [87] a scheme for the numerical computation of Evans functions is presented that relies on an appropriate boundary-value problem formulation. Several examples, including the computation of eigenvalues for a multi-dimensional problem are given using bvp5c, and bvp6c. The main advantage of the scheme is that the scheme is linear and scalable to large problems.

Zhao [88] developes approaches to solve the cavity expansion problem and used bvp6c and bvp4c. Their performance in solving cavity expansion problem is examined in comparison with multiple shooting and finite difference methods.

Wang et al. [89] worked on examples regarding the growth-induced deformations and instabilities in thin hyperelastic plates and used sixth order bvp solver.

Further remarks:

To install the code bvp6c on your computer, create an empty folder and copy the files of (bvp6c.zip bvp6c.m) into it. This archive is available from the web: http://read.pudn.com/downloads141/sourcecode/math/610474/bvp6c/bvp6c.m__.htm.

Note that an additional function file (bvp6c.zip odenumjac.m) is also needed. It can be downloaded as: http://read.pudn.com/downloads141/sourcecode/math/610474/bvp6c/odenumjac.m__.htm.

The code was developed by Nicholas Hale in his masters thesis, Imperial College, UK and is freely available from his web site (Stellenbosh University, South Africa): http://appliedmaths.sun.ac.za/ ~ nhale/publications/Hale2006.pdf.

Problem 10.8.2 Compare the run times of bvp4c, bvp5c and bvp6c BVP solvers using the same MATLAB script under the same circumstances (without providing analytical partial derivatives for the ODE and boundary conditions, without vectorization of the ODE, and using the same initial guess and the same number of points in the initial mesh), by imposing an absolute error tolerance value of 10^{-6} that apply to the corresponding components of the residual vector, while varying relative tolerance value from $10^{-1} \leq 10^{-k} \leq 10^{-12}$. Plot these elapsed time curves in the same figure as functions of relative tolerance (bvp456c_ThomasFermi.m).

Solution

Adaptive distribution of collocation nodes is the essential feature of these algorithms. The grid adaptation and error control are based on the residuals for the ODE and for the boundary conditions, however, the convergence speed towards the BVP solution mainly depends on the initial guess.

Note that a full code comparison is a broader subject and such a comparison is beyond the scope of this book. Thus the run time test that is performed here only demonstrates feasibility on a single problem, and do not represent code comparison issues in full.

We compare the performance of these codes by solving the Thomas-Fermi equation,

$$y'' = x^{-1/2} \cdot y^{3/2}, \quad x \in [0, \quad 1], \quad y(0) = 1, \quad y(1) = 0$$

This model problem illustrates a typical case observed in many tests.
Following is the edited output list of the MATLAB script (bvp456c_ThomasFermi.m).

```
Absolute tolerance=1e-6
Relative tolerance=10^k
```

k	bvp4c	bvp5c	bvp6c
-1	0.215773296983038	0.259012277461517	0.360633138842010
-2	0.300295123786550	0.325266725973315	0.426439708579116
-3	0.407935459478055	0.395024153022648	0.734856586424798
-4	0.710825078091421	0.542838516917789	0.838896650022763
-5	0.974806716293222	0.884672652341388	0.974895799422561
-6	1.400225047794946	1.108976582631581	1.168632721752335
-7	1.835339893190560	1.199207529618088	1.304563314089462
-8	2.304416757973453	1.158702621219766	1.404578216013863
-9	2.557379182545290	1.144278133129605	1.435822994695654
-10	2.923764135788963	1.197726368647336	1.612973213072477
-11	3.155868381934270	1.203830820874353	1.643654756484903
-12	3.727190511045692	1.423837003325633	1.945248359248861

Run time curves as functions of computational relative tolerance are displayed in Fig. 10.48.

```
%bvp456c_ThomasFermi.m   y"-y^(3/2)/sqrt(x)=0,  y(0)=1, y(1)=0
%Run time comparison for bvp4c, bvp5c and bvp6c
clc;clear;close;format long;
eps=1e-15;
ode = @(x,y)  [ y(2) ; y(1)^(3/2)/sqrt(x+eps)];
bc = @(ya,yb)[ ya(1)-1;  yb(1)];
solinit = bvpinit(linspace(0,1,11),[1 0]);
for k=1:12
options=bvpset('reltol',10^-k,'abstol',10^-6);
tic, sol4 = bvp4c(ode,bc,solinit,options),t4(k)=toc;
tic, sol5 = bvp5c(ode,bc,solinit,options),t5(k)=toc;
tic, sol6 = bvp6c(ode,bc,solinit,options),t6(k)=toc;
p(k)=k;
end
disp('bvp4c           bvp5c            bvp6c');
[t4' ,t5', t6']
plot(p,t4,'k',p,t5,'r',p,t6,'linewidth',2);grid;xlim([1 12]);
legend('bvp4c','bvp5c','bvp6c');xlabel('-k');ylabel('run time, s');
```

Fig. 10.48 Run time curves as functions of computational relative tolerance

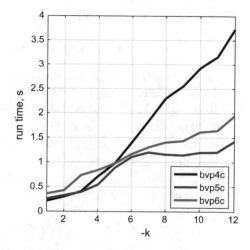

10.9 Exercises

1. Solve the Fisher equation (BVP) (bvp4c_Fisher.m) using the same set of parameters and default tolerances but without applying analytic Jacobian for the ODE. Draw a conclusion.

2. Compare and contrast the solutions of Fisher equation (BVP) presented in this book and in [22] considering the type of MATLAB functions used, guess functions, number of initial mesh points, overall length of the codes for performing the same task, and axes used for graphs.

3. The MATLAB code (bvp4c_Fisher.m) presented in this book uses x-axis limit values of $x = \pm 60$. Run the code with extended limit values. Observe that warning messages that are received beyond a certain value of parameter c indicates computational problems that are apparent in the curves displayed above a threshold on the x-axis. Find out these threshold x-axis values. Does it invalidate the solution of particular problem?

4. Type `edit shockbvp` in MATLAB command Window and examine demo problem named `shockbvp`.

5. Compute the value of parameter λ for which there is a solution of the nonlinear first order parametric BVP,

$$\varepsilon y' = \cos(x) - \frac{\lambda \cos^3(x)}{y}, \quad y\left(-\frac{\pi}{2}\right) = 0, \quad y\left(\frac{\pi}{2}\right) = 2.6.$$

For different values of ε, list first and last few terms of the solution. Plot the solution curve within the interval of definition. (bvp4c04.m).

6. In the solution of the BVP,

$$y'' + y = 1, \quad 0 \le x \le b, \quad y(0) = 0, \quad y(b) = 1, \quad y'(b) = 2,$$

where the value of right-end boundary b is unknown, solutions are influenced by the guessed values b_g for unknown boundary value. If the guessed value $0.09 \le b_g \le 0.767$, the unknown boundary value $b = 0.5236$ also yields $y'(0) \approx \sqrt{3}$, the value of derivative function at point $x = 0$. However, if the guessed value $0.79 \le b_g \le 2.767$, the unknown boundary value is computed as $b = 2.6180$ and $y'(0) \approx -\sqrt{3}$.
   What happens if $b_g > 2.76$ or $b_g < 0.09$ in this problem?
7. In the solution of the BVP,

$$y'' + y = 0, \quad y(0) = 0, \quad y(b) = 1, \quad y'(b) = 0.$$

where b is an unknown boundary point, determine the limits of possible initial guess values for the final boundary, b_g, or (xf_guess) to yield the same solution values. (bvp4c_unknownbndry2.m).
8. Use bvp4c and bvp5c solvers to solve the BVP [90],

$$y'' - 1.5y^2 = 0, \quad 0 \le x \le 1, \quad y(0) = 4, \quad y(1) = 1.$$

9. Use bvp4c solver to solve the nonhomogeneous steady Allen-Cahn equation,

$$\epsilon y'' + y(1 - y^2) = \sin(x), \quad 0 < x < 1, \quad \epsilon \in [0.1\ 0.01\ 0.001]$$

subject to boundary conditions, $y(0) = -1, y(1) = 1$
10. Use loops to modify the MATLAB script file (bvp4c14.m) for the numerical solution of steady Allen-Cahn equation when $\epsilon = 0.0001$, subject to boundary conditions, $y(0) = -1$, $y(1) = 1$, $0 < x < 1$.
   Solve following perturbation problems [91] using bvp4c and/or bvp5c.
11. $\epsilon y'' - y = 0$, $y(0) = 1$, $y(1) = 0$, $\epsilon \in [0.1, \quad 0.01, \quad 0.001]$
   Exact solution: $y(x) = (\exp(-x/\sqrt{\epsilon}) - \exp((x-2)/\sqrt{\epsilon})/(1 - \exp(-2/\sqrt{\epsilon}))$
12. $\epsilon y'' - y = 0$, $\quad y(0) = 1$, $\quad y(1) = 0$, $\quad \epsilon \in [0.1, \quad 0.01]$
   Exact solution: $y(x) = (1 - \exp((x-1)/\epsilon))/(1 - \exp(-1/\epsilon))$
13. $\epsilon y'' - xy' - y = -(1 + \epsilon \pi^2) \cos(\pi x) + \pi x \sin(\pi x)$,
   $y(-1) = y(1) = -1$, $\quad \epsilon \in [0.1, \quad 0.01, \quad 0.001]$
   Exact solution: $y(x) = \cos(\pi x)$
14. $\epsilon y'' - y = 0$, $y(0) = 1$, $y(1) = 2$, $\epsilon \in [0.2, 0.1, 0.01]$
   Exact solution: $(2 - \exp(-1/\epsilon) - \exp(-x/\epsilon))/(1 - \exp(-1/\epsilon))$
   Use MATLAB bvp4c and bvp5c solvers to solve the following BVPs using default settings.

15. $y' = \sin\theta$, $\theta' = M$, $\epsilon M' = -Q$,
    $\epsilon Q' = (y-1)\cos\theta - MT$, $T = \sec\theta + \epsilon Q\tan\theta$
    $y(0) = y(1) = 0$, $M(0) = M(1) = 0$.
    $\epsilon = 10^{-i}, i = 1, \ldots, 4$ [92]

16. $y^{(4)} = R(y'y'' - yy^{(3)})$, $y(0) = y'(0) = 0$, $y(1) = 1$, $y'(1) = 0$,
    $R = 10^i, i = 1, 2, 3$ [92].

17. $v^{(3)} + 2vv'' - (v')^2 + w^2 = \gamma^2$; $w'' + 2vw' - 2v'w = 0$
    $v(0) = v'(0) = v'(b) = 0$, $w(0) = 1, w(b) = \gamma, b = \infty.\gamma = 0, 5, 50, 400$ [92].

18. $\epsilon(1+x^2)yy'' - (1.2 - 2\epsilon x)yy' + (y'/y) + (2x/(1+x^2))(1 - 0.2y^2)$
    $= 0$, $y(0) = 0.92129, y(1) = 0.375$.
    $\epsilon = 5\times 10^{-i}$, $\epsilon = 10^{-i}$ $i = 1, \ldots, 4$ [92].

19. $\epsilon y'' + yy' - y = 0, y(0) = \alpha, y(1) = \beta$
    It is known that for small ϵ, $\epsilon > 0$, the behaviour of the solution depends on the
    used α and β. Following are in the collection used in [92].
    (a) $\alpha = -1/3, \beta = 1/3$, (b) $\alpha = 1, \beta = -1/3$, (c) $\alpha = 1, \beta = 1/3$,
    (d) $\alpha = 1, \beta = 3/2$, (e) $\alpha = 0, \beta = 3/2$, (f) $\alpha = -7/6, \beta = 3/2$.

20. $(\epsilon + x^2)y'' + 4xy' + 2y = 0$, $y(-1) = y(1) = 1/(1+\epsilon)$, $\epsilon \in [0.05, \quad 0.01]$
    Exact solution: $y(x) = 1/(\epsilon + x^2)$ [93].

21.
$$v^{(3)} + (1/2)(3-r)vv'' + r(v')^2 + w^2 - sv' = 1$$
$$w'' + (1/2)(3-r)vw' + (r-1)v'w - s(w-1) = 0$$
$$v(0) = v'(0) = w(0) = v'(b) = 0, \quad w(b) = 1, \ b = \infty.$$

It is known that numerical difficulty of this problem depends on the used couple
(r, s) of parameters, $(r, s) = (-0.1, 0.2), (0.2, 0.2), (0.05, 0.0), (0.5, 0.2)$.
Hint: Let b = 50 [94].

22.
$$\epsilon v^{(4)} + v^{(3)} + w' = 0$$
$$\epsilon w'' + vw' - v'w = 0$$
$$v(0) = v(1) = v'(0) = v'(1) = 0, w(0) = 1, w(1) = -1.$$

$\epsilon = 10^i, i = 0, \ldots, 4$[94].

23. $\epsilon y'' + \exp(y)y' - (\pi/2)\sin(\pi x/2)\exp(2y) = 0, y(0) = y(1) = 0$.
    $\epsilon = 10^{-i}$ $i = 1, \ldots, 5$ [93].

24. $\epsilon y'' + (y')^2 = 1$,
    $y(0) = 1 + \epsilon \ln\cos h(-0.745/\epsilon)$, $y(1) = 1 + \epsilon \ln\cos h(0.255/\epsilon)$
    $\epsilon = 0.2, 0.02, 0.002$ [93].

25. $\epsilon^2 y'' = y + y' - \exp(-2x/\epsilon)$, $y(0) = 1$, $y(1) = \exp(-1/\epsilon)$
    $\epsilon = 10^{-i}$ $i = 1, \ldots, 4$ [93].

26. $\epsilon y'' + y' + y^2 = 0$, $y(0) = 0$, $y(1) = 1/2$
    $\epsilon = 10^{-i}$ $i = 1, \ldots, 6$ [93].

27. (Radial stress distribution on a rotationally symmetric shallow membrane cap)

$$\left(x^3 y'\right)' = x^3 \left(\frac{1}{2} - \frac{1}{8} y^{-2}\right), \quad y'(0) = 0, \quad y(1) = 1$$

28. (Heat and mass transfer in a spherical catalyst) [95].

$$\left(x^2 y'\right)' = x^2 q^2 y \exp\left[\frac{rs(1-y)}{1+c(1-y)}\right], \quad y'(0) = 0, \quad y(1) = 1, \quad q = r = s = 1$$

29. Solve the following BVP [73],

$$\left(x^\alpha e^x y'\right)' = 5 e^x x^{\alpha+3} \frac{\left(5x^5 e^y - (\alpha+4) - x\right)}{4 + x^5},$$

$$y'(0) = 0, \quad y(1) + 5y'(1) = \log\left(\frac{1}{5}\right) - 5$$

Exact solution: $y = \log\left[1/\left(4 + x^5\right)\right]$

30. BVP with irregular singularity [96],

$$x^2 y'' - y' = 1, \quad y(0) = 1, \quad y(1) = 0$$

Exact solution: $y(x) = 1 - x$.

31. Cui and Geng [97]

$$x^2 y'' - xy' + y = 0, \quad 1 \le x \le 2, \quad y(1) = y(2) = 1$$

Exact solution: $y(x) = x - x\ln(x)/(2\ln(2))$

32. Ebaid and El Sayed [98, 99],

$$y'' + \frac{1}{2x} y' = e^y\left(\frac{1}{2} - e^y\right), \quad x \in [0, \quad 1], \quad y(0) = \ln(2), \quad y(1) = 0$$

Exact solution: $y(x) = \ln(2/(x^2 + 1))$

33. The nonhomogeneous Bessel equation [96, 97]

$$y'' + \frac{1}{x} y' + y = 4 - 9x + x^2 - x^3, \quad x \in [0, 1], \quad y(0) = y(1) = 0$$

34. A two parameter singularly perturbed problem of the form

$$\epsilon y'' + \mu a(x) y' - b(x) y = f(x)$$

$$y(0) = y_0, \quad y(1) = y_1, \quad 0 \le \mu \le 1, \quad 0 \le \epsilon \le 1$$

is known as a reaction-diffusion problem if $\mu = 0$, and convection-diffusion problem when $\mu = 1$ [100]. Solve this problem and plot the solution curves if

$$a(x) = (1+x)^2, \quad f(x) = 2x+1, \quad b(x) = 1$$
$$y(0) = y(1) = 0, \quad \mu = 0.1, \quad \epsilon = 0.02$$

35. Use bvp4c, bvp5c and bvp6c solvers to compute the solution of the fourth order BVP,

$$y^{(4)} + xy = -(8+7x+x^3)e^x, \quad y(0) = y(1) = 0, \quad y'(0) = 1, y'(1) = -e$$

Compare their run times and the number of function calls using the same number of initial mesh values. Determine the error in computing the value of $y(0.5)$ for each solver.
Exact solution: $y(x) = x(1-x)e^x$ [101].

36. Solve the 12th order BVP [102], $y^{(12)} = 2e^x \cdot y^2 + y'''$, subject to the boundary conditions,

$$y^{(2i)}(0) = 1, \quad y^{(2i)}(1) = e^{-1}, \quad i = 0,1,2,3,4,5$$

The exact solution is $y = e^{-x}$

37. Solve the system of ODEs [103],

$$\exp(x_1 + x_2') - (x_2')^2 - x_2^2 = 0$$
$$x_1' - x_2 = 0$$

subject to boundary conditions, $x_1(0) = 0$, $x_1(1) = \sin 1$ where $x_2(0)$ and $x_2(1)$ are not known.
Exact analytical solution: $x_1(t) = \sin t$, $x_2(t) = \cos t$, $0 \leq t \leq 1$

38. Solve some of the problems of this chapter using bvp6c solver. Note the number of function calls and elapsed time of computation.

39. Compare the run times of bvp4c, bvp5c and bvp6c BVP solvers using the same MATLAB script for solving the Thomas-Fermi equation by providing analytical partial derivatives for the ODE, without vectorization of the ODE, using the same number of points in the initial mesh and imposing an absolute error tolerance value of 10^{-6} that apply to the corresponding components of the residual vector, while varying relative tolerance value from $10^{-1} \leq 10^{-k} \leq 10^{-12}$. Plot the elapsed time curves in the same figure as functions of relative tolerance.

References

1. Kierzenka J, Shampine LF (2001) A BVP solver based on residual control and the MATLAB PSE. ACM TOMS 27(3):299–316
2. Gökhan FS (2011). Effect of the Guess function & continuation method on the run time of MATLAB BVP Solvers. In: Ionescu CM (ed) MATLAB. IntechOpen. https://doi.org/10.5772/19444
3. Hale N, Moore DR (2008) A sixth-order extension to the MATLAB package bvp4c of J. Kierzenka and L. Shampine. Report no 08/04A, Oxford University Computing Laboratory Numerical Analysis Group
4. Shampine LF, Gladwell I, Thompson S (2003) Solving ODEs with MATLAB. Cambridge University Press, New York
5. Hermann M, Saravi M (2016) Nonlinear ordinary differential equations: analytical approximation and numerical methods. Springer, Berlin (ex.4.15)
6. Bailey PB, Shampine LF, Wattman PF (1968) Nonlinear two point BVPs. Academic Press, London, pp 7–9
7. Abramowitz M, Stegun IA (1965) Handbook of mathematical functions. Dover Publications, New York (17.6)
8. Ascher UM et al (1995) Numerical solution of boundary value problems for ordinary differential equations. SIAM, Philadelphia
9. Bratu G (1914) Sur les equation integrals non-lineaires. Bull Math Soc France 42:113–142
10. Romero N (2015) Solving the one dimensional Bratu problem with efficient fourth order iterative methods. SeMA, Sociedad Española de Matemática Aplicada 71:1–14
11. Weibel ES (1959) On the confinement of a plasma by magnetostatic fields. Phys Fluids 2 (1):52–56
12. Gidaspow D, Baker BS (1973) A model for discharge of storage batteries. J Electrochem Soc 120(8):1005–1010
13. Roberts SM, Shipman JS (1976) On the closed form solution of Troesch's problem. J Comput Phys 21(3):291–304
14. Chang S-H (2010) Numerical solution of Troesch's problem by simple shooting method. Appl Math Comput 216(11):3303–3306
15. Vazquez-LH et al (2012) A general solution for Troesch's problem. Math Prob Eng. article ID 208375. http://dx.doi.org/10.1155/2012/208375
16. Zarebnia M, Sajjadian M (2012) The sinc–Galerkin method for solving Troesch's problem. Math Comput Model 56(9–10):218–228
17. Vazquez LH et al (2014) Direct application of Padé approximant for solving nonlinear differential equations. SpringerPlus 3:563. https://doi.org/10.1186/2193-1801-3-563
18. Saadatmandi A, Niasar TA (2015) Numerical solution of Troesch's problem using Christov Rational Functions. Comput Methods Differ Equ 3(4):247–257
19. Cengizci S, Eryilmaz A (2015) Successive complementary expansion method for solving Troesch's problem as a singular perturbation problem. Int J Eng Math. https://doi.org/10.1155/2015/949463
20. Fisher RA (1937) The wave of advance of advantageous genes. Ann Eugenics 7:353–369
21. Feng Z (2007) Traveling waves to a reaction–diffusion equation. Discrete Contin Dyn Syst Suppl, 382–390
22. Shampine LF, Gladwell I, Thompson S (2003) Solving ODEs with MATLAB. Cambridge University Press, New York, p 183
23. Ascher UM et al (1995) Numerical solution of boundary value problems for ordinary differential equations. SIAM, Philadelphia, p 13
24. Longuski JM, Guzman JJ, Prussing JE (1984) Optimal control with aerospace applications (Chap. 7.2). Springer, Berlin
25. Agrawal GP (2002) Fiber-optic communications systems, 3rd edn. John Wiley & Sons, Inc, London, pp 59–60

26. Levien R (2008) The elastica: a mathematical history. Tech. Rep. UCB/EECS-2008-103, EECS Department, University of California, Berkeley

27. Ferone V, Kawohl B, Nitsch C (2014) The elastica problem under area constraint. Math Annalen 365(3–4)

28. Lawden DF (1989) Elliptic Functions and Applications. Springer-Verlag Applied Mathematical Sciences 80, Berlin

29. Choi SUS (1995) Enhancing thermal conductivity of fluids with nanoparticles. In: Siginer DA, Wang HP (eds) Developments and applications of non-Newtonian flows, vol 231/MD 66. ASME, New York, pp 99–105

30. Ahmad A, Asghar S, Afzal S (2016) Flow of nanofluid past a Riga plate. J Mag Magnet Mater 402:44–48

31. Sheikholeslami M, Shamlooei M (2017) Fe3O4-H2O nanofluid natural convection in presence of thermal radiation. Int J Hydro Ener 42:5708–5718

32. Turkyilmazoglu M (2016) Performance of direct absorption solar collector with nanofluid mixture. Energy Conver Manage 114:1–10

33. Shahmohamadi H, Rashidi MM (2016) VIM solution of squeezing MHD nanofluid flow in a rotating channel with lower stretching porous surface. Adv Powd Tech 27:171–178

34. Bovand M, Rashidi S, Esfahani JA (2015) Enhancement of heat transfer by nanofluids and orientations of the equilateral triangular obstacle. Energy Conver Manage 97:212–223

35. Zheng L, Zhang C, Zhang X (2013) Flow and radiation heat transfer of a nanofluid over a stretching sheet with velocity slip and temperature jump in porous medium. J Franklin Inst 350(5):990–1007

36. Pantokratoras A, Magyari E (2009) EMHD free-convection boundary-layer flow from a Riga-plate. J Eng Math 64:303–315

37. Kumar CK, Bandari S (2014) Melting heat transfer in boundary layer stagnation point flow of a nanofluid towards a stretching/shrinking sheet. Canadian J Phys 92:1703–1708

38. Hayat T, Khan M, Khan MI, Alsaedi A, Ayub M (2017) Electromagneto squeezing rotational flow of Carbon (C)-Water (H2O) kerosene oil nanofluid past a Riga plate: a numerical study. PLoS One. 12(8):e0180976. https://doi.org/10.1371/journal.pone.0180976

39. Atlas M, Hussain S, Sagheer M (2018) Entropy generation and squeezing flow past a Riga plate with Cattaneo-Christov heat flux. Bull Pol Acad Sci Tech Sci 66(3):291–300

40. Cole JD (1968) Perturbation methods in applied mathematics. Blaisdell, Waltham, MA, pp 29–38

41. Allen S, Cahn JW (1979) A microscopic theory for antiphase boundary motion and its application to antiphase domain coarsening. Acta Metall 27:1084–1095

42. Kowalczyk M, Liu Y, Wei J (2015) Singly periodic solutions of the Allen-Cahn equation and the Toda lattice. Commun Partial Differ Equ 40(2):329–356

43. Huang R, Huang R, Ji S, Yin J (2015) Advances in difference equations. Springer Open J 2015:295. https://doi.org/10.1186/s13662-015-0631-3

44. Driscoll TB (2009) Learning MATLAB (Chap. 7.8). SIAM, Philadelphia, PA

45. Roberts SM, Shipman JS (1967) Continuation in shooting methods for two-point boundary value problems. J Math Anal Appl 18:45–58

46. Holt JF (1964) Numerical solution of nonlinear two-point boundary problems by finite difference methods. Comm ACM 7(6):366–373

47. Shampine LF, Gladwell I, Thompson S (2003) Solving ODEs with MATLAB. Cambridge University Press, Cambridge, pp 195–198

48. Ascher UM, Mattheij RMM, Russell RD, (1995) Numerical Solution of boundary value problems for ordinary differential equations. SIAM, p 23

49. Muir PH, Pancer RN, Jackson KR (2000) PMIRKDC: a parallel mono-implicit Runge Kutta code with defect control for boundary value ODEs. University of Toronto, Toronto, Ontario, Canada M5S 3G4

50. Keskin AU (2019) Ordinary differential equations for engineers, problems with MATLAB solutions. Springer, Berlin, p 181

51. Keskin AU (2019) Ordinary differential equations for engineers, problems with MATLAB solutions. Springer, Berlin, p 180
52. Keskin AU (2019) Ordinary differential equations for engineers, problems with MATLAB solutions. Springer, Berlin, p 187
53. Jamet P (1970) On the convergence of finite difference approximations to one dimensional singular boundary value problems. Numer Math 14:355–378
54. Gustafsson B (1973) A numerical method for solving singular boundary value problems. Numer Math 21:328–344
55. Cohen AM, Jones DE (1974) A note on the numerical solution of some singular second order differential equations. J Inst Math Appl 13:379–384
56. Reddien GW (1975) On the collocation method for singular two point boundary value problems. Numer Math 25:427–432
57. Kadalbajoo MK, Aggarwal VK (2005) Numerical solution of singular boundary value problems via Chebyshev polynomial and B-spline. Appl Math Comput 160:851–863
58. Lundqvist S, March NH (1983) Origins—the thomas–fermi theory. Theory of the inhomogeneous electron gas. Plenum Press, New York, pp 9–12
59. Fermi E (1928) Eine statistische Methode zur Bestimmung einiger Eigenschaften des Atoms und ihre Anwendung auf die Theorie des periodischen Systems der Elemente. Zeitschrift für Physik A Hadrons and Nuclei 48(1):73–79
60. Chawla M, Katti C (1985) A uniform mesh finite difference method for a class of singular two-point boundary value problems. SIAM J Numer Anal 1985:561–565
61. Chawla M, Subramanian R (1988) A new spline method for singular two-point boundary value problems. Int J Comput Math 24:291–310
62. Inc M, Ergut M, Cherruault Y (2005) A different approach for solving singular two-point boundary value problems. Kybernetes 34:934–940
63. Cen Z (2007) Numerical method for a class of singular non-linear boundary value problems using Greens functions. Int J Comput Math 84:403–410
64. Çağlar H, Çağlar N, Özer M (2009) B-spline solution of non-linear singular boundary value problems arising in physiology. Chaos Solitons Fractals 39:1232–1237
65. Yucel U, Sari M (2009) Differential quadrature method (DQM) for a class of singular two-point boundary value problem. Int J Comput Math 86(3):465–475
66. Kanth AR, Aruna K (2010) He's variational iteration method for treating nonlinear singular boundary value problems. Comput Math Appl 60:821–829
67. Wazwaz A, Rach R (2011) Comparison of the Adomian decomposition method and the variational iteration method for solving the Lane-Emden equations of the first and second kinds. Kybernetes 40:1305–1318
68. Ebaid A (2011) A new analytical and numerical treatment for singular two-point boundary value problems via the Adomian decomposition method. J Comput Appl Math 235:1914–1924
69. Singh R, Kumar J, Nelakanti G (2012) New approach for solving a class of doubly singular two-point boundary value problems using Adomian decomposition method. Adv Numer Anal 1–22
70. Danish M, Kumar S, Kumar S (2012) A note on the solution of singular boundary value problems arising in engineering and applied sciences: Use of OHAM. Comput Chem Eng 36:57–67
71. Singh R, Kumar J (2013) Solving a class of singular two-point boundary value problems using new modified decomposition method. ISRN Comput Math. article ID 262863, 1–11. http://dx.doi.org/10.1155/2013/262863
72. Roul P (2016) A new efficient recursive technique for solving singular boundary value problems arising in various physical models. Eur Phys J Plus 131:105
73. Thula K, Roul P (2018) A high-order b-spline collocation method for solving nonlinear singular boundary value problems arising in engineering and applied science. Mediterr J Math 15:176

74. Niu J, Xu M, Lin Y, Xue Q (2018) Numerical solution of nonlinear singular boundary value problems. J Comput Appl Math 331:42–51
75. Davis ME (1984). Numerical methods and modeling for chemical engineers. John Wiley and Sons, Inc, p 58
76. Carberry JJ (1976) Chemical and catalytic reaction engineering. McGrawHill, New York
77. Duggan R, Goodman A (1986) Pointwise bounds for a nonlinear heat conduction model of the human head. Bull Math Biol 48:229–236
78. Pandey RK, Singh AK (2009) On the convergence of a fourth-order method for a class of singular boundary value problems. J Comput Appl Math 224:734–742
79. Lin SH (1976) Oxygen diffusion in a spherical cell with nonlinear oxygen uptake kinetics. J Theor Biol 60(2):449–457
80. Jacobsen J, Schmitt K (2002) The Liouville-Bratu-Gelfand problem for radial operators. J Differential Equations 184:283–298
81. Huang S-Y, Wang S-H (2016) Proof of a conjecture for the one-dimensional perturbed Gelfand problem from combustion theory. Arch Ration Mech Anal 222(2):769–825
82. Hale N, Moore DR (2008) A sixth-order extension to the MATLAB package bvp4c of J. Kierzenka and L. Shampine. Oxford University Computing Laboratory, Report no. 08/04
83. Cash JR, Singhal A (1982) High order methods for the numerical solution of two-point boundary value problems. Behav Inf Technol 22:184–199
84. Cash JR, Moore DR (2004) High-order interpolants for solutions of two-point boundary value problems using MIRK methods. Comput Math Appl 48(10–11):1749–1763
85. Gokhan FS, Yilmaz G (2010) Numerical solution of Brillouin and Raman fiber amplifiers using bvp6c. COMPEL Int J Comput Math Electr 29(3):824–839
86. Hu X, Ning T, Pei L, Chen Q, Li J (2015) A simple error control strategy using MATLAB BVP solvers for Yb3+ -doped fiber lasers. Optik Int J Light Electron Opt. 126 (22):3446–3451
87. Barker B, Nguyen R, Sandstede B, Ventura N, Colin Wahl C (2018) Computing Evans functions numerically via boundary-value problems. Physica D 367:1–10
88. Zhao J (2011) A unified theory for cavity expansion in cohesive-frictional micromorphic media. Int J Solids Struct 48:1370–1381
89. Wang J, Steigmann D, Wang F-F, Daid H-H (2018) On a consistent finite-strain plate theory of growth. J Mech Phys Solids 111:184–214
90. Lesnic DA (2007) Nonlinear reaction-diffusion process using the adomian decomposition method. Internat Comm Heat Mass Trans 34:129–135
91. Hemker PW (1977) A Numerical study of stiff two point boundary value problems. Methematisch Centrum, Amsterdam
92. Cash JR, Wright MH (1991) A deferred correction method for nonlinear two point boundary value problems: implementation and numerical evaluation. SIAM J Numer Anal 12:971–989
93. Cash JR (1989) A comparison of some global methods for solving two-point boundary value problems. Appl Math Comput 31:449–462
94. Enright WH, Muir PH (1996) Runge-Kutta software with defect control for boundary value ODEs. SIAM J Sci Stat Comput 17:479–497
95. Kanth ASVR (2007) Cubic spline polynomial for nonlinear singular two point boundary value problems. Appl Math Comput 189:2017–2022
96. Chun C, Ebaid A, Lee MY, Aly E (2012) An approach for solving singular two-point boundary value problems: analytical and numerical treatment. ANZIAM J 53(E):E21–E43
97. Cui M, Geng F (2007) Solving Singular two-point boundary value problem in reproducing kernel space. J Comput Appl Math 205:6–15
98. Ebaid A (2010) Exact solutions for a class of nonlinear two-point boundary value problems: the decomposition method. Z Naturforsh A 65:145–150
99. El-Sayed SM (2002) Integral Methods for computing solutions of a class of singular two-point boundary value problems. Appl Math Comput 130:235–241

100. Shanthi V, Ramanujam N, Natesan S (2006) Fitted mesh method for singularly perturbed reaction-convection-diffusion problems with boundary and interior layers. J Appl Math Comput 22(1–2):49–65
101. Islam S, Tirmizi IA, Ashraf S (2006) A class of methods based on non-polynomial spline functions for the solution of a special fourth-order boundary-value problems with engineering applications. Appl Math Comput 174:1169–1180
102. Wazwaz AM (2000) Approximate solutions to boundary value problems of higher order by the modified decomposition method. Comput Math Appl 40:679–691
103. Dolezal J, Fidler J (1979) On the numerical solution of implicit two-point boundary value problems. Kybernetice 15(3):221–230

Index

Printed in the United States
By Bookmasters